MTP International Review of Science

Heterocyclic Compounds

MTP International Review of Science

Publisher's Note

The MTP International Review of Science is an important new venture in scientific publishing, which we present in association with MTP Medical and Technical Publishing Co. Ltd. and University Park Press, Baltimore. The basic concept of the Review is to provide regular authoritative reviews of entire disciplines. We are starting with chemistry because the problems of literature survey are probably more acute in this subject than in any other. As a matter of policy, the authorship of the MTP Review of Chemistry is international and distinguished; the subject coverage is extensive, systematic and critical; and most important of all, new issues of the Review will be published every two years.

In the MTP Review of Chemistry (Series One), Inorganic, Physical and Organic Chemistry are comprehensively reviewed in 33 text volumes and 3 index volumes, details of which are shown opposite. In general, the reviews cover the period 1967 to 1971. In 1974, it is planned to issue the MTP Review of Chemistry (Series Two), consisting of a similar set of volumes covering the period 1971 to 1973. Series Three is planned for 1976, and so on.

The MTP Review of Chemistry has been conceived within a carefully organised editorial framework. The over-all plan was drawn up, and the volume editors were appointed, by three consultant editors. In turn, each volume editor planned the coverage of his field and appointed authors to write on subjects which were within the area of their own research experience. No geographical restriction was imposed. Hence, the 300 or so contributions to the MTP Review of Chemistry come from many countries of the world and provide an authoritative account of progress in chemistry.

To facilitate rapid production, individual volumes do not have an index. Instead, each chapter has been prefaced with a detailed list of contents, and an index to the 10 volumes of the MTP Review of Organic Chemistry (Series One) will appear, as a separate volume, after publication of the final volume. Similar arrangements will apply to the MTP Review of subsequent series.

Organic Chemistry
Series One
Consultant Editor
D. H. Hey, F.R.S.
Department of Chemistry
King's College, University of London

Volume titles and Editors

1 **STRUCTURE DETERMINATION IN ORGANIC CHEMISTRY**
Professor W. D. Ollis, F.R.S.,
University of Sheffield

2 **ALIPHATIC COMPOUNDS**
Professor N. B. Chapman,
Hull University

3 **AROMATIC COMPOUNDS**
Professor H. Zollinger, *Swiss Federal Institute of Technology, Zurich*

4 **HETEROCYCLIC COMPOUNDS**
Dr. K. Schofield, *University of Exeter*

5 **ALICYCLIC COMPOUNDS**
Professor W. Parker, *University of Stirling*

6 **AMINO ACIDS, PEPTIDES AND RELATED COMPOUNDS**
Professor D. H. Hey, F.R.S. and Dr. D. I. John,
King's College, University of London

7 **CARBOHYDRATES**
Professor G. O. Aspinall, *Trent University, Ontario*

8 **STEROIDS**
Dr. W. F. Johns, *G. D. Searle & Co., Chicago*

9 **ALKALOIDS**
Professor K. F. Wiesner, F.R.S.,
University of New Brunswick

10 **FREE RADICAL REACTIONS**
Professor W. A. Waters, F.R.S.,
University of Oxford

INDEX VOLUME

Butterworth & Co. (Publishers) Ltd.

Physical Chemistry
Series One
Consultant Editor
A. D. Buckingham
Department of Chemistry
University of Cambridge

Volume titles and Editors

1 **THEORETICAL CHEMISTRY**
Professor W Byers Brown, *University of Manchester*

2 **MOLECULAR STRUCTURE AND PROPERTIES**
Professor G. Allen, *University of Manchester*

3 **SPECTROSCOPY**
Dr. D. A. Ramsay, F.R.S.C., *National Research Council of Canada*

4 **MAGNETIC RESONANCE**
Professor C. A. McDowell, F.R.S.C., *University of British Columbia*

5 **MASS SPECTROMETRY**
Professor A. Maccoll, *University College, University of London*

6 **ELECTROCHEMISTRY**
Professor J. O'M Bockris, *University of Pennsylvania*

7 **SURFACE CHEMISTRY AND COLLOIDS**
Professor M. Kerker, *Clarkson College of Technology, New York*

8 **MACROMOLECULAR SCIENCE**
Professor C. E. H. Bawn, F.R.S., *University of Liverpool*

9 **CHEMICAL KINETICS**
Professor J. C. Polanyi, F.R.S., *University of Toronto*

10 **THERMOCHEMISTRY AND THERMODYNAMICS**
Dr. H. A. Skinner, *University of Manchester*

11 **CHEMICAL CRYSTALLOGRAPHY**
Professor J. Monteath Robertson, F.R.S., *University of Glasgow*

12 **ANALYTICAL CHEMISTRY—PART 1**
Professor T. S. West, *Imperial College, University of London*

13 **ANALYTICAL CHEMISTRY—PART 2**
Professor T. S. West, *Imperial College, University of London*

INDEX VOLUME

Inorganic Chemistry
Series One
Consultant Editor
H. J. Emeléus, F.R.S.
Department of Chemistry
University of Cambridge

Volume titles and Editors

1 **MAIN GROUP ELEMENTS—HYDROGEN AND GROUPS I–IV**
Professor M. F. Lappert, *University of Sussex*

2 **MAIN GROUP ELEMENTS—GROUPS V AND VI**
Professor C. C. Addison, F.R.S. and Dr. D. B. Sowerby, *University of Nottingham*

3 **MAIN GROUP ELEMENTS—GROUP VII AND NOBLE GASES**
Professor Viktor Gutmann, *Technical University of Vienna*

4 **ORGANOMETALLIC DERIVATIVES OF THE MAIN GROUP ELEMENTS**
Dr. B. J. Aylett, *Westfield College, University of London*

5 **TRANSITION METALS—PART 1**
Professor D. W. A. Sharp, *University of Glasgow*

6 **TRANSITION METALS—PART 2**
Dr. M. J. Mays, *University of Cambridge*

7 **LANTHANIDES AND ACTINIDES**
Professor K. W. Bagnall, *University of Manchester*

8 **RADIOCHEMISTRY**
Dr. A. G. Maddock, *University of Cambridge*

9 **REACTION MECHANISMS IN INORGANIC CHEMISTRY**
Professor M. L. Tobe, *University College, University of London*

10 **SOLID STATE CHEMISTRY**
Dr. L. E. J. Roberts, *Atomic Energy Research Establishment, Harwell*

INDEX VOLUME

Organic Chemistry Series One

Consultant Editor
D. H. Hey, F.R.S.

MTP International Review of Science

Volume 4

Heterocyclic Compounds

Edited by **K. Schofield**
University of Exeter

Butterworths · London
University Park Press · Baltimore

THE BUTTERWORTH GROUP

ENGLAND
Butterworth & Co (Publishers) Ltd
London: 88 Kingsway, WC2B 6AB

AUSTRALIA
Butterworths Pty Ltd
Sydney. 586 Pacific Highway 2067
Melbourne: 343 Little Collins Street, 3000
Brisbane: 240 Queen Street, 4000

NEW ZEALAND
Butterworths of New Zealand Ltd
Wellington: 26–28 Waring Taylor Street, 1

SOUTH AFRICA
Butterworth & Co (South Africa) (Pty) Ltd
Durban: 152–154 Gale Street

ISBN 0 408 70278 8

UNIVERSITY PARK PRESS

U.S.A. and CANADA
University Park Press
Chamber of Commerce Building
Baltimore, Maryland, 21202

Library of Congress Cataloging in Publication Data

Schofield, Kenneth, 1921–
  Heterocyclic compounds.

  (MTP international review of science) (Organic
chemistry, series one, v. 4)
  1. Heterocyclic compounds. I. Title.
[DNLM: 1. Heterocyclic compounds.
QD 400 H589 1973]
QD251.2.O74 vol. 4 [QD400] 547′.008s [547′.59]
ISBN 0–8391–1032–4   72–11967

First Published 1973 and © 1973
MTP MEDICAL AND TECHNICAL PUBLISHING CO. LTD.
St. Leonard's House,
St. Leonard's Gate,
Lancaster, Lancs.
and
BUTTERWORTH & CO. (PUBLISHERS) LTD.

Filmset by Photoprint Plates Ltd., Rayleigh, Essex
Printed in England by Redwood Press Ltd., Trowbridge, Wilts
and bound by R. J. Acford Ltd., Chichester, Sussex

Consultant Editor's Note

The subject of Organic Chemistry is in a rapidly changing state. At the one extreme it is becoming more and more closely involved with biology and living processes and at the other it is deriving a new impetus from the extending implications of modern theoretical developments. At the same time the study of the subject at the practical level is being subjected to the introduction of new techniques and advancements in instrumentation at an unprecedented level. One consequence of these changes is an enormous increase in the rate of accumulation of new knowledge. The need for authoritative documentation at regular intervals on a world-wide basis is therefore self-evident.

The ten volumes in Organic Chemistry in this First Series of biennial reviews in the MTP International Review of Science attempt to place on record the published achievements of the years 1970 and 1971 together with some earlier material found desirable to assist the initiation of the new venture. In order to do this on an international basis Volume Editors and Authors have been drawn from many parts of the world.

There are many alternative ways in which the subject of Organic Chemistry can be subdivided into areas for more or less self-contained reviews. No single system can avoid some overlapping and many such systems can leave gaps unfilled. In the present series the subject matter in eight volumes is defined mainly on a structural basis on conventional lines. In addition, one volume has been specially devoted to methods of structure determination, which include developments in new techniques and instrumental methods. A further separate volume has been devoted to Free Radical Reactions, which is justified by the rapidly expanding interest in this field. If there prove to be any major omissions it is hoped that these can be remedied in the Second Series.

It is my pleasure to thank the Volume Editors who have made the publication of these volumes possible.

London   D. H. Hey

Preface

Heterocyclic systems form a majority of the several thousand ring-systems known to organic chemistry. Even an extended review cannot deal with more than a fraction of them. In this volume the major families are dealt with, though many minor systems receive incidental mention.

Most of the work reviewed relates to the chemistry of heteroaromatic systems, and of these a fairly uniform kind of treatment is given. This treatment is less appropriate to such a diverse group as the Oxygen Heterocycles (Chapter 7), and inapplicable to Reduced Heterocycles (Chapter 10). The Macromolecular Heterocyclic Compounds (Chapter 11) include individual compounds of the greatest importance, and the study of this family is now a major activity in the field.

Despite the limitation imposed by the unavoidable need to be selective, it is hoped that this review mirrors accurately the important interests and achievements of those working at present on Heterocyclic Organic Chemistry.

Exeter   K. Schofield

Contents

1
Small (3- and 4-Membered) and Medium (Especially 7-Membered) Rings

D. R. MARSHALL
University College of North Wales, Bangor

1.1 THREE-MEMBERED RINGS

1.1.1 Ring synthesis

Established methods of ring synthesis[1] are generally used, but new methods have been found, especially for nitrogen compounds. Thus nitrenes will add to multiple bonds, and generation of nitrenes by oxidation of N-aminophthalimide and related compounds (e.g. by lead tetra-acetate) in the presence of olefins gave aziridines[2]. With acetylenes the nitrenes gave $2H$-azirines[3], a reaction discussed in Section 1.1.2.4. If the nitrene function is generated within the olefin by decomposition of a vinyl azide, cyclisation takes place and the unsaturation is retained. The azide group can be introduced by

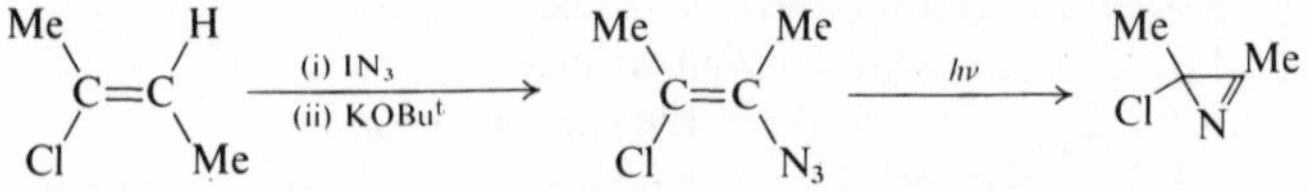

addition of iodine azide, as in the preparation of a 2-chloro-$2H$-azirine[4].

A novel thi-irane synthesis also utilises elimination of nitrogen, but to give an ylid which underwent conrotatory cyclisation (Section 1.1.3.1). As the ylid largely retained its configuration, the *trans*-diethylthiadiazoline

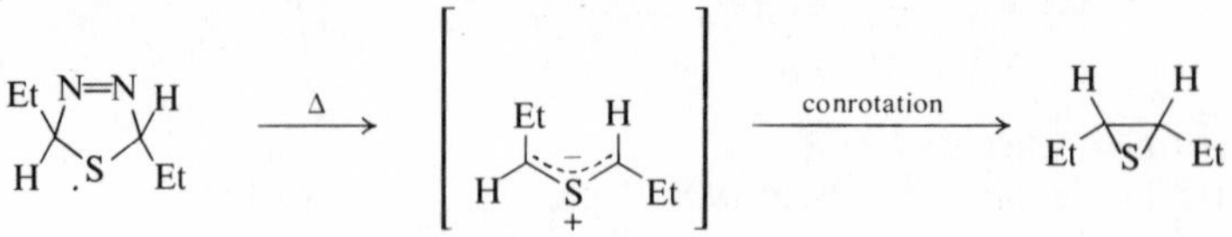

giving a product containing 93 % of *cis*-diethylthi-irane, cyclisation may have been rapid and not significantly reversible[5]. Azoxyalkanes, which can be

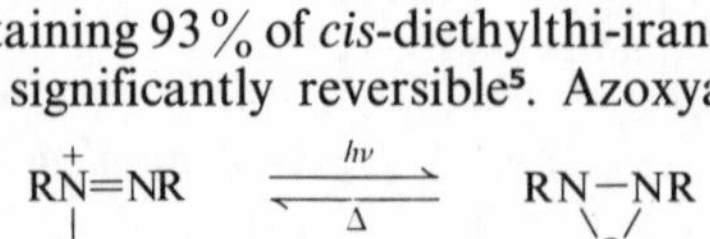

classed as ylids, have been cyclised to oxadiaziridines, in which the ring contains no carbon. The compounds readily reverted to the azoxyalkanes[6].

1.1.1.1 α-Lactams and related compounds

α-Lactones, the classic three-membered cyclic intermediates in reactions, are unstable, but α-lactams have been found to be more stable. It is well known that the presence of substituents favours ring closure, and it also stabilises the ring. α-Lactams can be obtained by treatment of α-bromo-amides with base, and 1,3-di-t-butylaziridin-2-one is quite stable at room temperature, decomposes slowly at 140 °C, and is attacked very slowly by nucleophiles[7]. This synthesis may provide a useful route to amino acids, for the lactams are alkylated at oxygen to imino-ethers by triethyloxonium fluoborate and the ethers are hydrolysed by bicarbonate to amino acids[8].

Cyclisation of α-bromoamidines with base gives aziridine imines analogous

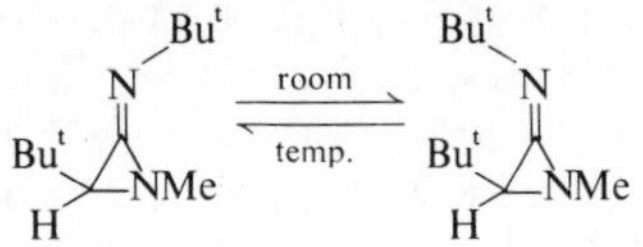

(1) 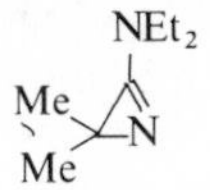   (2)

to α-lactams ('α-lactamidines'?), e.g. (1)[9], while a vinyl azide route gave the isomeric aminoazirine structure (2)[10].

1.1.2 Structural and theoretical studies

1.1.2.1 Absolute configurations

Spectroscopic methods are generally used to determine relative stereo-chemistry of reactants and products, but the absolute stereochemistry of enantiomers may not be known. This has been determined for 2-phenylaziri-dine[11]. Iodine isocyanate (another useful reagent for introduction of nitrogen functions) added to styrene gave a racemic product, converted by addition of menthol to diastereomeric urethanes, which were separated. Cyclisation

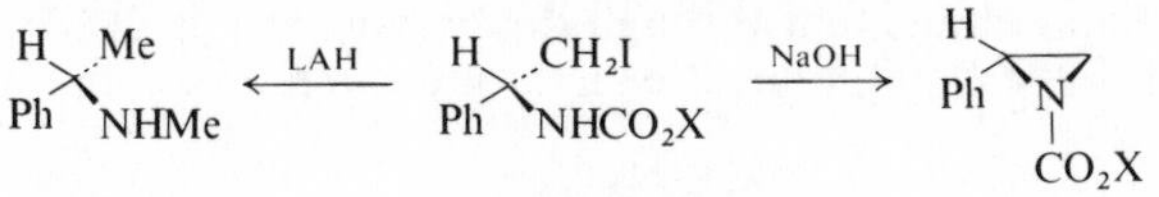

S(−)   (X = menthyl)   R(−)

with base gave the aziridines, while reduction gave N-methyl-α-phenyl-ethylamines related to known enantiomers.

1.1.2.2 Nitrogen inversions and invertomers

Though nitrogen usually inverts rapidly, in small rings inversion may be so slow as to allow separation of stable invertomers. The inductive and electro-static effects of an electronegative atom next to the inverting nitrogen appear to be required. The first stable invertomers were N-chloroaziridines[12, 13], and others such as N-chloro-trans-2-phenyl-3-benzoylaziridine[14] have since been separated. N-alkoxyaziridines also show slow inversion[15].

The adjacent heteroatom may be in the ring, as in oxaziridines and diaziri-dines. Thus invertomers of 2,3-dimethyl-3-benzyloxaziridine and 1,3-dimethyl-2,3-dibenzyldiaziridine have been isolated, as well as various others[16]. In the pair of diaziridines the two inverting groups are always *trans* to one another, though it is thought that the two nitrogen atoms do not invert synchronously.

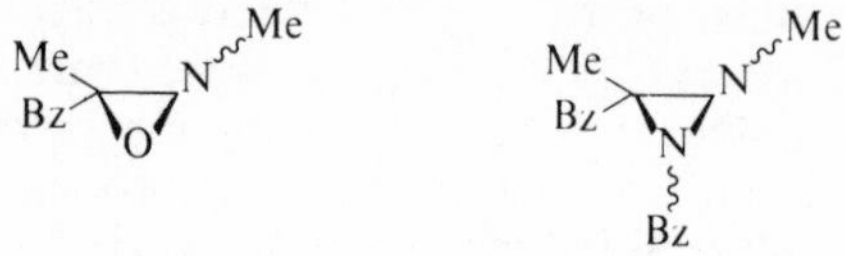

The inversion barriers for several oxaziridines and diaziridines lie in the region of 110–135 kJ mol^{-1} (with, reasonably, entropy barriers near zero), well above the minimum needed to stabilise invertomers[16]. For many other compounds barriers are in the range 40–80 kJ mol^{-1}, which is too small to allow separation of invertomers, but large enough to produce n.m.r. coalescence temperatures up to *c.* 125 °C [17]. Similar inversion rates are found in dialkyldiaziridinones, for which the high i.r. carbonyl frequencies (up to 1880 cm^{-1}) suggest little amide conjugation and hence pyramidal nitrogen[18]. (A similar effect is found in the four-membered 1,2-diazetidinones[19].) How-

ever, hetero-substituents such as chlorine on carbon may lower inversion barriers through double bond–no bond resonance[17].

1.1.2.3 Conjugative properties

In small heterocyclic rings distortion has effects very similar to those found in cyclopropane. This extends to conjugative effects, for a ^{19}F n.m.r. study has shown that conjugation through the oxirane ring is 26% as effective as through a double bond, almost exactly the same as for the cyclopropane

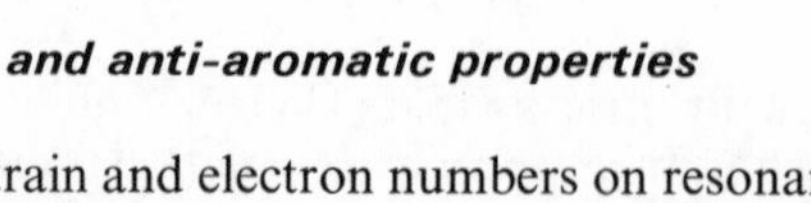

ring[20]. A related observation is that in diazirine and 3,3-dimethyldiazirine the unshared electron pairs are only *c.* 50% localised[21].

1.1.2.4 Aromatic and anti-aromatic properties

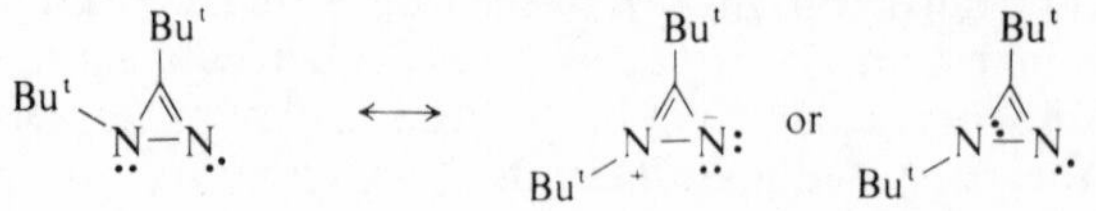

The effects of ring strain and electron numbers on resonance are of considerable interest. The amidines (1) and (2) (Section 1.1.1.1) would, but for the ring structure, be strong bases forming stable cations. Unfortunately their pK values are not known, but (2) immediately dimerises in acid, contradicting cation stability[10]. The behaviour of the unknown 1*H*-diazirine ring would be interesting, as it might show some amidine-like stability, or the instability of a 4-π-electron anti-aromatic system. Its synthesis might be eased by t-butyl substituents (cf. lactams).

Dual possibilities also exist for 2,3-diphenylthi-irene 1-oxide[22], which might be predominantly olefinic, or anti-aromatic or aromatic. In fact it is

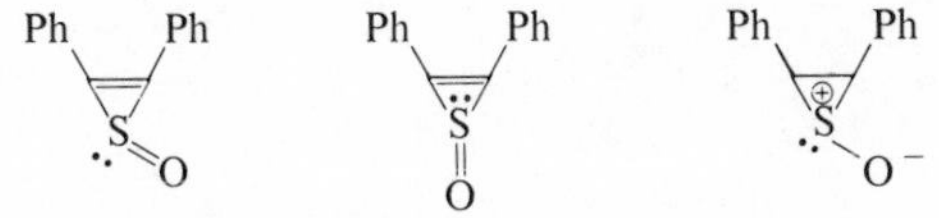

more stable than the corresponding sulphone and its spectroscopic properties indicate conjugation, showing that it adopts the third, most stable, structure. Its stability contrasts with the thermal instability of saturated thi-irane S-oxides and aziridine N-oxides, which readily cleave to olefins[23, 24].

Dialkyl-2-chloro-2H-azirines do not appear to ionise extensively to aromatic azapropenium ions, but the ethyl methyl derivative isomerises via

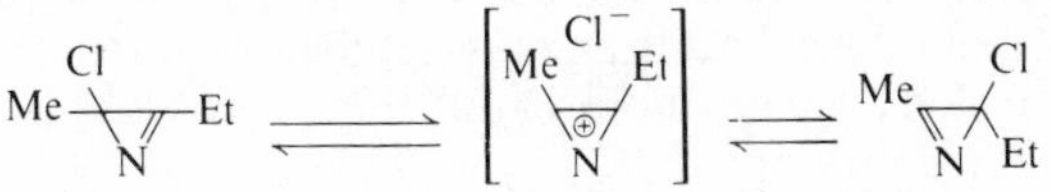

a polar transition state which may be markedly delocalised[4].

Anti-aromatic properties[25] are expected for 1H-azirines, the lower vinylogues of pyrroles. They are isoelectronic with cyclopropyl anions, for which an anti-aromatic energy of c. 400 kJ mol^{-1} (depending on the conformation assumed) is calculated[26]. Attempted preparations have failed, and typically give 2H-azirines instead, this behaviour being explained by anti-aromatic instability[3]. The intermediacy of a 1H-azirine has been demonstrated, however, by pyrolysing two isomeric phthalimidotriazoles. Loss of

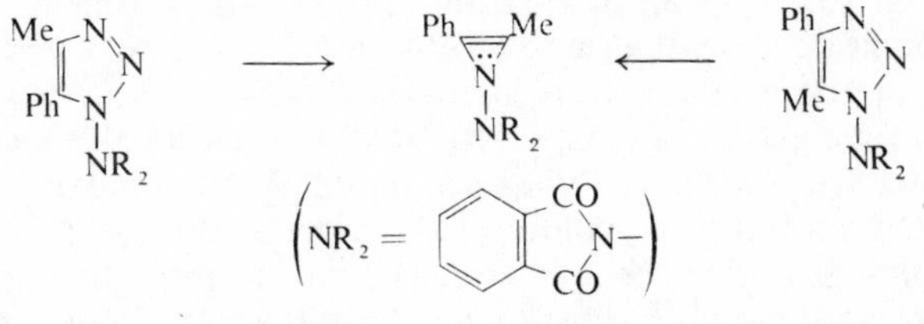

nitrogen led to nitrenes, which underwent cyclisations and rearrangements, but identical product mixtures were obtained from both triazoles, showing a common intermediate. This was clearly unstable, and was most probably the 1H-azirine[27].

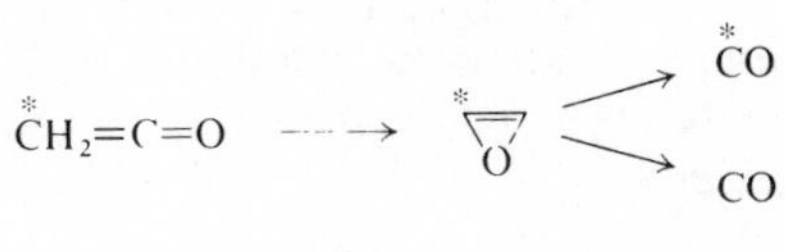

Oxirenes and thi-irenes should be similarly anti-aromatic. Oxirenes have been identified in photolyses of α-diazoketones and alkyl diazoacetates[28], and have been postulated as reaction intermediates, e.g. in the photolysis of ^{14}C labelled ketene, when carbon scrambling is observed[29]. Thi-irenes have also been postulated as intermediates, as in the photolysis of a mesoionic

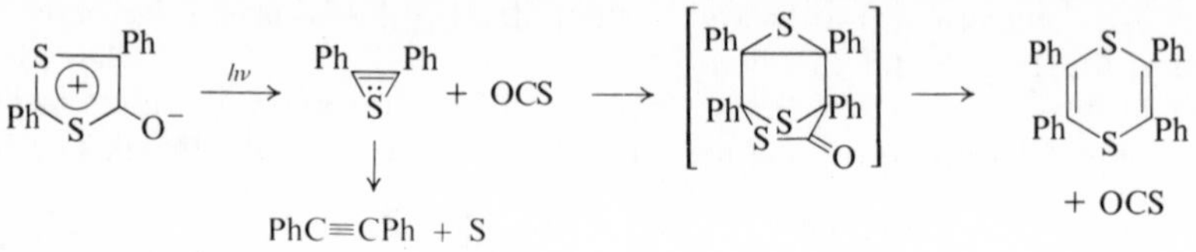

$$PhC{\equiv}CPh + S$$

dithiolium oxide[30].

1.1.3 Ring-modifying processes

1.1.3.1 Ylid formation and cyclo-addition

Aziridines undergo thermolytic or photolytic C—C bond cleavage to give dipolar ylids, thermolysis being conrotatory and photolysis disrotatory, in accordance with the Woodward–Hoffmann rules[31]. Cleavage is first order and reversible, and internal rotation of the ylid allows *cis–trans* equilibration of aziridines, which can be more rapid than cyclo-addition with added

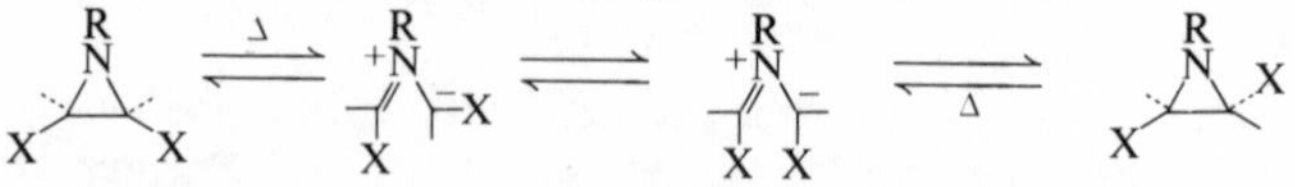

$$(X = CO_2Me)$$

dipolarophile[32]. Recyclisation (over a few seconds) of these ylids when produced by flash photolysis showed activation entropies of -134 J mol^{-1}K^{-1} (to *cis*-aziridine) and -88 J mol^{-1}K^{-1} (to *trans*-aziridine), showing the transition states to be cyclic[33]. Analogous behaviour is shown by oxiranes such as *cis*- and *trans*-dicyanostilbene oxides[34], while the stereochemistry of the cycloadducts from 2-cyano-*trans*-stilbene oxide and dimethyl fumarate show that here too thermolysis is conrotatory[35].

Whereas at room temperature oxiranes are photolysed to ketones and carbenes, in rigid glasses of ethanol, etc. at 77 K ylids are also formed, giving rise to photochromism[36]. The ylids are immobilised, however, and cannot isomerise. On warming the colours fade, more rapidly in the presence of dipolarophiles. Aziridines behave similarly, but the nitrogen ylids are more stable, even, with suitable substituents, in solution at room temperature.

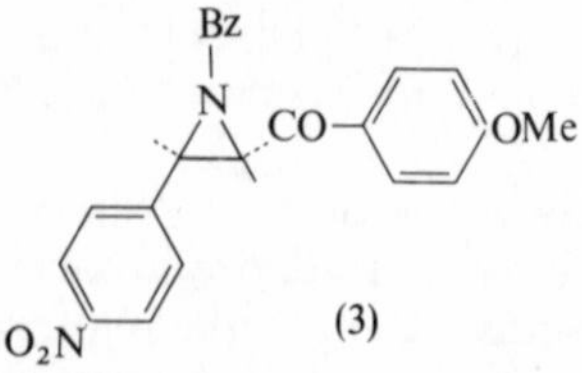

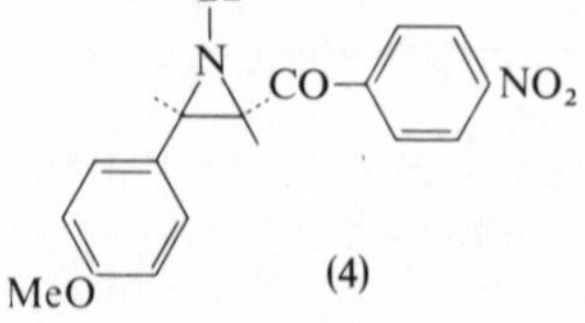

Thus ylids from (3) (blue) and (4) (red) then have half lives of minutes and hours, respectively. These ylids gave adducts with dipolarophiles, or were reconverted to the parent aziridines by irradiation with visible light.

Dipolarophiles may be electrophilic olefins, carbonyl groups, azo groups, nitroso groups, or imines[37].

1.1.3.2 Bicyclic compounds

When a second ring is fused to the 2,3-positions of an aziridine ring, it is geometrically impossible for conrotatory ylid formation to take place unless the second ring is large, though disrotatory photolysis is not hindered. This is well shown by an elegant example in which aziridine and dipolarophile

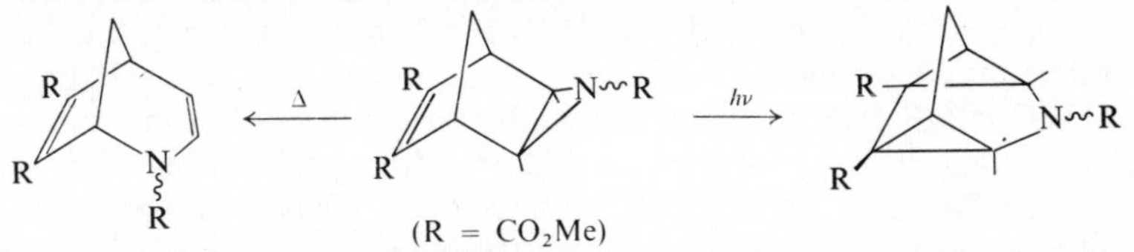

$$(R = CO_2Me)$$

are in the same molecule[38]. Photolysis and cyclo-addition are normal, but it is apparent that thermolysis breaks a C—N bond to give a *nitrenium* ylid *not* stabilised by mesomerism, which then undergoes a 1,2-shift, or that an equivalent synchronous rearrangement takes place. Similar C—N thermolysis may account for formation of the same stable ylid both by photolysis and by thermolysis of an indano[1,2-*b*]aziridine[39]. An apparently discordant observation is that an aziridinosuccinimide readily underwent parallel photolysis to an ylid trapped by dimethylacetylene dicarboxylate, but did

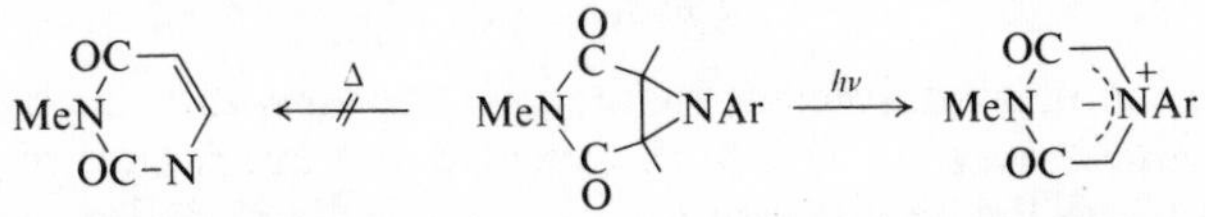

not undergo thermolysis at all, even at 180 °C[40]. This can be explained, however, by the poor migratory aptitude of the carboxamido group.

Abnormal photolysis is also found, in which the stereochemistry of ylid adducts suggests that the photolysis was *conrotatory*[36]. This is thought to happen when a photo-excited aziridine is converted, before cleavage, by intersystem crossing to a vibrationally excited ground state, equivalent to thermal excitation, and that this ground state undergoes normal conrotatory opening.

Thermolysis of C—N bonds can be the preferred process even in monocyclic aziridines. Several *N*-aminoaziridines have been found to cleave readily and reversibly to nitrenes and olefins. In the presence of another added olefin the alternative aziridine is also formed on recombination[41]. (Cf. Section 1.1.1.) A related process is the thermolysis of hydrazones ob-

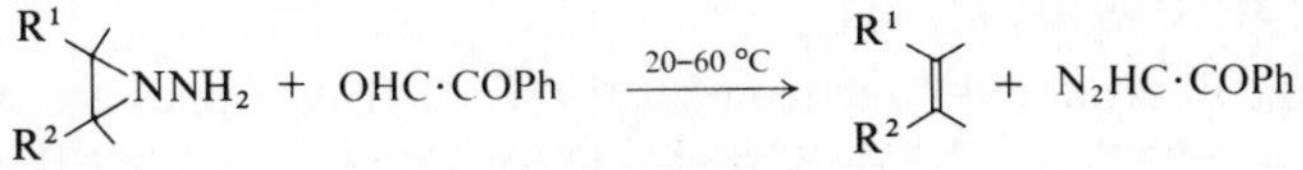

tained from *N*-aminoaziridines and phenylglyoxal[42]. In these reactions the stereochemistry of the aziridine is preserved in the olefin formed.

1.1.3.3 Nucleophilic ring-opening

The relative reactivities of the aziridine ring and the carbonyl group have been studied by allowing *N*-ethoxycarbonylaziridine to react with a series of nucleophiles[43]. The general trend is for strong nucleophiles, including the fluorenide carbanion[44], to attack the carbonyl group, while weaker nucleophiles attack the ring.

Under basic conditions nucleophiles such as amines preferentially attack 2-alkylaziridines at the less substituted carbon atom in an S_N2 step, though some 'abnormal' attack in the other direction also occurs. It had earlier been assumed that abnormal attack was by an S_N1 mechanism, but as abnormal attack is disfavoured by solvent polarity, it is probably S_N2 also[45]. S_N2 attack in both directions is also found in the boron trifluoride-catalysed aminolysis of 2-phenylaziridine[46], although aziridine–BF_3 complex formation might be expected to favour S_N1 cleavage adjacent to the phenyl group, since aziridinium *cations* are known to undergo S_N1 cleavage[47]. In intramolecular reactions abnormal reaction may be dominant[45].

When an aziridine N-substituent carries an α-heteroatom, ring expansion can take place. Thus 2,3-disubstituted N-acylaziridines give Δ^2-oxazolines.

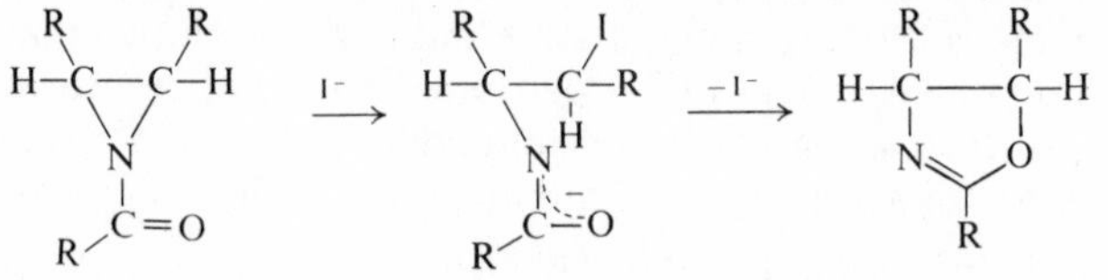

When the reaction is catalysed by iodide ions, the *trans*-aziridines give mainly *trans*-oxazolines, showing that both cleavage and cyclisation proceed with inversion, though the *cis*-compounds are less selective[48]. Iodide ion, a potent nucleophilic catalyst, also takes part in the production of an intriguing fused ring system. Reaction of aziridine with (remarkably) bis(2-chlorotetrafluoro-

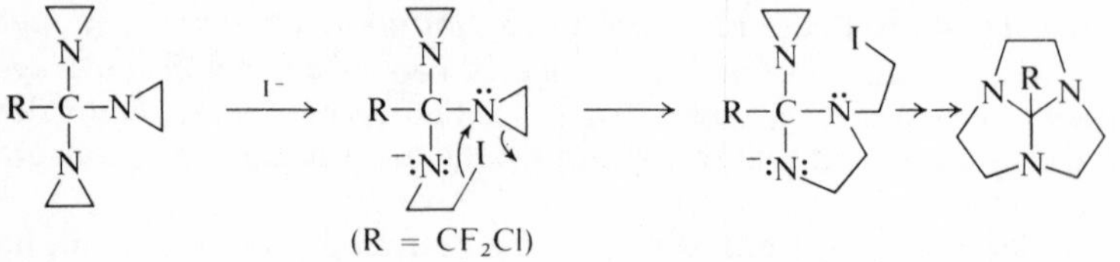

(R = CF_2Cl)

ethyl)disulphide gave the trisaziridine, which by iodide catalysis was rearranged to the strain-free tricyclic product[49].

1.1.3.4 3H-Diazirines

Among the unsaturated three-membered rings 3-H-diazirines are strikingly stable. The 3-methyl-3-ω-hydroxylalkyl derivatives can be converted in side chain reactions to the tosylates, and to the carboxylic acids and various derivatives of the acids without affecting the ring[50]. This stability is due at least in part to the symmetry-disallowed nature of ring thermolysis, whether to diazoalkane or to nitrogen and carbene[51]. In agreement, photolysis takes place readily. The compounds are readily accessible through reactions such as iodine oxidation of diaziridines[50], or oxidative cyclisation of amidines[52].

Thermolysis of 3-vinyl-3H-diazirines is allowed, however, and these compounds are much more reactive than the simple azirines. Thus methylvinylazirine isomerises to 3-methylpyrazole almost quantitatively in a few days at room temperature, via a dipolar intermediate which can be trapped as an ester by benzoic acid[51]. Formation of a carbonium ion adjacent to the

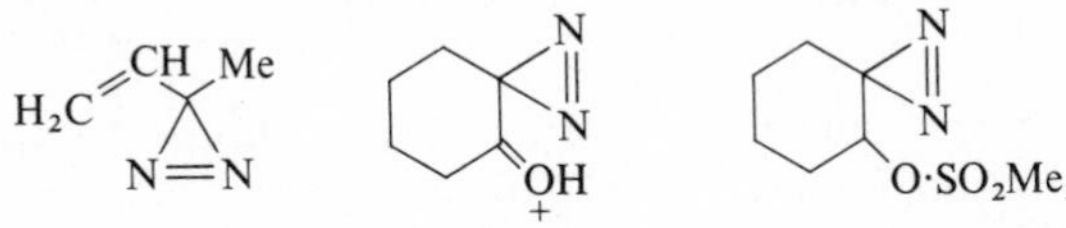

azirine ring also induces loss of nitrogen, as in the treatment of an α-ketone with acid, or an α-methanesulphonate ester with water alone[51].

1.2 FOUR-MEMBERED RINGS

As most of the work reported here is synthetic, it is arranged according to the types of ring system and heteroatom, save for Section 1.2.1.4.

1.2.1 Monocyclic compounds

Although unusual rings such as an inorganic diazadiphosphetidine[53] and an oxagermetane[54] have been reported, most recent work is on four-membered rings containing one atom of oxygen, sulphur or nitrogen.

1.2.1.1 Oxygen-containing rings

In recent years the preparation of oxetanes by photocyclo-addition of ketones to olefins (the Paterno–Büchi reaction) has been studied extensively[55]. Carbonyl n–π* triplet states are often involved, but singlet states are important in reactions of alkanones. Thus addition of cyclobutanone to butadiene or piperylene is regio- and stereo-specific, leading to the conclusion that addition is through a singlet[56]. This was originally ascribed to disturbance of the carbonyl energy levels in cyclobutanone by ring strain, but a detailed study of additions to alkanones[57] has shown that addition of acetone to dicyanoethylene occurs only through the ketone n–π* singlet, so addition of singlet alkanones appears to be general. Addition to vinyl ethers takes place through both singlets and triplets, but while singlet addition to give oxetanes is stereospecific, addition of triplets is not, because the triplet, biradical intermediate must undergo relatively slow spin inversion before yielding any product. This allows internal rotation to occur, whereas singlet intermediates cyclise more rapidly in a more concerted fashion[57].

A similar reaction of aromatic esters such as dimethyl terephthalate may possibly be through a carbonyl π–π* singlet[58].

This kind of addition is not restricted to olefins, for furans and thiophens will also undergo additions at the 2,3-positions[59].

Sensitised photo-addition of singlet molecular oxygen gives dioxetanes, though these cannot always be isolated. Thus thujopsene reacted only at the double bond, and lack of accompanying deuterium exchange suggested a

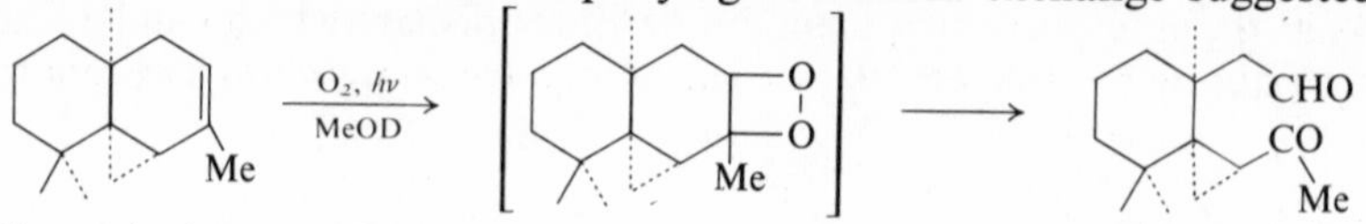

1,2-dioxetane intermediate[60]. However, sensitised photo-oxidation at $-78\,°C$ yields stable 1,2-dioxanes from alkoxyethylenes[61, 62]. On warming, the products are cleaved to carbonyl compounds. Correspondingly, cyclic

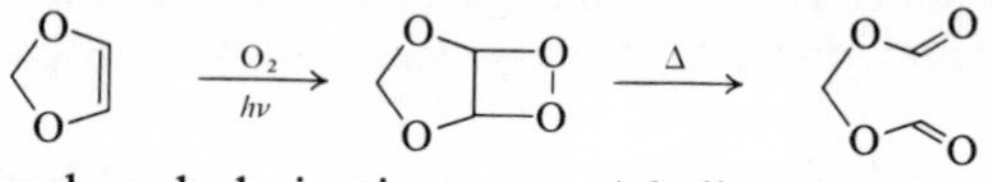

olefins give dicarbonyl derivatives, e.g. 1,3-dioxole gives bisformyloxy-methane[62].

1.2.1.2 Sulphur-containing rings

Thietanes are the best known of these[63]. A general preparation is the fusion of 1,3-dioxan-2-ones with potassium thiocyanate. The high stereospecificity supports the postulated mechanism of successive S_N2 displacements[64]. A

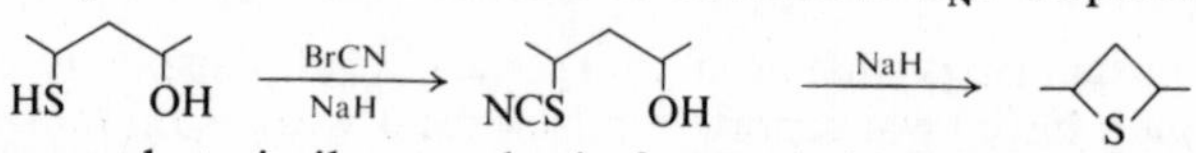

new and somewhat similar synthesis forms a thiocyanate from cyanogen bromide and effects ring closure with sodium hydride, but in this case both cis- and trans-products are formed[65]. Another new preparation uses removal of sulphur (by tris(diethylamino)phosphine) from a dithiolane ring. An

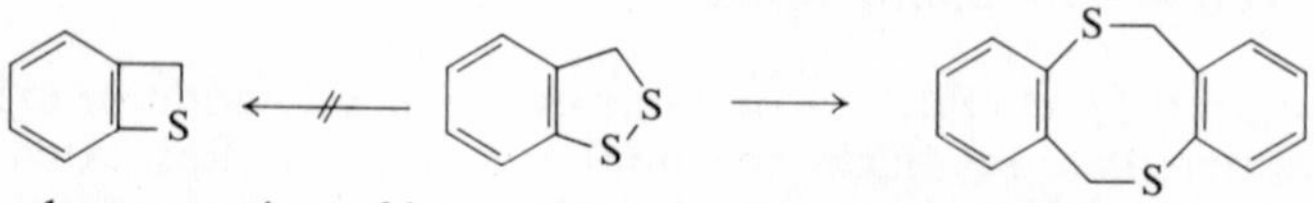

attempted preparation of benzothiete by this method failed, however[66].

The thietane ring is a boat with substituents (or electron pairs) axial or equatorial. Microwave, dipole moment, and n.m.r. measurements of the bond angles agree, and show the usual equatorial preference of substituents[67–69]. Unpublished x-ray work shows that the azetidine ring has a similar conformation[70].

Thietanes are readily oxidised to their sulphoxides and sulphones. In these, protons adjacent to sulphur are acidic, and an interesting series of base-promoted epimerisations and rearrangements of 2,4-diphenylthietane and its oxides has been observed[69, 71]. The course of reaction is highly dependent on conditions. Methanolic methoxide epimerises the trans (a,e) isomer to the cis (e,e). Stronger bases in aprotic media produce the confor-

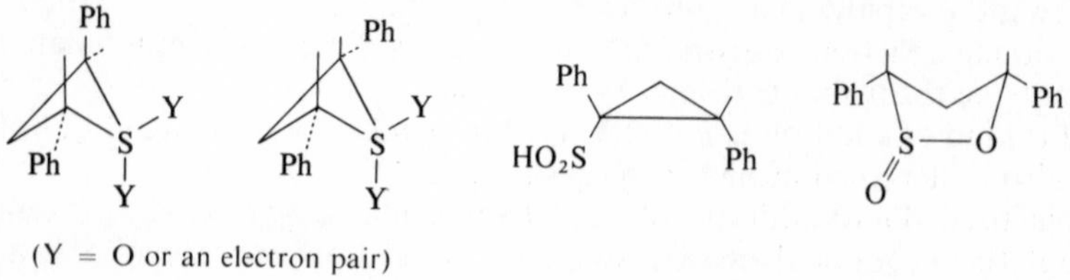

(Y = O or an electron pair)

mationally more stable anion from either isomer, and this rearranges. Thus with ethylmagnesium bromide, the sulphones give selectively *trans*-1,2-diphenylpropanesulphinic acid, while with potassium t-butoxide in DMF the sulphoxides give selectively the *cis*-diphenyl product (which dispropor-tionates). With magnesium bromide in t-butyl alcohol, on the other hand, the *cis*- and *trans*-sulphones give the 3,5-diphenyl-1,2-oxathiolane 2-oxides, a rearrangement hitherto effected only by pyrolysis or in mass spectrometry.

Pyrolysis also converts thietane sulphones into cyclopropanes, but unselectively[65, 69]. However, less vigorous desulphurisation of 2,4-dimethyl-thietane *S*-methyl sulphonium salts with butyl-lithium in THF produced 90% inversion of the 2,4-dimethyl geometrical configuration[65]. Thietanes themselves are difficult to desulphurise.

An interesting synthesis gives an unsaturated thiete dioxide, which will

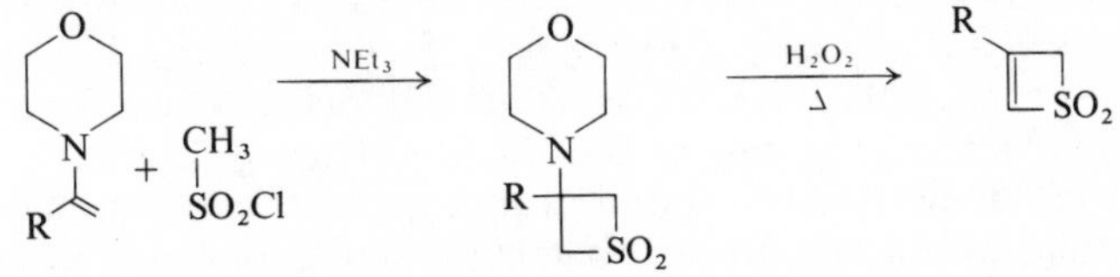

add bromine but not amines, (though it was hydrolysed by hydroxide)[72]. This suggests that this little-known system is not conjugated, which is reason-able, as the unsaturated ring must be virtually planar and the dihedral angle between the olefinic and S=O bonds high.

1.2.1.3 Nitrogen-containing rings

Saturated rings are again the best known of this group (e.g. Ref. 1). The less common unsaturated azetines have been prepared by general methods such as ring expansion through carbenes or nitrenes. Thus the starting point may be an azirine and the reaction analogous to expansion of a pyrrole to a

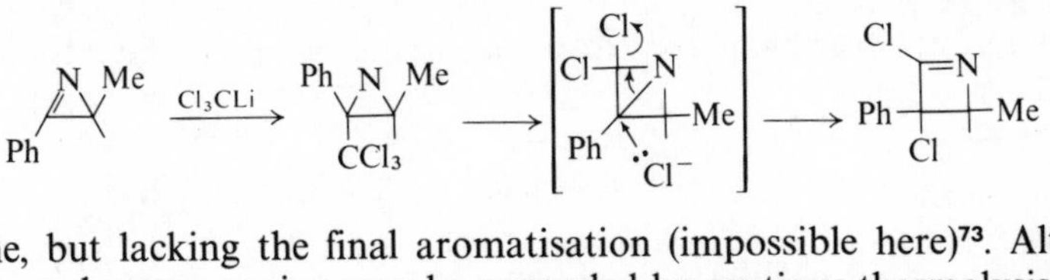

pyridine, but lacking the final aromatisation (impossible here)[73]. Alterna-tively, a cyclopropane ring may be expanded by cautious thermolysis of an

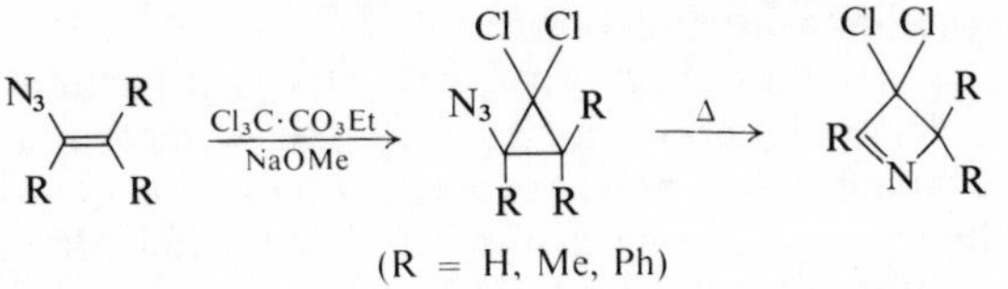

(R = H, Me, Ph)

explosive azide[74]. Both methods give Δ^1-azetines, compounds which are readily hydrated by acid to β-lactams[73], and reduced by lithium aluminium hydride to azetidines[74], as are β-lactams, for which diborane reduction can also be used[75]. An attempt to prepare an azetine by silver decyanation of 2-cyano-N-t-butylazetidine in ethanol gave the 2-ethoxyazetidine[76].

1.2.1.4　Cyclo-addition of heterocumulenes

The $[2+2]$ cyclo-addition of heterocumulenes is a versatile method for synthesis of heterocyclic four-membered rings, the classic example being the dimerisation of ketene. Since Staudinger's early work much information has

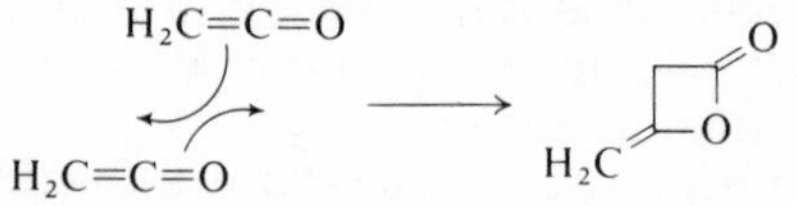

accumulated[77], and new results continue to appear. Thus addition of ketenes to aldehydes or ketones gives 2-oxetanones (β-lactones) as with ketene itself (e.g. Ref. 78), though the opposite direction of addition to give 3-oxetanones is not observed. Similarly, imines give 2-azetidinones (β-lactams).

With an unsymmetrical ketene and an arylimine without *ortho*-substituents only the *trans*-product is formed, but if the aryl group carries an *ortho*-

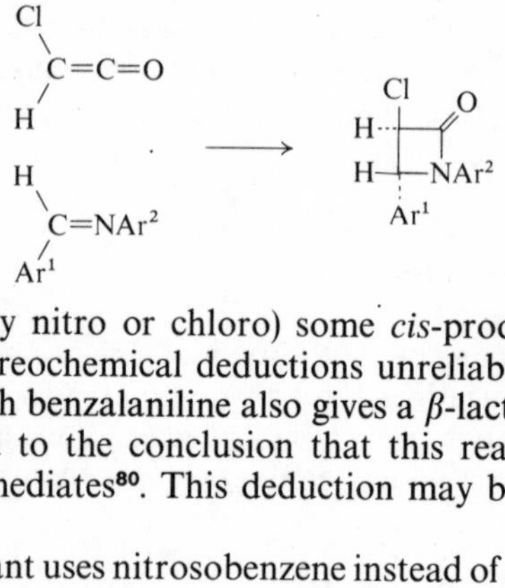

substituent (especially nitro or chloro) some cis-product is formed[79]. This variability makes stereochemical deductions unreliable. Thus reaction of a carbonyl chloride with benzalaniline also gives a β-lactam, and formation of some cis-product led to the conclusion that this reaction did not always involve ketene intermediates[80]. This deduction may be correct in this case, but could be wrong.

An interesting variant uses nitrosobenzene instead of a carbonyl compound

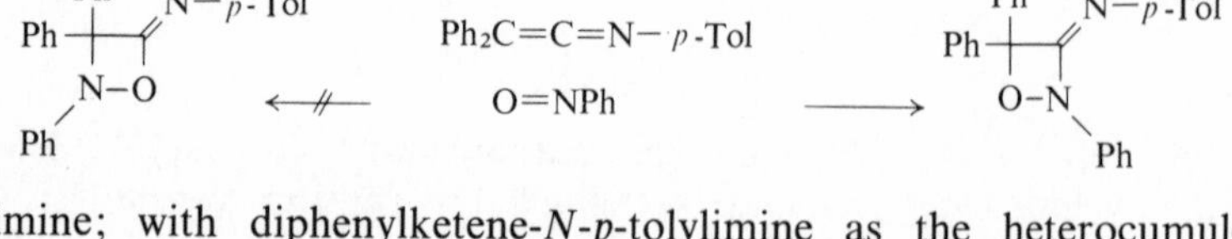

or imine; with diphenylketene-N-p-tolylimine as the heterocumulene, addition is in the opposite direction[81].

Isocyanates, commonly used, give with olefins β-lactams in which the whole of the lactam function comes from the heterocumulene. Chloro-sulphonyl isocyanate is very reactive, needing only mild conditions, and the adducts lose the chlorosulphonyl group readily by mild hydrolysis, providing a potentially useful synthetic route[82]. In additions of tosyl isocyanate to vinyl ethers entropies of activation of the order of $-190 \, \text{J mol}^{-1} \text{K}^{-1}$ indicate cyclic transition states. The *trans*-products are more stable than the *cis*, with which they equilibrate, possibly through open chain dipolar intermediates[83]. A formally analogous but intramolecular cyclisation has pro-

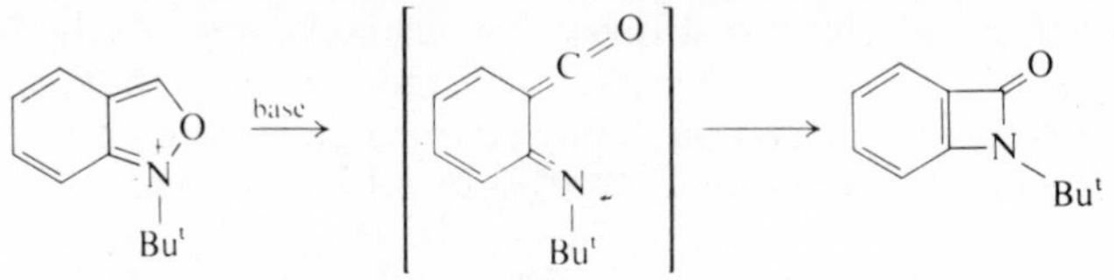

duced the first stable benzazetinone[84].

When both reactants are heterocumulenes, the product has two exocyclic double bonds. Diazetidines, oxazetidines, and thiazetidines are available from carbodi-imides and isocyanates or isothiocyanates[85]. The products undergo thermolysis to the starting materials or to alternative heterocumulenes, thus providing synthetically useful exchange reactions leading to new heterocumulenes. In some cases, when four-centre intermediates could not be isolated, though exchange products were formed, six-centre intermediates, also known, cannot be ruled out.

1.2.2 Bicyclic antibiotics

By far the most important groups of four-membered heterocyclic compounds are the penicillins and cephalosporins. Their great value lies in their specific

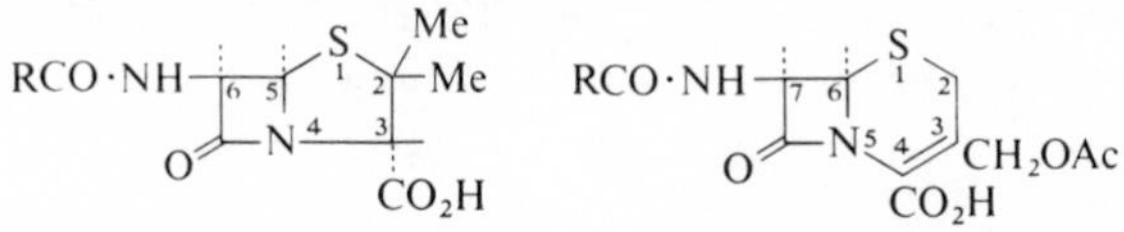

action in inhibiting bacterial cell wall synthesis, arising from their β-lactam structure. This, it is thought, mimics the substrate of a particular transpeptidase and so inhibits this enzyme. A very wide range of structures is effective, and the compounds are toxic only to bacteria, not to higher animals. When bacterial strains acquire tolerance to particular antibiotics of this class by producing β-lactamases, new effective variants can be introduced by changing the substituent at the penicillin C-6-amino group, or the acyloxy group of cephalosporins.

Initially new penicillins could only be produced by fermentation, but enzymic deacylation of the C-6-amino group (giving penicillanic acid) allows many new semisynthetic derivatives to be made, especially from penicillins G (R = $PhCH_2$) and V (R = $PhOCH_2$). This deacylation can now also be carried out chemically (e.g. Ref. 86). Total synthesis was first achieved in 1959.

The structure of cephalosporin C [R = $HO_2C \cdot CH(NH_2) \cdot (CH_2)_3$] was established in 1961, and total synthesis achieved in 1966. The cephalosporins are very valuable clinically, as they are active against many penicillin-resistant bacteria, and are less likely than penicillins to induce allergic reactions.

The importance of these compounds has given rise to a vast body of work, described in recent monographs[87, 88], and more succinctly in other articles (e.g. Ref. 89).

All operations on β-lactams are complicated by the sensitivity of the strained ring system, particularly to hydrolysis. Interestingly, monocyclic β-lactams do not seem to undergo basic attack any more rapidly than open chain amides[90]. Penicillins, however, are more strained than monocyclic β-lactams by about 25 kJ mol^{-1} and cephalosporins by a lesser amount.

1.2.2.1 Stereochemistry

Correct stereochemistry is essential for biological activity, though the overall molecular shape is clearly less important. Sweet and Dahl have determined by x-ray measurements the structures of representative $Δ^2$- and $Δ^3$-cephalosporins and compared them with penicillins[91]. One key point is the bridgehead nitrogen atom of the bicyclic system, the amide nitrogen of the β-lactam. In the inactive $Δ^2$-cephalosporins the nitrogen configuration is virtually planar, but in all the active compounds it is markedly pyramidal. Correspondingly, the active compounds show longer N—CO bonds, shorter lactam C=O bonds, and higher lactam carbonyl i.r. frequencies. All this is consistent with lack of amide resonance in the active β-lactams, leading to enhanced carbonyl reactivity (which can be observed in the form of readier base hydrolysis.) This reactivity contributes to their biological activity.

A second key point is C-6 in the penicillins, C-7 in the cephalosporins, at which it is apparently necessary to have an α-hydrogen atom and a β-substituent.

1.2.2.2 Synthesis of the β-lactam ring: epimerisation

Recently, research has been directed towards manipulation of the sulphur-containing ring, but attention is still paid to new β-lactam syntheses. In one sequence D-pipecolic acid was converted via N-acylation into a diazoester,

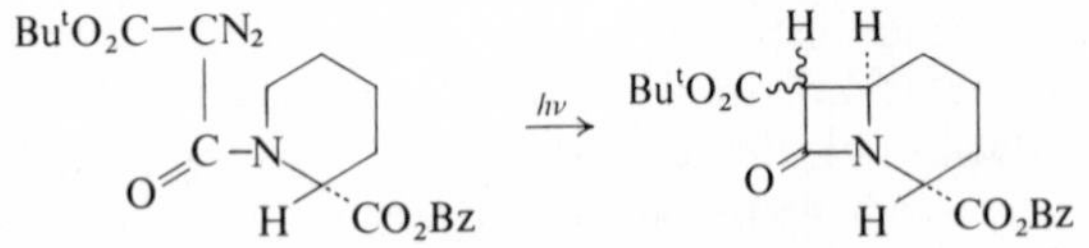

which was cyclised photolytically via the carbene, both the 6α- and 6β-epimers being formed. The 6β-epimer was converted via 6β-azide and 6β-isocyanate into the 6β-phenylacetamido-derivative (cf. penicillin G), and the benzyl ester was cleaved by hydrogenolysis[92]. Though biologically inactive, the product is an analogue of the antibiotics with the correct stereochemistry, and this synthetic route could lead to useful antibiotics.

One difficulty with syntheses of the β-lactam ring is that the active 6β-amino compounds have the unnatural D-amino acid configuration at C-6. In this synthesis the 6β-ester obtained on cyclisation was readily epimerised by base to the natural, but inactive, L-form, which is generally the more stable.

Another recent synthesis indeed produces 6-*epi*-penicillins[93]. It is clearly

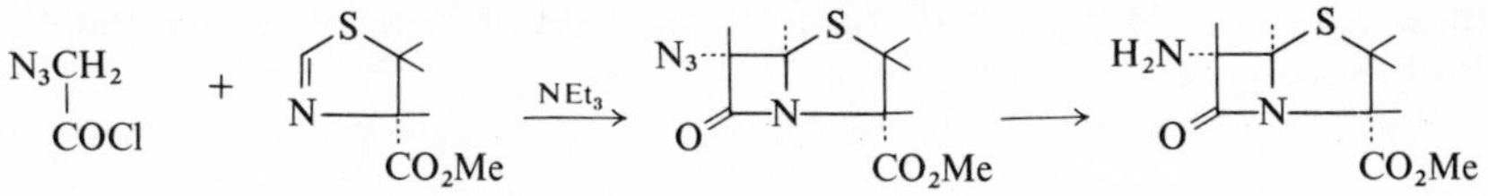

important to be able to epimerise such products.

Numerous epimerisations have been observed[89], usually 6β to 6α, often accompanied by deuteration in suitable media. Monocyclic β-lactams have been epimerised by the action of a fairly strongly basic amidine, though triethylamine was not effective[94]. Proton ionisation can account for such observations, but for other epimerisations induced by sodium hydride Wolfe has suggested the further intermediacy of a ring-opened β-elimination

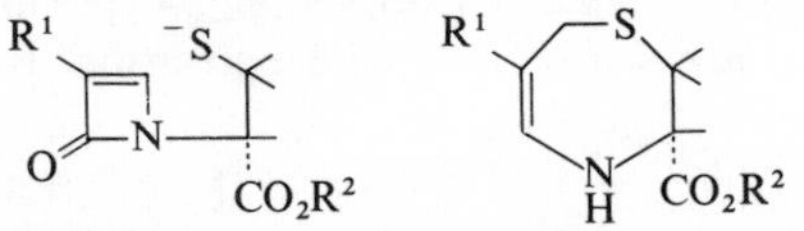

product[95], which could also account for the base catalysed formation of thiazepines from penicillins[96]. Recently amidine catalysis has been used to equilibrate the 6-epimers of penicillanic acid derivatives[97], which should provide a means of converting synthetic 6α-products into active 6β-compounds. An alternative reagent, avoiding basic conditions is N,O-bistrimethyl-silylacetamide[98].

Various workers have introduced other substituents into the lactam rings of the antibiotics. Two recent examples introduce 6α-alkyl groups without altering the 6β-amino-function[99, 100], but the products appear to be inactive.

1.2.2.3 Modification of the second ring

Modifications to the rings and their immediate substituents generally destroy antibacterial properties. An elegant programme of work to define more clearly the structural requirements for antibacterial activity has been reported by Heusler[101]. It is known that the lactam ring must be intact. The thiazolidine ring of penicillins was severed at the various possible points, with suitable structural fragments left attached to the molecule. All were inactive, suggesting that the second, fused ring is also essential. This ring was then modified for comparative studies, and 'benzylcephalocillin' and .

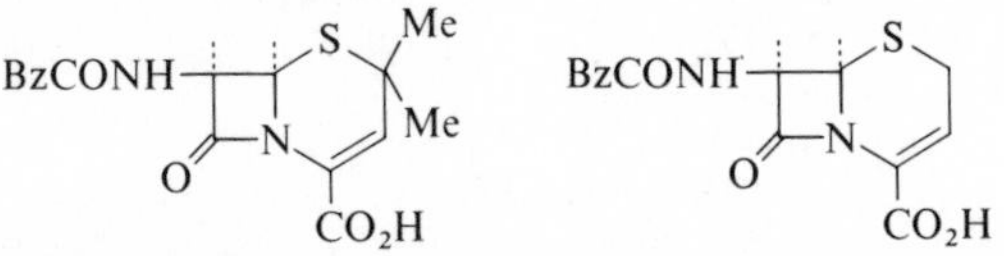

'bis-nor-cephalocillin' were both found to be active. This is in accord with the stereochemical findings discussed in Section 1.2.2.1.

Because penicillins G and V are readily available, conversion of penicillins into cephalosporins is a potentially attractive route to new semi-synthetic antibiotics. The conversion can be achieved through penicillin sulphoxides, readily obtained by oxidation. The β-sulphoxide is the more stable when (as usually) there is a 6β-NH group available for chelation, and is produced from the α-sulphoxide by heating, the reverse isomerisation being photocatalysed[102]. Both isomerisations proceed by S-1—C-2 cleavage[103].

With *p*-toluenesulphonic acid as catalyst the sulphoxides rearrange to

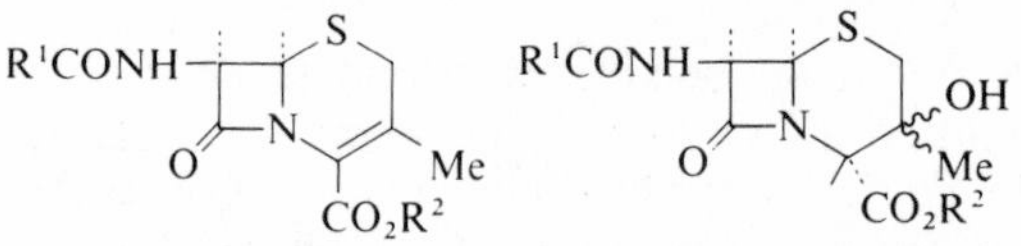

desacetoxycephalosporins, while with sulphuric acid they give in addition the corresponding epimeric hydrates[104]. With acetic anhydride the sulphoxides give a mixture of epimeric acetoxypenicillins, the desacetoxycephalosporin, and the acetate of the desacetoxycephalosporin hydrate. The cephalosporin derivatives are formed by incorporation into the ring of the

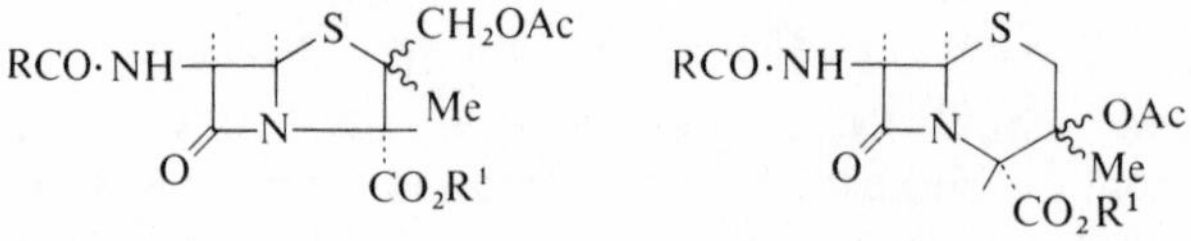

substituent carbon *cis* to the S-oxide group. This rearrangement has been used to synthesise cephalosporins[103]. The acetoxypenicillin from rearrangement of methyl methylpenicillanate β-sulphoxide was oxidised again and,

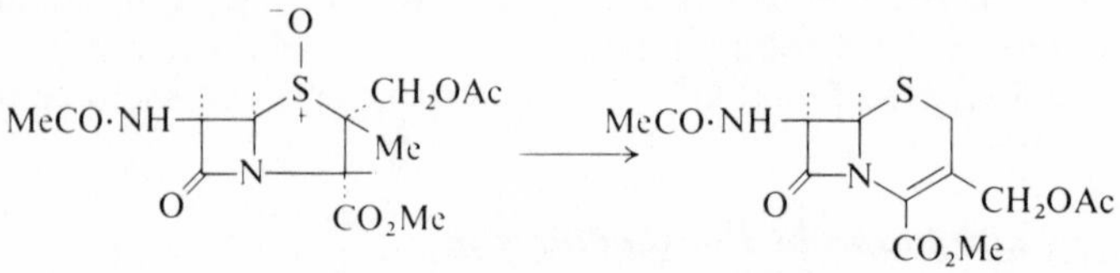

after separation, the correct sulphoxide was again rearranged, giving the cephalosporin. This conversion is much shorter than others reported.

Other recent work on modification of antibiotic structures also involves penicillin sulphoxides. The sulphenic acid obtained by heating a sulphoxide was trapped by conversion into a thioether, and the C-2—C-3 fragment together with its substituents was then removed completely by addition of diazomethane followed by base treatment, or reduction[105].

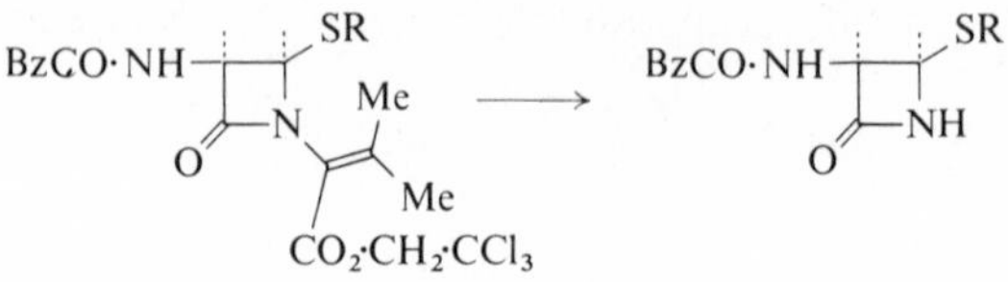

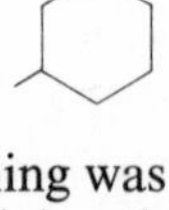

In another method ring opening was by isobutyl thiol, giving a disulphide. This was degraded as before, giving the isobutyldithiolactam. Alternatively, the initial disulphide was desulphurised before degradation[106].

Methods, such as these, of removing the thiazolidine ring should allow re-synthesis to give new materials.

1.3 SEVEN-MEMBERED RINGS

The greater ease of synthesis and greater stability of nitrogen heterocyclic compounds has led to attention being concentrated on them, rather than on oxygen or sulphur heterocyclic compounds.

1.3.1 Monocyclic fully-conjugated compounds

Much recent interest in these compounds is due to their reactivity and readiness to undergo skeletal transformations.

1.3.1.1 Synthesis

1H-Azepines are readily accessible through expansion of suitable six-membered rings. Thus photolysis of ethyl azidoformate in benzene gives 1-ethoxycarbonyl-1H-azepine through nitrene insertion[107], the rate-determining step being azidoformate decomposition[108]. With substituted benzenes the reaction is unselective[107], but isomeric products have been separated successfully[109]. The necessary nitrenoid species have also been generated intramolecularly by reduction of nitrobenzenes with triethyl phosphite (leading to azepinylphosphonates)[110]. Alternatively, nitrogen can be inserted as an ylid, either intermolecularly through action of the nitrosobenzene–triethyl phosphite adduct on benzene (giving 1-phenyl-1H-azepine)[111], or intramolecularly by photolysis of 1-ethoxycarbonylaminopyridinium ylid (giving this time a diazepine, 1-ethoxycarbonyl-1H-1,2-diazepine)[112]. Intramolecular carbonium ion insertion into a dihydropyridine has also been used[113].

Five-membered rings too can be expanded. Photolysis of pyrrole with dimethyl acetylenedicarboxylate gives 3,4-bis(methoxycarbonyl)-1H-azepine via 2,3-addition[114], while indole reacts similarly without irradiation[115]. Yet another route involves addition of cyclopentadienones (with loss of carbon monoxide) with 2H-azirines[116].

In contrast to azepines and diazepines, oxepins[117, 118] and thiepins[119] are less widely accessible (and less stable) and so less well known.

1.3.1.2 Structure and theoretical studies

In agreement with a common extension of Hückel's rule, x-ray analysis shows that these molecules are non-planar polyenes in boat conformations[120], which may invert quite slowly[121]. 1H-Azepines have little tendency to

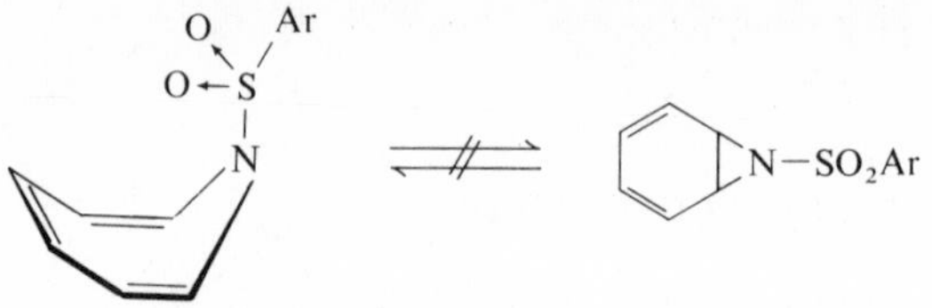

tautomerise to azanorcaradienes[107], in contrast to oxepins, in which rapid

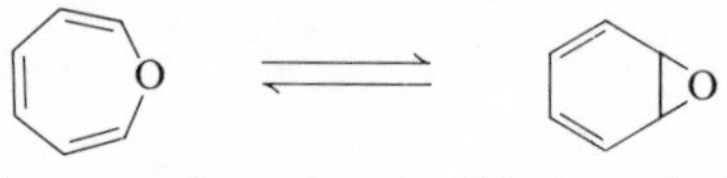

valence tautomerisation may be extensive[118], though there is some evidence for such tautomerisation of azepines[122].

In view of this it is not surprising that calculations of the ground state energies of azepine, oxepin, thiepin, and their benzo- and dibenzo-derivatives show them to have no resonance stabilisation in the seven-membered rings[123].

1.3.1.3 Intramolecular ring modification

This lack of resonance stability renders the fully conjugated rings very reactive. Thus, they quite generally undergo photochemical transannular bond formation giving a [3.2.0]bicyclic structure with the heteroatoms in the five-membered ring. Examples are oxepins and azepines[124], benzoxepins[125], 1,2-diazepines[126], and 1,2-diazepin-4-ones[127]. Reversion to the monocyclic

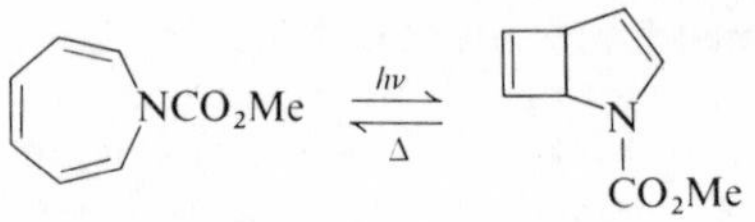

structures may occur by simple thermolysis[124] or may be attainable only indirectly[126] or not at all[125].

A different kind of transannular rearrangement of azepines gives fulvenes

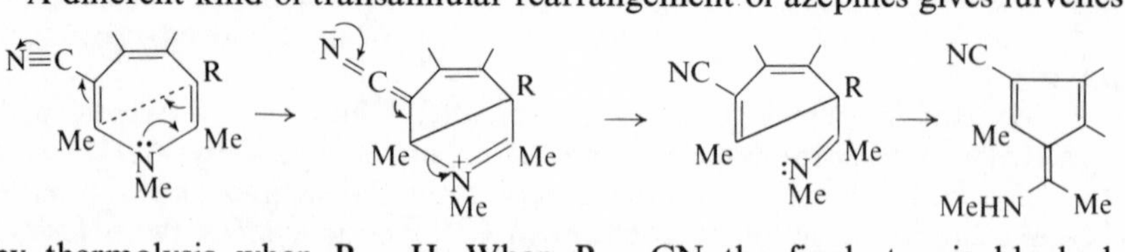

by thermolysis when R = H. When R = CN the final step is blocked, supporting the proposed mechanism[128].

1.3.1.4 Cyclo-addition with dienophiles

Most work has used carbonyl substituted compounds in order to reduce polyene activity, which may otherwise be inconveniently high, especially with azepines. Thus N-methylazepine undergoes very rapid (6+6) dimerisation at the 2,7-positions above 0 °C[129]. Additions to N-ethoxycarbonylazepines are generally allowed thermal (4+2) cyclo-additions at the 2,5-positions by dienophiles such as tetracyanoethylene[108, 109, 130] and 1,4-phthalazinedione[131]. These azepines can also act as 2π-donors (at the 4,5-positions) in allowed thermal cyclo-additions to dienes[130], reversing their role.

N-Ethoxycarbonylazepines will also undergo thermal (6+2) cyclo-additions of nitrosobenzene[130] and diethyl *trans*-azodicarboxylate (in contrast to *cis*-dienophiles)[131]. As these are disallowed as concerted processes, they are probably two-step reactions.

1.3.1.5 Metal complexes

The tendency to react as dienes at the 2,5-positions recurs in the formation of complexes with tricarbonyliron. 1-Methoxycarbonyl-1*H*-azepine[132], 3-acetyl-1*H*-azepine[133], 1-isopropoxycarbonyl-1,2-diazepine[134], and several 3,5,7-triphenyl-1,2-diazepines[135] have all been shown by x-ray measurements to form complexes in which the iron is bonded effectively to a butadiene unit. This makes an angle of close to 140 °C with the rest of the ring. Some complexes are different, however. A rhodium–triphenyldiazepine complex has

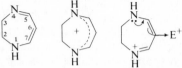

the metal atom bonded to one nitrogen atom[136], while the same diazepine reacts with di-iron nonacarbonyl by N—N bond fission, giving a complex with two N—Fe—N bridges[137].

It is clear that non-planar polyene structures restrict conjugation to produce predominantly diene behaviour in these analogues of cyclo-octatetraene.

1.3.2 2,3-Dihydro-1,4-diazepines

In great contrast to the fully conjugated rings are the partly conjugated 2,3-dihydro-1,4-diazepines. These are vinylogous amidines, and like amidines

are strongly basic, forming symmetrical mesomeric cations. The resonance energy of the cations[138] is such (c. 80 kJ mol^{-1}) that they are effectively devoid of olefinic properties, and instead undergo electrophilic substitution (at position 6) like benzenoid compounds. For this reason they have been described as *quasi-aromatic* (as have various metal acetylacetonates with similar properties[139]), or more recently *meneidic*[140] (signifying 'retention of type' in Robinson's phrase[141]).

Examples of substitution reactions carried out with these compounds are nitration[142], halogenation[143], deuteration[144, 145], azo-coupling[146], reduction of nitro-derivatives to amines[142], and the Sandmeyer reaction upon the amines[142].

It is not essential for the ring to be seven-membered, as any molecular skeleton suitable for holding the conjugated system would serve, but the diazepine rings are readily synthesised from 1,2-diamines and 1,3-dicarbonyl compounds. Indeed, substituents which can be introduced by reaction can alternatively be present in the carbonyl precursor[147].

1.3.3   1,5-Benzodiazepines and analogues

1.3.3.1   Chemical character

These compounds are alternatively named as 2,3-benzo-derivatives of the 2,3-dihydro-1,4-diazepines, and are synthesised similarly[148]. For this reason

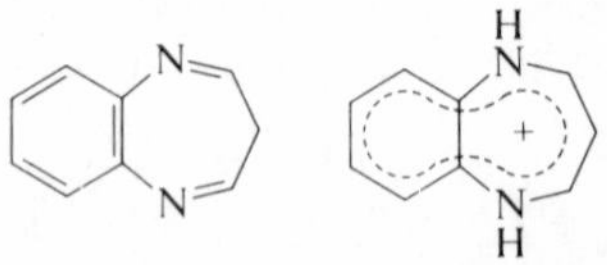

they were the first seven-membered heterocyclic rings to be examined[149]. The bases are not aromatic, the major base tautomer being not even fully conjugated. Protonation affords purple cations, indicating full peripheral conjugation of a 12π-system, but this is not very stable. The pK is low[150], basic decomposition is facile[149], and recent attempts at nitration[151] and acetylation[152] failed to give diazepine products. However, the cations are stable in acidic solutions.

Analogous 2,3-benzo-1,4-thiazepines readily extrude sulphur, or otherwise undergo ring contraction, under basic conditions[153].

Another interesting analogue is an isoelectronic, fully conjugated 2,3-furazano-1,4-diazepinium salt, but its reactions have not yet been examined[154].

1.3.3.2   Metal complexes

A blue-black ferrous sulphate complex of 2,4-dimethyl-1,5-benzodiazepine has long been known[155], and a series of sulphate[156] and halide[156, 157] complexes of this diazepine with cobalt, manganese, zinc, cadmium, copper and antimony have now been prepared and examined. In the halides the metals are in tetrahedral environments, suggesting MX_4^{2-}, the sulphates appear to contain octahedral $M(H_2O)_6$, and the spectra (apart from metal absorptions) resemble those of the diazepinium cations. Thus in no compound does the heterocycle seem to be co-ordinated, in direct contrast to the polyolefinic compounds discussed earlier. This again suggests that, although both types of heterocycle contain $4n$ π-electrons, not $(4n+2)$, these benzodiazepinium cations are significantly stabilised by delocalisation.

1.3.4 Pharmacologically active diazepines

Since 1960, when chlordiazepoxide (2-methylamino-5-phenyl-7-chloro-3*H*-1,4-benzodiazepine 4-oxide) was introduced as a tranquilliser, a large number of seven-membered heterocyclic compounds have been synthesised with a view to discovering other useful drugs. Most are 1,4-benzodiazepines[148, 158]. Compounds are now known which are variously hypnotics, central muscle relaxants, anticonvulsants, coronary vasodilators, etc. Some seven benzodiazepines are now in clinical use.

Synthesis of a variety of compounds has allowed structural aids to activity to be better defined[158]. An electron-withdrawing substituent at C-7 is needed,

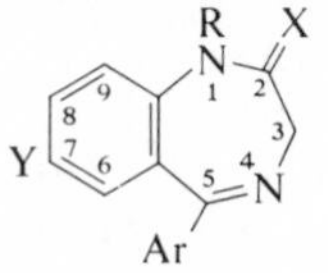

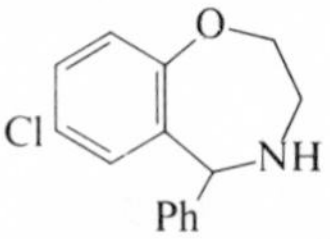

and chlorine is generally used, though a nitro-group is effective, as in nitrazepam (R = H, X = O, Ar = Ph, Y = NO$_2$). Other desirable factors are an *ortho* (but not *para*) halogen in Ar, and also R = Me, but these are not essential, as the benzoxazepine shown possesses hypotensive activity[159]. However, an easily formed metabolite of diazepam (R = Me, X = O, Ar = Ph, Y = Cl) is the *des*-methyl derivative, and as this may therefore be the active agent, other (and less readily metabolised) R groups have been examined[160-162]. The N-4 nitrogen seems essential, as the C-4 analogue of diazepam is inactive[163].

A different approach is suggested by the observation that diazepam and diphenylhydantoin, both effective antiepileptics, have very different chemical structures but rather similar spatial properties, indicating that synthesis of *stereo*-analogues may be fruitful (cf. penicillins and cephalosporins)[164].

These benzodiazepines are generally synthesised by ring enlargement of appropriate 2-(α-chloroalkyl)quinazoline 3-oxides through treatment with amines or hydroxide ions[158] (cf. ring expansion to azepines) though other ring enlargements (such as a Beckmann rearrangement) can be used. The substituents can often be modified as required after ring synthesis, though structures carrying groups difficult to introduce may require more elaborate methods, as when a pyrrolino analogue of diazepam was synthesised recently from 5-chloroindoline in seven steps[165]. Such syntheses generally employ standard methods.

A range of dibenzo-compounds has also been examined recently, such as the dibenzthiepin octoclothepin and derivatives[166], and various dibenzo-

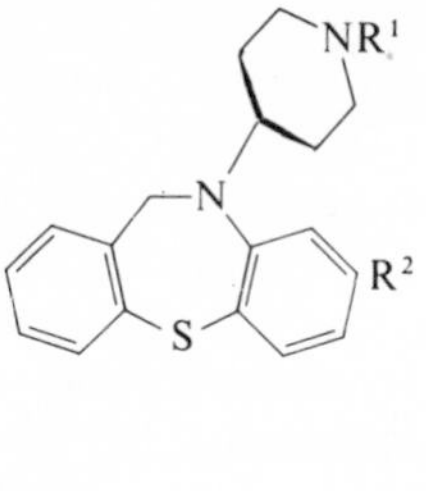

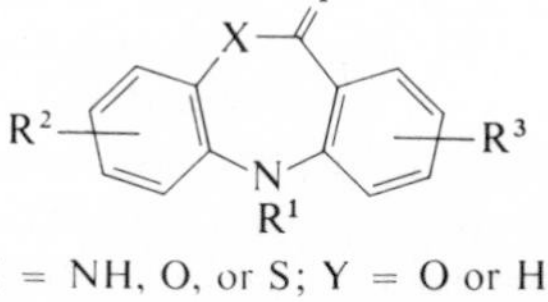

(X = NH, O, or S; Y = O or H$_2$)

diazepines[167-169] oxazepines[170], and thiazepines[170]. The properties claimed range from anti-depressant to anti-allergy and (potentially) anti-malarial.

1.4 EIGHT- AND NINE-MEMBERED RINGS

Nitrogen (though not oxygen or sulphur) can form a neutral, fully conjugated eight-membered ring. Azacyclo-octatetraene derivatives have been prepared by standard ring expansions through azides[171] and oximes[172] but one recent synthesis involves an interesting series of cyclo-additions and fragmentations[173]. Formation of the azocine ring depends on ready valence isomerisation, which also appears similarly in the fully conjugated system and has

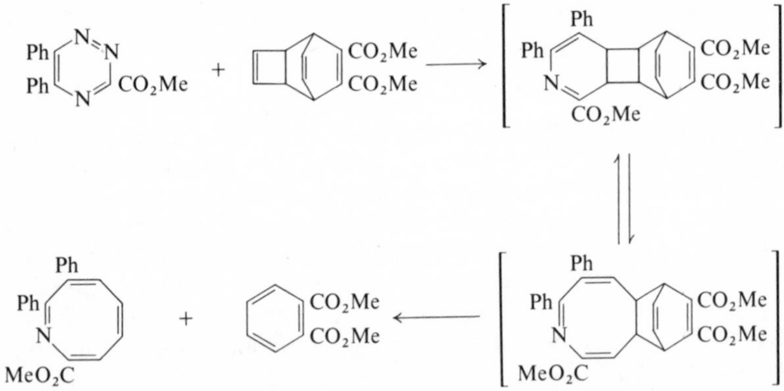

again been utilised synthetically to prepare azocines from [4.2.0]bicyclic precursors[174]. In the fully conjugated azocines this tautomerisation is an equilibrium, which is normally almost entirely in favour of the azocine form. However, fusion of a tri-, tetra-, or penta-methylene bridge across the 2,7-positions (giving azapropellanes) so strains the tub-shaped azocine that azabicyclo-octatriene forms are preferred[174]. This parallels the known behaviour of cycloheptatrienes, for example.

Electrochemical[175] or potassium metal[176] reduction of azocines gives their planar di-anions. Like cyclo-octatetraene di-anion, these 10π systems are delocalised, stable, do not rearrange readily, and support a diamagnetic ring current, in agreement with Hückel's rule, and in contrast to the poly-olefinic character of the parent diazocines.

Thermal valence tautomerisation has also yielded 2,3-benzo-derivatives of 10-electron systems of nitrogen[177] and sulphur[178], but these (like the oxygen

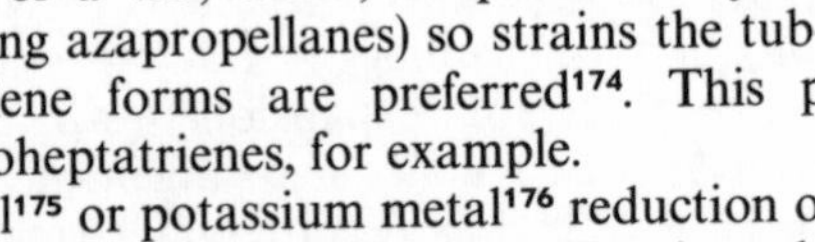

analogue[179]) do not support diamagnetic ring currents and appear to be non-aromatic. Upon photo-irradiation they isomerise again to the [4.2.0] bicyclic systems, an interconversion very like that shown by azepines.

An aromatic sulphur derivative has been prepared, however, by fusing three tetrafluorocyclobutene rings to a 1,4-dithiocin ring. The compound shows an aromatic ring current similar in magnitude to that in the analogous

benzene derivative[180]. It seems likely that steric factors are mainly responsible, since the four-membered rings will tend to force on the dithiocin ring the bond angles needed for planarity. It would be very interesting to know the properties of the analogous dihydrodiazocines.

It appears to be rather easier to achieve a neutral aromatic nine-membered ring. Azonin itself has been prepared[181], and is very reactive, but quite stable in an inert environment and shows an aromatic ring current[182]. Of various heteronins now known, it appears[182] that oxonin, N-ethoxycarbonylazonin, N-acetylazonin and also a substituted thionin[183], are not aromatic because the heteroatom is insufficiently able to provide an electron pair to the π-system, while azonin, its potassium salt, and its N-methyl- and N-ethyl-derivatives are aromatic because the electrons are available.

In view of the known properties of carbocyclic 10π-electron systems, it seems likely that azonins are the largest rings able to show aromatic properties without also having 'inside' protons.

References

1. E.g. Weissberger, A., ed. (1964). *Heterocyclic Compounds with Three- and Four-membered Rings*. (New York: Interscience); Schmitz, E. (1967). *Dreiringe mit Zwei Heteroatomen*. (Berlin, Heidelberg, New York: Springer-Verlag); Dermer, O. C. and Ham, G. E. (1969). *Ethyleneimine and Other Aziridines*. (New York: Academic Press)
2. Anderson, D. J., Gilchrist, T. L., Horwell, D. C. and Rees, C. W. (1970). *J. Chem. Soc. C*, 576
3. Anderson, D. J., Gilchrist, T. L. and Rees, C. W. (1969). *Chem. Commun.*, 147
4. Ciabattoni, J. and Cabell, L. (1971). *J. Amer. Chem. Soc.*, **93**, 1482
5. Kellogg, R. M., Wassenaar, S. and Buter, J. (1970). *Tetrahedron Lett.*, 4689
6. Greene, F. D. and Hecht, S. S. (1970). *J. Org. Chem.*, **35**, 2482
7. Sheehan, J. C. and Beeson, J. H. (1967). *J. Amer. Chem. Soc.*, **89**, 362
8. Sheehan, J. C. and Nafissi-v, M. M. (1970). *J. Org. Chem.*, **35**, 4246
9. Quast, H. and Schmitt, E. (1970). *Angew. Chem. Int. edn.*, **9**, 381
10. Rens, M. and Ghosez, L. (1970). *Tetrahedron Lett.*, 3765
11. Fujita, S., Imamura, K. and Nozaki, H. (1971). *Bull. Chem. Soc. Japan*, **44**, 1975
12. Brois, S. J. (1968). *J. Amer. Chem. Soc.*, **90**, 506, 508
13. Felix, D. and Eschenmoser, A. (1968). *Angew. Chem. Int. edn.*, **7**, 224
14. Padwa, A. and Battisti, A. (1971). *J. Org. Chem.*, **36**, 230
15. Brois, S. J. (1970). *J. Amer. Chem. Soc.*, **92**, 1079
16. Mannschreck, A. and Seitz, W. (1971). *23rd International Congress of Pure and Applied Chemistry*, Vol. 2, 309. (London: Butterworths)
17. Anderson, D. J. and Gilchrist, T. L. (1971). *J. Chem. Soc. C*, 2273
18. Greene, F. D., Stowell, J. C. and Bergmark, W. G. (1969). *J. Org. Chem.*, **34**, 2254
19. Fahr, E., Rohlfing, W., Thiedemann, R., Mannschreck, A., Rissmann, G. and Seitz, W. (1970). *Tetrahedron Lett.*, 3605
20. Pews, R. G. and Ojha, N. D. (1970). *Chem. Commun.*, 1033
21. Haselbach, E., Heilbronner, E., Mannschreck, A. and Seitz, W. (1970). *Angew. Chem. Internat. edn.*, **9**, 902

22. Carpino, L. A. and Chen, H. W. (1971). *J. Amer. Chem. Soc.*, **93**, 785
23. Heine, H. W., Myers, J. D. and Peltzer, E. T. (1970). *Angew. Chem. Int. edn.*, **9**, 374
24. Baldwin, J. E., Bhatnagar, A. K., Chun-Choi, S. and Shortridge, T. J. (1971). *J. Amer. Chem. Soc.*, **93**, 4082
25. Breslow, R. (1968). *Angew. Chem. Int. edn.*, **7**, 565
26. Clark, D. T. (1969). *Chem. Commun.*, 637
27. Gilchrist, T. L., Gymer, G. E. and Rees, C. W. (1971). *Chem. Commun.*, 1519
28. Thornton, D. E., Gosavi, R. K. and Strausz, O. P. (1970). *J. Amer. Chem. Soc.*, **92**, 1768
29. Russell, R. L. and Rowland, F. S. (1970). *J. Amer. Chem. Soc.*, **92**, 7508
30. Kato, H., Kawamura, M., Shiba, T. and Ohta, M. (1970). *Chem. Commun.*, 959
31. Huisgen, R., Scheer, W. and Ham, G. E. (1967). *J. Amer. Chem. Soc.*, **89**, 1753
32. Huisgen, R. and Maeder, H. (1971). *J. Amer. Chem. Soc.*, **93**, 1777
33. Hermann, H., Huisgen, R. and Maeder, H. (1971). *J. Amer. Chem. Soc.*, **93**, 1779
34. Hamberger, H. and Huisgen, R. (1971). *Chem. Commun.*, 1190
35. Dahmen, A., Hamberger, H., Huisgen, R. and Markowski, V. (1971). *Chem. Commun.*, 1192
36. Trozzolo, A. M. and DoMinh, T. (1971). *23rd International Congress of Pure and Applied Chemistry*, Vol. 2, 251, (London: Butterworths)
37. Huisgen, R., Martin-Ramos, V. and Scheer, W. (1971). *Tetrahedron Lett.*, 477
38. Klaus, M. and Prinzbach, H. (1971). *Angew. Chem. Int. edn.*, **10**, 273
39. Lown, J. W. and Matsumoto, K. (1971). *J. Org. Chem.*, **36**, 1405
40. Huisgen, R. and Maeder, H. (1969). *Angew. Chem. Int. edn.*, **8**, 604
41. Gilchrist, T. L., Rees, C. W. and Stanton, E. (1971). *J. Chem. Soc. C*, 988
42. Mueller, R. K., Felix, R. D., Schreiber, J. and Eschenmoser, A. (1970). *Helv. Chim. Acta*, **53**, 1479
43. Hassner, A. and Kascheres, A. (1970).*Tetrahedron Lett.*, 4623
44. Stamm, H. (1971). *Tetrahedron Lett.*, 1205
45. Gaertner, V. (1971). *J. Heterocyclic Chem.*, **8**, 177, 519
46. Moehrle, H. and Feil, R. (1971). *Tetrahedron*, **27**, 1033
47. Crist, D. R. and Leonard, N. J. (1969). *Angew. Chem. Int. edn.*, **8**, 962
48. Foglia, T. A., Gregory, L. M. and Maerker, G. (1970). *J. Org. Chem.*, **35**, 3779
49. Lautenschlaeger, F. (1970). *J. Heterocyclic Chem.*, **7**, 1413
50. Church, R. F. R. and Weiss, M. J. (1970). *J. Org. Chem.*, **35**, 2465
51. Schmitz, E. (1971). *23rd International Congress of Pure and Applied Chemistry*, vol. 2, 283, (London: Butterworths)
52. Frey, H. M. and Liu, M. T. H. (1970). *J. Chem. Soc. A*, 1916
53. Nixon, J. F. and Wilkins, B. (1970). *Z. Naturforsch.*, **25**, 649
54. Massol, M., Mesnard, D., Barrau, J. and Satge, J. (1971). *Compt. Rend., Ser. C*, **272**, 2081
55. Arnold, D. R. (1968). *Adv. Photochem.*, **6**, 301
56. Dowd, P., Gold, A. and Sachdev, K. (1970). *J. Amer. Chem. Soc.*, **92**, 5725
57. Turro, N. J. (1971). *Pure Appl. Chem.*, **27**, 679
58. Shigemitsu, Y., Katsuhara, Y. and Odaira, Y. (1971). *Tetrahedron Lett.*, 2887
59. Rivas, C., Velez, M. and Crescente, O. (1970). *Chem. Commun.*, 1474
60. Ito, S., Takeshita, H. and Hiraina, M. (1971). *Tetrahedron Lett.*, 1181
61. Foote, C. S. (1971). *Pure Appl. Chem.*, **27**, 635
62. Schaap, A. P. (1971). *Tetrahedron Lett.*, 1757
63. Sander, M. (1966). *Chem. Rev.*, **66**, 341
64. Paquette, L. A. and Freeman, J. P. (1970). *J. Org. Chem.*, **35**, 2249
65. Trost, B. M., Schinski, W. L., Chen, F. and Mantz, I. B. (1971). *J. Amer. Chem. Soc.*, **93**, 676
66. Harpp, D. N. and Gleason, J. G. (1970). *J. Org. Chem.*, **35**, 3259
67. Kumakura, S., Shimozawa, T., Ohnishi, Y. and Ohno, A. (1971). *Tetrahedron*, **27**, 767
68. Siegl, W. O. and Johnson, C. R. (1971). *Tetrahedron*, **27**, 341
69. Dodson, R. M., Jancis, E. H. and Klose, G. (1970). *J. Org. Chem.*, **35**, 2520
70. Gaertner, V. R. (1970). *J. Org. Chem.*, **35**, 3952
71. Dodson, R. M. *et al., J. Org. Chem.*, **36**, 2693, 2698, 2703
72. Siegl, W. O. and Johnson, C. R. (1970). *J. Org. Chem.*, **35**, 3657
73. Hassner, A., Currie, J. O., Jr., Steinfeld, A. S. and Atkinson, R. F. (1970). *Angew. Chem. Internat. edn.*, **9**, 731
74. Levy, A. B. and Hassner, A. (1971). *J. Amer. Chem. Soc.*, **93**, 2051

75. Wells, J. N. and Tarwater, O. R. (1970). *J. Pharm. Sci.*, **60**, 156
76. Masuda, T., Chinone, A. and Oiita, M. (1970). *Bull. Chem. Soc. Japan*, **43**, 3281
77. Ulrich, H. (1967). *Cycloaddition Reactions of Heterocumulenes*, (New York: Academic Press)
78. Brady, W. T. and Patel, A. D. (1971). *J. Heterocyclic Chem.*, **8**, 739
79. Nelson, D. A. (1971). *Tetrahedron Lett.*, 2543
80. Bose, A. K., Spiegelman, G. and Manhas, M. S. (1971). *Tetrahedron Lett.*, 3167
81. Barker, M. W. and Gill, G. T. (1970). *J. Heterocyclic Chem.*, **7**, 1203
82. Bestian, H. (1971). *Pure Appl. Chem.*, **27**, 611
83. Effenberger, F., Prossel, G. and Fischer, P. (1971). *Chem. Ber.*, **104**, 2002
84. Olofson, R. A., Van der Meer, R. K. and Stournas, S. (1971). *J. Amer. Chem. Soc.*, **93**, 1543
85. Ulrich, H. (1971). *23rd International Congress of Pure and Applied Chemistry*, vol. 2, 265. (London: Butterworths)
86. Fosker, G. R., Hardy, K. D., Nayler, J. H. C., Seggery, P. and Stove, E. R. (1971). *J. Chem. Soc. C*, 1917
87. Bose, A. K. and Manhas, M. S. (1969). *Synthesis of Penicillin, Cephalosporin C, and Analogs*, (New York: Dekker)
88. Manhas, M. S. and Bose, A. K. (1971). *β-Lactams Natural and Synthetic*, (New York: Wiley-Interscience)
89. Lucke, J. L. and Balavoine, G. (1971). *Bull. Soc. Chim. France*, 2733
90. Washkuhn, R. J. and Robinson, J. R. (1971). *J. Pharm. Sci.*, **60**, 1168
91. Sweet, R. M. and Dahl, L. F. (1970). *J. Amer. Chem. Soc.*, **92**, 5489
92. Brunwin, D. M., Lowe, G. and Parker, J. (1971). *Chem. Commun.*, 865
93. Bose, A. K., Spiegelman, G. and Manhas, M. S. (1968). *J. Amer. Chem. Soc.*, **90**, 4506; (1971). *J. Chem. Soc. C*, 2468
94. Bose, A. K., Narayanan, C. S. and Manhas, M. S. (1970). *Chem. Commun.*, 975
95. Wolfe, A. *et al.*, (1968). *Chem. Commun.*, 242. (1970). *Chem. Commun.*, 1067
96. Kovacs, O. K. J., Elkström, B. and Sjöberg, B. (1969). *Chem. Commun.*, 1863
97. Jackson, J. R. and Stoodley, R. J. (1971). *Chem. Commun.*, 647
98. Gutowski, G. E. (1970). *Tetrahedron Lett.*, 1779
99. Boehme, E. H. W., Applegate, H. E., Toeplitz, B., Dolfini, J. E. and Gougoutas, J. Z. (1971). *J. Amer. Chem. Soc.*, **93**, 4324
100. Kaiser, G. V., Ashbrook, C. W. and Baldwin, J. E. (1971). *J. Amer. Chem. Soc.*, **93**, 2342
101. Heusler, K. (1971). *23rd International Congress of Pure and Applied Chemistry*, vol. 3, 87, (London: Butterworths)
102. Archer, R. A. and DeMarco, P. V. (1969). *J. Amer. Chem. Soc.*, **91**, 1530
103. Spry, D. O. (1970). *J. Amer. Chem. Soc.*, **92**, 5006
104. Gutowski, G. E., Foster, B. J., Daniels, C. J., Hatfield, L. D., and Fisher, J. W. (1971). *Tetrahedron Lett.*, 3433
105. Barton, D. H. R., Greig, D. G. T., Sammes, P. G. and Taylor, M. V. (1971). *Chem. Commun.*, 845
106. Barton, D. H. R., Sammes, P. G., Taylor, M. V., Cooper, C. M., Hewitt, G., Looker, B. E. and Underwood, W. G. E. (1971). *Chem. Commun.*, 1137
107. Paquette, L. A. (1971). *Angew. Chem. Int. edn.*, **10**, 11
108. Sasaki, T., Kanematsu, K. and Kakelu, A. (1970). *Bull. Chem. Soc. Japan*, **43**, 2893
109. Photis, J. M. (1970). *J. Heterocyclic Chem.*, **7**, 1249; (1971). *ibid.*, **8**, 167
110. Cadogan, J. I. G., Sears, D. J., Smith, D. M. and Todd, M. J. (1969). *J. Chem. Soc. C*, 2813
111. Sundberg, R. J. and Smith, R. H. (1971). *Tetrahedron Lett.*, 267
112. Gletter, R., Schmidt, D. and Streith, J. (1971). *Helv. Chim. Acta*, **54**, 1645
113. Van-Bergen, T. J. and Kellog, R. M. (1971). *J. Org. Chem.*, **36**, 978
114. Gandhi, R. P. and Chadha, V. K. (1971). *Ind. J. Chem.*, **9**, 305.
115. Acheson, R. M. and Bridson, J. N. (1971). *Chem. Commun.*, 1225
116. Anderson, D. J. and Hassner, A. (1971). *J. Amer. Chem. Soc.*, **93**, 4339
117. Vogel, E., Böll, W. A. and Günther, H. (1965). *Tetrahedron Lett.*, 609
118. Vogel, E. and Günther, H. (1967). *Angew. Chem. Int. edn*, **6**, 385
119. Hoffmann, J. M. and Schlessinger, R. H. (1970). *J. Amer. Chem. Soc.*, **92**, 5263
120. Paul, I. C., Johnson, S. M., Paquette, L. A., Barrett, J. H. and Haluska, R. J. (1968). *J. Amer. Chem. Soc.*, **90**, 5023

121. Svanholm, U. (1971). *Acta Chem. Scand.*, **25**, 640
122. Prinzbach, H., Stusche, D. and Kitzing, R. (1970). *Angew. Chem. Int. edn.*, **9**, 377
123. Dewar, M. J. S. and Trinajstic, N. (1970). *Tetrahedron*, **26**, 4269
124. Paquette, L. A. and Barrett, J. H. (1966). *J. Amer. Chem. Soc.*, **88**, 1718
125. Hofmann, H. and Hofmann, P. (1971). *Tetrahedron Lett.*, 4055
126. Kan, G., Thomas, M. T. and Snieckus, V. (1971). *Chem. Commun.*, 1022
127. Moore, J. A., Volker, E. J. and Kopay, C. M. (1971). *J. Org. Chem.*, **36**, 2676
128. Mahendran, M. and Johnson, A. W. (1971). *J. Chem. Soc. C*, 1237
129. Göttlicher, S. and Habermehl, G. (1971). *Chem. Ber.*, **104**, 524
130. Paquette, L. A., Kuhla, D. E., Barrett, J. H. and Leichter, L. M. (1969). *J. Org. Chem.*, **34**, 2888
131. Murphy, W. S. and McCarthy, J. P. (1970). *Chem. Commun.*, 1129
132. Johnson, S. M. and Paul, I. C. (1970). *J. Chem. Soc. B*, 1783
133. Waite, M. G. and Sim, G. A. (1971). *J. Chem. Soc. A*, 1009
134. Allman, R. (1970). *Angew. Chem. Int. edn.*, **9**, 958
135. Carty, A. J., Kan, G., Madden, D. P., Snieckus, V., Stanton, M. and Birchall, T. (1971). *J. Organometallic Chem.*, **32**, 241
136. Smith, R. A., Madden, D. P., Carty, A. J. and Palenik, G. J. (1971). *Chem. Commun.*, 427
137. Carty, A. J., Madden, D. P., Mathew, M., Palenik, G. J. and Birchall, T. (1970). *Chem. Commun.*, 1664
138. Lloyd, D. and Marshall, D. R. (1972). *Chem. Ind.*, 335
139. Collman, J. P. (1963). *Reactions of Coordinated Ligands*, 78, *Advan. Chem. Ser.*, Washington, D. C. (American Chemical Society)
140. Lloyd, D. and Marshall, D. R. (1971). *The Jerusalem Symposia on Quantum Chemistry and Biochemistry*, vol. 3, 85. (Jerusalem: The Israel Academy of Sciences and Humanities)
141. Armit, J. W. and Robinson, R. (1925). *J. Chem. Soc.*, **127**, 1604
142. Gorringe, A. M., Lloyd, D. and Marshall, D. R. (1970). *J. Chem. Soc. C*, 617
143. Barnett, C., Lloyd, D., Marshall, D. R. and Mulligan, L. A. (1971). *J. Chem. Soc. B*, 1529
144. Barnett, C. and Warkentin, J. (1968). *J. Chem. Soc. B*, 1572
145. Butler, A. R., Lloyd, D. and Marshall, D. R. (1971). *J. Chem. Soc. B*, 795
146. Grant, E. M., Lloyd, D. and Marshall, D. R. (1970). *Chem. Commun.*, 1320
147. E.g. Reichardt, C. (1971). *Annalen*, **746**, 207
148. Archer, G. A. and Sternbach, L. H. (1968). *Chem. Rev.*, **68**, 747
149. Thiele, J. and Steimmig, G. (1907). *Chem. Ber.*, **40**, 955
150. Schwarzenbach, G. and Lutz, K. (1940). *Helv. Chim. Acta*, **23**, 1147
151. Levshina, K. V., Glazyrina, L. P., and Safonova, T. S. (1970). *Khim. Geterotsikl. Soedin.*, 1133
152. Levshina, K. V., Glazyrina, L. P. and Safonova, T. S. (1970). *Khim. Geterotsikl. Soedin.*, 1135
153. Wilhelm, M. and Schmidt, P. (1970). *Helv. Chim. Acta*, **53**, 1697
154. Gasco, A., Rua, G., Menziani, G. M. and Tappi, G. (1970). *J. Heterocyclic Chem.*, **7**, 131
155. Emmert, B. and Gottschneider, H. (1933). *Ber.*, **66**, 1871
156. Hunter, P. W. W. and Webb, G. A. (1971). *J. Inorg. Nucl. Chem.*, in the press
157. Ouchi, A., Takeuchi, T., Nakatani, M. and Takahashi, Y. (1971). *Bull. Chem. Soc. Japan*, **44**, 434
158. Sternbach, L. H. (1971). *Angew. Chem. Int. edn.*, **10**, 34
159. Walker, G. N. and Smith, R. T. (1971). *J. Org. Chem.*, **36**, 305
160. Gilman, N. W. and Sternbach, L. H. (1971). *J. Heterocyclic Chem.*, **8**, 297
161. Lamdan, S., Gaozza, C. H., Sicardi, S. and Izquierdo, J. A. (1970). *J. Med. Chem.*, **13**, 742
162. Robichaud, R. C., Gylys, J. A., Sledge, K. L. and Hillyard, I. W. (1970). *Arch. Int. Pharmacodyn. Ther.*, **185**, 213
163. Loev, B., Goodman, M. M., Zirkle, C. and Macko, E. (1970). *Arzneim. Forsch.*, **20**, 974
164. Camerman, A. and Camerman, N. (1970). *Science*, **168**, 1457
165. Hester, J. B., Rudzik, A. D. and Veldcamp, W. (1970). *J. Med. Chem.*, **13**, 827
166. Jilek, J. O., Pomykacek, J., Metysova, J. and Protiva, M. (1971). *Coll. Czech. Chem. Commun.*, **36**, 2226

167. Greig, M. E., Gibbons, A. J. and Young, G. A. (1971). *Arch. Int. Pharmacodyn. Ther.,* **190,** 299
168. Davis, M., Knowles, P., Sharp, B. W. and Walsh, R. J. A. (1971). *J. Chem. Soc. C,* 2449
169. Lindt, S., Lauener, H. and Eichenberger, E. (1971). *Farmaco, Ed. Prat.,* **26,** 585
170. Yale, H. L., Beer, B., Pluscec, J. and Spitzmiller, E. R. (1970). *J. Med. Chem.,* **13,** 713
171. Looker, J. J. (1971). *J. Org. Chem.,* **36,** 2681
172. Coates, R. M. and Johnson, E. F. (1971). *J. Amer. Chem. Soc.,* **93,** 4016
173. Elix, J. A., Wilson, W. S. and Warrener, R. N. (1970). *Tetrahedron Lett.,* 1837
174. Paquette, L. A., Kakihana, T., Hansen, J. F. and Philips, J. C. (1971). *J. Amer. Chem. Soc.,* **93,** 152
175. Anderson, L. B., Hansen, J. F., Kakihana, T. and Paquette, L. A. (1971). *J. Amer. Chem. Soc.,* **93,** 161
176. Paquette, L. A., Hansen, J. F. and Kakihana, T. (1971). *J. Amer. Chem. Soc.,* **93,** 168
177. Shue, H. J. and Fowler, F. W. (1971). *Tetrahedron Lett.,* 2437
178. Coffen, D. L., Poon, Y. C. and Lee, M. L. (1971). *J. Amer. Chem. Soc.,* **93,** 4627
179. Schroth, W. and Werner, B. (1967). *Angew. Chem. Int. edn.,* **6,** 697
180. Riley, M. O. and Park, J. D. (1971). *Tetrahedron Lett.,* 2871
181. Anastassiou, A. G. and Gebrian, J. H. (1970). *Tetrahedron Lett.,* 825
182. Anastassiou, A. G., Eachus, S. W., Cellura, R. P. and Gebrian, J. H. (1970). *Chem. Commun.,* 1133
183. Garratt, P. J., Holmes, A. B., Sondheimer, F. and Vollhardt, K. P. C. (1970). *J. Amer. Chem. Soc.,* **92,** 4492

2
Nitrogen Heterocyclic Molecules, Part 1
Pyrroles, Indoles, Isoindoles, Indolizines and Cyclazines

R. ALAN JONES
University of East Anglia, Norwich

2.1 PYRROLES, INDOLES, ISOINDOLES, INDOLIZINES AND CYCLAZINES

During the period covered by this review a comprehensive text on the chemistry of indoles[1] and a review of the physico-chemical properties of pyrroles[2] have been published. Both survey the literature up to 1968. This review covers the most significant developments in the chemistry of simple pyrroles and their benzo derivatives up to the end of 1971.

2.1.1 Ring synthesis

2.1.1.1 Pyrroles

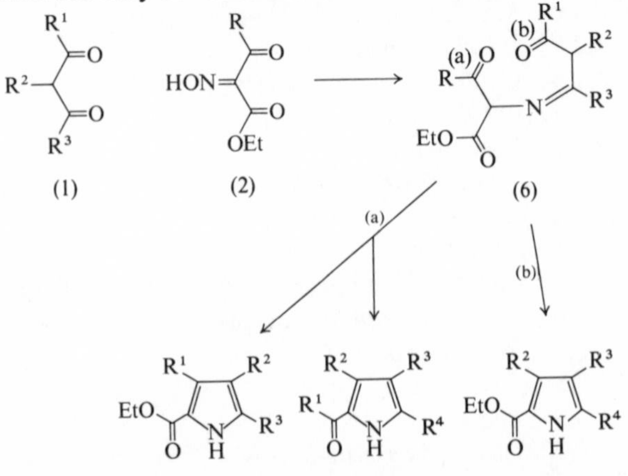

The classic Knorr pyrrole synthesis has again been studied in detail. Oxygen appears to be important to the reaction and the low yields sometimes encountered probably result from denitrosation of the intermediate nitrosocarbonyl compound[3]. The modified reaction involving condensation of β-diketones (1) with nitrosoketoacetates (2) can theoretically lead to three products (3–5) via (6). The normal product (3) is usually found to be impure and the contaminant has now been shown to be the acetylpyrrole (4)[4]. Such an observation can only be rationalised if one considers that the enol form of carbonyl (a) in (6), leading to (3) and (4), is more stable than that of carbonyl group (b). Although a pyrrole of type (5, $R^1 = R^2 = R^4 = Me$) has been isolated[5] from the condensation of (2, $R = Me$) with the diketone (1, $R^2 = H$, $R^3 = R^4 = Me$), such compounds are more likely to be formed from keto-acetates having electron-donating R groups[4]. Steric factors appear to be negligible[4]. Under similar conditions β-dicarbonyl compounds with cyano-acetic esters gave 2-cyanopyrroles with lesser amounts of the 2-ester[6].

A general synthesis of pyrroles based upon the Knorr reaction has been described by Pleininger[7] and a useful one step conversion of β-dicarbonyl compounds into pyrroles using hydroxylamine O-sulphonic acid has been reported[8].

Considerable interest[9-13] has also been shown in a modification of the Fischer conversion of hydrazones into indoles (see below) for the synthesis of pyrroles. Azines (7) formed from hydrazine or its N-methyl and N,N'-dimethyl derivatives, cyclise under a variety of conditions to give the pyrroles (8). Side reactions can, however, produce significant amounts of the pyrazolines (9)[9-11]. It is also thought that thermolysis in a mass spectrometer of

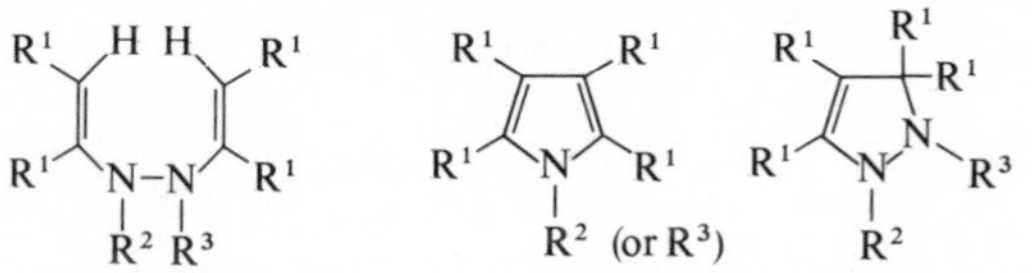

(7)   (8)   (9)

the readily attainable 1,2,4,5-tetramethylperhydro-1,2,4,5-tetrazine gives the pyrrole (8, $R^2 = Me$, $R^1 = H$) via (7, $R^2 = R^3 = Me$, $R^1 = H$)[14].

Further examples of the now well established addition reaction of alkynes to 1,3-dipolar systems to form five-membered heterocycles have been described for the synthesis of pyrroles. From azlactones[15] and mesoionic compounds[16, 17], e.g. (10, $R = H$ or Me), pyrroles of the type (11, $R^1 = H$ or Me) are formed. Indolizines have been prepared by a similar route[18].

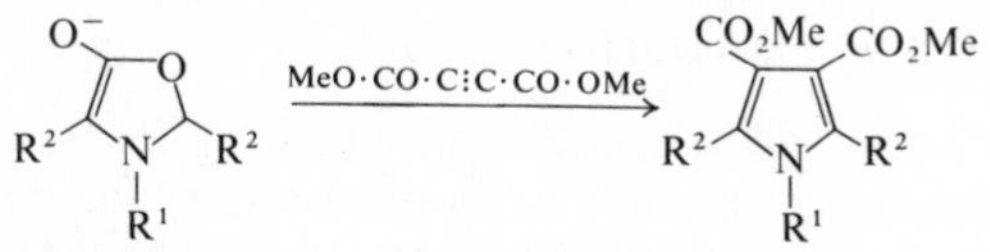

(10)   (11)

An interesting new synthesis of pyrrole via an initial 1,3-dipolar addition of an alkoxyalkene to a sulphonylazide probably follows the route (12 → 13)[19]. Several other novel syntheses have been reported. Two research groups have converted α,β-unsaturated derivatives of aziridine into pyrroles[20, 21]. Thus, (14) rearranged when heated to give the 2,3-dihydropyrrole (15) which was readily dehydrogenated to the pyrrole[20], whilst the preparation of 2-phenyl-indole[21] from N-phthalimide-2,3-diphenylaziridine proceeds via 2,3-diphenylazirine[22]. High yields of pyrroles have also been isolated from the thermal rearrangement of O-vinylketoximes[23], obtained from the oxime and acetylene and it appears that, by starting from the oxime of cyclohexanone, there may be a viable route to indoles. Ring expansion of 3-acylazetidines under photolytic conditions has been shown to give 3-alkylpyrroles[24].

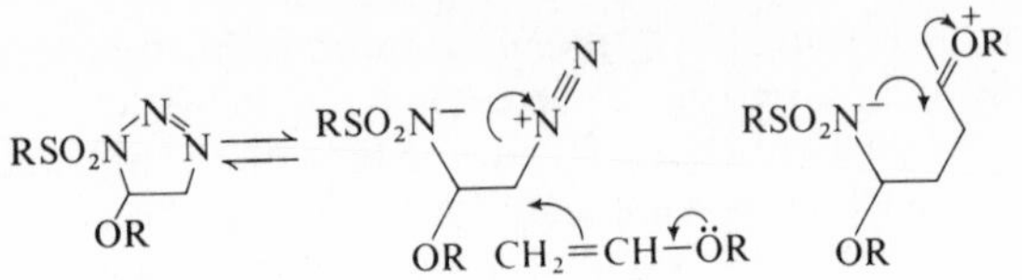

Mechanistic studies of the established synthesis of N-arylpyrroles by the Diels–Alder addition of nitrosobenzenes to butadienes have now shown that only the adduct (16) rearranges to the pyrrole, whilst (17) gives 3-anilino-6-arylpyr-2-ones[25]. The orientation of the addition is controlled by the polarisation of the butadiene system. Thus, in the addition of nitrosobenzene to

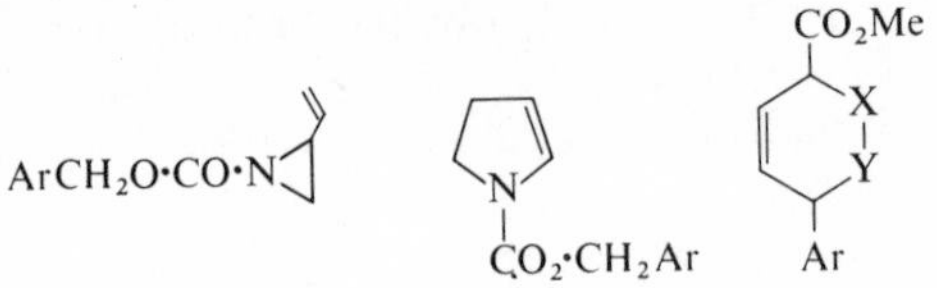

1-aryl-4-carbomethoxybuta-1,3-diene, (17) is favoured when the aryl group is substituted in the *para*-position by electron withdrawing groups. A re-examination of the by-products of the standard procedure for the preparation of pentaphenylpyrrole from tetraphenylcyclopentadienone has provided evidence for a reaction mechanism involving a 1,2-oxazine intermediate similar to (16)[26].

Widespread interest in cyclophanes, annulenes and helicenes has been enhanced by the incorporation of the pyrrole ring into such systems using conventional ring syntheses[27–30].

2.1.1.2 Indoles and isoindoles

The Fischer reaction is still one of the preferred methods for the preparation of indoles[31–34] and the optimum conditions necessary for the cyclisation of acetaldehydehydrazone to indole have been studied in detail[35]. It appears, however, that in many instances the yields of 2-arylindoles from the Bischler cyclisation of ω-arylaminoketones[36] are superior to those from the corresponding Fischer reaction.

N-Methylphenylhydrazines and hydrazobenzene eliminate methylamine or aniline during Fischer cyclisation to give indoles unsubstituted on nitrogen (19, $R^2 = H$)[37]. An alternative cyclisation similar to that observed with N,N'-dialkylazines (see above) gave pyrazolines (9, $R^2 = H$, $R^3 = $ Me or Ph). Irrespective of the cyclisation conditions, migration of the methoxy group of 2-methoxy-phenylhydrazones (18, $R^1 = $ 2OMe, $R^2 = H$) resulted in the formation of 5-methoxyindoles as well as the 7-methoxy isomer[38], whilst the unexpected by-product (20) was isolated[39] from the cyclisation of 2,5-dimethoxyphenylhydrazones.

Further studies[40] into the mechanism of the Nenitzescu synthesis of hydroxy-indoles have provided evidence for the postulated bicyclic quinone-imine intermediate[41, 42]. Both 5- and 6-hydroxyindoles have been isolated using Nenitzescu's method[43, 44], the ratio of the isomers depending largely upon the solvent[44]. In acetic acid 4,5-dihydroxy derivatives were formed[45]. 4,5-Disubstituted indoles, as well as 4-monosubstituted indoles, have also been synthesised[46, 47] from the readily attainable 4-oxo-4,5,6,7-tetrahydro-indoles[48–51].

The established acid-catalysed conversion of N-2-bromo-allylaniline into 2-methylindole[52] has been subjected to further examination[53] using ^{14}C-labelled compounds but still no clear distinction, to explain the apparent migration of the methyl group, can be made between a Claisen type of re-arrangement and an alternative concerted migration of the arylamino group with the elimination of the bromide ion. Simple $3 \rightarrow 2$ valence isomerism is definitely excluded.

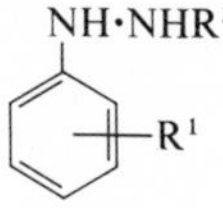

(18) 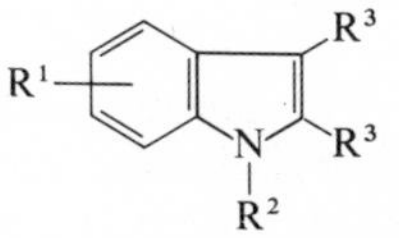   (19) 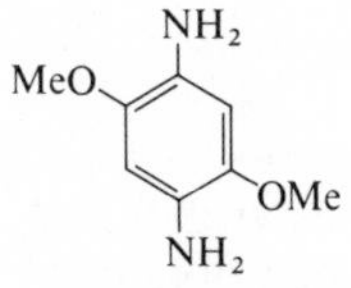   (20)

Photochemical rearrangements of N-methyl-N-phenylglycine esters[54] (21a) and of 2-(2-nitrophenyl)ethanol[55] (21b) leading respectively to the oxindoles (23a and b) proceed through related intermediates (22a and b) and an analogous direct conversion of α-(N-alkylanilino)ketones into N-alkyl-indoles[56] may also go through an intermediate of type (22c, R^1 = NHalkyl, R^2 = alkyl). It is equally conceivable that the cyclisation of o-nitroaceto-phenone to oxindole[57] under Willgerodt reaction conditions proceeds via a similar type of intermediate. All of these reactions have a close similarity to the photochemically-induced formation of 1-hydroxy-3,3-dimethyloxindoles

$C_6H_5 \cdot N(Ph) \cdot CH_2 \cdot COR$
(21a)

$o\text{-}NO_2 \cdot C_6H_4 \cdot CH_2 \cdot CH_2OH$
(21b)

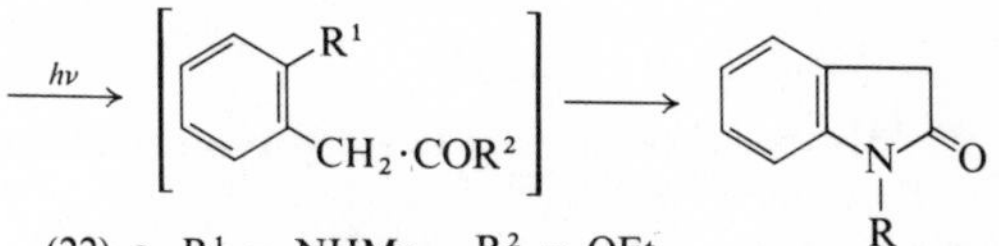

(22) a, R^1 = NHMe, R^2 = OEt
     b, R^1 = NHOH, R^2 = OH   (23) a, R = Me
     c, R^1 = NHAlkyl, R^2 = Alkyl   b, R = OH

from o-nitro-t-butyl benzene[58], the mechanism of which has recently been investigated. Several synthetic routes to indole involve photochemical or thermal decomposition of aryl azides[59] and α-azidocinnamates[60]. Although one usually assumes a nitrene intermediate for such reactions[61], the azido-cinnamates may form intermediary azirines (cf. Ref. 22).

Nucleophilic attack by sulphonium ylides upon o-acylanilines (24 → 25) gives a reactive intermediate in a potentially useful general route to sub-stituted indoles[62]. Arsonium ylides[63] and methylsulphinyl anions[64] have also been used.

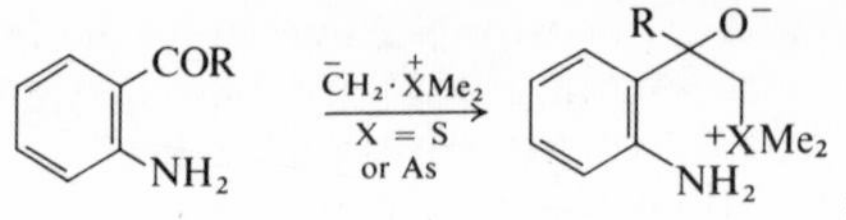

Few new preparative procedures have been described for isoindoles. The relatively stable 2-substituted-1,3-dicarbomethoxyisoindoles have been prepared by the oxidation of the corresponding isoindolines with chloranil[65], whilst 1,2-dibenzoylcyclohexa-1,4-diene has been converted by the Paal–Knorr reaction into 2,3-diphenylisoindole[66].

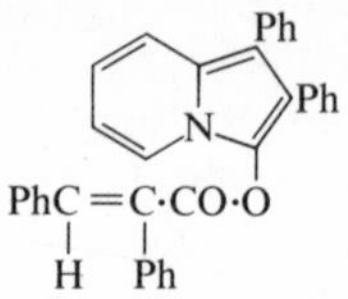

Photochemical cyclisation of 1-benzyl-2-bromopyridinium salts gave pyrido[2,1-a]isoindoles (benzo[c]indolizines)[67].

2.1.1.3 Indolizines and cyclazines

Further examples of the standard 1,3-dipolar addition reaction of alkynes and alkenes to pyridinium N-ylides to give indolizines have been described[68-70] and the product obtained by Breslow[71] from the reaction of diphenylcyclo-propenone with pyridine has been unequivocally identified[72] as the diphenyl-acrylic ester (26).

An alternative synthesis of indolizines from pyridines has been illustrated[73] by the preparation of 3-amino- and 3-hydroxy-indolizines by the reaction of 2-acylpyridines with aryl aldehydes in the presence of ammonium acetate or perchloric acid respectively. The initial step of the reaction is presumably the formation of the azachalcone.

Using a procedure based upon Hafner's synthesis[74] of polycyclic non-benzenoid aromatic systems, Leaver[75] converted 3H-pyrrolizine into cyclo[3,2,2]azine by two routes and prepared cyclopenta[h]cyclo[4,2,2]azine by the reaction of 3H-pyrrolizine with the dimethylimmonium salt derived from 2-formyl-6-dimethylaminofulvene.

2.1.2 Theoretical studies and physico-chemical properties

2.1.2.1 Molecular orbital calculations

Continued effort has been made to correlate the observed preferential reaction of electrophiles at the α-position of pyrroles with the calculated electron densities of the α- and β-positions and MO calculations have been used to determine the ground-state geometry and electron distribution[76-81]. The ground-state resonance energies have also been recalculated for indole, isoindole and carbazole[79] and the π-electron distribution of indolizine, as calculated by a simple Huckel MO approach[82], has been shown to agree with the observed reactivities of the various positions to electrophilic attack whilst more sophisticated calculations have been used to rationalise the ^{13}C magnetic resonance data[83].

Experimental observations of ^{14}N quadrupole coupling constants for pyrrole have been satisfactorily interpreted using a simple valence-bond approach[84] and a theoretical study of the electronic transitions of mono-substituted pyrroles has been undertaken[85].

The observed dipole moment of 1-trimethylsilylpyrrole and values estimated from Pariser–Pople–Parr calculations appear to be consistent with d_π–p_π overlap between the Si and N nuclei[86]. Similar d_π–p_π conjugation in 2-pyrrylphosphonium salts is also possible[87].

2.1.2.2 Hydrogen bonded complexes

Dielectric absorption measurements of the pyrrole:pyridine system have now established[88] beyond doubt the size and shape of the complexed species. It has been suggested[89], however, that the solvent effect upon the v(NH) frequency of the associated pyrrole is frequently compatible with a 'multi-molecular complex'.

Although the need for caution has been indicated in the correlation of the enthalpy of association of pyrrole with Lewis bases with the associated v(NH) frequency shifts, a recent study[90] has shown that there is a linear relationship between the two sets of data. The enthalpy and entropy changes accompanying dimerisation of 2-formylpyrrole through hydrogen bonds has also been estimated from infrared absorption data[91, 92].

An application of ^{14}N magnetic resonance spectroscopy to a study of the interaction of pyrrole and indole with Lewis bases showed a direct linear correlation of the observed downfield shifts (1–15 p.p.m.) of the ^{14}N signals with the shifted hydrogen bonded NH proton resonances[93].

2.1.2.3 N.M.R. measurements

The ^{1}H magnetic resonance spectrum of pyrrole has been subjected to rigorous analysis[94–96] both at 60 and 100 MHz; in particular, a sub-spectral analysis[95] of the 100 MHz spectrum of [^{15}N]pyrrole provided accurate values of the coupling constants. Chemical shift data and coupling constants for several monosubstituted pyrroles[97–100], and indoles[101] and 4,5,6,7-tetra-fluoroindoles[102] have also been carefully analysed.

Re-examination[103] of the ^{1}H magnetic resonance data for trialkyl-pyrroles in strong acid confirmed the accepted view[2] that protonation occurs at the α-position but, contrary to earlier observations for 2-acetyl-pyrrole[2], the same workers suggested that 3-nitro- and 3-acetyl-compounds are also protonated at the α-position. A recent interpretation[104] of the ^{1}H magnetic resonance signals for N-indolyl metallic derivatives indicated that whereas the Li$^+$ salt exists as a solvent separated ion pair, the K$^+$ and Na$^+$ salts are contact-ion pairs. The absence of time averaging of the signals of indolyl magnesium bromide in the presence of indole, observed with the other indolyl salts, was taken as evidence for a particularly strong association between the Mg and N nuclei.

The ^{13}C magnetic resonance signals for indole and several methylindoles[105] and for indolizine[83] have been unambiguously assigned and, although there are several anomalies, the effect of the methyl groups upon the ring carbon resonance signals of the indoles were found to be additive.

^{13}C Magnetic resonance measurements have been used successfully[106] to confirm a proposed biosynthetic route for prodigiosin.

2.1.2.4   Infrared spectroscopy

The fine structure of the NH vibrational band of pyrrole in the vapour phase has attracted further attention[107] and the influence of solvents, particularly those capable of hydrogen bonding, upon the anharmonicity of $v(NH)$ has been discussed in detail[108]. In continuation of the investigation of the electronic effect of substituents upon the pyrryl NH frequency, cyanopyrroles have been studied[109] and the interesting observation has been reported[110] that cyanopyrroles react with halogen acids to give 6-amino-6-halogeno-2-aza-fulvenium salts (27, R = H, X = halogen). Complex salts are formed with stannous chloride and may well explain the low susceptibility of cyano-pyrroles to the Stephen reduction.

Recent photoelectron spectroscopic studies[111] have required minor reassignments to be made to the vibrational modes of the pyrrole ring[2] and attempts have been made to calculate the force constants for these vibrations.

2.1.2.5   Dipole moments

From a comparison of the observed dipole moments of indole and its 5- and 6-halogenated derivatives[112], the direction of the dipole moment for indole was found to subtend the short axis by c. 50 degrees[113]. In making the calculations no account was taken of possible conjugative interaction between the substituent and the heterocyclic ring. The result is compatible, however, with the evaluation of the dipole moments of methylindoles in the ground and excited singlet state[114] which showed that, compared with the cases of indole and its 1-, 2-, and 3-methyl derivatives, there is a smaller variation between μ_g and μ_e for 5- and 7-methylindole indicative of a greater polarisation along the long axis in the ground state of the latter compounds.

Dipole moments of several substituted pyrroles have been measured or re-measured[115] and analysis of the data for 2-acylpyrroles confirms the greater stability of the syn conformation (28) (see below). It was also concluded from the dielectric relaxation times for these compounds that rotation of the substituent was negligible.

2.1.2.6   Mass spectrometric measurements

An important fragmentation[116] of naturally occurring indoles having the 3-CH_2R substituent leads to the loss of the R group to give a base peak at $m/e = 130$, often regarded as the quinolinium ion, whilst fragmentation of isomeric hydroxyskatoles is controlled by the hydroxyl group and, owing to substituent randomisation, the spectra of the isomers are virtually indistinguishable[117]. A similarity in the spectra of methylindolizines has also led to the suggestion that a common ion $[M - H^{\cdot}]^+$ is formed[118]. The formation of such an ion can be envisaged through the expansion of either the five- or six-membered rings by the incorporation of the substituent, followed by ring opening of the bicyclic system to give a ten-membered ring, which may subsequently form the quinolinium ion[118]. Further fragmentation of the $[M - H^{\cdot}]^+$ ion is certainly similar to that expected of the quinolinium ion. The appearance of a peak with $m/e = 141 [M - OH^{\cdot} - Me^{\cdot}]^+$ in the spectrum of 3-acetyl-4-methylindolizine was rationalised with the formation of the cyclo[3,2,2]azine ion[118].

π-Pyrrylmanganese carbonyl complexes $(\pi$-pyrr$)Mn(CO)_3$, lose the carbonyl groups in consecutive steps[119] and the resulting ion $m/e = 121$ fragments further in a manner analogous to that of azaferrocene[120]. Complexes of the type $(\pi$-pyrr$)Mn(CO)_2MPh_3$ (M = P, As or Sb) also lose carbon monoxide readily but the resulting ion then follows two fragmentation pathways giving $[(\pi$-pyrr$)Mn]^+$ and $[Ph_3M - Mn]^+$; the ratio depending upon the stability of the M−Mn system[119].

2.1.2.7   Ionisation measurements

A measure of the transmission of substituent effects across indole[121] and pyrrole[122] rings has been obtained from the ionisation constants of the 2-carboxylic acids and a high correlation of the σ_p values for 5- and 6-substituents and σ_m values for 4- and 7-substituents, according to the equation $pK_{a'} = 2.56 \Sigma \sigma - 2.55$, was found for the protonation of indole in the 3-position[123].

Although protonation of indole and pyrrole occurs almost exclusively on carbon[2], it has been possible to estimate the nitrogen basicity of these compounds by extrapolation of basicity data for N-methylcarbazole[124] which is unusual in its preferential protonation of the nitrogen atom in strong acids. An estimated value of -10 was given for the nitrogen protonation of pyrrole.

2.1.2.8   Conformational analysis

The controversial problem of the conformational preference of 2-acetyl- and 2-formyl-pyrrole $(28 \rightleftharpoons 29, R^1 = H, R^2 = Me$ or H) (see, for example, Refs. 2, 115, 125–127) and the furan and thiophen analogues has now been resolved. Measurement of the Nuclear Overhauser Effect (NOE) on the resonance signals of the 3-proton of 4-bromo-2-formyl-1-methylpyrrole upon irradiation of the formyl proton has established[128] that the *syn* conformation is preferred by a factor of 9.3:1 over the *anti* conformation at room temperature. This ratio is considerably smaller than that proposed for 2-formyl-1-methylpyrrole[125] and is in complete contrast to the suggestion by Swedish workers[127] that the *anti* conformation $(29, R^1 = Me, R^2 = H)$ is the more stable. The latter work appears to be at fault in the method of preparation of the compound, as all attempts to reproduce the n.m.r. data failed except when 2-formylpyrrole (5% w/w) was added to the 2-formyl-1-methylpyrrole[126]. Dipole moment measurements[115, 126] confirm the earlier conformational assignments[125]. Attempts to use NOE to establish the preferred conformation of 2-acetyl-1,3-dimethylpyrrole failed[129] owing, most probably, to the distance between the acetyl and the 1- and 3-methyl groups.

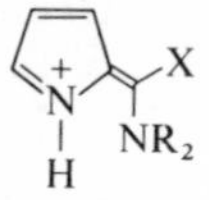

Related conformational studies of the 6-aryl-6-dimethylamino-2-azafulvenes[130], the bases of the salts (27, R = Me, X = aryl) isolated from the Vilsmeier–Haack benzoylation of pyrrole, have established the conformational preference of the 6-substituents. The energy barrier to rotation about the C-6—N bond of c. 62 kJ mol^{-1} showed no direct dependence upon the electronic character of the p-substituent on the aryl ring.

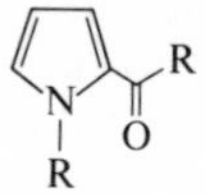

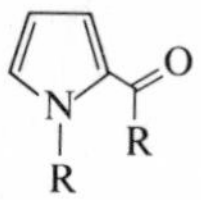

Hückel MO calculations indicate[131] that in the preferred conformations of 1- and 3-phenylpyrrole, the two rings are twisted out of co-planarity by c. 40 and 30 degrees respectively, whereas for 2-phenylpyrrole the co-planar system is the most stable. Although reasonable, there is little evidence, as yet, to support these calculations.

2.1.3   Reactions of the aromatic rings

Electrophiles normally attack the α-position of pyrrole and, although the sensitivity of the pyrrole ring to substituent effects appears to be less than that of furan or thiophene[132], both electronic and steric factors control the orientation of attack upon substituted pyrroles. Similarly, although electrophilic attack on indole occurs preferentially on the heterocyclic ring[1], activation of the benzene ring by electron-donating substituents results in substitution of other sites.

2.1.3.1   Halogenation

Bromination of 1-methylindole by molecular bromine in acetic acid gives, as the major product, 2,3,5,6-tetrabromo-1-methylindole and, as a secondary product, the polybromo-oxindole (30, R = Me, X^1 = X^2 = Br)[133]. A similar compound (30, R = X^2 = H, X^1 = Cl) is the major product of the chlorination of indole with N-chloramines under acidic conditions[134] and 3,3-dihalogeno-oxindoles have also been isolated from the halogenation of several other substituted indoles[135–137]. Thus, although deactivation of the heterocyclic ring by electron-withdrawing substituents normally leads to electrophilic substitution of the benzene ring[1], chlorination of 3-acyl- and 3-carboxy-indoles gave (31, X^1 = X^2 = Cl) and (31, X^1 = Cl, X^2 = H) respectively[135, 137]. An elucidation of the reaction mechanism[135, 136] was greatly aided by the isolation and identification of the oxindole (31, X^2 = Cl, X^1 = CO_2Me) from the chlorination of 2-carbomethoxyindole.

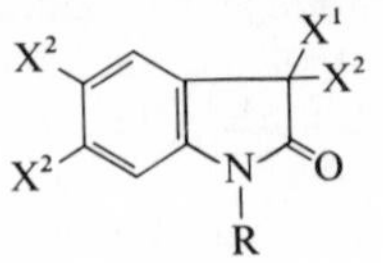

(30) 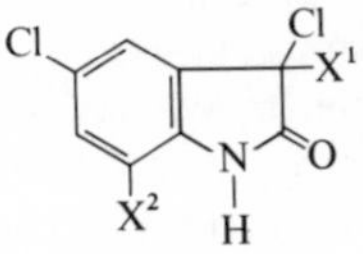   (31)

In general, pyrroles react quite normally with halogens. Pyrroles, mono-substituted with electron-withdrawing groups in the 2-position, 'direct' the halogen into the 4-position[138-140]. With an excess of halogen further substitution generally yields 3,4- and 4,5-dihalogenopyrroles[140], although 3,5-dihalogeno compounds have also been isolated[138]. Even pyrroles having two electron-withdrawing substituents react readily with sulphuryl chloride or pivalyl chloride to give fully chlorinated products[141, 142].

The behaviour of pyrroles with halogens in the presence of thiourea and thiols is anomalous. Under these conditions the halogen preferentially oxidises the sulphur compounds which subsequently react with the pyrrole to give pyrrylthiols[143, 144]. Somewhat more surprising is the report that iodine added to indolizine instead of giving the expected substitution product[145].

2.1.3.2   Nitration and nitrosation

3-Nitropyrrole, which is the minor product of the nitration of pyrrole[146], can be obtained in a higher yield by the nitration of 1-acetylpyrrole and the subsequent removal of the acetyl group[147]. Although 2-methylpyrrole is preferentially nitrated in the 5-position[148], electron-withdrawing substituents in the 2-position promote nitration of the 4-position, whilst similar groups in the 3-position lead to nitration of the 5-position[147, 149].

Nitrosation of 2,5-disubstituted pyrroles has long been thought to give β-nitroso derivatives, but a recent investigation[150] has shown that the initial nitrosation product of 2-methyl-5-phenylpyrrole rearranged under acidic conditions to give 3-benzoyl-5-methylisoxazole[150].

2.1.3.3   Alkylation

The introduction of a t-butyl group into the pyrrole ring has been accomplished using t-butyl acetate in the presence of concentrated sulphuric acid[151] and also with t-butyl chloride in the presence of a Lewis acid[152]. The latter reaction has also been applied to the synthesis of 3-t-butylindole when a by-product is the 1,3-di-t-butylindole[153]. The presence of Lewis acids also causes the isomerisation of 3-t-butylpyrrole to the thermodynamically more stable 2-isomer[152]. An efficient procedure has been described for the preparation of 3-tritylindole using trityl chloride in the presence of pyridine[154], and benzylation can be effected with benzyl alcohols or amines and BF_3[155].

The low yields usual for the Mannich addition of primary amines and formaldehyde with pyrrole have been improved[156] by varying the reaction conditions; the yields are considerably higher with bulky amines. A re-examination of the acid-catalysed reaction of pyrroles with formaldehyde to give dipyrromethanes has shown that under suitable conditions tris-pyrryltrismethanes (32) were formed[157], whilst from the reaction of pyrroles with acetone a series of novel spiro compounds, e.g. (33), have been isolated[158]. With indole the initial hydroxymethylation occurs at the 3-position. If this position is blocked, the unstable 1-hydroxymethyl compound is formed[159] but subsequently rearranges to give the 2-isomer, which reacts further to give either 2,2'-bisindolylmethane or a trisindolyltrismethane system[160]. The presence of oxygen, however, can cause the formation of the fully aromatic system (34)[161].

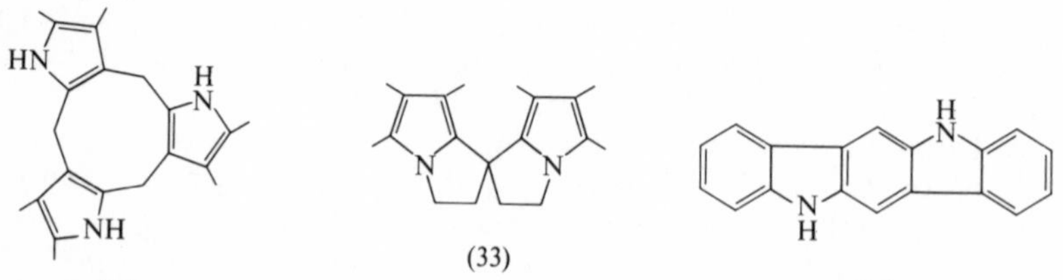

(32) (33) (34)

Activation of the benzene ring of indole to electrophilic attack results in the preferential, and reversible[162], hydroxymethylation of the 4-position, in the case of 5-hydroxyindole, and the 5- and 7-positions, in the case of 6-hydroxy-indole[163]. Should these positions be blocked, however, substitution reverts to the 3-position.

Nucleophilic substitution of a pyrrole ring is extremely rare, but one case has recently been described[164]. 2-Acetyl-5-nitropyrrole reacts normally with ethylene oxide to give (35) but subsequent nucleophilic displacement of the nitro group results in the formation of the bicyclic system (37). An analogous reaction has been found to occur with 3-substituted indolyl magnesium bromide in which nucleophilic attack of the alkoxy anion upon the $C=N$ bond of the intermediate 3,3-disubstituted 3H-indole gave the 2,3,3a,8a-tetrahydro[2,3-b] indole system[165].

1-Substituted pyrroles react normally to give a mixture of 2- and 3-(2-hydroxyethyl)pyrrole[166].

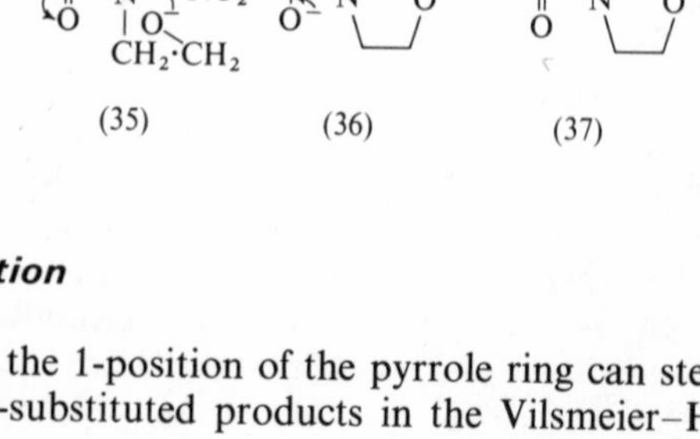

(35) (36) (37)

2.1.3.4 Acylation

Substituents on the 1-position of the pyrrole ring can sterically control the ratio of α- to β-substituted products in the Vilsmeier–Haack formylation reaction[167]. Thus, although 1-methylpyrrole is formylated exclusively at the 2-position, the predominant formylation product of 1-t-butylpyrrole is the 3-isomer. Steric effects are of lesser importance in the formylation of 2-substituted indoles[168] and of indolizines[169]; 2-t-butylindole, for example, has been observed to give high yields of the 3-formyl derivative[170]. The electronic effect of 1-substituents upon the orientation of formylation of 1-substituted pyrroles is relatively insignificant[167], but, as might be expected, a 4-alkoxy group on an indole system promotes formylation of the benzene ring[171]. 2-Oxo-Δ^3- and 2-oxo-Δ^4-pyrrolines, the tautomeric forms of 2-hydroxypyrrole, both give formyl-2-halogenopyrroles under Vilsmeier–Haack conditions[172, 173].

The Vilsmeier–Haack reaction was also used in the preparation of 3-chloro-1-(2-pyrrolyl)propane which, remarkably, in the presence of sodium hydride, gave 2-pyrrolylcyclopropane to the complete exclusion of 8-oxo-5,6,7,8-tetrahydroindolizine[174].

An interesting variation on the Vilsmeier–Haack reaction, which has some potential for the direct insertion of a carboxyl function into pyrrole and indole systems, utilises dithiocarbamic esters[175] (38 → 40). The initial product (40) is readily converted into carboxamides, carboxylic esters, etc.

$$R_2N{\cdot}CS{\cdot}SR \xrightarrow{COCl_2} R_2\overset{+}{N}{=}C\overset{Cl^-}{\underset{Cl}{\overset{SR}{\diagup}}} \xrightarrow{ArH} Ar{-}\underset{\overset{+}{N}R_2}{\overset{SR}{\diagdown}}$$

(38)   (39)   (40)

Ar = 2-pyrryl or
3-indolyl

The long-standing problem concerned with the structure of the benzoylation product of tetra-alkylpyrroles is now resolved[176–178]. 2,3,4,5-Tetramethylpyrrole reacts with two molecules of benzoyl chloride to give (41, $R^1 = R^2 =$ Me)[176, 177]. Surprisingly, only a single product (41, $R^1 = Et, R^2 = Me$) was isolated from the benzoylation of phyllopyrrole and, contrary to earlier findings, 2,5-dimethylpyrrole was reported[177] to give (41, $R^1 = R^2 = H$). The same authors[179], however, found that phthalic anhydride reacted with 2,5-dimethylpyrrole to give the 3-(2-carboxyphenyl) derivative.

(41)   (42)   (43)

Acylation of indoles with acyl chlorides in the presence of pyridine is unsatisfactory (see below), although 3-trichloroacetyl derivatives have been prepared by this method[180]. 3-Trifluoroacetylation of indole is also simple[181] and the facile cleavage of the CF_3 group provides a valuable route to 3-carboxyindoles but, in general, acylation of indole with acyl chlorides or anhydrides yields bis- and tris-indolyl systems[182–184].

The cyclic pyrroloketones (e.g. 42) which are important models in the conformational analysis of acylpyrroles (see above) and also as precursors for the novel 1-azapentalene anion[185] have been prepared by direct cyclisation of the pyrrylpropionic acids[185, 186].

2.1.3.5　Reactions with pyridinium and pyrylium ions

The failure of acyl chlorides in pyridine to acylate indoles is due to the preferential nucleophilic attack by indole at the 4-position of the 1-acyl-pyridinium ion to give 1-acyl-1,4-dihydro-4(3-indolyl)pyridine which, during the work-up procedure, loses a proton and the acyl group giving 3-(4-pyridyl)indole[180, 187]. An analogous reaction has been observed with pyrylium salts[188] and the 4-(3-indolyl)pyrylium salt has been converted into the corresponding phosphabenzene by standard procedures[189].

Indoles, oxindoles and 2-oxopyrrolines also react in a similar manner with quinoline 1-oxides and pyridine 1-oxides[187, 190, 191] in the presence of acylating agents, and nucleophilic attack by indole on the activated ring of 2-nitrophenazine 10-oxide yields 1-(3-indolyl)-2-nitrophenazine[192].

2.1.3.6　Reactions with aryl azides

Continued interest has been shown in the reactions of toluene-p-sulphonyl azides and picryl azide with indoles[193-197] and indolizines[198, 199]. The formation of both toluene-p-sulphonamidoindoles and the azoindoles can be rationalised in terms of electrophilic attack at the 3-position of the indole ring followed by cleavage to give the 3-indolyl diazonium ion. In a similar reaction of pyrryl magnesium bromide with toluene-p-sulphonyl azide, the heteroaryl diazonium ion is probably a precursor of the azapyrrole formed[200].

2.1.3.7　Reactions of the NH group

Two research groups[201-203] have described the preparation of 1-pyrrylboranes and the reduction of indole by diborane in the presence of a proton source is known to proceed via the initial formation of a N–B adduct[204]. 1-Pyrryl-phosphorus compounds have also been synthesised and their reactions with hydroxy compounds have been studied[205].

Pyrrole reacts with thallium(I) ethoxide to form a stable thallium(I) salt which is superior to pyrrylpotassium as a reagent for the preparation of 1-substituted pyrroles[206]. The derivatives of pyrrole with chromium or manganese carbonyl compounds have been characterised[207, 208] as π-complexes and similar complexes are formed with nickel and titanium ions[209, 210].

The acylation of pyrryl magnesium halides to give 2-acylpyrroles has been re-examined to determine whether 1-acylpyrroles are intermediates of the reaction[211]. The formation of only 1-(3,4,5-trideuteriopyrrolyl)pyrrole from the reaction of 1-carbomethoxypyrrole with tetradeuteriopyrryl magnesium bromide precludes the initial acylation of the nitrogen atom to give a symmetrical intermediate followed by migration to the 2-position. A detailed analysis of the products of the methylation of tetramethylpyrryl magnesium bromide has also revealed that, contrary to earlier reports, both the hitherto unknown pentamethyl-β-pyrrolenine and the α-pyrrolenine are formed to an equivalent extent[212].

Pyrryl-2-thiocarboxylic esters have been isolated in good yield from the reaction of alkyl thiocarbonates with pyrryl magnesium bromide[213] but, as might be expected, the sodium salts of indole reacted with carbon disulphide to give both 1- and 3-indolyldithiocarboxylic acids[214]. Pyrryldithiocarboxylic acids have also been prepared by the $AlCl_3$ catalysed Friedel–Crafts reaction of pyrroles with carbon disulphide[215].

As a protective substituent for the NH group of pyrrole, the 1-benzyloxymethyl derivative, formed by the reaction of benzyl chloromethyl ether with the pyrryl anion, is superior to the benzyl derivative[216] in many reactions and yet it is conveniently removed by catalytic hydrogenation followed by treatment of the resultant 1-hydroxymethyl compound with base. It has been suggested[217] that the toluene-p-sulphonyl group is the most suitable protective substituent for the indolyl NH.

2.1.3.8 Oxidation

The initial monomeric oxidation products of pyrrole with hydrogen peroxide have been identified as 2-oxopyrrolines[218], their yield increasing as the pH of the reaction medium is increased (cf. Ref. 219). The dimeric and trimeric oxidation products have also been characterised.

Photo-oxidation of pyrrole and of C-alkylpyrroles in methanol produced 5-hydroxy- (or 5-methoxy-)-2-oxo-Δ^3-pyrroline and the corresponding maleimides via the peroxide (43)[220, 221]. A similar intermediate has been postulated for the singlet oxygen oxidation of 1-benzoyl-2,3,4,5-tetraphenylpyrrole[222]. The acyclic products from tetraphenylpyrroles have been explained[222, 223] by a rearrangement of the peroxide (43) to a 2,3-peroxide and subsequent cleavage of the 1,2- or 2,3-bond. The singlet oxygen oxidation products of 1-phenylpyrroles have also been studied[224] and a complete analysis of the electrochemical oxidation products of tetraphenylpyrrole in nitromethane has revealed ring expansion to a tetraphenylpyridine as an important reaction pathway[225].

The majority of oxidation reactions of indole proceed through the initial formation of 3-peroxy-3H-indole derivatives[226–232], which in the absence of a 3-substituent give indoxyls[230–232]. Further reaction leads to the formation of 3-indolyl(2-indol-3-ones) and other dimeric products[231, 233]. o-Acylanilines, produced by singlet-oxygen oxidation of indoles[234], may result from the formation of a 2,3-peroxide analogous to that formed with pyrrole, whilst the oxidation of ethylindoles to acetylindoles suggests preferential attack by the peroxy radicals on the ethyl group[235]. Oxidation of isoindoles is more akin to that of pyrrole with the initial formation of a 1,4-peroxide similar to (43) which then breaks down to give a phthalimide derivative[236, 237].

2.1.3.9 Reduction

Catalytic hydrogenation of indoles over Raney nickel generally proceeds through the 2,3-dihydroindole, which can be isolated. Substitution at the 1-position appears to reduce this selectivity and, in general, whilst substituents on the benzene ring have little effect upon the reduction, the rate of hydrogenation is reduced by 2- and 3-substituents[238]. Conversely, the benzene ring of 5- and 6-methoxyindoles was reduced by lithium in liquid ammonia, but attempts to reduce 7-methoxyindole failed[239]. Under similar conditions the six-membered ring of indolizine was reduced to give 5,6-dihydro- and 5,6,7,8-tetrahydroindolizine[240].

2.1.3.10   Diels–Alder reaction

In the past the Diels–Alder addition reaction involving pyrroles has met with varied success. The most recent report[241], however, showed that with the active dienophile, hexafluorobicyclo[2,2,0]hexadiene, pyrrole forms both a mono-adduct (44) and a di-adduct (45), the stereochemistry of which was assigned on the basis of ^{19}F n.m.r. chemical shifts.

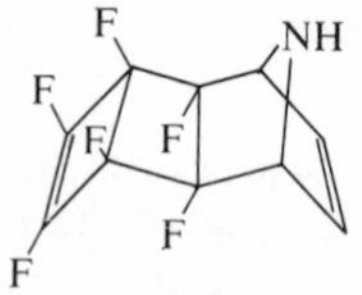

(44) 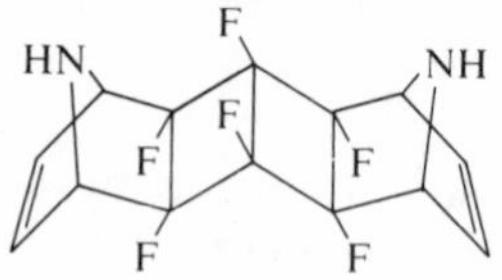   (45)

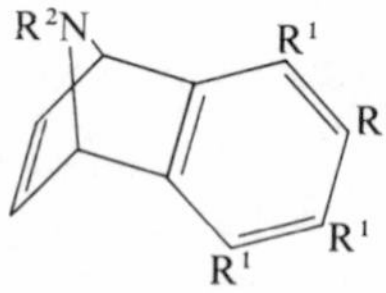

(46)

Under photolytic conditions, instead of a normal Diels–Alder addition reaction, dimethyl acetylenedicarboxylate has been found[242] to give 3,4-dicarbomethoxyazepine. This compound differs from that obtained by Prinzbach[243] from a sigmatropic rearrangement of the normal 2+4 Diels–Alder addition and must result from an initial 2+2 cyclo-addition of the acetylenic ester to the 2,3-bond of pyrrole. A similar 2+2 cyclo-addition has been observed [244] between 3-dialkylaminoindoles and acetylenic esters to give benzo-azepines.

In contrast to the normal acid catalysed rearrangement of benzo-7-azanorbornadienes to 1-aminonaphthalenes, the photolysis products of (46, R^1 = H, R^2 = CO$_2$Et) have been identified[245] as azepines (cf. Ref. 243) and 1-(aminomethylene)indenes. An unusual pathway has also been observed[246] in the thermal cleavage of (46, R^1 = F, R^2 = Me) to give 2-(2,3,4,5-tetrafluorophenyl)-1-methylpyrrole.

As is to be expected, acetylenic esters readily form typical Diels–Alder adducts with 1,3,4,7-tetramethylisoindole but the further addition of a second molecule of the acetylenic ester gave a significant yield of a 3a,9a-dihydrobenz[g]indole tetra-ester[236]. The corresponding mono-adduct of 1,2,3,4,7-pentamethylisoindole is stable in the absence of dimethyl acetylenedicarboxylate but, when heated in its presence gave methylamine and 2,3-dicarbomethoxy-1,4,5,8-tetramethylnaphthalene[247].

2.1.4   Photochemistry and ring modifying processes

2.1.4.1   Photochemical reactions

Following upon Bryce-Smith's earlier reports[248, 249] of the photolysis of pyrrole in benzene, the photoaddition of naphthalene and pyrrole has been studied in detail[250, 251]. From the quenching effect of pyrrole on the fluorescence of naphthalene it was concluded that a charge-transfer exiplex of singlet napthalene and pyrrole was formed[251]. Subsequent proton transfer from the pyrryl NH group to naphthalene and addition of the pyrryl anion to the naphthalene cation gave 2-(2-pyrryl)-1,2-dihydronaphthalene and 1-(2-pyrryl)-1,4-dihydronaphthalene. This mechanism is consistent with the absence of reaction with 1-methylpyrrole[251] and the incorporation of deuterium into the naphthalene ring when 1-deuteriopyrrole was used[250]. The more stable 2-substituted compound was preferentially formed as the polarity of the solvent was increased[251].

Although well known for furans and thiophens, the first example of 2 ⟶ 3 valence isomerisation of pyrroles has been reported for 2-cyanopyrrole[252]. Irradiation of 2-nitropyrrole in acetone, however, gave 3-hydroximino-2-oxo-Δ^4-pyrroline[253], possibly via the isomeric nitrite and, instead of rearrangement, methanol added across the 3,4-bond during the photolysis of 2,2,5-trialkyl-2H-pyrroles in methanol[254].

Friedel–Crafts alkylation of indole (see above) with methyl chloroacetate gave all seven isomers under photolytic conditions and, upon irradiation, methyl 1-indolylacetate rearranged to the 6- and 3-indolyl esters[255]. On irradiation, 2-styrylindole[256] and its aza analogues[257] cyclised to the corresponding benzo- and pyrido-[c]carbazoles.

2.1.4.2 Ring expansion and homopyrroles

The mechanism of the ring expansion of pyrroles, when treated with chloroform and sodium ethoxide, to give chloropyridines is still a matter of dispute. It is now apparent that the 2-dichloromethyl-2H-pyrrole is not an intermediate in the formation of the 3-chloropyridine[258, 259] but it is evident that the 2H-pyrroles do undergo ring expansion under strongly basic conditions, possibly via the bicyclic aziridinopyrrole[259] or the homopyrrole (47, X = Cl, R = H)[260].

The synthesis of 2,3-homopyrrole has been investigated by two research groups. Fowler[261] has shown that 1-carbomethoxypyrrole reacts with diazomethane in the presence of cuprous chloride to give three products (47, X = H, R = CO$_2$Me; 48; 49). Ring expansion of the homopyrrole occurred

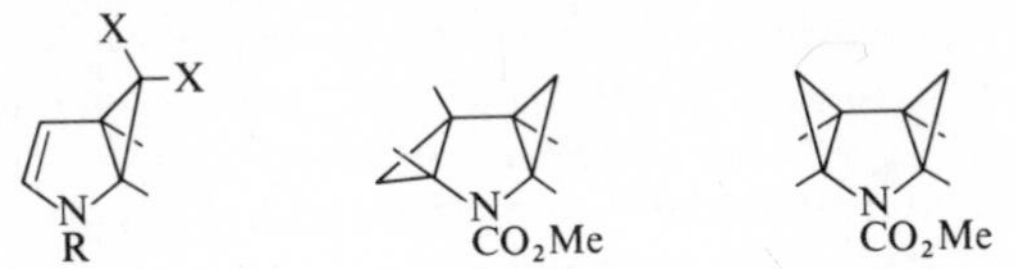

(47)   (48)   (49)

thermally at 285 °C to give 1-carbomethoxy-1,2-dihydropyridine[261] and a similar reaction of the two isomers (50, R^1 = H, R^2 = CO$_2$Et and R^1 = CO$_2$Et, R^2 = H) gave 2,5-dicarboethoxy-1-carbomethoxy-1,2-dihydro-3,6-dimethylpyridine[262]. Photolysis of the dihydropyridine led to the reformation of the two homopyrroles suggesting that there is a common acyclic triene intermediate which undergoes electrocyclic closure to the dihydropyridine and an intramolecular 2+4 cyclo-addition to give the homopyrrole.

Diels–Alder addition of dimethyl acetylene dicarboxylate to 1-carbo-methoxyhomopyrrole gave (51, R = H)[261]. In a similar reaction the *exo*-isomer was formed by the homopyrrole diester (50) to the complete exclusion of the *endo*-isomer[261]. Such selectivity is consistent with a disrotatory ring-opening of the homopyrrole followed by a 2+4 cyclo-addition from the unhindered side of the dipolar intermediate.

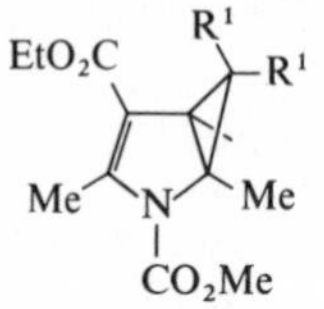

(50) 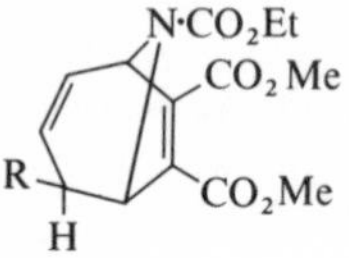   (51) 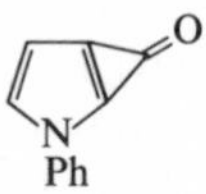   (52)

2.1.5 Properties of functional groups

2.1.5.1 Acyl compounds

It has been proposed[263] that the anomalous reaction of the oxime of 3-acetyl-1-methylindole with toluene-*p*-sulphonyl chloride in which 3-acetylindole is regenerated via its *N*-chloroimine derivative is due to the formation of a spiro-aziridine during the anchimerically-assisted elimination of the toluene-*p*-sulphonyl group. The corresponding Beckmann rearrangement of ω-indolyl-acetophenones with the preferential formation of the indolylacetanilide appears to be controlled by intramolecular hydrogen bonding[264]. The alternative rearrangement product, the *N*-(indolylmethyl)acetamide, was obtained by the Schmidt reaction[264]. The Arndt–Eistert reaction and analogous rearrangements of 2-indolyl diazoketones have also been studied in detail[265].

The Wittig reaction involving acylpyrroles is now well established and has recently been used in the synthesis of 3-oxopyrrolizines[266]. Indolylethenes have also been prepared by the alternative procedure starting from indolylmethyl triphenylphosphonium salts[267]. In the aldol condensation of 3-formylindole with methyl ketones a phosphorane was used in an unusual role as the base; apparently no Wittig-reaction products were detected[268].

Treibs has described the synthesis of two novel heterocyclic systems. The base-catalysed reaction of 2-formylpyrrole with cyclopentadiene gave 6-(2-pyrryl)fulvene[269], whilst 1,3-dipolar addition to a 2-pyrrylmethylazide gave a pyrryl triazoyl methane[270]. Comparatively standard reactions were used in the conversion of acylpyrroles into 1,5- and 2,6- diazaindenes[271, 272] and 2,5,6-triazaindene derivatives[272, 273].

2.1.5.2 Alkyl and aryl compounds

2-(1-Pyrryl)anilines have been converted by standard reactions into a series of novel pyrrolo[2,1-*c*]benzazines[274, 275] and a new route into the 9*H*-pyrrolo[1,2-*a*]indole ring system has been devised using an internal Mannich-type reaction of 2-(1-pyrryl)benzaldehyde with dialkylamines[276].

In a reaction, which has some analogy with the peroxide induced oxidation of the ethyl group of 2-ethyl-3H-indoles[235], irradiation of 2,3-dimethylindole in acetic acid gave 2-formyl-3-methylindole in moderate yield[277].

2.1.5.3 Carboxylic acid derivatives

Detailed studies of the kinetics and the mechanism of the decarboxylation of 2-carboxypyrrole have shown[278] that at low pH values the decarboxylation of the 2-protonated species is the rate-determining step, whereas, at lower hydrogen ion concentrations, the overall rate of decarboxylation decreases and is controlled by the protonation step. In decarboxylation and in the direct elimination of carbalkoxy and acyl groups from pyrroles phosphoric acid is the reagent of choice[279] and the optimum conditions for the basic decarboxylation of 2-carboxyindoles have also been reported[280].

The comparative merits of the conversion of 2-carbethoxyindole into 2-formylindole by the McFadyen–Stevens reaction or by reduction via the carbinol have been discussed[281] and Raney-nickel-catalysed reduction of S-ethyl pyrryl-2-thiocarboxylates has been shown to be selective under specified conditions giving the aldehyde, carbinol or methyl compounds[282], whilst Huang–Minlon conditions completely removed the thiocarboxylate group and has been used successfully in the synthesis of 3-substituted pyrroles[283].

The Curtius rearrangement of 5-methyl-2-pyrrolylazides followed the normal route to give 2-aminopyrrole derivatives but, owing to their high susceptibility to oxidation, attempts to isolate the free amines resulted in the formation of di-imidoporphins[284].

Several interesting attempts, based upon the successful pyrolytic decomposition of phthalic anhydride[285], have been made to prepare the benzyne analogue of pyrrole. Pyrolysis of the anhydride of 1-phenylpyrrole-2,3-dicarboxylic acid gave the bicyclic ketone (52) which, instead of losing carbon monoxide, dimerised to give the bispyrroloquinone[286], whilst thermal rearrangement of the analogous ketone obtained from the 3,4-dicarboxylic acid gave furano[2,3-c]quinoline[286].

2.1.5.4 Oxo compounds and related derivatives

By integrating the steps of the reaction of isatin with a Grignard reagent and LiAlH$_4$ in a 'one pot' process, good yields of 3-substituted indoles have been obtained[287]. In similar reactions concomitant decarboxylation of 2-carboxy-2-substituted indol-3-ones during reduction with NaBH$_4$ gave 2-substituted indoles[288] and it has been found that the reduction of 2-benzylisoindol-1-one did not proceed beyond the formation of the isoindole when reduced with sodium bis(2-methoxyethoxy)aluminium hydride[289]. Catalytic reduction of isatin in the presence of dimethylamine gave 3-dimethylamino-oxindole[290].

The established reactivity of the enolate anion of indolones and of oxo-pyrrolines with electrophiles has been investigated further. Both the syn- and anti-benzylidene derivatives obtained from the base-catalysed condensation of 1-methyloxindole with benzaldehyde have been characterised[291] and the complex products of the reaction of acetone with 2-oxo-Δ^3-pyrroline have been shown to result from the initial condensation at the 5-position[292]. All attempts to monoalkylate 1-methyloxindole in the 3-position with alkyl halides and sodium hydride have resulted in dialkylation[293] and, although the structural evidence was incomplete, 1-bromo-2-chloroethane was thought to give the spirocyclopropyl compound[294]. In this connection it is interesting to find that, although 1-chloro-1-methylbut-3-ene gave the 2-alkylated derivative of 2-carbethoxyindol-3-one, allyl toluene-p-sulphonate gave the O-allyl compound[295], which thermally rearranged to the C-allyl derivative. The analogous Claisen rearrangements of 2-allylthioindoles to give the 3-allyl-indol-2-thiones[296, 297] and of 1-allylindoles to the 2-allylindoles[298] have also been reported. The rearrangements of the O- and S-ethers to indolones and indolthiones suggest a greater thermodynamic stability of the latter systems but spectroscopic studies of the tautomerism of the 2-mercaptindole $\rightleftharpoons$ indole-2-thione system showed that in chloroform both tautomeric forms were present in comparable amounts whilst, as might be expected, in a more polar solvent the thiol predominated[299].

References

1. Sundberg, R. J. (1970). *The Chemistry of Indoles*, (New York: Academic Press)
2. Jones, R. A. (1970). *Advan. Heterocyclic Chem.*, **11**, 383
3. Castro, A. J., Giannini, D. D. and Greenlee, W. F. (1970). *J. Org. Chem.*, **35**, 2815
4. Harbuck, J. W. and Rapoport, H. (1971). *J. Org. Chem.*, **36**, 853
5. Roomi, M. W. and MacDonald, S. F. (1970). *Can. J. Chem.*, **48**, 1689
6. Skylar, E. Y. and Evstigneeva, R. P. (1970). *Zhur. Obshch. Khim.*, **40**, 1365
7. Pleininger, H. and Husseini, H. (1970). *Synthesis*, **11**, 587
8. Tamura, Y., Kato, S. and Ikeda, M. (1971). *Chem. and Ind. (London)*, 767
9. Chapelle, J-P., Elguero, J., Jacquier, R. and Tarrago, G. (1970). *Bull. Soc. Chim. Fr.*, 3147
10. Fritz, H. and Uhrhan, P. (1971). *Ann.*, **744**, 81
11. Stapfer, C. H. and D'Andrea, R. W. (1970). *J. Heterocyclic Chem.*, **7**, 651
12. Sucrow, W. and Chondromatidis, G. (1970). *Chem. Ber.*, **103**, 1759
13. Chapelle, J-P., Elguero, J., Jacquier, R. and Tarrago, G. (1971). *Bull. Soc. Chim. Fr.*, 280
14. Sucrow, W., Bethke, H. and Chondromatidis, G. (1971). *Tetrahedron Letters*, 1481
15. Bayer, H. O., Gotthardt, H. and Huisgen, R. (1970). *Chem. Ber.*, **103**, 2356
16. Huisgen, R., Gotthardt, H., Bayer, H. O. and Schaefer, F. C. (1970). *Chem. Ber.*, **103**, 2611
17. Potts, K. T., Husain, S. and Husain, S. (1970). *Chem. Commun.*, 1360
18. McEwen, W. E., Mineo, I. C. and Shen, Y. H. (1971). *J. Amer. Chem. Soc.*, **93**, 4479; McEwen, W. E., Kanitkar, K. B. and Hung, W. M. (1971). *J. Amer. Chem. Soc.*, **93**, 4484
19. Khayat, M. A. R. and Al-Isa, F. S. (1970). *Tetrahedron Letters*, 1351
20. Mente, P. G. and Heine, H. W. (1971). *J. Org. Chem.*, **36**, 3076
21. Gilchrist, T. L., Rees, C. W. and Stanton, E. (1971). *J. Chem. Soc. C*, 3036
22. Bowie, J. H. and Nussey, B. (1970). *Chem. Commun.*, 1565
23. Sheradsky, T. (1970). *Tetrahedron Letters*, 25
24. Padwa, A., Albrecht, F., Singh, P. and Vega, E. (1971). *J. Amer. Chem. Soc.*, **93**, 2928
25. Kresze, G., Saitner, H., Firl, J. and Kosbahn, W. (1971). *Tetrahedron*, **27**, 1941
26. Ranganathan, S. and Kar, S. K. (1971). *Tetrahedron Letters*, 1855
27. Wasserman, H. H. and Bailey, D. T. (1970). *Chem. Commun.*, 107
28. Fujita, S., Kawaguti, T. and Nozaki, H. (1971). *Tetrahedron Letters*, 1119
29. Paudler, W. W. and Stephen, E. A. (1970). *J. Amer. Chem. Soc.*, **92**, 4468

30. Teubner, H-J. and Vogel, L. (1970). *Chem. Ber.*, **103**, 3319
31. Shukri, J., Alazawe, S. and Al-Tai, A. S. (1970). *J. Ind. Chem. Soc.*, **47**, 123
32. Gupta, D. R. and Ojha, A. C. (1971). *J. Ind. Chem. Soc.*, **48**, 295
33. Chastrette, F. (1970). *Bull. Soc. Chim. Fr.*, 1151
34. Keglevic, D. and Goles, D. (1970). *Croat. Chem. Acta*, **42**, 513
35. Kanerov, V. Y., Suvorov, N. N., Starostenko, N. E. and Okinikov, Y. G. (1970). *Zhur. Fiz. Khim.*, **44**, 1583
36. Buu-Höi, N. P., Saint-Ruf, G., Deschamps, D., Bigot, P. and Hieu, H-T. (1971). *J. Chem. Soc. C*, 2606
37. Chapelle, J-P., Elguero, J., Jacquier, R. and Tarrago, G. (1970). *Bull. Soc. Chim. Fr.*, 240
38. Ishii, H., Murakami, Y., Suzuki, Y. and Ikeda, N. (1970). *Tetrahedron Letters*, 1181
39. Ishii, H., Murakami, Y., Tani, S., Abe, K. and Ikeda, N. (1970). *Yakagaku Zasshi*, **90**, 724
40. Raileanu, D., Palaghita, M. and Nenitzescu, C. D. (1971). *Tetrahedron*, **27**, 5031
41. Raileanu, D. and Nenitzescu, C. D. (1965). *Rev. Roumain. Chem.*, **10**, 339
42. Allen, G. R. and Weiss, M. J. (1966). *Chem. and Ind. (London)*, 117
43. Poletto, J. F. and Weiss, M. J. (1970). *J. Org. Chem.*, **35**, 1190
44. Eiden, F. and Kuckländer, U. (1971). *Arch. Pharm.*, **304**, 57
45. Kuckländer, U. (1971). *Arch. Pharm.*, **304**, 602
46. Remers, W. A., Roth, R. H., Gibs, G. J. and Weiss, M. J. (1971). *J. Org. Chem.*, **36**, 1232
47. Remers, W. A. and Weiss, M. J. (1971). *J. Org. Chem.*, **36**, 1241
48. Friary, R. J., Franck, R. W. and Tobin, J. F. (1970). *Chem. Commun.*, 283
49. Roth, H. J. and Hagen, H-E. (1971). *Arch. Pharm.*, **304**, 70
50. Roth, H. J. and Hagen, H-E. (1971). *Arch. Pharm.*, **304**, 73
51. Julia, M. and Pascal, Y. R. (1970). *Chim. Ther.*, **5**, 279
52. Degutis, J. and Barkauskas, V. (1969). *Khim. Geterotsikl. Soedin.*, 1003; *Chem. Abstr.*, **73**, 35147
53. George, C., Gill, E. W. and Hudson, J. A. (1970). *J. Chem. Soc. C*, 74
54. Arora, K. J. S., Dirania, M. K. M. and Hill, J. (1971). *J. Chem. Soc. C*, 2865
55. Bakke, J. (1970). *Acta Chem., Scand.*, **24**, 2650
56. Hill, J. and Townend, J. (1970). *Tetrahedron Letters*, 4607
57. Stefanescu, P. N. (1971). *Rev. Chim. (Bucharest)*, **22**, 370
58. Dopp, D. (1971). *Chem. Ber.*, **104**, 1043
59. Cliff, G. R. and Jones, G. (1971). *J. Chem. Soc. C*, 3418
60. Hemetsberger, H., Knittel, D. and Weidmann, H. (1970). *Monatsh. Chem.*, **101**, 161
61. Barton, D. H. R., Sammes, P. G. and Weingarton, G. G. (1971). *J. Chem. Soc. C*, 721
62. Bravo, P., Gaudiano, G. and Umani-Ronchi, A. (1970). *Gazz. Chim. Ital.*, **100**, 652
63. Bravo, P., Gaudiano, G., Ponti, P. P. and Zubiani, M. G. (1970). *Tetrahedron Letters*, 4535
64. Bravo, P., Gaudiano, G. and Ponti, P. P. (1971). *Chem. and Ind. (London)*, 253
65. Cignarella, G. and Saba, A. (1970). *Ann. Chim. (Rome)*, **66**, 765
66. White, J. D., Mann, M. E., Kirshenbaum, H. D. and Mitra, A. (1971). *J. Org. Chem.*, **36**, 1048
67. Bradsher, C. K. and Voigt, C. F. (1971). *J. Org. Chem.*, **36**, 1603
68. Sasaki, T. and Yoshioka, T. (1971). *Bull. Chem. Soc. Jap.*, **44**, 803
69. Douglass, J. E. and Wesolosky, J. M. (1971). *J. Org. Chem.*, **36**, 1165
70. Fröhlich, J. and Kröhnke, F. (1971). *Chem. Ber.*, **104**, 1621
71. Breslow, R., Eicher, T., Krebs, A., Peterson, R. A. and Posner, J. (1965). *J. Amer. Chem. Soc.*, **87**, 1320
72. Lown, J. W. and Matsumoto, K. (1971). *Can. J. Chem.*, **49**, 1165
73. Fröhlich, J. and Kröhnke, F. (1971). *Chem. Ber.*, **104**, 1629, 1645
74. see e.g. Hafner, K., Häfner, K. H., König, C., Kreuder, M., Ploss, G., Schulz, G., Sturm, E. and Vöpel, K. H. (1963). *Angew. Chem. Internat. edn.*, **2**, 123
75. Jessep, M. A. and Leaver, D. (1970). *Chem. Commun.*, 790
76. Palmer, M. H. and Gaskell, A. J. (1971). *Theoret. Chim. Acta*, **23**, 52
77. Hammond, H. A. (1970). *Theoret. Chim. Acta*, **18**, 239
78. O'Sullivan, P. S., de la Vega, J. and Hameka, H. F. (1970). *Chem. Phys. Letters*, **5**, 576
79. Dewar, M. J. S., Harget, A. J., Trinajstić, N. and Worley, S. D. (1970). *Tetrahedron*, **26**, 4505

80. Häfelinger, G. (1970). *Chem. Ber.*, **103**, 2902
81. Heidrich, D. and Scholz, M. (1970). *Monatsh. Chem.*, **101**, 1394
82. Engewald, W. Mühlstädt, M. and Weiss, C. (1971). *Tetrahedron*, **27**, 851
83. Pugmire, R. J., Robins, M. J., Grant, D. M. and Robins, R. K. (1971). *J. Amer. Chem. Soc.*, **93**, 1887
84. Eletr, S. (1970). *Molec. Phys.*, **18**, 119
85. Marey, T. and Arriau, J. (1971). *Compt. Rend.*, **272C**, 851
86. Nagy, J. and Hencsei, P. (1970). *J. Organometallic Chem.*, **24**, 603
87. Goetz, H. and Marshner, F. (1971). *Tetrahedron*, **27**, 3581
88. Tucker, S. W. and Walker, S. (1970). *J. Phys. Chem.*, **74**, 1270
89. Pham-Van-Huong and Lassegues, J. C. (1970). *Spectrochim. Acta*, **26**, 269
90. Nozai, M. S. and Drago, R. S. (1970). *J. Amer. Chem. Soc.*, **92**, 7086
91. Chadwick, D. J. and Meakins, G. D. (1970). *Chem. Commun.*, 637
92. Jung, A. (1970). *Method Phys. Anal.*, **6**, 54
93. Saito, H. and Nukado, K. (1971). *J. Amer. Chem. Soc.*, **93**, 1072
94. Fukui, H., Shimokawa, S. and Sohma, J. (1970). *Molec. Phys.*, **18**, 1217
95. Rahkamaa, E. (1970). *Molec. Phys.*, **19**, 727
96. Rahkamaa, E. (1971). *Z. Naturforsch., A*, **26**, 1187
97. Shimokawa, S., Fukai, H. and Sohma, J. (1970). *Molec. Phys.*, **19**, 695
98. Gagnaire, D., Ramasseul, R. and Rassat, A. (1970). *Bull. Soc. Chim. Fr.*, 415
99. Roomi, M. W. and Dugas, H. (1970). *Can. J. Chem.*, **48**, 2303
100. Fukui, H., Shimokawa, S., Sohma, J. Iwadare, T. and Esumi, N. (1971). *J. Molec. Spectrosc.*, **39**, 521
101. Lallemand, J-Y and Bernath, T. (1970). *Bull. Soc. Chim. Fr.*, 4091
102. Berus, E. I., Barkhash, V. A. and Molin, Y. N. (1970). *Zhur. Strukt. Khim.*, **11**, 819. *(English Edn.*, 759)
103. Melent'eva, T. A., Filipporva, T. M., Kazanskaya, L. V., Kustanovich, I. M. and Bevezovskii, V. M. (1971). *Zhur. Obshch. Khim.*, **41**, 179
104. Reinecke, M. G., Sebastian, J. F., Johnson, H. W. and Pyun, C. (1971). *J. Org. Chem.*, **36**, 3091
105. Parker, R. G. and Roberts, J. D. (1970). *J. Org. Chem.*, **35**, 996
106. Cushley, R. J., Anderson, D. R., Lipsky, S. R., Sykes, R. J. and Wasserman, H. H. (1971). *J. Amer. Chem. Soc.*, **93**, 6284
107. Eloranta, J. K. (1970). *Z. Naturforsch. A*, **25**, 1296
108. Foldes, A. and Sandorfy, C. (1970). *Can. J. Chem.*, **48**, 2197
109. Elsom, L. F. and Jones, R. A. (1970). *J. Chem. Soc. B*, 79
110. Skylar, E. Y., Evstigneeva, R. P. and Preobrazhenskii, N. A. (1970). *Zhur. Obshch. Khim.*, **40**, 1877
111. Derrick, P. J., Asbrink, L., Edqvist, O. and Lindholm, E. (1971). *Spectrochim. Acta*, **27A**, 2525
112. Scott, D. W. (1971). *J. Molec. Spectrosc.*, **37**, 77
113. Weiler-Feilchen, H., Pullman, A., Berthod, H. and Geissner-Prettre, C. (1970). *J. Molec. Struct.*, **6**, 297
114. Abribat, G. and Viallet, P. (1970). *Compt. Rend.*, **271C**, 1029
115. Cumper, C. W. N. and Wood, J. W. M. (1971). *J. Chem. Soc. B*, 1811
116. Jamieson, W. D. and Hutzinger, O. (1970). *Phytochemistry*, **9**, 2029
117. Marchelli, R., Jamieson, W. D., Safe, S. H., Hutzinger, O. and Heacock, R. A. (1971). *Can. J. Chem.*, **48**, 2197
118. Jones, G. and Stanyer, J. (1970). *Org. Mass Spectrom.*, **3**, 1489
119. King, R. B. and Efraty, A. (1970). *Org. Mass. Spectrom.*, **3**, 1227
120. Catoliotti, R., Foffani, A. and Pignataro, S. (1970). *Inorg. Chem.*, **9**, 2594
121. Kost, A. N., Minkin, V. I., Sagitullin, R. S., Gorbunov, V. I. and Sadekov, I. D. (1970). *Zhur. Org. Khim.*, **6**, 845; *Chem. Abstr.*, **73**, 13956
122. Fringnelli, F., Marino, G. and Savelli, G. (1970). *Tetrahedron*, **26**, 5815
123. Aten, W. C. and Büchel, K. H. (1970). *Z. Naturforsch., B*, **25**, 961
124. Chen, H. J., Hakka, L. E., Hinman, R. L., Kresge, A. J. and Whipple, E. B. (1971). *J. Amer. Chem. Soc.*, **93**, 5102
125. Jones, R. A. and Wright, P. H. (1968). *Tetrahedron Letters*, 5495
126. Candy, C. F. (1971). Ph.D. Thesis, University of East Anglia
127. Arlinger, L., Dahlqvist, K. I. and Forsen, S. (1970). *Acta Chem. Scand.*, **24**, 672

128. Roques, B., Janreginberry, C., Fournie-Zaluski, M. C. and Combrisson, S. (1971). *Tetrahedron Letters*, 2693
129. Meinwald, J. and Ottenheym, H. C. J. (1971). *Tetrahedron*, **27**, 3307
130. Candy, C. F. and Jones, R. A. (1971). *J. Chem. Soc. B*, 1405
131. Galasso, V. and De Alti, G. (1971). *Tetrahedron*, **27**, 3307
132. Clementi, S. and Marino, G. (1970). *Chem. Commun.*, 1642
133. Da Settimo, A. and Nannipieri, E. (1970). *J. Org. Chem.*, **35**, 2546
134. Snieckus, V. and Mei-Sie Lin. (1970). *J. Org. Chem.*, **35**, 3994
135. Muchowski, J. M. (1970). *Can. J. Chem.*, **48**, 422
136. Bass, R. J. (1971). *Tetrahedron*, **27**, 3263
137. Bergmann, J. (1971). *Acta Chem. Scand.*, **25**, 2865
138. Birchale, G. R. and Rees, A. H. (1971). *Can. J. Chem.*, **49**, 919
139. Janreginberry, C., Fournie-Zaluski, M-C. and Chevallier, J-P. (1971). *Compt. Rend.*, **273C**, 276
140. Farnier, M. and Fournari, P. (1971). *Compt. Rend.*, **273C**, 919
141. Motekaitis, R. J., Heinert, D. H. and Martell, A. E. (1970). *J. Org. Chem.*, **35**, 2504
142. Matsui, M. and Okada, K. (1970). *Agr. Biol. Chem. (Tokyo)*, **34**, 648
143. Harris, R. L. N. (1970). *Aust. J. Chem.*, **23**, 1199
144. Beveridge, S. and Harris, R. L. N. (1971). *Aust. J. Chem.*, **24**, 1229
145. Treibs, W. (1970). *Naturwissenschaften*, **57**, 306
146. Cooksey, A. R., Morgan, K. J. and Morrey, D. P. (1970). *Tetrahedron*, **26**, 5101
147. Morgan, K. J. and Morrey, D. P. (1971). *Tetrahedron*, **27**, 254
148. Sonnet, P. E. (1970). *J. Heterocyclic Chem.*, **7**, 399
149. Sunder, S. and Blanton, C. D. (1970). *J. Chem. Eng. Data*, **15**, 592
150. Spiro, V., Ajello, E. and Petruso, S. (1971). *Ann. Chim. (Rome)*, **48**, 546
151. Treibs, A. and Schulze, L. (1970). *Liebigs Ann. Chem.*, **739**, 222
152. Anderson, H. J. and Huang, C. W. (1970). *Can. J. Chem.*, **48**, 1550
153. Tarnopolskii, Y. I., Denisovich, L. I. and Golikov, A. A. (1970). *Khim. Geterotsikl. Soedin*, 620; *Chem. Abstr.*, **73**, 98728
154. Buckus, P. and Raguotiene, N. (1970). *Khim. Geterotsikl. Soedin*, 1056; *Chem. Abstr.*, **75**, 48805
155. Decodts, G., Wakselman, M. and Vilkas, M. (1970). *Tetrahedron*, **26**, 3313
156. Raines, S. and Kovacs, C. A. (1970). *J. Heterocyclic Chem.*, **7**, 233
157. Treibs, A., Kreuzer, F. H. and Häberle, N. (1970). *Liebigs Ann. Chem.*, **733**, 37
158. Treibs, A., Jacob, K., Kreuzer, F. H. and Tribollet, R. (1970). *Liebigs Ann. Chem.*, **742**, 107
159. Wolinsky, J. and Sundeen, J. E. (1970). *Tetrahedron*, **26**, 5427
160. Bergmann, J., Högberg, S. and Lindström, J-O. (1970). *Tetrahedron*, **26**, 3347
161. Bergmann, J. (1970). *Tetrahedron*, **26**, 3353
162. Monti, S. A. and Castillo, G. D. (1970). *J. Org. Chem.*, **35**, 3764
163. Monti, S. A. and Johnson, W. O. (1970). *Tetrahedron*, **26**, 3685
164. Vecchietti, V., Dradi, E. and Lauria, F. (1971). *J. Chem. Soc. C*, 2554
165. Onaka, T. (1971). *Tetrahedron Letters*, 4391
166. Ponomarev, A. A., Skvortsov, I. M. and Lenin, V. M. (1970). *Khim. Geterotsikl. Soedin*, 1339; *Chem. Abstr.*, **74**, 64151
167. Candy, C. F. and Jones, R. A. (1970). *J. Chem. Soc. C*, 2563
168. Horning, D. E. and Muchowski, J. M. (1970). *Can. J. Chem.*, **48**, 193
169. McKenzie, S. and Reid, D. H. (1970). *J. Chem. Soc. C*, 145
170. Kamenov, L. and Yudin, L. G. (1970). *Khim. Geterotsikl. Soedin*, 923
171. Seeman, F., Wiskott, E., Niklaus, P. and Troxler, F. (1971). *Helv. Chim. Acta*, **54**, 2411
172. v. Dobeneck, H. and Messerschmitt, T. (1971). *Liebigs Ann. Chem.*, **751**, 32
173. Messerschmitt, T., v. Speck, U. and v. Dobeneck, H. (1971). *Liebigs Ann. Chem.*, **751**, 50
174. Cooper, G. H. (1971). *J. Org. Chem.*, **36**, 2897
175. Harris, R. L. N. (1970). *Tetrahedron Letters*, 5217
176. Roomi, M. W. (1970). *Tetrahedron*, **26**, 4243
177. Treibs, A. and Jacob, K. (1970). *Liebigs Ann. Chem.*, **733**, 27
178. Roomi, M. W. (1971). *Experientia*, **27**, 505
179. Treibs, A. and Jacob, K. (1970). *Liebigs Ann. Chem.*, **740**, 196
180. Bergmann, J. (1970). *J. Heterocyclic Chem.*, **7**, 1071

181. Katner, A. S. (1970). *Org. Prep. Proc.*, **2**, 297
182. Kametani, T. and Suzuki, T. (1970). *Yakugaku Zasshi*, **90**, 771
183. Bergmann, J. (1971). *J. Heterocyclic Chem.*, **8**, 329
184. Zhungietu, G. I., Sukhanyuk, B. P. and Protsap, G. A. (1970). *Khim. Geterotsikl. Soedin,* 1058; *Chem. Abstr.*, **75**, 48814
185. Volz, H., Zingibl, U. and Messner, B. (1970). *Tetrahedron Letters*, 3593
186. Volz, H. and Draese, R. (1970). *Tetrahedron Letters*, 4917
187. Denbel, H., Wolkenstein, D., Jokisch, H., Messerschmitt, T., Brodka, S. and v. Dobeneck, H. (1971). *Chem. Ber.*, **104**, 705
188. Zhungietu, G. P., Shantsevoi, I. V. and Sukhanyuk, B. P. (1971). *Zhur. Vses. Khim. Obshchest.*, **16**, 232
189. U.S.S.R. patent. (1971). 287 936. *Chem. Abstr.*, **75**, 36347
190. Hamana, M. and Kumadaki, I. (1970). *Chem. Pharm. Bull. (Tokyo)*, **18**, 1743, 1822
191. Kametani, T. and Suzuki, T. (1971). *J. Chem. Soc. C*, 1053
192. Pietra, S. and Casiraghi, G. (1970). *Gazz. Chim. Ital.*, **100**, 128
193. Bailey, A. S., Warr, W. A., Allison, G. B. and Prout, C. K. (1970). *J. Chem. Soc. C,* 956
194. Bailey, A. S., Scattergood, R., Warr, W. A., Cameron, T. S. and Prout, C. K. (1970). *Tetrahedron Letters*, 2979
195. Bailey, A. S., Scattergood, R. and Warr, W. A. (1971). *J. Chem. Soc. C*, 2479
196. Tickle, I. J. and Prout, C. K. (1971). *J. Chem. Soc. C*, 3401
197. Bailey, A. S., Scattergood, R. and Warr, W. A. (1971). *J. Chem. Soc. C*, 3796
198. Bailey, A. S., Brown, B. R. and Churn, M. C. (1971). *J. Chem. Soc. C*, 1590
199. Collona, M., Greci, L. and Padovano, G. (1970). *Atti Accad. Sci. Ist. Bologna, Classe Sci. Fis. Rend.*, **7**, 84; *Chem. Abstr.*, **75**, 76510
200. Yoshida, Z., Hashimoto, H. and Yoneda, S. (1971). *Chem. Commun.*, 1344
201. Bellut, H. and Köster, R. (1970). *Liebigs Ann. Chem.*, **738**, 86
202. Szarvas, P., Emri, J. and Gyori, B. (1970). *Acta Chim. (Budapest)*, **64**, 203
203. Szarvas, P., Emri, J. and Gyori B. (1971). *Acta Chim. (Budapest)*, **70**, 1
204. Monti, S. A. and Schmidt, R. R. (1971). *Tetrahedron*, **27**, 3331
205. Burgada, R. and Bernard, D. (1971). *Compt. Rend.*, **273C**, 164
206. Candy, C. F. and Jones, R. A. (1971). *J. Org. Chem.*, **36**, 3993.
207. Öfele, K. and Dotzauer, E. (1971). *J. Organometallic Chem.*, **30**, 211
208. Coleman, K. J., Davies, C. S. and Gogan, N. J. (1970). *Chem. Commun.*, 1414
209. Tille, D. (1970). *Z. Naturforsch., B*, **25**, 1358
210. Wenschuh, E. and Meibner, I. (1970). *Z. Chem.*, **10**, 2
211. Loader, C. E. and Anderson, H. J. (1971). *Can. J. Chem.*, **49**, 1064
212. Wong, J. L. and Ritchie, M. H. (1970). *Chem. Commun.*, 142
213. Loader, C. E. and Anderson, H. J. (1971). *Can. J. Chem.*, **49**, 45
214. Kobayashi, G., Matsuda, Y., Natzuki, R. and Tominaga, Y. (1970). *Yakugaku Yasshi*, **90**, 1251
215. Treibs, A. and Friess, R. (1970). *Liebigs Ann. Chem.*, **737**, 179
216. Anderson, H. J. and Groves, J. K. (1971). *Tetrahedron Letters*, 3165
217. Bowman, R. E., Evans, D. D. and Islip, P. J. (1971). *Chem. Ind. (London)*, 33
218. Bocchi, V., Chierici, L., Gardini, G. P. and Mondelli, R. (1970). *Tetrahedron*, **26**, 4073
219. Lilie, J. (1971). *Z. Naturforsch., B*, **26**, 197
220. Quistad, G. B. and Lightner, D. A. (1971). *Chem. Commun.*, 1099
221. Quistad, G. B. and Lightner, D. A. (1971). *Tetrahedron Letters*, 4417
222. Ranjon, A. (1971). *Bull. Soc. Chim. Fr.*, 2068
223. Rio, G. and Lecas-Nawrocka, A. (1971). *Bull. Soc. Fr.*, 1723
224. Franck, R. W. and Auerbach, J. (1971). *J. Org. Chem.*, **36**, 31
225. Libers, M. and Caullet, C. (1971). *Bull. Soc. Chim. Fr.*, 1947
226. Nakagawa, M. and Hino, T. (1970). *Tetrahedron*, **26**, 4491
227. Dmitrienko, G. I. (1971). *Can. J. Chem.*, **49**, 3642
228. Robinson, B. and Zubair, M. U. (1971). *J. Chem. Soc. C*, 976
229. Nakagawa, M., Yamaguchi, H. and Hino, T. (1970). *Tetrahedron Letters*, 4035
230. Hino, T., Nakagawa, M., Hashizuma, T., Yamaji, N. and Miwa, Y. (1970). *Tetrahedron Letters*, 2205
231. Dolby, J. J. and Rodia, R. M. (1970). *J. Org. Chem.*, **35**, 1493
232. Kanaoka, Y., Aiura, M. and Hariya, S. (1971). *J. Org. Chem.*, **36**, 458

233. Iddon, B., Phillips, G. O., Robbins, K. E. and Davies, J. V. (1971). *J. Chem. Soc. B*, 1887
234. Evans, N. A. (1971). *Aust. J. Chem.*, **24**, 1971
235. Kanaoka, Y., Miyashita, K. and Yonemitzu, O. (1970). *Chem. Pharm. Bull. Japan*, **18**, 634
236. Bender, C. O., Bonnett, R. and Smith, R. G. (1970). *J. Chem. Soc. C*, 1251
237. Kricka, L. J. and Vernon, J. M. (1971). *J. Chem. Soc. C*, 2667
238. Tóth, T. and Gerecs, A. (1971). *Acta Chim. (Budapest)*, **67**, 229
239. Remers, W. A., Gibs, G. J., Pidacks, C. and Weiss, M. J. (1971). *J. Org. Chem.*, **36**, 279
240. Cliff, G. R., Jones, G. and Stanyer, J. (1971). *J. Chem. Soc. C*, 3426
241. Barlow, M. G., Haszeldine, R. N. and Hubbard, R. (1970). *J. Chem. Soc. C*, 90
242. Gandhi, R. P. and Chadha, V. K. (1971). *Ind. J. Chem.*, **9**, 305
243. Prinzbach, H., Fuchs, R. and Kitzing, R. (1968). *Angew. Chem. Internat. edn.*, **7**, 67
244. Lin, M-S. and Snieckus, V. (1971). *J. Org. Chem.*, **36**, 645
245. Kaupp, G., Perreten, J., Leute, R. and Prinzbach, H. (1970). *Chem. Ber.*, **103**, 2288
246. Coe, P. L. and Uff, A. J. (1971). *Tetrahedron*, **27**, 4065
247. Kricka, L. J. and Vernon, J. M. (1971). *Chem. Commun.*, 942
248. Bellas, M., Bryce-Smith, D. and Gilbert, A. (1967). *Chem. Commun.*, 263
249. Bryce-Smith, D., Clarke, M. T., Gilbert, A., Klunklin, G. and Manning, C. (1971). *Chem. Commun.*, 916
250. McCullough, J. J., Huang, C. W. and Wu, W. S. (1970). *Chem. Commun.*, 1368
251. McCullough, J. J. and Wu. W. S. (1971). *Tetrahedron Letters*, 3951
252. Hiraoka, H. (1970). *Chem. Commun.*, 1306
253. Hunt, R. and Reid, S. T. (1970). *Chem. Commun.*, 1576
254. Patterson, J. M., Beine, R. L. and Boyd, M. R. (1971). *Tetrahedron Letters*, 3923
255. Naruto, S. and Yonemitsu, O. (1971). *Tetrahedron Letters*, 2297
256. de Silva, O. and Snieckus, V. (1971). *Synthesis*, 254
257. Husson, H-P., Thal, C., Potier, P. and Wenkert, E. (1970). *J. Org. Chem.*, **35**, 442
258. Nicoletti, R. and Forcellese. M. L. (1965). *Gazz. Chim. Ital.*, **95**, 83
259. Jones, R. L. and Rees, C. W. (1969). *J. Chem. Soc. C*, 2255
260. Gambacorta, A., Nicoletti, R. and Forcellese, M. L. (1971). *Tetrahedron*, **27**, 985
261. Fowler, F. W. (1971). *Angew. Chem. Internat. edn.*, **10**, 135
262. Biellmann, J. F. and Goeldner, M. P. (1971). *Tetrahedron*, **27**, 2957
263. Cohen, M. (1970). *Tetrahedron Letters*, 2165
264. Rosenmund, P., Sauer, D. and Trommer, W. (1970). *Chem. Ber.*, **103**, 496
265. Bhandari, K. S. and Snieckus, V. (1971). *Can. J. Chem.*, **49**, 2354
266. Flisch, W. and Neumann, U. (1971). *Chem. Ber.*, **104**, 2170
267. Eenkhoorn, J. A., de Silva, O. S. and Snieckus, V. (1970). *Chem. Commun.*, 1095
268. Orlova, I. A., Turchin, K. F. and Semenov, N. N. (1971). *Zhur. Org. Khim.*, **7**, 598
269. Treibs, A. and Häberle, N. (1970). *Liebigs Ann. Chem.*, **739**, 220
270. Treibs, A. and Jacob, K. (1970). *Liebigs. Ann. Chem.*, **737**, 176
271. Bisagin,É., Bourzat, J-D. and André-Louisfert, J. (1970). *Tetrahedron*, **26**, 2087
272. Kreher, R. and Vogt, G. (1970). *Angew. Chem. Internat. edn.*, **9**, 955
273. Gillis, B. T. and Valentour, J. C. (1971). *J. Heterocyclic Chem.*, **8**, 13
274. Cheeseman, G. W. H. and Rafiq, M. (1971). *J. Chem. Soc. C*, 2732
275. Cheeseman, G. W. H., Rafiq, M., Roy, P. D., Turner, C. J. and Boyd, C. V. (1971). *J. Chem. Soc. C*, 2018
276. Raines, S., Chai, S. Y. and Plopoli, F. P. (1971). *J. Org. Chem.*, **36**, 3992
277. Mudry, C. A. and Arasca, A. R. (1971). *Chem. and Ind. (London)*, 1038
278. Dunn, G. E. and Lee, G. K. J. (1971). *Can. J. Chem.*, **49**, 1032
279. Treibs, A. and Schulze, L. (1971). *Liebigs Ann. Chem.*, **739**, 225
280. Bowman, R. E. and Islip, P. J. (1971). *Chem. and Ind. (London)*, 154
281. Bhat, G. A. and Siddappa, S. (1971). *J. Chem. Soc. C*, 178
282. Bullock, E., Chen, T-S., Loader, C. E. and Wells, A. E. (1970). *Can. J. Chem.*, **48**, 1651
283. Groves, J. K., Anderson, H. J. and Nagy, H. (1971). *Can. J. Chem.*, **49**, 2427
284. Treibs, A. and Grimm, D. (1971). *Liebigs Ann. Chem.*, **752**, 44
285. Fields, E. K. and Meyerson, S. (1965). *Chem. Commun.*, 475
286. Cava, M. P. and Bravo, L. (1970). *Tetrahedron Letters*, 4631
287. Bergmann, J. (1971). *Acta Chem. Scand.*, **25**, 1277
288. Pleininger, H. and Sirowej, H. (1971). *Chem. Ber.*, **104**, 1869

289. Garmaise, D. L. and Ryan, A. (1970). *J. Heterocyclic Chem.*, **7,** 413
290. Waite, D. (1970). *J. Chem. Soc. C*, 550
291. Daisley, R. W. and Walker, J. (1971). *J. Chem. Soc. C*, 3357
292. Bocchi, V. and Gardini, G. P. (1971). *Tetrahedron Letters*, 211
293. Daisley, R. W. and Walker, J. (1971). *J. Chem. Soc. C*, 1375
294. Horning, D. E., Lacasse, G. and Muchowski, J. M. (1971). *Can. J. Chem.*, **49,** 246
295. Pleininger, H., Sirowej, H. and Rau, D. (1971). *Chem. Ber.*, **104,** 1863
296. Bycroft, B. W. and Laudon, W. (1970). *Chem. Commun.*, 168
297. Bycroft, B. W. and Laudon, W. (1970). *Chem. Commun.*, 967
298. Casnati, G. and Pochini, A. (1970). *Chem. Commun.*, 1328
299. Hino, T., Nakagawa, M., Suzuki, T., Takeda, S., Kano, N. and Ishii, Y. (1971). *Chem. Commun.*, 836

3
Nitrogen Heterocyclic Molecules, Part 2 Azoles and Benzazoles

M. R. GRIMMETT
University of Otago, Dunedin, New Zealand

3.1   INTRODUCTION

This chapter surveys the advances during 1970–1971 in the chemistry of pyrazoles, imidazoles, triazoles, tetrazoles, and their simple benz-derivatives. The naming and numbering of compounds will follow ordinary practice except in one point; in discussions other than those concerned with tautomerism, only one of two possible tautomers will be named, e.g. 4-methylimidazole rather than 4(5)-methylimidazole. This usage carries no implication that the tautomer actually named is necessarily the dominant form.

3.2 RING SYNTHESIS

3.2.1 Pyrazoles

Considerable use has been made of the reaction of hydrazines with 1,3-difunctional compounds[1-13]. With unsaturated β-diketones the products are alkenylpyrazoles[1], while with acetylenic carbonyl compounds mono- and 1,2-di-substituted hydrazines form 1-substituted pyrazoles and allow the formation of isomers previously difficult to obtain[2]. Most 1,1-di-substituted hydrazines react with α-benzoylbenzylcyanide with the elimination of one substituent and the formation of 1-alkyl-5-amino-3,4-diphenylpyrazole[3]. The eliminated alkyl group is found either as a quaternary hydrazinium ion ($N \rightarrow N$-transalkylation), or as an alkyl derivative of α-benzoylbenzylcyanide ($N \rightarrow O$- and $N \rightarrow C$-transalkylation). When the hydrazines are unsymmetrically substituted the alkyl groups are split off at different rates. In the presence of alkoxides, α-acetylenic esters or nitriles react with phenylhydrazine to form 3-hydroxy-1-phenylpyrazoles[4]. When excess alkali is present, the amino group attacks the ester or nitrile function to form a hydrazide which cyclises to the hydroxypyrazole[5]. Some success has been achieved in the synthesis of 4-hydroxypyrazoles from 1,2,3-triones or 2-acetoxyl-1,2,3-triones and hydrazines[6,7]. Lower yields of these hydroxypyrazoles are obtained from the reaction of chalcone oxide with toluene-p-sulphonylhydrazide in sodium ethoxide, followed by chromic oxide oxidation of the pyrazolinol formed[6]. Alkylbut-1-en-3-ynyl ethers and N,N-dialkylbut-1-en-3-ynylamines react with methylhydrazine hydrochloride to form mixtures of 1,3- and 1,5-dimethylpyrazoles[8]. The intermediates formed when nicotinoyl- and benzoyl-hydrazines react with acetylacetone have been shown to be cyclic monohydrazones (i.e. 1-acyl-5-hydroxy-2-pyrazolines) which are readily dehydrated to pyrazoles[9]. This is in contrast to the acyclic intermediate formed in the reaction of acetylacetone with hydrazine. Substituted 5-amino-pyrazoles result when β-cyanoalkylhydrazones are cyclised

Figure 3.1

with basic catalysts[14]. In a new synthetic route to pyrazoles[15], conjugated diacetylenic compounds condense with hydrazine hydrate at 80–100 °C. Reaction of phenylhydrazine with the hydrogen sulphates of 3-aryl-1,2-dithiolium (1) gives a mixture of 3- and 5-aryl-1-phenylpyrazoles[16]. (See Figure 3.1.) With 3,5-diaryl-1,2-dithiolium perchlorate two isomeric 3,5-

diaryl-1-phenylpyrazoles are formed[16]. Good yields (c. 60%) of pyrazoles result from the interaction of hydrazines and 3-acetyl-1,4,5,6-tetrahydropyridine[17]. The condensation of 1,4-di-anions of acetophenone phenylhydrazone (and related compounds) with a variety of aromatic esters, followed by acid cyclisation provides a further route to pyrazoles[18, 19]. Arylamides of arylsulphonylcyanothioacetic acid are cyclised with hydrazine to 3-arylamino-4-arylsulphonyl-5-aminopyrazole[20]. The synthesis of pyrazoles and 1,3-difunctional compounds has been discussed[21, 22].

Transformations of other heterocyclic molecules into pyrazoles include the 1,3-dipolar cyclo-addition reactions of N-acylsydnone imines with dimethylacetylenedicarboxylate[23]. Similarly, the in situ formation of dipolar intermediates has been employed[24], e.g. the direct reaction of N-nitrosarcosine with acetylene dicarboxylic ester in the presence of acetic anhydride. Thermolysis or photolysis of N-phenylsydnones with acetylenedicarboxylates results in pyrazole formation[25]. Both 2-phenyl-4,6- and 2-phenyl-4,5-dichloro-3(2H)-pyridazinone are converted in alkali into 2-phenyl-5-hydroxypyrazole-3-carboxylic acid[26], possibly through a hetaryne intermediate. Ring contraction to pyrazole carboxylic acids also occurs in the alkaline reaction of the corresponding unchlorinated pyridazinones[27]. The oxidative photolysis of 3-substituted Δ^2-pyrazolines yields pyrazoles in 33–82% yields[28], but the unsubstituted compounds give rise to β-anilinopropionitriles.

Pyrazole syntheses involving the 1,3-dipolar addition of aliphatic diazo compounds to unsaturated compounds have received considerable attention. Diazomethane adds exclusively to the α,β-triple bond of diynes, ArC⋮C·C⋮C·COR, to form mixtures of isomeric pyrazoles[29], while with acetylenic carbohydrate derivatives 4-glycosylpyrazoles result[30]. The cyclo-addition of diazomethane to unsymmetrical difunctionalised acetylenes[31] occurs with such an orientation as to suggest that the addition always occurs according to the expected polarisation of the triple bond by substituents. In a similar test[32] of von Auwers orientation rule[33] the addition of diazomethane and diazopropene to mono- and di-substituted acetylenic compounds shows that, in contrast to diazomethane, diazopropene reacts only with carbonyl-conjugated acetylenes, and then only in one of the two possible orientations[32] (see Figure 3.2). A product ratio analysis has been made for the reaction of

$$CH_2\!:\!CH\cdot\overset{-}{C}\!-\!\overset{+}{N}_2 \;+\; O_2N\!\!\left\langle\bigcirc\right\rangle\!\!-\!C\!\vdots\!C\cdot CHO \;\longrightarrow$$

Figure 3.2

diazomethane with methyl cis- and trans-β-chloroacrylates[34]. When diphenyldiazomethane and cyclo-octyne interact, the product, 3,3-diphenyl-4,5-cyclo-octanopyrazolenine, undergoes a van Alphen rearrangement to 1,5-diphenyl-3,4-cyclo-octanopyrazole when heated in acetic acid[35]. Carbonyl-stabilised sulphur ylides react with 1,3-dipoles, e.g. phenacylsulphonium ylids give high yields of pyrazoles with nitrile imines[36, 37].

Pyrazoles are formed in a dark reaction during the photolysis of some unsaturated tosylhydrazones[38]. In a novel reaction, α-keto-ester 2-carbo-alkoxymethylhydrazones undergo a Dieckmann reaction to form 4-hydroxy-pyrazole-5-carboxylates[39]. Although the usual products are pyrroles, the cyclisation of bis-en-hydrazines to pyrazolines can occur, and if a suitable leaving group is present, β-elimination can give rise to pyrazoles or pyrazolium species[40].

3.2.2 Indazoles

In a reaction resembling earlier alkaline cyclisations of benzyl compounds with o-nitro substituents, N,N-dimethyl-2,4-dinitrobenzylamine forms a 2-methylisoindazole[41]. Aromatic diazonium carboxylates have been isolated as intermediates in the Jacobson indazole synthesis[42]. In the thermal reaction of N-substituted-N-arylcarbamoylazides, a Curtius rearrangement results in the formation of 1-substituted-3-hydroxyindazoles[43]. Tropone phenylhydrazone is transformed photochemically into 2-phenylindazole (13%), and by lead tetra-acetate oxidation into the same product (20%)[44]. It is possible to dehydrogenate tetrahydroindazoles and benzodihydroindazoles to the fully aromatic compounds[45].

3.2.3 Imidazoles

The synthesis and properties of imidazoles have been reviewed[46]. An improved synthesis of 2-imidazol-ylethanol has been reported[47] from but-2-yne-1,4-diol, with the product being isolated and purified as its methoxy derivative. As an extension of earlier studies of imidazole formation by the action of ammonia on reducing oligosaccharides, Richards[48] has developed what appears to be a simple, sensitive method for linkage identification in polysaccharides. The action of ammonia on periodate-oxidised polysaccharides yields 2-imidazol-ylethanol from $1 \rightarrow 4$-linked, imidazole and 4-methylimidazole from $1 \rightarrow 6$-linked, and no imidazoles from $1 \rightarrow 3$-linked

$$\begin{array}{ccccc} R^1{-}C{=}O & & NR^2 & & R^1{-}NR^2 \\ | & + & || & \xrightarrow{-H_2O} & \diagup \diagdown \\ S{=}CNH_2 & & CHR^3 & & HS{\diagdown_N\diagup}R^3 \end{array}$$

Figure 3.3

reducing sugars. The reaction of α-halogenocarbonyl compounds with guanidines produces 2-amino-1-(2-thiazolyl)imidazoles[49]. In a reaction which resembles the Marckwald synthesis, 1,2,5-trisubstituted imidazole-4-thiols are derived from the exothermic combination of N-unsubstituted α-oxothionamides and aldimines (see Figure 3.3)[50]. In preparing mesoionic derivatives of 4-aminoimidazole, Chinone[51] formed 4-amino-1,2,3-triphenyl-imidazolium salts from amidines and halogenoacetonitriles. The condensation of α-aminonitriles with thioimidates forms α-amidinonitriles which can cyclise, depending on the reaction conditions and on the natures of the substituents, to 4-amino- or 4-amidino-imidazoles[52]. A Claisen-type rearrangement of the adduct (2) of an aromatic amidoxime and a propiolate ester provides a new and direct route to imidazole-2-carboxylates (see Figure 3.4)[53]. Pyrolysis of the adduct (2) in refluxing diphenyl ether gives

Figure 3.4

61–72% of imidazoles. The 'no-mechanism' rearrangement proposed involves three heteroatoms in the rearrangement ring[53]. A new, one-step synthesis of triarylimidazoles and triarylimidazolines has been developed from the reaction of benzylamine and derivatives with carbon tetrachloride in the presence of catalytic amounts of a metal carbonyl[54]. A free-radical mechanism has been proposed for the reaction[54]. Cyclisation of compounds PhCH: NC(CN):C(R)·SMe with dry HCl forms chloroimidazoles[55]. When diaminomaleonitrile is heated with a lower alkyl orthoformate in alcoholic solution, high yields of 4,5-dicyanoimidazoles are formed[56].

Interest in the chemistry of imidazole N-oxides and -hydroxides has increased in recent years. The compounds may be prepared from α-oximinoketones and α-dioximes[57], from α-oximinoketones and aldimines[58], or from α-oximinoketones, ammonia (or amines), and aldehydes[59, 60]. When the aldehyde used is formaldehyde, the product is the imidazolin-2-one which Lettau[61] considers to be formed through rearrangement of the N-oxide, but this seems unlikely in view of the stability (under the reaction conditions) of imidazole N-oxides unsubstituted at C-2[62].

The conversion of other heterocycles into imidazoles has received attention. In the synthesis of 2-arylimidazoles[63], selenium dehydrogenation of imidazolines has proved valuable. Hydrazine reacts with oxazoles in the presence of Brönsted acids to form 1-aminoimidazoles[64], while the 1,3-dipolar cyclo-addition of 5-hydroxy-3-methyl-2,4-diphenyloxazolium hydroxide inner salt with alkyl cyanides yields imidazoles[65]. Phenylisocyanate converts oxazole N-oxides into dihydroimidazoles[66]. Imidazole and 2-cyanoimidazole have been isolated in 15% yields from a ring transformation of 2-chloropyrazine by potassium amide in liquid ammonia[67]. Whereas benzamidine reacts with 2-amino-3-phenacyl-1,3,4-oxadiazolium bromides to form 1-acylamino-2-benzimidoylamino-4-arylimidazoles[68, 69], replacement of benzamidine with ammonia in the reaction results in the formation of 2-aminoimidazoles[69].

3.2.4 Benzimidazoles

The synthetic procedures of choice leading to benzimidazoles have involved variations on the cyclisation of o-phenylenediamine derivatives. Thus have been prepared 2-styryl-[70], 2-substituted-[71,72], and 2-substituted-amino-1,3-diphenylbenzimidazoles[73], and quaternary salts of 1,3-diphenylbenzimidazoles[74]. Kiffer[75] has described the synthesis of 2-aminobenzimidazoles through desulphurisation of 1-(2-aminophenyl)-3-phenylthioureas (prepared by condensation of phenylisothiocyanate with o-phenylenediamines). Depending on the nature of the substitution these thioureas can be decomposed thermally to 2-benzimidazolinethiones. When, however, the o-phenylenediamine is substituted on nitrogen by a β-dimethylamino function, reaction with phenylisothiocyanate leads directly to the 2-(N-substituted)benzimidazolinethione. In a study of the formation of benzimidazoles by the peroxyacid-catalysed cyclisation of o-acylamino-N,N-dialkylanilines, Meth-Cohn[76] has shown that the reaction proceeds via a N-oxide intermediate, as it proved possible to prepare this N-oxide and cyclise it under the reaction conditions. A corrected mechanism has been proposed[77] for the formation of 2-cyclopropylbenzimidazole from o-phenylenediamine and 2-bromocyclobutane. An improvement in the reduction of o-nitroanilines using hydrazine hydrate in ethanol with Raney nickel has increased yields of benzimidazoles to c. 65%[78], and has been applied to the synthesis of N-substituted-2-mercaptobenzimidazoles, using carbon disulphide as the cyclising agent. The reaction of triethyl phosphite with N-benzylidene-2-nitroaniline in boiling t-butylbenzene is reported[79] to give better yields (c. 47%) of 2-phenylbenzimidazole than the standard procedure from o-phenylenediamine.

Although the oxidation of the parent compounds with organic peracids has not proved fruitful[80], benzimidazole N-oxides may be prepared by the action of boiling hydrochloric acid on o-nitro-N,N-dialkylanilines[81]. Prolonged treatment, however, results in the preferential formation of a chloro-substituted benzimidazole[81]. The most useful method of synthesis of benzimidazole N-oxides seems to be the reductive ring-closure of N-acyl-o-nitroaniline derivatives with ammonium sulphide[82], although N-2,4-dinitrophenylamino acids have been converted into compounds designated as 6-nitrobenzimidazole 1-oxides[83,84], but which probably exist mainly as the hydroxy tautomers.

Pyrolysis of pyrrolidinobenzenesulphonylazide forms, among other products, a 1,2-disubstituted benzimidazole by a Curtius-type rearrangement[85]. The preparation has been reported[86] of 5,6-dehydro-1,3-dimethylbenzimidazole by oxidation of 5-amino-1,2-dimethyltriazolo[4,5-d]benzimidazole, or by aprotic diazotisation of the appropriate 6-aminobenzimidazole-5-carboxylic acid.

3.2.5 Triazoles

3.2.5.1 1,2,3-Triazoles

Use has been made of the 1,3-dipolar addition of azides to acetylenic compounds in the formation of 1-N-glycosyl-[87], 4-phenyl-[88], and 5-diethylamino-1-sulphonyl-1,2,3-triazoles[89]. Syntheses involving condensation of arylazides with active methylene compounds[90] include the preparation of 1,2,3-triazoles in 60% yields from the reaction of azide ion with α,β-unsaturated nitro compounds and nitriles[91] (a reaction in which, unexpectedly, considerable quantities of symmetrical triarylbenzenes were also formed), and the synthesis of 1-vinyl-1,2,3-triazoles[92]. The addition of diazomethane to certain nitriles produces 2-methyl-4-substituted-1,2,3-triazoles[93]. A new general method of synthesis is available through the interaction of α-keto phosphorus ylids with arylazides in refluxing benzene[94-97], which produces 54–91% yields of 1-aryl-5-substituted products. The reaction is accelerated by electron-withdrawing substituents on the ylid. Low entropies of activation[95], and kinetic data[96] support a concerted cyclo-addition as the first step in the reaction (see Figure 3.5). The method has been extended[97] to the synthesis of

Figure 3.5

N-2-acyl- and N-2-carbethoxy-1,2,3-triazoles, and shown to proceed through the corresponding N-1-substituted triazoles (cf. ref. 96), which rearrange by an intermolecular process under the influence of the ylid as base. The use of super-dry solvents is reported to prevent hydrolysis and raise yields[97]. The reactions of the product of oxidising benzil osazone with sodium ethoxide and iodine[98, 99], which is converted by heating or ultraviolet irradiation into 2,4,5-triphenyl-1,2,3-triazole, have been rationalised[100] by postulation of a mesionic intermediate ((3) Figure 3.6) which loses phenyl-nitrene to form the triazole. 1,5-Dipolar cyclisations (which do not take place in the thermolysis of vinyl azides in neutral or protic solvents because the loss of a nitrogen molecule occurs faster) can be carried out in strongly

Figure 3.6

(i) 85%
(ii) 14%

basic media wherein the resonance-stabilised triazolylanion is formed, e.g. α-azidostyrene yields 5-phenyl-1,2,3-triazole[88]. Cyclisation of ethyl N-methyl-N-arylazoamino-acetates with thionyl chloride forms anhydro 3-aryl-4-hydroxy-1-methyl-1,2,3-triazolium hydroxides[101]. Phenylglyoxal dihydra-zones are oxidised with iodine and mercuric oxide to 1,2,3-triazoles[102] (cf. Refs. 98–100). The compound prepared from 2,3-dioxobutane-bis-benzoylhydrazone with potassium hexacyanoferrate, previously considered to be dihydrotetrazine[103], has now been shown[104] to have the structure of a mesoionic triazole.

3.2.5.2   1,2,4-Triazoles

Substituted thioureas, isothioureas, and compounds containing the thioureido moiety as part of a heterocyclic structure, react at 100 °C with hydrazine hydrate to form 4-amino-5-mercapto-1,2,4-triazoles[105]. When aniline is added to 4-(p-substituted phenyl)-1,1,4-trihalogenodiazabutadienes, high yields of 3-amino-4,5-diaryl-1,2,4-triazoles result[106]. In confirmation of the suggestion that 1-halogenomethylamino-2,3-diphenylguanidines are intermediates in this reaction, it was found that 1-bromo-(p-substituted phenyl)-methyleneamino-2,3-diphenylguanidines form high yields of triazoles when stirred with triethylamine[106]. The Michael adducts of hydrazine and amino-methylenemalonates rearrange, depending on the leaving group, to form 4-substituted-1,2,4-triazoles or 4-carbethoxy-5-hydroxy-4H-pyrazole[107]. Hydroxylamine reacts with N-acylamidrazones (4) to form 4-hydroxy-1,2,4-triazoles[108] (see Figure 3.7). Other cyclisation reactions which lead to

$$R^1C\underset{NH}{\overset{OR^3}{<}} + R^2C\underset{NH\cdot NH_2}{\overset{O}{<}} \longrightarrow R^1C\underset{NH}{\overset{NH-NH}{<}}\underset{O}{CR^2} \xrightarrow{NH_2OH\cdot HCl} R^1C\underset{NOH}{\overset{NH-NH}{<}}\underset{O}{CR^2}$$

(4)

$$\underset{OH}{\overset{N-N}{R^1\left(\!\!\!\underset{N}{\right)}\!\!\!R^2}}$$

Figure 3.7

substituted 1,2,4-triazoles include the formic acid-induced cyclisation of 2-hydrazinobarbituric acids to triazolobarbituric acids[109], the reaction of N-aminoamidines with isocyanide chlorides in boiling toluene[110], the cyclisation by heat or aqueous alkali of N-(1,2,4-oxadiazol-3-yl)-N'-aryl-formamidines to 3-acylamino-1-aryl-1,2,4-triazoles[111], the oxidative cyclisation with alkaline ferricyanide of 4-substituted semicarbazones to 3,4-disubstituted-5-hydroxy-1,2,4-triazoles[112], the alkaline cyclisation of 1-amino-4-aroylthiosemicarbazides and their hydrazones to 3-aryl-5-mercapto-1,2,4-triazoles[113], and the cyclisation of 1-(N,N-dialkyloxamoyl)-4-methyl-semicarbazides to 1,2,4-triazoles and 1,2,4-triazines[114]. Triazoles unsubstituted on nitrogen have been prepared by the photolytic desulphurisation of 1,2,4-triazoline-3-thiones[115], while oxidation of the latter compounds forms the corresponding triazole sulphonic acids[116].

A number of other heterocyclic systems may be transformed into 1,2,4-triazoles. Mesoionic 1,3,4-thiadiazoles (5) rearrange to the isomeric triazoles (6) either by heating in ethanol, or by treatment with phenylisothiocyanate

$$MeN\overset{Ph}{\underset{N}{\bigoplus}}\underset{NPh}{S} \xrightarrow[\text{or (ii) PhNCS}]{\text{(i) Heat in EtOH}} Me\cdot N\overset{Ph}{\underset{N}{\bigoplus}}\underset{S}{NPh}$$

(5)   (6)

Figure 3.8

in benzene at room temperature[117] (see Figure 3.8). A wide variety of primary amines and ammonia react with 1,3,4-oxadiazolium salts to form 1,2,4-triazoles and -triazolium salts[118]. The intermediate oxadiazolines or amidrazones resulting from attack of the amine at C-2 of the oxadiazolium ring may have been isolated, and their subsequent fate has been shown to depend on the nature of the amine and on the reaction conditions. In a reaction analogous to the formation of 3-o-hydroxyphenyl-1,2,4-triazoles from 1,3-benzoxazin-4-ones[119], 2-aryl-2-oxazolin-4-ones form 5-substituted-3-hydroxymethyl-1,2,4-triazoles with hydrazine[120]. With two moles of dicarbethoxydi-imine, the inner salt of 5-hydroxy-3-methyl-2,4-diphenyloxazolium hydroxide is transformed into 4-methyl-3,5-diphenyl-1,2,4-triazole[65]. In an unprecedented ring contraction of 4,6-disubstituted pyrimidines by the action of hydrazine to 3-methyl-1,2,4-triazole[121], evidence is presented that the C(2)—N(3 or 1)—C(4 or 6)—C(5) fragment of the pyrimidine is involved in the triazole ring formation. Low yields of 3,5-diphenyl-1,2,4-triazole result from the photolysis of 5-phenyltetrazole[122]. In a new reaction N-tetrazol-5-yl p-substituted benzhydrazidic bromides and chlorides are converted in aqueous ethanol (via azocarbonium ion intermediates) into 3-azido-5-aryl-1,2,4-triazoles in 60–90% yields[123].

3.2.6   Benzotriazoles

Benzotriazoles have been prepared by reactions of o-phenylenediamines with nitrous acid[124, 125] and hydrazine hydrate[124]. Tetrafluorobenzotriazol-1-ol is obtained by the reaction of pentafluoronitrobenzene with hydrazine hydrate[124]. 2-Substituted benzotriazoles have been prepared by a multistep synthesis from o-nitroanilines[126].

3.2.7   Tetrazoles

Methods of synthesis of tetrazoles (and triazoles) have been reviewed[127]. Nitrous acid cyclises 2-hydrazinobarbituric acids to tetrazolo[5,4-a]-barbituric acids[109]. Bis-formazans are smoothly oxidised with concentrated nitric acid to the bis-tetrazolium salts[128].

3.3   STRUCTURE AND THEORETICAL STUDIES

There has been considerable application of theoretical studies to the azole field in recent years. Calculations[129] of electron densities, dipole moments and energies of formation give values which reflect the decrease in azole stability as the numbers of nitrogen atoms increase. When SCF calculations of π-electron distributions were made[130] for the ground states of azoles and benzazoles, the following orders of electrophilic substitution were predicted: pyrazole 4 > 5 > 3; imidazole 5 > 2 > 4; benzimidazole 7 > 6 > 5 > 4; 2H-indazole 3 > 5 > 7 > 4; 1H-indazole 7 > 5. It is admitted that these calculations do not allow for the tautomeric equivalence of the 3- and 5-positions of pyrazole, 4- and 5-positions of imidazole etc., but they do predict the possibility of attack at the 3 (or 5)-positions in pyrazole and they are in accordance with the known orientations of electrophilic attack in benzimidazole and indazole. Whereas the calculations predict initial electrophilic substitution at C-5 in 1,2,3-triazole, bromination gives the 4,5-dibromo compound and the observation[130] that the carbon atoms are π-deficient in 1,2,4-triazole corresponds to the known resistance of that compound to halogenation, sulphonation and alkylation. Although electrophilic attack in $1H$-benzotriazole is predicted to occur at C-4 and C-7[130], nitration and chlorination form 4,7-dinitro- and 4,5,6,7-tetrachloro-benzotriazole respectively. Semi-empirical LCAO calculations for azoles, introducing σ-electrons, indicate that the charges are weak except for those on NH nitrogen atoms, and that the σ-dipole moments are close to those of lone pairs[131]. These workers[131] concluded that it is unsuitable to take σ-polarity into account with regard to the approximations used in the π-calculations relative to these heterocycles. Calculations of thermodynamic pK values, ΔH^0 and ΔS^0 values have been made for some azoles[132]. Theoretical calculations and physical measurements have been brought to bear on a number of more specific problems in the field.

3.3.1 Pyrazoles

Bond energy calculations for the different excited states of pyrazole show that, as in the ground state, the N—N bond is the weakest[133]. Considerable effort has been expended in studies of the tautomerism of hydroxy- and amino-pyrazoles. Using a semi-empirical SCF molecular-orbital π-approximation, heats of atomisation have been calculated for the hydroxy and amino compounds[134], as well as for the '-one' and '-imino' tautomers. From

Figure 3.9

the differences in these values, the order of tautomer stability for amino-pyrazoles has been predicted as $(8) > (7) > (9)$ (see Figure 3.9), whereas pyrazolones (CH-form; (10)) were predicted to be more stable than the hydroxy tautomers (OH-form: (11)). (See Figure 3.10.) However, from the natural-abundance ^{13}C n.m.r. spectra of some 1-phenyl-2-pyrazolin-5-ones the com-

CH-form   OH-form   NH-form

(10)   (11)   (12)

Figure 3.10

pounds were shown to exist in the OH form (11), while the 3-amino derivatives existed in the CH-form (10) under the experimental conditions (in dimethyl sulphoxide)[135]. In non-polar solvents, or solvents of low polarity, 2-pyrazolin-5-ones generally exist as the CH tautomers (10)[136], but frequently a different tautomer or a mixture of tautomers is obtained as the polarity increases.

In the solid state, because of the possibility of strong hydrogen bonding, additional tautomers are possible. Electron-attracting substituents in the 3- or 4-positions favour the OH-form (11); electron-releasing groups favour the CH-form (10). In the absence of substituents, or with groups of weak electronic effect, which tautomer is preferred depends largely on the medium[136]. Theoretical studies[137-139] applied to pyrazolone tautomerism (and also to the anions and cations) largely confirm the previous work but suggest that the OH-form (11) is least stable in the absence of specific interactions, and that the most polar structure (NH-form; (12)) would be expected to be predominant in polar solvents. Infrared studies on the solid state[140] have proved useful in tautomerism studies, but it should be emphasised that in view of the considerable effects of the medium on tautomeric equilibria, any conclusions from infrared, ultraviolet and proton magnetic resonance spectra can only relate to the medium in which the study is being made. Designation[7] of the preferred structure of 4-hydroxy-3,5-dimethylpyrazole, while not necessarily incorrect, might be criticised on the above grounds. The prototropic tautomerism of 3-substituted pyrazoles has been reviewed[141].

Total electron densities, charge distributions, dipole moments, ionisation potentials, and proton affinities have been calculated for a series of alkyl-pyrazoles using the CNDO/2 method[142]. The results indicate that in the parent molecule C-4 is the most negative, and that in methylpyrazoles, C-methyl groups withdraw electron density from the σ-system, but increase π-electron density, whereas a N-methyl group does the opposite. In the cation, a methyl group in any position increases σ-electron density.

The crystal structure of pyrazole has been re-examined[143] showing greater uniformity in the lengths of the five covalent bonds than had been thought, and suggesting that some re-interpretation of the molecular structure in terms of resonance structure is necessary. Parallel studies[144, 145] of the molecular structure of 1,1-dimethyl-3-phenylpyrazolium 5-oxide are consistent with a significant contribution to the resonance hybrid of an acyclic ketene form. From measurements of the angles made by the two rings in 1-phenylpyrazoles[146, 147] the influence of steric hindrance and electron delocalisation has been evaluated. The values obtained (e.g. 1-phenyl-pyrazole, 11; 3-methyl-1-phenylpyrazole, 9; 3,5-dimethyl-1-phenylpyrazole, 37; 5-methyl-1-phenylpyrazole, 45 degrees) point to a significant steric effect of a methyl group at C-5. Studies[148] of the conformation of 1-substituted pyrazoles by depolarised Rayleigh scattering show that the 3- and 5-positions can be characterised individually.

3.3.2 Imidazoles and benzimidazoles

Radical reactivity indices calculated for the 4- and 5-positions of 1,2-dimethyl-imidazole suggest that the 5-position should be the more reactive[149]. Charge density calculations and correlations between the charges and ^{1}H n.m.r. chemical shifts have been made for benzo-substituted 2-chloro-1-methyl-benzimidazoles[150]. Studies of crystal structure have been made for imidazole[151, 152] at room temperature (comparing the results with those obtained[153] at −115 °C), 4,5-di-t-butylimidazole[154], 4-acetylamino-2-bromo-5-isopropyl-1-methylimidazole[155] and histamine phosphate[156, 157]. Histamine phosphate was shown to exist in the crystalline state entirely as the *trans*-rotamer[156, 157], whereas in D_2O solution[158] the univalent cation has approximately equal proportions of the *trans*- and *gauche*-rotamers. Examination of the crystal structure and ^1H n.m.r. spectrum of 1,3-dimethyl-2(3H)-imidazole-thione[159] has been interpreted as showing partial double bond

(13) (14) (15)

Figure 3.11

character in the N—C—N system, but no aromaticity. Structure (13) (see Figure 3.11) is preferred to (14) or (15)[159]. In view of the variable bond lengths derived from crystal structure studies of imidazole itself[151−153], and as the crystal structure data and ^1H n.m.r. results advanced[159] to support what is essentially an isolated resonance structure appear to conflict to some degree, it would be premature to draw any firm conclusions as to the correct structure.

Ionisation constants have been determined[160] for 1-methyl-, 1-benzyl- and 1-phenylimidazole and benzimidazole, and it has been demonstrated that in 1-phenylimidazole, the phenyl group departs from the plane of the heterocyclic ring by *c.* 40 degrees.

There has been a number of studies of tautomerism in imidazole systems[134, 161], including a review[162]. Coburn[163] has suggested that some of the reactions of 2-aminoimidazole can be explained by its existence as the imino-tautomer in dimethyl formamide, but, of course, conclusions drawn from chemical reactions do not provide a sound basis for allocation of structure in a tautomeric system.

3.3.3 Triazoles and tetrazoles

The microwave spectrum of 1,2,4-triazole in the vapour phase shows that the molecule is planar and exists predominantly in the unsymmetrical tautomeric form[164], while the dipole moment obtained (2.72 D) agrees quite well with values obtained by other methods. By means of SCF calculations, ionisation potentials, electron affinities and dipole moments have been determined for triazoles and tetrazoles[165].

Theoretical studies[134] suggest that 5-aminotetrazole should exist as such in the crystalline state and in solution as it is more stable than the tautomeric imine. Experimental evidence[166−171] leaves the matter in dispute. The ionisation potentials of benzotriazole and indazole have been calculated[172].

3.4 PHYSICAL PROPERTIES

3.4.1 Dipole moments

Dipole moments have been determined for benzimidazolones[173], 3-nitro-1,2,4-triazoles[174], 1,2,4-triazole[164], alkylpyrazoles[142] and pyrazolones[138].

3.4.2 Ionisation constants

A study has been made[175] of the influence of substituents on the pK values (determined both experimentally and using the Hammett equation) of benzimidazoles, indazoles and benzotriazoles. The acidities of benzimidazoles are affected more by substituents in the 2-position than by those in the benzene ring. Measurement of the pK_a values of 1-substituted-4- and 1-substituted-5-nitroimidazoles[176] shows that the 5-nitro substituent causes a greater increase in basicity (compared with the N-unsubstituted compound) because the interaction between the nitro group and N-1 will decrease the electron-withdrawing ability of N-1, and also influence the interaction between N-1 and N-3. The ionisation constants of aminoimidazoles[177], and of some pyrazoles[178] have been determined (see also Section 3.3).

3.4.3 Spectroscopic properties

3.4.3.1 Ultraviolet spectra

Comparatively little use has been made of ultraviolet spectroscopy as a structural tool for azoles during the period surveyed by this article. It has been applied in studies of benzimidazole[179], substituted 2,4,5-triarylimidazolyl radicals[180] and nitro derivatives of 1,2,4-triazole[181].

3.4.3.2 Infrared spectra

Further study[182] of the infrared spectra of imidazole and 1-methylimidazole shows that the imidazole spectrum is most sensitive to temperature change between 100 and $-160\,°C$. This involves a shift in the band at $918\,cm^{-1}$ corresponding to the non-planar deformational vibrations of N—H, and may be accounted for by a decrease in N—H bond length with decrease in temperature causing an increase in bond strength. The observation that the band at $618\,cm^{-1}$ is unaffected by temperature suggests that it is probably incorrectly assigned[183] to a N—H vibration. A new matrix isolation device developed[184] to study hydrogen-bonded, high-boiling organic compounds allows the isolation of monomeric species in an inert gas, e.g. argon, and has been applied to a study of imidazole and pyrazole. A number of new band assignments have been made for these heterocycles using the device[184]. Further infrared studies have been carried out on benzimidazolones[173],

4-methylimidazole[185] and salts[186], salts of imidazole and 1,3-dideuterio-imidazole[187], and benzylated aminotetrazoles[186].

3.4.3.3 Raman spectra

Applications in the imidazole field have been studied[187, 188].

3.4.3.4 Nuclear magnetic resonance

Studies[150, 189] of the ^{1}H n.m.r. spectra of 2-chlorobenzimidazoles have been applied to prototropic tautomerism. The distinction between 1- and 2-methyl-4-nitroindazoles also relies on the method and the ^{1}H n.m.r. spectrum of the latter supports the contribution of a dipolar structure to the resonance hybrid[190]. A detailed study of the spectra of 1-arylazoles and the corresponding compounds nitrated in the benzene ring has been applied[191, 192] to determination of the conformations of these molecules. Deshielding (c. 0.1 p.p.m.) of the ortho-protons of the aryl ring results from the ring current of the heterocyclic nucleus. A further deshielding effect (c. 0.7 p.p.m.) is caused by the sp^2 nitrogen atom at the 2-position affecting specifically the opposite ortho-proton, and this permits assignment of the conformations of 1-(2′,4′-dinitrophenyl)pyrazole and 1-pyridazinopyrazoles. The position of the amino group in aminopyrazoles may be determined by examination of solvent effects on the ^{1}H n.m.r. spectrum[193]. The extent of formation of 1,4- and 1,5-disubstituted triazoles from the cyclo-addition of vinyl- and β-iodo-azide with acetylenic esters has been determined by ^{1}H n.m.r. methods[194]. A phenyl group located in a sterically-hindered position on the triazole resonates as a more or less sharp singlet[97]. The ^{13}C magnetic resonance spectrum has been obtained[195] for benzimidazole and the chemical shifts have been compared with protonation parameters. Further applications of ^{1}H n.m.r. have been made to structural problems[126, 196-199], and tautomerism studies[135, 159, 163, 166-171, 189, 200].

3.4.3.5 Electron spin resonance

Progress has been made in the interpretation of the e.s.r. spectrum of 2,4,5-triphenylimidazolyl radicals by examining the spectra of the partially and completely deuterated species and by the application of simple molecular orbital calculations[201]. The e.s.r. spectra of 1-hydroxy-2,4,5-triphenyl-imidazolyl and the corresponding 3-oxide radicals have been recorded and interpreted[202].

3.4.4 Mass spectrometry

From a study of deuterated derivatives, the mass-spectral fragmentation pattern for pyrazole has been determined[203, 204]. The predominant process is loss of HCN from the molecular ion by cleavage of the N—N bond (which is known[133] to be the weakest bond) with formation of a fragment radical-ion, $C_2H_3N^{+\cdot}$. Subsequent, or preceding, loss of a hydrogen radical, gives rise to an ion, $C_2H_2N^+$. The loss of HCN takes place with high specificity from the 3(5)-position[204]. A second prominent process is the loss of a nitrogen molecule, after initial loss of $H^\cdot$ or a substituent, giving rise to a species, $C_3H_2R^+$, which may be a cyclopropenyl ion. Some loss of acetylene is evident[203]. The fragmentation process is strongly influenced by substituents, e.g. 4-nitropyrazole acts like an aromatic nitro compound; 1-methylpyrazole loses a $CH_2CN^\cdot$ radical[203]. Elimination of nitrogen has been noted also in the mass spectra of diphenyl- and triphenyl-1,2,3-triazoles[205], benzotriazole[206], 1,2,4-triazole 4-oxide[207] and 5-aminotetrazole[208]. Although skeletal rearrangements are not common in the mass spectra of azoles, a major cleavage in the fragmentation pattern of 1-(benzylideneamino)benzotriazole involves rearrangement[209]. Benzotriazoles substituted on N-1 commonly lose the substituent during fragmentation[206]. Studies of the mass spectra of triazolinones[210], 1-chlorobenzotriazole[211] and mesoionic 1,2,4-triazoles[117] have been reported. The use of mass spectrometry of N-phenylosatriazoles[212] and their acetates[213] in oligosaccharide sequencing has been studied. By adding a fused pyrazole ring to cyclic ketones it is possible to study the fragmentation of medium sized ketones[214] (6- to 10-membered rings).

3.5 SUBSTITUTION REACTIONS

3.5.1 Electrophilic substitution

3.5.1.1 Pyrazoles

Electrophilic substitution on nitrogen occurs in the synthesis of 1-acyl-pyrazoles from pyrazoles and acid chlorides[215], the arylation of 1-acyl-pyrazoles[216], pyrazoles and aminopyrazoles[217], and in the formation of 1-deuteriopyrazole[218]. A comprehensive study has been made[219] of the Ullmann condensation of azoles with aryl halides. Cyanogen bromide, phenylisocyanate, and propargyl bromide react with the pyrazolyl anion (as sodium pyrazole) to form the 1-cyano-, 1-phenylcarbamoyl- and 1-propargyl derivatives[220]. The 4-chloro- and 4-bromo-pyrazole sodium salts, however, react with cyanogen bromide to form the 1,1'-ketiminobipyrazole[220]. Iodination of the silver salt of 4-iodo-3,5-dimethylpyrazole results in replacement of the silver by iodine[221].

Electrophilic attack at ring carbon atoms includes the direct carboxylation of 5-aminopyrazoles to the 4-carboxy derivatives using aqueous potassium bicarbonate[222], and the Friedel-Crafts substitution of 1,3-dialkyl-5-chloro-pyrazole with aroyl halides in refluxing 1,1,2,2-tetrachloroethane[223]. This latter reaction fails when the aroyl halide is susceptible to the action of aluminium chloride or to self-aroylation.

Methods have been devised for the synthesis of all possible deuterated derivatives of pyrazole[218]. While exchange at C-4 requires heating with D_2O at 200 °C for 12 h, the 3- and 5-hydrogen atoms can only be exchanged in alkaline medium, reflecting the usual substitution behaviour of pyrazole. From a study of the kinetics of deuteration of pyrazole[224] the mechanism of 4-substitution is believed to involve the general acid-catalysed formation of sigma intermediates from the molecular and conjugate base forms of pyrazole. As the rate of deuteration at the 3(5)-position exhibited neither buffer nor pD dependence, the behaviour is considered to be consistent with the formation of an ylide intermediate from nucleophilic attack on the conjugate acid of pyrazole by OD^- anions. Exchange occurs at the 4-position most rapidly at $pH \geqslant 6$.

A kinetic study has been carried out of the aqueous bromination of pyrazole, 1-methyl- and 3,5-dimethyl-pyrazoles[178]. Substitution occurs as expected at C-4 with 1:1 stoichiometry. The observed rates may be interpreted in terms of direct attack by molecular bromine on neutral substrates. The high reactivity of 3,5-dimethylpyrazole is discussed[178]. A study of the halogenation of silver[221, 225] and sodium salts[220] of pyrazoles shows that the silver salt is halogenated initially at C-4, then at C-3(5), and finally forms 3,4,5-trihalogenopyrazoles[225]. In addition, chlorination produces side products associated with the condensation of several pyrazole molecules. Whereas 3-methyl- and 3,5-dimethylpyrazole silver salts are halogenated at C-4, the 4-methyl-compound is iodinated at C-3. Bromination of the 4-methyl- or 3,5-dimethylpyrazole salts again results in the formation of products composed of two or more heterocyclic nuclei[221]. When the iodination of pyrazole and 4-methylpyrazole is followed by deiodination with tritium, 4-[3H]pyrazole and 4-methyl-3-[3H]pyrazole are formed[226].

Nitration of pyrazole normally occurs at C-4 unless this position is blocked, in which case substitution is possible in the 3- and 5-positions using nitric acid in acetic anhydride[227, 228], in sulphuric acid[229, 230], and with 70% nitric acid[217, 230]. Thus nitration of 4-bromo-1-methylpyrazole forms the 3,5-dinitro-compound[230]. Nitration of 1-picrylpyrazole with mixed acids yielded the 4-nitro-compound[217], but it did not prove possible to nitrate 4-picryl-pyrazole[229]. 1-Methyl-4-picrylpyrazole, however, in refluxing 90% nitric acid forms the 5-nitro-product, and with mixed acids the 3,5-dinitropyrazole results[229]. When the pyrazole is substituted by phenyl, nitration in sulphuric acid or with fuming nitric acid usually results in p-nitrophenylpyrazoles[10,227,231] owing to nitration of the conjugate acid of 1-phenylpyrazole. On the other hand, nitration in acetic anhydride involves the free base which is attacked at C-4 of the heterocyclic ring (unless that is blocked[228]). A study of the kinetics of mononitration in 70% nitric acid of 4-(2',4'-dinitrophenyl)1-methyl-pyrazole[232] has been interpreted as indicating that, although both 3- and 5-nitro-products are formed, the latter decomposes at a rate faster than its formation.

3.5.1.2 Indazoles

While 2-methylindazole reacts with butyl-lithium to form the 3-lithio derivative, 1-methylindazole is lithiated on the methyl function[233]. 2-Glycosylindazoles have been prepared by electrophilic substitution reactions[197], sometimes using HgCN as a hydrogen halide acceptor[234].

3.5.1.3   Imidazoles

Electrophilic substitution reactions of imidazoles have been reviewed[46]. Methods are described[235] for the selective alkylation of azoles by following the sequence: (a) acylation, (b) alkylation with oxonium or carboxonium reagents and (c) deacylation with alcohol or water. Under these conditions the most sterically-hindered product is formed in good yield (e.g. 1-ethyl-5-phenylimidazole; 86%). Other reactions involving electrophilic attack at the multiply-bonded nitrogen include alkylation with Mannich bases[236], phosphorylation in alkaline media with phosphoryl chloride[237] and trifluoromethylsulphonation[238] to form the imidazolide which is a convenient reagent for the introduction of the 'triflate' group, and the formation of 1:1 adducts with 1,3-dipolar compounds[239]. Kinetic studies have been made of the reaction of imidazole with acrylonitrile[240] and with picryl chloride[241]. Although the imidazole-catalysed hydrolysis of esters can be classified as an electrophilic attack on the multiply-bonded ring nitrogen, the results for 1970 have been summarised[242] and space does not permit more than the listing of recent references[243, 244] to the field. The imidazole-catalysed isomerisation of penicillins into penicillenic acids[245] probably involves an initial nucleophilic attack of imidazole on penicillin.

The kinetics of deuteration of imidazole have been examined at various pD values for both the 2- and 4(5)-positions[246]. As in the case of pyrazole[224], parallel rate determining proton abstractions from the conjugate acid of imidazole by OD^- (see Figure 3.12) and D_2O leading to an ylide intermediate

(16)

Figure 3.12

(16) account for the pD rate profile for 2-substitution while an additional path involving proton-abstraction from the imidazole neutral molecule accounts for the 4-substitution profile.

From a study of base catalysis in the iodination of imidazole[247] it is considered that the rate-determining step is a nucleophilic attack of the base on the hydrogen atom being abstracted from the sigma complex and that the acceleration follows the nucleophilicity rather than the basicity of the catalyst. The iodinating agent is believed[227] to be of the general form, BI_2, where B is the base.

In the reaction of Mannich bases with imidazoles, substitution can occur at positions 1,2,4 or 5 depending on the reaction conditions[236]. In acid media the attack is at the multiply-bonded nitrogen, but in basic conditions the principal attack is at nitrogen or C-4. The N-substitution is reversible and hence C-substituted product accumulates. The carboxylation of 5-aminoimidazoles occurs in 40–50% yields to give the 4-carboxyl compounds[222]. The first electrophilic acylation at imidazole ring carbons has been reported[248, 249] in the reaction of 1-methylimidazole with benzoyl chloride in acetonitrite containing triethylamine. The observation that only 2-substituted product is obtained, as is also the case with metallation[250], is of interest in view of the orientation of nitration of 1-methylimidazole, which yields 1-methyl-4-nitro- and 1-methyl-5-nitro-imidazoles in the ratio of 2.6:1 [251]. Although no theoretical predictions have been made concerning the relative reactivities of ring carbons in 1-methylimidazole, chemical shift data[252] suggest that the order of reactivity for electrophilic substitution should by 5 > 4 > 2, on the basis that higher τ values mean higher electron densities. Recent work, though, indicates that all of the ring positions of imidazole and 1-methylimidazole are reactive. Both 1-methyl-4- and 1-methyl-5-chloro-imidazole are nitrated in sulphuric acid[253], while exhaustive nitration of imidazole in sulphuric acid yields in succession 4-mono, 4,5-di- and 2,4,5-trinitroimidazole[254]. Nitration of 4-substituted imidazoles with dinitrogen tetroxide in acetonitrile yields a mixture of 5- and 2-nitro derivatives[255].

3.5.1.4 Benzimidazoles

Bromination in acetic acid of 2-amino-1-methylbenzimidazole occurs at C-6 [256]. In alkaline solution 1,3-dimethylbenzimidazolium trichloromercurate salts form 2-chloromercury derivatives by electrophilic attack on the zwitterion formed by the loss of the 2-H [257].

3.5.1.5 Triazoles

Electrophilic attack by methyl propiolate on 3-amino-1,2,4-triazole occurs at both N-2 and N-4 [258], whereas acetylation of the compound yields the 1-acetyl-5-amino-1,2,4-triazole by attack at N-2 only[198]. In the case of 1,2,3-triazole, substitution at N-1 [259] and at N-1 and N-2 [260, 261] occurs. The halogenation of 1-hydroxy-2-phenyl-1,2,4-triazole takes place at the 5-position[108]. In the bromination of 1,2-disubstituted-1,2,3-triazolium salts[262] with N-bromoacetamide in alkali, the reaction involves attack of the bromonium ion on the triazole anion. Deuterium exchange-rate measurements predict much faster bromination at C-5 than at C-4, a result which is confirmed experimentally. While nitric acid converts 3-amino-1,2,4-triazole into the 3-nitramino derivative, phenyltriazoles form p-nitrophenyltriazoles[263].

3.5.1.6 Benzotriazoles

In the alkylation of benzotriazoles with chloroacetic acid in ethanolic alkali the 2-substituted product predominates[126]. Methylation of 5-methylbenzotriazole yielded the 1-methyl-(27%), 2-methyl-(45%), and 3-methyl-(45%) products[126]. Products of 1-substitution predominate in vinylation reactions with acetylene and cuprous chloride catalyst[264] and with Mannich bases[265].

3.5.1.7 Tetrazoles

Alkylation of 5-substituted tetrazoles produces mixtures of 1- and 2-alkyl-derivatives[166, 266].

3.5.2 Nucleophilic substitution

3.5.2.1 Pyrazoles and indazoles

In a study of base-catalysed interhalogenation and cine-substitution of bromopyrazolium salts, Begtrup[267] has shown that nucleophilic attack of hydroxide ion on 1,2-dimethyl-3,4-di- and 3,4,5-tri-bromopyrazolium tosylate results in the formation of 1,2-dimethyl-4-bromopyrazol-4-in-3-one and its 5-bromo derivative. Chlorine or bromine can replace iodine in the silver salts of 4-iodo-3,5-dimethylpyrazole[221] and pyrazole-3-diazonium chloride undergoes nucleophilic displacement with sodium azide to form the hitherto unknown 3-azidopyrazole[268]. Metallation and deuterium exchange reactions, while often depending on initial nucleophilic attack, have been classified as electrophilic substitutions.

3.5.2.2 Imidazoles

The reactions of acetylimidazolium ion with amines have been shown[269, 270] to depend largely on amine basicity. Attempts to prepare 1-methyl-4- and 1-methyl-5-piperidinoimidazole by nucleophilic substitution for halogen failed[271], but 1-methyl-2-bromoimidazole could be converted into the 2-piperidino compound by heating with piperidine at 200 °C for 60 h [271]. The fusion of 1-substituted imidazoles at 160–400 °C with powdered potassium hydroxide results in the formation of imidazolones[272]. The reaction of 5-bromo-1-methylimidazole with potassium amide in liquid ammonia does not result in nucleophilic substitution, but rather in migration and elimination of halogen to give a mixture of 4-bromo-1-methylimidazole and 1-methylimidazole[273].

3.5.2.3 Benzimidazoles

Substitution in the 2-position of benzimidazoles is easier, and has been studied more, than in the case of imidazoles. Direct amination with sodamide forms 2-aminobenzimidazoles, particularly when there are substituents present having a large electron-donor effect[274]. As with imidazoles, direct hydroxylation is possible[272]. The displacement of chlorine on C-2 occurs with monoacetylethylenediamine[73] and thiophenol[275, 276]. When o-nitro-N,N-dialkylanilines are subjected to prolonged treatment with hot hydrochloric acid, 6-chlorobenzimidazoles are formed[81]. This reaction is reported to involve a nucleophilic substitution with elimination of OH^- from an N-oxide intermediate—a type of S_N2' reaction. Other nucleophiles such as CN^-, SCN^- and N_3^- take part in the reaction, but with Br^-, I^- and $C_2H_5S^-$ reduction of the N-oxide is the only reaction[81].

3.5.2.4 Triazoles

Although 5-nitro substituents in 1,2,4-triazoles may be replaced by chlorine or bromine, 3-nitro substituents are unaffected[277]. Furthermore, chloro and bromo substituents on C-5 may be displaced by alkoxyl and alkylamine groups. With concentrated hydrochloric acid and urea, 5-amino-1-methyl-3-nitro-1,2,4-triazole is converted into the 3-chloro compound[277]. When hydrazine is the nucleophile, 1-methyl-3,5-dinitro-1,2,4-triazole is converted into a mixture of 5-hydrazino- and 5-amino-1-methyl-3-nitro-1,2,4-triazole[278]. In a series of publications concerned with the bromination in the presence of hydroxide or alkoxide ions of 1,3-dimethyl-1,2,3-triazolium salts, Begtrup[262, 279, 280] has shown that bromo groups in the 4- and 5-positions can be replaced by methoxyl[262, 280] and hydroxyl[262]. The usual products are triazolones formed from the methoxyl compounds by dequaternisation involving specific loss of the O-methyl group[262]. The migration of the amino function to C-5 when 1-substituted amino-4-phenyl-1,2,3-triazoles are boiled with concentrated hydrochloric acid[281] could be considered to be an intramolecular nucleophilic substitution.

3.5.2.5 Benzotriazoles

Amination of 5-nitro-,1- and 2-methyl-5-nitro-, and 2-phenyl-5-nitro-benzotriazoles results in the formation of the 4-amino compounds[282].

3.5.3 Radical substitution

There has been little study of this type of substitution in the azole field. Although placement in the Photochemistry Section might seem more logical, the photo-addition of acetone to 1- and 2- and 1,2-dimethyl-imidazole[149] will be discussed here in view of the theoretical considerations involved. The α-hydroxyalkylimidazoles formed result from selective attack of excited carbonyl oxygen at C-5 of 1-methylimidazole. In the case of 2-methylimidazole the products are the 4-mono-(8%) and the 4,5-disubstituted (14.5%)

(17)

Figure 3.13

compounds. Imidazole does not react, reflecting the requirement for more electron-rich substrates. The formation of mainly 4-substituted products is a consequence of the difference in reactivity of the 4- and 5-positions, and also the difference in stabilities of the radical intermediates derived from carbonyl oxygen attack at C-4 or C-5. A suggested reaction mechanism is shown in Figure 3.13. Simple Hückel calculations of radical reactivity indices showed that the 5-position is the more reactive, and further calculations showed that the biradical intermediate (17) is more stable than the one substituted at C-4 [149].

When nitrosaminotetrazoles are refluxed in benzene, the NHNO function is displaced to form the phenyl substituted product[283] by a Gomberg-Bachman reaction.

3.6 RING MODIFYING PROCESSES

3.6.1 Pyrazoles

Although 1,4-dihydroxypyrazoles are oxidised by peracids to 1,3,4-oxadiazin-6-one-4-oxides[284], and 3-aminopyrazole is degraded on treatment with peroxytrifluoroacetic acid[285], 5-amino-1-methylpyrazole gave 1-methyl-5-nitropyrazole 2-oxide[285], and 1-methylpyrazole can also be converted into the 2-oxide with hydrogen peroxide[286]. Pyrazolium salts are reduced by metal hydrides to 3-pyrazolines, pyrazolidines and alkylhydrazines[287]. The novel cleavage of 3-methyl-4-nitroso-1,5-diphenylpyrazole by triethylphosphite to phenyl(phenylimino)acetonitrile is thought to proceed via a nitrene intermediate[288].

3.6.2 Indazoles

Oxidation of 1- and 2-aminoindazoles with lead tetra-acetate results in ring expansion to 1,2,3-benzotriazines[289], whereas it was expected by analogy with the corresponding aminotriazoles[290] that nitrogen would be expelled from the intermediate nitrenes formed in the reaction. Under the influence of organosodium compounds 1-benzyl-3-bromoindazole is ring-opened to the nitrile of N-benzylanthranilic acid[291]. In the presence of hydride ions 3-amino-1-benzylindazole rearranges to form 68 % of 4-amino-2-phenyl-quinazoline[292].

3.6.3 Imidazoles and benzimidazoles

The ring cleavage of imidazoles under strongly basic conditions has been known for many years. Cleavage of 4-amino-1,2,3-triphenylimidazolium by sodium hydroxide forms N-cyanomethyl-N,N'-diphenylbenzamidine[51]. In aqueous media 1,3-dimethyl- and 1,2,3-trimethylbenzimidazolium salts form stable ammonium hydroxides with bases, but in non-aqueous media, although the reactivity is lower than in analogous series, hydride ion can lead to the formation of an ylide, or a methylene base[293]. The corresponding 1,3-diphenyl compounds are cleaved in aqueous alkali and eliminate a hydracid molecule very readily in non-aqueous media with the formation of bis-(1,3-diphenyl)-2-benzimidazolinylidine[294]. On treatment with hydrogen peroxide in alkali, imidazoles are converted into 1,2,4-oxadiazoles[295].

3.6.4 Triazoles and benzotriazoles

The oxidation of 1-aminobenzotriazole provides a convenient route to benzyne, and this has now been complemented by the process of deoxygenation of 1-nitrosobenzotriazole with diethylphosphite[290]. Oxidative fragmentation of 1-amino-1,2,3-triazoles by a concerted process with lead tetra-acetate leads to nitrogen and nitriles[296]. Reduction with lithium aluminium hydride of 1,2- and 1,3-dimethylbenzotriazolium salts produces o-phenylenediamines[297]. The pyrolysis of triazoloarenes results in ring contraction via a Wolff rearrangement of intermediate 1,3-biradicals[298], e.g. benzotriazole forms 1-cyanopentadiene and aniline. Pyrolysis of 1-phthalimido-1,2,3-triazoles results in the formation of $2H$-azirines with $1H$-azirines as intermediates[299].

3.6.5 Tetrazoles

Tetrazolylhydrazones react with chlorine in 90% aqueous acetic acid to form trichlorodiazabutadienes, compounds in which three halogens may be replaced sequentially by nucleophilic reagents[300]. The gas-phase pyrolysis of pentamethylenetetrazole results in the elimination of nitrogen to form pent-4-enylcyanamide[301].

3.7 PHOTOCHEMISTRY

The photo-isomerisation of five-membered heterocyclic molecules has been reviewed[302].

3.7.1 Pyrazoles

Spiropyrazoles are photolysed to cyclopropabenzenes. The formation of these products involves a n→ π* excitation, followed by ring enlargement to indazoles which eliminate nitrogen[303].

3.7.2 Indazoles

The photo-isomerisation of indazoles to benzimidazoles is a characteristic reaction analogous to the conversion of pyrazoles into imidazoles[304, 305]. While 2-alkylindazoles form 1-alkylbenzimidazoles[133], 3-substituted indazoles isomerise to 2-substituted benzimidazoles[306].

3.7.3 Imidazoles

Although the photo-isomerisation of imidazoles to pyrazoles is unknown, there is a photo-equilibrium between 1,4- and 1,2-dimethylimidazole[305]. In the chemiluminescence reaction of 2,4,5-triphenylimidazole, and in its photochemical oxidation in the presence of ammonia, 2,4,5-triphenyl-1,3,5-triazine is formed[307]. The flash photolysis of hexa-arylbi-imidazole produces imidazolyl radicals[308], which have been shown[180] to be more nearly planar than the parent dimers; *ortho*-substituents in the aryl rings decrease the radical stability. The radicals oxidise electron-rich substrates by rapid electron abstraction from tertiary amines, iodide and metal ions and by hydrogen abstraction from phenols, mercaptans, secondary amines and active methylene compounds[309]. Studies[310, 311] have been made of the photo-oxidation of *leuco*-triphenylmethane dyes by these radicals. The photo-chemical decomposition of imidazole diazonium fluoroborates in fluoroboric acid results in the formation of 2- and 4-fluoroimidazoles[312].

3.7.4 Benzimidazoles

Photolysis in methanol of 1,2-polymethylene-[81] and 1,2-dialkyl-benzimida-zole 3-oxides[313] produces benzimidazolones among the reaction products. Both 1-methyl-2-benzimidazolethione and 1-methyl-2-methylthiobenzimi-dazole are desulphurised on photolysis[314], even though the latter compound might be expected to rearrange as under thermolysis conditions.

3.7.5 Triazoles and benzotriazoles

A characteristic transformation in the photolysis of triazoles[315] and benzo-triazoles[316-318] is the loss of nitrogen, a process which may result in the for-mation of a cyclic molecule, e.g. indole[316] or carbazole[318]. The effects of solvent, concentration and frequency have been examined[317] for the photo-lysis of 1-acylbenzotriazoles, in which system it was discovered that the diradicals formed by nitrogen elimination can be stabilised by (a) addition of hydrogen or of solvent molecules, (b) ring-closure accompanied by hydrogen shift or (c) polymerisation. The photolytic transformation of 5-chloro-1-(*p*-chlorophenyl)benzotriazole (18) yields a symmetrical car-bazole (20) which confirms the intermediacy of a diradical (19) and the

Figure 3.14

operation of a 1,5-cyclisation[318]. (See Figure 3.14). Whereas the thermolysis of 1-benzoylbenzotriazole yields 2-phenylbenzoxazole, photolysis in benzene and acetonitrile forms *o*-phenylbenzanilide[318]. The photolytic transforma-tions of 4-(*p*-nitrophenyl)imino-[315] and 4-acylamino-1,2,4-triazolium ylids[319] have been studied. In the photolysis of 1-hydroxy-1,2,4-triazole in methanol the products include 1,2,4-triazole, 1,2,4-triazol-3-one, and 4-hydroxy-3-hydroxymethyl-1,2,4-triazole[320].

3.7.6 Tetrazoles

Among the products of ultraviolet irradiation of 5-phenyltetrazoles are nitrogen and triazoles[122]. It has been demonstrated using ^{15}N-labelling techniques that the nitrogen is expelled exclusively from positions 3 and 4[122]. As a consequence of the probable existence in solution of the tetrazole as a hydrogen-bonded dimer, 1,2-dihydro-3,6-diphenyl-1,2,4,5-tetrazine is formed, and from this compound are derived 3,5-diphenyl-1,2,4-triazole and benzonitrile[122]. The photolysis of tetrazolide anions (which cannot associate through hydrogen bonding) does not form tetrazines, but rather two equivalents of nitrogen and the products derived from a phenylcarbene intermediate[321]. Benzil osazone is a product of the photolysis of 2,5-diphenyltetrazole[322].

3.8 FUNCTIONAL GROUP PROPERTIES

3.8.1 Acyl, aroyl and formyl substituents

Acyl and formyl functions on imidazoles and pyrazoles condense with amino compounds and other nucleophiles in a normal manner. The reaction of hydrazine hydrate with 1-substituted-imidazole-4,5-dialdehyde results in cyclisation to imidazo[4,5-d]pyridazines[323], while 4-acetyl-3-methyl-1-phenylpyrazolin-5-one reacts with aromatic aldehydes to form 4-cinnamoyl derivatives, and with amines and hydrazines to yield 4-acetylimino compounds[324]. Irradiation of 4-keto-1,2,3-triazoles in isopropanol forms pinacols[325]. In the solid state 5-aryl-1,2,4-triazole-3-carbaldehydes have been found to dimerise to carbonyl-free hemiaminals, although they exist as monomers in solution[326, 327].

3.8.2 Alkyl, arylalkyl and aryl substituents

Condensation of trichloroacetaldehyde with 1,2-dimethylbenzimidazole in the presence of zinc chloride, followed by treatment with alkali, results in the formation of 1-methylbenzimidazole-2-acrylic acid[328]. Lithiation of 1,3,5-trimethylpyrazole forms the 1-lithiomethyl compound, while a 1-phenyl substituent is substituted in the *ortho*-position[329]. The preferential attack at a 1-methyl substituent may be explained by the inductive effect of the ring nitrogen atoms contributing to the increased 'acidity' of the proton substituted. Nitroaryl substituents on nitrogen in 1,2,4-triazole are removed quantitatively with aqueous ammonia or ammonium carbonate[330]. Halogenation of 1-vinylimidazole and 1-vinylbenzimidazole is believed to occur through initial complex formation followed by some addition at the exocyclic double bond[331]. In line with the difficulty of addition to vinylimidazoles are the reactions of 1-vinylbenzotriazoles. Bromine reacts with simultaneous complex formation and some addition; chlorine forms mainly the addition product; hydrogen halides form only complexes[332]. The vinyl function may be reduced by hydrogenation over ethanolic Raney nickel[264].

3.8.3 Amino and aminoalkyl substituents

With aqueous acetic anhydride, 2-aminoethylbenzimidazole forms the amide in high yield[333]. The intramolecular Mannich reaction of 2-alkylaminomethylbenzimidazoles has been reported[334]. The Dimroth rearrangement of 4-amino-3-benzyl-1,2,3-triazoles and their 5-substituted derivatives in alkali results in the formation of the 4-benzylamino products[281]. The reaction is retrogressive in neutral solvents[281], and also occurs with 4-substituted-5-amino-1-phenyl-1,2,3-triazoles[335]. By means of successive treatments with sodium in liquid ammonia, and air oxidation, 1-benzyl-2-aminobenzimidazole is converted into 2,2′-azobenzimidazole (32 %), 2-nitrobenzimidazole (1–2 %), and 2-aminobenzimidazole (40 %). The last compound is transformed by potassium in liquid ammonia in the presence of ferric nitrate and platinum black into the azo compound (36 %) and the 2-nitrobenzimidazole (10 %)[336]. 5-Aminopyrazoles may be converted into the azides which lose nitrogen above room temperature[337]. Treatment of 1-substituted-5-aminotetrazoles with nitrous acid produces the nitrosaminotetrazoles[338].

3.8.4 Carboxyl substituents and derivatives

Heating above the melting point provides a smooth method of decarboxylation of 1,2,3-triazole-4-carboxylic acids[92]. A kinetic study[339] of the decarboxylation of some 1-glycosyl-5-aminoimidazole-4-carboxylic acids has been interpreted as a first-order decarboxylation of both the acid and the zwitterion, whereas the anion is stable and is not decarboxylated. An intramolecular condensation of 5-amino-1-carboxyalkyl-2-methyl-4-nitroimidazoles results in lactam formation[340]. In an unusual reaction PCl_5 is reported[341] to convert 1-methylimidazole-4,5-dicarboxylic acid into 1-methyl-4-trichloromethylimidazole-5-carboxylic acid chloride.

3.8.5 Halogeno and halogenoalkyl substituents

Butyl-lithium reacts with 1-methyl-3-chloromethylindazole to form the 3-pentyl compound[291]. A highly efficient reagent for the nuclear chlorination of carbazole and derivatives is 1-chlorobenzotriazole, which acts as a source of electrophilic chlorine[342].

3.8.6 Hydroxy and oxide substituents

The N-oxide function in imidazole N-oxides may be removed with zinc in acetic acid[343], and by heating[62]. Dehydrogenation of 1-hydroxy-2,4,5-triphenylimidazole and the 3-oxide produces the 1-oxide and 1,3-dioxide radicals respectively[202]. When 3,5-dimethyl-4-hydroxypyrazole is treated with dimethylsulphate, the 4-methoxy product results[7]. Reduction to the oxygen-free compound occurs in the reactions of 4-hydroxy-1,2,4-triazole with zinc and acetic acid, phosphoryl chloride, or phosphorus trichloride[108].

3.8.7 Nitrogen substituents: nitro, diazo and cyanoalkyl

Selective hydrogenation of 4-nitro-1,2,3-triazoles over palladium–carbon gives the amino compounds in greater than 90% yields[259]. Reaction of 3-diazopyrazole with excess diazomethane yields 3-(1-tetrazolyl)pyrazole and pyrazolyl[5,1-c]triazole[268]. With sodium azide, the diazo compound is converted into the azido function[268]. In an unprecedented reaction, 3-cyanomethyl-1,2,4-triazole reacts with aqueous hydrazine at 210 °C to form 3-methyl-1,2,4-triazole[121].

3.8.8 Thiol, thione and related functions

Imidazoles with thiol substituents in the 4-position can be oxidised to form the bis(4-imidazolyl)disulphides which may be cleaved by hydrogen sulphide[50]. With 15% alkaline hydrogen peroxide at 80–90 °C the thiol is oxidised to the sulphonic acid[50]. Oxidation of 1,2,4-triazoline-3-thiones also forms the corresponding sulphonic acid[116]. A number of azole thiones, mercapto-alkylazoles and disulphides are desulphurised on photolysis[116, 314]. With chlorine in carbon tetrachloride bis-(1,3,5-trisubstituted pyrazol-4-yl)-disulphides are converted into the pyrazole sulphenyl chlorides[344]. Mercapto-imidazoles[345] and -benzimidazoles[346] readily form S-glycosides.

3.8.9 Miscellaneous

Imidazole arynes have been surveyed[347], but there has been criticism[271] of their reported formation from 4- and 5-halogenoimidazoles with lithium piperidide in piperidine. The use of azoles in peptide synthesis as terminating agents[348], and as carboxyl group activators[349, 350] has been reported. Although there has been a multitude of references to azoles and benzazoles in metal-organic complexes, lack of space prevents any discussion of this material.

References

1. Elguero, J., Gelin, R., Gelin, S. and Tarrago, G. (1970). *Bull. Soc. Chim. Fr.*, 231
2. Coispeau, G., Elguero, J. and Jacquier, R. (1970). *Bull. Soc. Chim. Fr.*, 689
3. Broser, W. and Bollert, U. (1971). *Chem. Ber.*, **104**, 2053
4. Nakamura, N., Kishida, Y. and Ishida, N. (1971). *Chem. Pharm. Bull. (Tokyo)*, **19**, 1389
5. Al-Jallo, H. N. (1970). *Tetrahedron Lett.*, 875
6. Nye, M. J. and Tong, W. P. (1970). *Can. J. Chem.*, **48**, 3563
7. Dittli, C., Elguero, J. and Jacquier, R. (1971). *Bull. Soc. Chim. Fr.*, 1038
8. Lisitsyn, E. A., Belyaeva, A. N., Malaeva, A. K., Maretina, I. A. and Petrov, A. A. (1970). *J. Org. Chem., U.S.S.R.* **6**, 438
9. Hedbom, C. and Helgstrand, E. (1970). *Acta Chem. Scand.*, **24**, 1744
10. Comrie, A. M. (1971). *J. Chem. Soc. C.*, 2807
11. Ioffe, B. V. and Zelenina, N. L. (1970). *Khim. Geterotsikl Soedin.*, 1414; *Chem. Abstr.* (1971). **74**, 53634
12. Aspart-Pascot, L., Lematre, J. and Sournia, A. (1971). *Complt. Rend. Acad. Sci. Ser. C.*, **272**, 103; *Chem. Abstr.* (1971). **74**, 87888
13. Ismailov, A. G. and Rustamov, M. A. (1970). *Azerb. Khim. Zh.*, **1–2**, 124; *Chem. Abstr.* (1971). **74**, 41986
14. Beyer, H. (1970). *Z. Chem.*, **10**, 386
15. Darbinyan, E. G., Mitardzhyan, Yu. B. and Matsoyan, S. G. (1970). *Arm. Khim. Zh.*, **23**, 640; *Chem. Abstr.* (1971). **74**, 76362
16. Bergeon, M. T., Metayer, C. and Quiniou, H. (1971). *Bull. Soc. Chim. Fr.*, 917
17. Quinn, L. D. and Pinion, D. O. (1970). *J. Org. Chem.*, **35**, 3134
18. Foote, R. S., Beam, C. F. and Hauser, C. R. (1970) *J. Heterocycl. Chem.*, **7**, 589

19. Beam, C. F., Foote, R. S. and Hauser, C. R. (1971). *J. Chem. Soc. C*, 1658
20. Neplyuev, V. M., Usenko, Yu. N., Dubenko, R. G. and Pel'kis, P. S. (1970). *Khim. Geterotsikl. Soedin.*, 1194; (1971). *Chem. Abstr.* **75**, 5775
21. Coispeau, G. and Elguero, J. (1970). *Bull. Soc. Chim. Fr.*, 2717
22. Elguero, J. (1971). *Bull. Soc. Chim. Fr.*, 1925
23. Potts, K. T., Husain, S. and Husain, S. (1970). *Chem. Commun.*, 1360
24. Potts, K. T. and Singh, U. P. (1969). *Chem. Commun.*, 66
25. Gotthardt, H. and Reiter, F. (1971). *Tetrahedron Lett.*, 2749
26. Maki, Y., Beardsley, G. P. and Takaya, M. (1971). *Tetrahedron Lett.*, 1507
27. Maki, Y. and Takaya, M. (1971). *Chem. Pharm. Bull. (Tokyo)*, **19**, 1635
28. Schrader, L. (1971). *Tetrahedron Lett.*, 2977
29. Stephan, E., Vo-Quang, L. and Vo-Quang-Yen. (1971). *Compl. Rend. Acad. Sci. Ser. C*, **272**, 1731
30. Garcia-Lopez, M. T., Garcia-Munoz, G. and Madronero, R. (1971). *J. Heterocycl. Chem.*, **8**, 525
31. Bastide, J. and Lematre, J. (1971). *Bull. Soc. Chim. Fr.*, 1336
32. Manecke, G. and Schenck, H.-U. (1971). *Chem. Ber.*, **104**, 3395
33. Von Auwers, K. and Ungemach, O. (1933). *Chem. Ber.*, **66**, 1205
34. Witiak, D. T. and Sinha, B. K. (1970). *J. Org. Chem.*, **35**, 501
35. Wittig, G. and Hutchison, J. J. (1970). *Liebigs Ann. Chem.*, **741**, 89
36. Hayashi, Y. and Oda, R. (1969). *Tetrahedron Lett.*, 853
37. Hayashi, Y., Watanabe, T. and Oda, R. (1970). *Tetrahedron Lett.*, 605
38. Dürr, H. (1970). *Chem. Ber.*, **103**, 369
39. Farkas, J. and Flegelova, Z. (1971). *Tetrahedron Lett.*, 1591
40. Fritz, H. and Uhrhan, R. (1971). *Liebigs Ann. Chem.*, **744**, 81
41. Patey, A. L. and Waldron, N. M. (1970). *Tetrahedron Lett.*, 3375
42. Rüchardt, C. and Chuan Cheng Tan. (1970). *Chem. Ber.*, **103**, 1774
43. Kametani, T., Sota, K. and Shio, M. (1970). *J. Heterocycl. Chem.*, **7**, 807
44. Tezuka, T., Yanagi, A. and Mukai, T. (1970). *Tetrahedron Lett.*, 637
45. Piozzi, F. and Passannanti, S. (1970). *Gazz. Chim. Ital.*, **100**, 639
46. Grimmett, M. R., *Advan. in Heterocycl. Chem.*, **12**, 103. (New York: Academic Press)
47. Bloemhoff, W. and Kerling, K. E. T. (1970). *Rec. Trav. Chim.*, **89**, 1181
48. Richards, E. L. (1970). *Aust. J. Chem.*, **23**, 1033
49. Beyer, H. and Schmidt, S. (1971). *Liebigs Ann. Chem.*, **748**, 109
50. Asinger, F., Sans, A., Offermanns, H., Krings, P. and Andree, H. (1971). *Liebigs Ann. Chem.*, **744**, 51
51. Chinone, A., Sato, S. and Ohta, M. (1971). *Bull. Chem. Soc. Japan*, **44**, 826
52. Julia, M. and Huynh Dinh Tam. (1971). *Bull. Soc. Chim. Fr.*, 1303
53. Heindel, N. D., (1971). *Tetrahedron Lett.*, 1439
54. Mori, Y. and Tsuji, J. (1971). *Tetrahedron*, **27**, 4039
55. Hartke, K. and Seib, B. (1971). *Pharmazie*, **25**, 517
56. Yamada, Y., Zama, T. and Kumashiro, I. (1971). *Japanese Pat.* 71 04,373 [(1971). *Chem. Abstr.*, **74**, 125693]
57. Towliati, H. (1970). *Chem. Ber.*, **103**, 3952
58. Lettau, H. (1970). *Z. Chem.*, **10**, 431
59. Volkamer, K. and Zimmermann, H. (1969). *Chem., Ber.*, **102**, 4177
60. Akagane, K., Allan, F. J., Allan, C. G., Friberg, T., Muircheartaigh, S. O. and Thomson, J. B. (1969). *Bull. Chem. Soc. Japan*, **42**, 3204
61. Lettau, H. (1970). *Z. Chem.*, **10**, 462
62. Ferguson, I. J., (1971). *University of Exeter*. Personal communication
63. Klem, R. F., Skinner, H. F. and Isensee, R. W. (1970) *J. Heterocycl. Chem.*, **7**, 403
64. Hafner, W. and Prigge, H. (1970). *Ger. Offen.*, 1 923 643 [(1971). *Chem. Abstr.*, **74**, 22838]
65. Brunn, E. Funke, E., Gotthardt, H. and Huisgen, R. *Chem. Ber.*, **104**, 1562
66. Goto, Y., Honjo, N. and Yamazaki, M. (1970). *Chem. Pharm. Bull. (Tokyo)*, **18**, 2000
67. Lont, P. J., Van der Plas, H. C. and Koudijs, A. (1971). *Rec. Trav. Chim.*, **90**, 207
68. Hetzheim, A. and Manthey, G. (1970). *Chem. Ber.*, **103**, 2845
69. Hetzheim, A. Pusch, H. and Beyer, H. (1970). *Chem. Ber.*, **103**, 3533
70. Balazs, M. K. and Furst, A. (1970). *J. Heterocycl. Chem.*, **7**, 1381
71. Buechel, K. H. (1970). *Z. Naturforsch. B*, **25**, 945
72. Smith, J. G. and Ho, I. (1971). *Tetrahedron Lett.*, 3541

73. Härter, H. P., Stauss, U. and Schindler, O. (1971). *Helv. Chim. Acta,* **54,** 2114
74. Bourson, J. (1970). *Bull. Soc. Chim. Fr.,* 1867
75. Kiffer, D. (1970). *Bull. Soc. Chim. Fr.,* 2377
76. Meth-Cohn, O. (1971). *J. Chem. Soc. C,* 1356
77. De Selms, R. C. (1970). *Tetrahedron Lett.,* 3001
78. Terent'ev, A. P., Il'ina, I. G., Rukhadze, E. G. and Vorontsova, I. G. (1970). *J. Gen. Chem. U.S.S.R.,* **40,** 1592
79. Cadogan, J. I. G., Marshall, R., Smith, D. M. and Todd, M. J. (1970). *J. Chem. Soc. C,* 2441
80. Kew, D. J. and Nelson, P. F., (1962). *Aust. J. Chem.,* **15,** 792
81. Fielden, R., Meth-Cohn, O. and Suschitzky, H. (1970). *Chem. Commun.,* 1658
82. Takahashi, S. and Kano, H. (1963). *Chem. Pharm. Bull. (Tokyo),* **11,** 1375
83. Neadle, D. J., and Pollitt, A. J. (1967). *J. Chem. Soc. C,* 1764
84. Ljublinskaya, L. A. and Stepanov, V. M. (1971). *Tetrahedron Lett.,* 4511
85. Martin, J., Meth-Cohn, O. and Suschitzky, H. (1971). *Chem. Commun.,* 1319
86. Perera, R. C., and Smalley, R. K. (1970). *Chem. Commun.,* 1458
87. Harmon, R. E., Earl, R. A. and Gupta, S. K. (1971). *J. Org. Chem.,* **36,** 2553
88. Woerner, F. P. and Reimlinger, H. (1970). *Chem. Ber.,* **103,** 1908
89. Regitz, M. and Himbert, G. (1970). *Tetrahedron Lett.,* 2823
90. L'abbé, G. (1969). *Chem. Rev.,* 345
91. Zefirov, N. S., Chapovskaya, N. K. and Kolesnikov, V. V. (1971). *Chem. Commun.,* 1001
92. L'abbé, G. and Hassner, A. (1970). *J. Heterocycl. Chem.,* **7,** 361
93. Stewart, J. M., Clark, R. L. and Pike, P. E. (1971). *J. Chem. Eng. Data,* **16,** 98
94. Ykman, P., L'abbé, G. and Smets, G. (1970). *Tetrahedron Lett.,* 5225
95. Ykman, P., L'abbé, G. and Smets, G. (1971). *Tetrahedron,* **27,** 845
96. Harvey, G. R. (1966). *J. Org. Chem.,* **31,** 1587
97. Ykman, P. L'abbé, G. and Smets, G. (1970). *Tetrahedron,* **27,** 5623
98. Woodward, R. B. and Wintner, C. (1969). *Tetrahedron Lett.,* 2697
99. Wintner, C. (1970). *Tetrahedron Lett.,* 2275
100. Angadiyavar, C. S., Sukumaran, K. B. and George, M. V. (1971). *Tetrahedron Lett.,* 633
101. Potts, K. T. and Husain, S. (1970). *J. Org. Chem.,* **35,** 3451
102. Khadem, H. E., Shaban, M. A. and Nassr, M. A. (1970). *J. Chem. Soc. C,* 2167
103. Stollé, R. (1926). *Chem. Ber.,* **59,** 1742
104. Petersen, S. and Heitzer, H. (1970). *Angew. Chem. Internat. Ed. Engl.,* **9,** 67
105. Isaacs, N. W. and Kennard, C. H. L. (1970). *Chem. Commun.,* 631; (1971). *J. Chem. Soc. B,* 1270
106. O'Halloran, J. K. and Scott, F. L. (1971). *Chem. Commun.,* 426
107. Gupta, C. M., Bhaduri, A. P. and Khanna, N. M. (1970). *Tetrahedron,* **26,** 3069
108. Becker, H., Görmer, G. and Timpe, H-J. (1970). *J. Prakt. Chem.,* **312,** 610
109. Spassov, W. S. and Raikov, Z. (1971). *Z. Chem.,* **11,** 422
110. Ollis, W. D. and Ramsden, C. A. (1971). *Chem. Commun.,* 1224
111. Ruccia, M., Vivona, N. and Cusmano, G. (1971). *J. Heterocycl. Chem.,* **8,** 137
112. Husain, S., Srinivasan, V. R. and Surendra Nath, T. G. (1971). *Indian J. Chem.,* **9,** 642
113. Kurzer, F. (1970). *J. Chem. Soc. C,* 1813
114. Pesson, M. and Antoine, M. (1970). *Bull. Soc. Chim. Fr.,* 1590, 1599
115. Blackman, A. J. (1970). *Aust. J. Chem.,* **23,** 631
116. Blackman, A. J. and Polya, J. B. (1970). *J. Chem. Soc. C,* 2403
117. Ollis, W. D. and Ramsden, C. A. (1971). *Chem. Commun.,* 1222
118. Boyd, G. V. and Summers, A. J. H. (1971). *J. Chem. Soc. C,* 409
119. Mustafa, A. and Hassan, A. E. A. A. (1957). *J. Amer. Chem. Soc.,* **79,** 3846
120. Browne, E. J., Nunn, E. E. and Polya, J. B. (1970). *J. Chem. Soc. C,* 1515
121. Van der Plas, H. C. and Jongejan, H. (1970). *Rec. Trav. Chim.,* **89,** 680
122. Scheiner, P. and Dinda, J. F. (1970). *Tetrahedron,* **26,** 2619
123. Scott, F. L., Cronin, D. A. and O'Halloran, J. K. (1971). *J. Chem. Soc. C,* 2769
124. Birchall, J. M., Haszeldine, R. N. and Kemp, J. E. G. (1970). *J. Chem. Soc. C,* 1519
125. Kamel, M., Ali, I. M. and Kamel, M. M. (1970). *Liebigs Ann. Chem.,* **733,** 115
126. Schellhammer, C. W., Schroeder, J. and Joop, N. (1970). *Tetrahedron,* **26,** 497
127. Avramenko, L. F., Vilenskii, Yu. B., Ivanov, B. M., Ol'shevskaya, I. A., Pochinok, V. Ya., Skripnik, L. I., Federova, L. N. and Federova, N. P. (1970). *Usp. Nauchn. Fotogr.,* **14,** 5 [(1971). *Chem. Abstr.,* **74,** 55101]

128. Stashkevich, V. V., Pilyugin, G. T. and Stashkevich, O. M. (1970). *J. Gen. Chem. U.S.S.R.*, **40**, 178
129. Roche, M. and Pujol, L. (1971). *J. Chim. Phys.*, **68**, 465
130. Kamiya, M. (1970). *Bull. Chem. Soc. Japan*, **43**, 3344
131. Roche, M. and Pujol, L. (1970). *Bull. Soc. Chim. Fr.*, 273
132. Hansen, L. D. and Baca, E. J. (1970). *J. Heterocycl. Chem.*, **7**, 991
133. Labhart, H., Heinzelmann, W. and Dubois, J. B. (1970). *Pure Appl. Chem.*, **24**, 495
134. Bodor, N., Dewar, M. J. S. and Harget, A. J. (1970). *J. Amer. Chem. Soc.*, **92**, 2929
135. Feeney, J., Newman, G. A. and Pauwels, P. J. S., (1970). *J. Chem. Soc. C*, 1842
136. Newman, G. A. and Pauwels, P. J. S. (1970). *Tetrahedron*, **26**, 1571
137. Deschamps, J., Sauvaitre, H., Arriau, J., Maquestiau, A., van Haverbeke, Y. and Jacquerye, R. (1971). *Tetrahedron Lett.*, 2929
138. Deschamps, J., Arriau, J. and Parmentier, P. (1971). *Tetrahedron*, **27**, 5779, 5795
139. Arriau, J., Chaillet, M. and Deschamps, J. (1971). *Tetrahedron*, **27**, 5807
140. Newman, G. A. and Pauwels, P. J. S., (1970). *Tetrahedron*. **26**, 3429
141. Katritzky, A. R. (1970). *Chimia*, **24**, 134
142. Burton, R. E. and Finar, I. L. (1970). *J. Chem. Soc. B*, 1692
143. Berthou, J., Elguero, J. and Rerat, C. (1970). *Acta Crystallogr.* **B26**, 1880
144. De Camp, W. H. and Stewart, J. M. (1970). *J. Heterocycl. Chem.* **7**, 895
145. Henery-Logan, K. R. and Keiter, E. A. (1970). *J. Heterocycl. Chem.*, **7**, 923
146. Rioux, J.-P., Clement, C., Jacquier, R. and Tarrago, G. (1970). *Bull. Soc. Chim. Fr.*, 2144
147. Tabak, S., Grandburg, J. I. and Kost, A. N. (1966). *Tetrahedron*, **22**, 2703
148. Rioux, J. -P. and Clement, C. (1970). *Bull. Soc. Chim. Fr.*, 2139
149. Matsuura, T., Banba, A. and Ogura, K. (1971). *Tetrahedron*, **27**, 1211
150. Dembech, P., Seconi, G., Vivarelli, P., Schenetti, L. and Taddei, F. (1971). *J. Chem. Soc. B*, 1670
151. Omel'chenko, Yu. A. and Kondrashev, Yu. D. (1971). *Kristallografiya*, **16**, 115; (1971). *Chem. Abstr.* **74**, 147539
152. Will, G. (1969). *Z. Krystallogr.*, **129**, 211
153. Martinez-Carrera, S. (1966). *Acta Crystallogr.*, **20**, 783
154. Visser, G. J. and Vos, A. (1971). *Acta Crystallogr.*, **B27**, 1802
155. van Remoortere, F. P. and Boer, F. P. (1971). *J. Chem. Soc. B*, 976
156. Veidis, M. V. and Palenik, G. J. (1969). *Chem. Commun.*, 196
157. Veidis, M. V., Palenik, G. J., Schaffrin, R. and Trotter, J. (1969). *J. Chem. Soc. A*, 2659
158. Casy, A. F. and Ison, R. R. (1970). *Chem. Commun.*, 1343
159. Ansell, G. B., Forkey, D. M. and Moore, D. W. (1970). *Chem. Commun.*, 56
160. Pozharskii, A. F., Chegolya, T. N. and Simonov, A. M. (1968). *Khim. Geterotsikl. Soedin.*, **4**, 503
161. Jacquier, R., Lacombe, J. -M. and Maury, G. (1971). *Bull. Soc. Chim. Fr.*, 1040
162. Khristich, B. I. (1970). *Khim. Geterotsikl. Soedin.*, 1683
163. Coburn, M. D. and Newman, P. N. (1970). *J. Heterocycl. Chem.*, **7**, 1391
164. Bolton, K., Brown, R. D., Burden, F. R. and Mishra, A. (1971). *Chem. Commun.*, 873
165. Carbo, R. and Fraga, S. (1970). *An. Fis.*, **66**, 401 [(1971). *Chem. Abstr.*, **74**, 130 493]
166. Scott, F. L. and Tobin, J. C. (1971). *J. Chem. Soc. C*, **703**
167. Fletcher, I. J. and Katritzky, A. R. (1970). *Chem. Commun.*, 706
168. Bianchi, G. and Katritzky, A. R. (1971). *Chem. Commun.*, 846
169. Butler, R. N. (1969). *Chem. Commun.*, 405
170. Butler, R. N. (1970). *J. Chem. Soc. B*, 138
171. Butler, R. N. (1970). *Chem. Commun.*, 1096
172. Yoshida, Z. and Kobayashi, T. (1971). *Theoret. Chim. Acta*, **20**, 216
173. Kalmykov, V. V., Blokhin, V. E. and Aglitskaya, K. V. (1969). *Tr. Voronezh. Gos. Univ.*, **73**, 67 [(1971). *Chem. Abstr.*, **74**, 47 608]
174. Pevzner, M. S., Federova, E. Ya. Shokor, I. N. and Bagal, L. N. (1971). *Khim. Geterotsikl. Soedin.*, **7**, 275
175. Aten, W. C. and Buechel, K. H. (1970). *Z. Naturforsch.* **B25**, 961
176. Blazevic, N., Kajfez, F. and Sunjic, V. (1970). *J. Heterocycl. Chem.*, **7**, 227
177. Litchfield, G. J. and Shaw, G. (1971). *J. Chem. Soc.* **C**, 817
178. Boulton, B. E. and Coller, B. A. W. (1971). *Aust. J. Chem.*, **24**, 1413
179. Gordon, R. D. and Yang, R. F. (1970). *Can. J. Chem.*, **48**, 1722
180. Cescon, L. A., Coraor, G. R., Dessauer, R., Silversmith, E. F. and Urban, E. J. (1971). *J. Org. Chem.*, **36**, 2262

181. Bagal, L. I. and Pevzner, M. S. (1971). *Khim. Geterotsikl. Soedin.*, **7**, 272
182. Kanaskova, Yu. D., Sukhorukov, B. I., Pentin, Yu. A. and Komarovskaya, G. V. (1970). *Bull. Acad. Sci. U.S.S.R.*, 1637
183. Cordes, M. and Walter, J. L. (1968). *Spectrochim. Acta*, **24A**, 24
184. King, S. T. (1970). *J. Phys. Chem.*, **74**, 2133
185. Wolff, H. and Wolff, E. (1971). *Spectrochim. Acta*, **27A**, 2109
186. Bellocq, A.-M. and Garrigou-Lagrange, C. (1970). *J. Chim. Phys.*, **67**, 951
187. Bellocq, A.-M. and Garrigou-Lagrange, C. (1970). *J. Chim. Phys.*, **67**, 1091
188. Colombo, L., Furic, K. and Kirin, D. (1971). *J. Molec. Spectrosc.*, **39**, 217
189. Benassi, R., Lazzeretti, P., Schenetti, L., Taddei, F. and Vivarelli, P. (1971). *Tetrahedron Lett.*, 3299
190. Diaz, E., Maldonado, L. A. and Ortega, A. (1970). *Spectrochim. Acta.*, **26A**, 284
191. Elguero, J., Jacquier, R. and Mondon, S. (1970). *Bull. Soc. Chim. Fr.*, 1346
192. Elguero, J., Jacquier, R. and Tarrago, G. (1970). *Bull. Soc. Chim. Fr.*, 1345
193. Elguero, J., Jacquier, R. and Mondon, S. (1970). *Bull. Soc. Chim. Fr.*, 4436
194. L'abbé, G., Galle, J. E. and Hassner, A. (1970). *Tetrahedron Letters*, 303
195. Pugmire, R. J. and Grant, D. M. (1971). *J. Amer. Chem. Soc.*, **93**, 1880
196. Garner, G. V., Meth-Cohn, O. and Suschitzky, H. (1971). *J. Chem. Soc. C*, 1234
197. Revankar, G. R. and Townsend, L. B. (1970). *J. Heterocycl. Chem.*, **7**, 1329
198. Coburn, M. D., Loughran, E. D. and Smith, L. C. (1970). *J. Heterocycl. Chem.*, **7**, 1149
199. Smith, R. F., Deutsch, J. L., Almeter, P. A., Johnson, D. S., Roblyer, S. M. and Rosenthal, T. C. (1970). *J. Heterocycl. Chem.*, **7**, 671
200. Bekarek, V. and Slouka, J. (1970). *Collect. Czech. Chem. Commun.*, **35**, 2936
201. Natsuko Cyr, Wilks, M. A. J. and Willis, M. R. (1971). *J. Chem. Soc. B*, 404
202. Volkamer, K. and Zimmermann, H. (1970). *Chem. Ber.*, **103**, 296
203. Van Thuijl, J., Klebe, K. J. and Van Houte, J. J. (1970). *Org. Mass Spectrom.*, **3**, 1549
204. Van Thuijl, J., Klebe, K. J. and Van Houte, J. J. (1971). *J. Heterocycl. Chem.*, **8**, 311
205. Compernolle, F. and Dekeirel, M. (1971). *Org. Mass Spectrom.*, **5**, 427
206. Lawrence, R. and Waight, E. S. (1970). *Org. Mass Spectrom.*, **3**, 367
207. Becker, H. G. O., Beyer, D. and Timpe, H.-J. (1970). *J. Prakt. Chem.*, **312**, 869
208. Brady, L. E. (1970). *J. Heterocycl. Chem.*, **7**, 1223
209. Rapp, U., Staab, H. A. and Wünsche, C. (1971). *Tetrahedron*, **27**, 2679
210. Kametani, T., Hirata, S., Shibuya, S. and Shio, M. (1971). *Org. Mass. Spectrom.*, **5**, 117
211. Thomas, C. B. (1970). *Org. Mass. Spectrom.*, **3**, 1523
212. Chishov, O. S., Kochetkov, N. K., Malysheva, N. N. and Shiyonok, A. I. (1971). *Org. Mass. Spectrom.*, **5**, 481
213. Chishov, O. S., Kochetkov, N. K., Malysheva, N. N., Shiyonok, A. I. and Chashchin, V. L. (1971). *Org. Mass. Spectrom.*, **5**, 1145
214. Audier, H.-E., Bottin, J., Fetizon, M. and Tabet, J.-C. (1971). *Bull. Soc. Chim. Fr.*, 2911
215. Barthel, J. and Schmeer, G. (1970). *Liebigs Ann. Chem.*, **738**, 195
216. Chiriac, C. and Zugravescu, I. (1970). *Rev. Roum. Chim.*, **15**, 1201
217. Coburn, M. D. (1970). *J. Heterocycl. Chem.*, **7**, 345
218. Elguero, J., Jacquier, R., Pellegrin, V. and Tabacik, V. (1970). *Bull. Soc. Chim. Fr.*, 1974
219. Khan, M. A. and Polya, J. B. (1970). *J. Chem. Soc. C*, 85
220. Reimlinger, H., Noels, A., Jadot, J. and van Overstraeten, A. (1970). *Chem. Ber.*, **103**, 1954
221. Reimlinger, H., Noels, A. and Jadot, J. (1970). *Chem. Ber.*, **103**, 1949
222. Cusack, N. J., Shaw, G. and Litchfield, G. J. (1971). *J. Chem. Soc. C*, 1501
223. Butler, D. E. and De Wald, H. A. (1971). *J. Org. Chem.*, **36**, 2542
224. Chung Wu, E. and Vaughan, J. D. (1970). *J. Org. Chem.*, **35**, 1146
225. Reimlinger, H., Noels, A., Jadot, J. and Van Overstraeten, A. (1970). *Chem. Ber.*, **103**, 1942
226. Gosztonyi, T., Carnmalm, B. and Sjoberg, B. (1970). *Acta Chem. Scand.*, **24**, 3078
227. Barry, W. J., Birkett, P. and Finar, I. L. (1969). *J. Chem. Soc. C*, 1328
228. Habraken, C. L., Cohen-Fernandes, P. and van Erk, K. C. (1970). *Tetrahedron Lett.*, 479
229. Coburn, M. D. (1970). *J. Heterocycl. Chem.*, **7**, 707
230. Coburn, M. D. (1971). *J. Heterocycl. Chem.*, **8**, 153
231. Finar, I. L. and Hurlock, R. J. (1957). *J. Chem. Soc.*, 3024
232. Coburn, M. D. (1971). *J. Heterocycl. Chem.*, **8**, 293
233. Tertov, B. A. and Onishchenko, P. P. (1970). *Khim. Geterotsikl. Soedin.*, 1435 [*Chem. Abstr.* (1971). **75**, 5785]

234. Alonso, G., Garcia-Muñoz, G. and Madroñero, R. (1970). *J. Heterocycl. Chem.*, **7**, 1435
235. Olofson, R. A. and Kendall, R. V. (1970). *J. Org. Chem.*, **35**, 2246
236. Stocker, F. B., Kurtz, J. L., Gilman, B. L. and Forsyth, D. A. (1970). *J. Org. Chem.*, **35**, 883
237. Guibe-Jampel, E., Wakselman, M. and Vilkas, M. (1971). *Bull. Soc. Chim. Fr.*, 1308
238. Effenberger, F. and Mack, K. E. (1970). *Tetrahedron Lett.*, 3947
239. Ruccia, M., Vivona, N., Cusmano, G. and Marino, M. L. (1970). *Gazz. Chim. Ital.*, **100**, 358
240. Yamada, F. and Fujimoto, Y. (1971). *Bull. Chem. Soc. Japan*, **44**, 533
241. Minetti, R. and Bruylants, A. (1970). *Bull. Cl. Sci. Acad. Roy. Belg.*, **56**, 1047
242. Tillett, J. G. (1970). *Annual Reports*, **67B**, 66, 71, 75, 87. (London: The Chemical Society)
243. Kunitake, T. and Shinkai, S. (1971). *J. Amer. Chem. Soc.*, **93**, 4247, 4256
244. Blyth, C. A. and Knowles, J. R. (1971). *J. Amer. Chem. Soc.*, 93, 3021
245. Bundegaard, H. (1971). *Tetrahedron Lett.*, 4613
246. Vaughan, J. D., Mughrabi, Z. and Chung Wu, E. (1970). *J. Org. Chem.*, **35**, 1141
247. Schutte, L. and Havinga, E. (1970). *Tetrahedron*, **26**, 2297
248. Regel, E. and Buechel, K. H. (1970). *Ger. Offen.*, 1 926 206 [(1971). *Chem. Abstr.* **74**, 31 754]
249. Regel, E. and Buechel, K. H. (1971). *Ger. Offen.*, 1 956 711 [(1971). *Chem. Abstr.* **75**, 49 086]
250. Tertov, B. A. and Burykin, V. V. (1970). *Khim. Geterotsikl. Soedin.*, 1554. [(1971). *Chem. Abstr.*, **74**, 76 466]
251. Hazeldine, C. E., Pyman, F. L. and Winchester, J. (1924). *J. Chem. Soc.*, 1431
252. Barlin, G. B. and Batterham, T. J. (1967). *J. Chem. Soc. B*, 516
253. Shimada, K., Kuriyama, S., Kanazawa, T., Satoh, M. and Toyoshima, S. (1971). *J. Pharm. Soc. Japan*, **91**, 221
254. Novikov, S. S., Khmelnitskii, L. I., Lebedev, O. V., Sevastyanova, V. V. and Yepishina, L. V. (1970). *Chem. Heterocycl. Compds. U.S.S.R.*, 503
255. Novikov, S. S., Khmelnitskii, L. I., Lebedev, O. V., Sevastyanova, V. V. and Yepishina, L. V. (1970). *Chem. Heterocycl. Compds. U.S.S.R.*, 669
256. Shiokawa, Y. and Ohki, S. (1971). *Chem. Pharm. Bull. (Tokyo).*, **19**, 401
257. Cooksey, C. J., Dodd, D. and Johnson, M. D. (1971). *J. Chem. Soc. B*, 1380
258. Reimlinger, H. and Peiren, M. A. (1970). *Chem. Ber.*, **103**, 3266
259. Neumann, P. N. (1970). *J. Heterocycl. Chem.*, **7**, 1159
260. Oliver, J. E. and Stokes, J. B. (1970). *J. Heterocycl. Chem.*, **7**, 961
261. Neumann, P. N. (1971). *J. Heterocycl. Chem.*, **8**, 51
262. Begtrup, M. and Poulsen, K. V. (1971). *Acta. Chem. Scand.*, **25**, 2087
263. Kröeger, T. C. F. and Miethchen, R. (1969). *Z. Chem.*, **9**, 378
264. Shostakovskii, M. F., Skvortsova, G. G., Domnina, E. S. and Makhno, L. P. (1970). *Khim. Geterotsikl. Soedin.*, 1289 [(1971). *Chem. Abstr.* **74**, 141 657]
265. Zelnik, R. and Strehlan, F. (1971). *Experientia*, **27**, 20
266. Einberg, F. (1970). *J. Org. Chem.*, **35**, 3978
267. Begtrup, M. (1970). *Acta Chem. Scand.*, **24**, 1819
268. Reimlinger, H. and Merenyi, R. (1970). *Chem. Ber.*, **103**, 3284
269. Oakenfull, D. G. and Jencks, W. P. (1971). *J. Amer. Chem. Soc.*, **93**, 178
270. Oakenfull, D. G., Salvesen, K. and Jencks, W. P. (1971). *J. Amer. Chem. Soc.*, **93**, 188
271. de Bie, D. A., van der Plas, H. C. and Guertsen, G. (1971). *Rec. Trav. Chim.*, **90**, 594
272. Kashparov, I. S. and Pozharskii, A. F. (1971). *Khim. Geterotsikl. Soedin*, 124 [(1971). *Chem. Abstr.* **75**, 35, 922]
273. De Bie, D. A. and van der Plas, H. C. (1969). *Rec. Trav. Chim.*, **88**, 1246
274. Pozharskii, A. F., Simonov, A. M., Mar'yanovskii, V. M. and Zinchenko, R. P. (1970). *Khim. Geterotsikl. Soedin*, 1060. [(1971). *Chem. Abstr.* **75**, 48 982]
275. Seconi, G., Vivarelli, P. and Ricci, A. (1970). *J. Chem. Soc. B*, 254
276. Dembech, P., Ricci, A., Seconi, G. and Vivarelli, P. (1971). *J. Chem. Soc. B*, 557
277. Bagal, L. I., Pevzner, M. S., Samarenko, V. Ya and Egorov, A. P. (1970). *Khim. Geterotsikl. Soedin.*, 1701. [(1971). *Chem. Abstr.* **74**, 99 948]
278. Bagal, L. I., Pevzner, M. S., Samarenko, V. Ya. and Egorov, A. P. (1970). *Khim. Geterotsikl. Soedin.*, 997. [(1971). *Chem. Abstr.* **74**, 76 376]
279. Begtrup, M. (1971). *Acta. Chem. Scand.*, **25**, 795
280. Begtrup, M. (1971). *Acta. Chem. Scand.*, **25**, 803

281. Albert, A. (1970). *J. Chem. Soc. C*, 230
282. Kamel, M. M., Abdul Hamid, M. M. and Kamel, M. (1971). *Liebigs Ann. Chem.*, **746**, 76
283. Scott, F. L., Lambe, T. M. and Tobin, J. C. (1971). *Chem. Commun.*, 411
284. Freeman, J. P., Surkey, D. L. and Kassner, J. E. (1970). *Tetrahedron Lett.*, 3797
285. Coburn, M. D. (1970). *J. Heterocycl. Chem.*, **7**, 455
286. Parnell, E. W. (1970). *Tetrahedron Letters*, 3941
287. Elguero, J., Jacquier, R. and Tizane, D. (1970). *Bull. Soc. Chim. Fr.*, 1121
288. Wright, J. B. (1969). *J. Org. Chem.*, **34**, 2474
289. Adams, D. J. C., Bradbury, S., Horwell, D. C., Keating, M., Rees, C. W. and Storr, R. C. (1971). *Chem. Commun.*, 828
290. Campbell, C. D. and Rees, C. W. (1969). *J. Chem. Soc. C*, 742
291. Tertov, B. A., Onishchenko, P. P. and Kazanbieva, M. A. (1970). *J. Org. Chem. U.S.S.R.*, **6**, 2147
292. Finch, N. and Gschwend, H. W. (1971). *J. Org. Chem.*, **36**, 1463
293. Bourson, J. (1971). *Bull. Soc. Chim. Fr.*, 152
294. Bourson, J. (1971). *Bull. Soc. Chim. Fr.*, 3541
295. Van Meeteren, H. W. and Van der Plas, H. C. (1969). *Rec. Trav. Chim.*, **88**, 204
296. Sakai, K. and Anselme, J.-P. (1970). *Tetrahedron Lett.*, 3851
297. Rudaya, L. I. and El'tsov, A. V. (1970). *J. Org. Chem. U.S.S.R.*, **6**, 2150
298. Wentrup, C. and Crow, W. D. (1970). *Tetrahedron*, **26**, 3965
299. Gilchrist, T. L., Gymer, G. E. and Rees, C. W. (1971). *Chem. Commun.*, 1519
300. Scott, F. L., Donovan, J. and O'Halloran, J. K. (1970). *Tetrahedron Lett.*, 4079
301. Wentrup, C. (1971). *Tetrahedron*, **27**, 1281
302. Lablache-Combier, A. and Remy, M. (1971). *Bull. Soc. Chim. Fr.*, 679
303. Dürr, H. and Schrader, L. (1970). *Chem. Ber.*, **103**, 1334
304. Beak, P., Meisel, J. L. and Messer, W. R. (1967). *Tetrahedron Lett.*, 5315
305. Beak, P. and Messer, W. R. (1969). *Tetrahedron*, **25**, 3287
306. Dubois, J. P. and Labhart, H. (1969). *Chimia*, **23**, 109
307. Maeda, K. and Hayashi, T. (1971). *Bull. Chem. Soc. Japan*, **44**, 533
308. Riem, R. H., MacLachlan, A., Coraor, G. R. and Urban, E. J. (1971). *J. Org. Chem.*, **36**, 2272
309. Cescon, L. A., Coraor, G. R., Dessauer, R., Deutsch, A. S., Jackson, H. L., MacLachlan, A., Marcali, K., Potrafke, E. M., Read, R. E., Silversmith, E. F. and Urban, E. J. (1971). *J. Org. Chem.*, **36**, 2267
310. MacLachlan, A. and Riem, R. H. (1971). *J. Org. Chem.*, **36**, 2275
311. Cohen, R. L. (1971). *J. Org. Chem.*, **36**, 2280
312. Cohen, L. A. and Kirk, K. L. (1971). *J. Amer. Chem. Soc.*, **93**, 3060
313. Ogata, M., Matsumoto, H., Takahashi, S. and Kano, H. (1970). *Chem. Pharm. Bull. (Tokyo)*, **18**, 964
314. El'tsov, A. V. and Krivozheiko, K. M. (1970). *J. Org. Chem. U.S.S.R.*, **6**, 637
315. Bird, C. W., Wong, D. Y., Boyd, G. V. and Summers, A. J. H. (1971). *Tetrahedron Lett.*, 3187
316. Burgess, E. M., Carithers, R. and Cullagh, Mc.L. (1968). *J. Amer. Chem. Soc.*, **90**, 1923
317. Meier, H. and Menzel, I. (1970). *Liebigs Ann. Chem.*, **739**, 56
318. Ohashi, M., Tsujimoto, K. and Yonezawa, T. (1970). *Chem. Commun.*, 1089
319. Becker, H. G. O., Beyer, D. and Timpe, H.-J. (1970). *Z. Chem.*, **10**, 264
320. Becker, H. G. O., Böttcher, H., Fischer, G., Rückauf, H. and Saphon, S. (1970). *J. Prakt. Chem.*, **312**, 586
321. Scheiner, P. (1971). *Tetrahedron Lett.*, 4489
322. Angadiyavar, C. S. and George, M. V. (1971). *J. Org. Chem.*, **36**, 1589
323. Schubert, H. and Rudorf, W. D. (1971). *Z. Chem.*, **11**, 175
324. Mustafa, A., Fleifel, A. M., Ali, M. I. and Hassan, N. M. (1970). *Liebigs Ann. Chem.*, **739**, 75
325. Van Thielen, J., Van Thien, T. and De Schryver, F. C. (1971). *Tetrahedron Lett.*, 3031
326. Browne, E. J. (1970). *Tetrahedron Lett.*, 943
327. Browne, E. J. (1971). *Aust. J. Chem.*, **24**, 393
328. Popov, I. I., Simonov, A. M. and Kolodyazhaaya, S. N. (1971). *Khim. Geterotsikl Soedin.*, 1566
329. Micetich, R. G. (1970). *Can. J. Chem.*, **48**, 2006
330. Henning, G. and Wolf, F. (1971). *Z. Chem.*, **11**, 153

331. Skvortsova, G. G., Domnina, E. S., Glazkova, N. P., Chipanina, N. N. and Shergina, N. I. (1971). *J. Gen. Chem. U.S.S.R.*, **41**, 620
332. Skvortsova, G. G., Domnina, E. S., Makhno, L. P., Frolov, Yu. L., Voronov, V. K., Chipanina, N. N. and Shergina, N. I. (1971). *Bull. Acad. Sci. U.S.S.R.*, 2570
333. Chub, N. K., Tsupak, E. P., and Simonov, A. M. (1970). *Khim. Geterotsikl. Soedin.*, 1393; [(1971). *Chem. Abstr.*, **74**, 76 368]
334. Ishiwata, S. and Shiokawa, Y. (1970). *Chem. Pharm. Bull. (Tokyo).*, **18**, 1245
335. Sutherland, D. R. and Tennant, G. (1971). *J. Chem. Soc. C*, 706
336. Pozharskii, A. F., Zvezdina, E. A., Andreichikov, Yu. P., Simonov, A. M., Anisimova, V. A. and Popova, S. F. (1970). *Khim. Geterotsikl. Soedin.*, 1267. [(1971). *Chem. Abstr.* **75**, 5804]
337. Smith, P. A. S., Breen, G. J. W., Hajek, M. K. and Awang, D. V. C. (1970). *J. Org. Chem.*, **35**, 2215
338. Tobin, J. C., Butler, R. N. and Scott, F. L. (1970). *Chem. Commun.*, 112
339. Litchfield, G. J. and Shaw, G. (1971). *J. Chem. Soc. B*, 1474
340. Sunjic, V., Fajdiga, T. and Japelj, M. (1970). *J. Heterocycl. Chem.*, **7**, 211
341. Mitsuhashi, K., Takahashi, K., Zaima, T. and Asahara, T. (1971). *Kogyo Kagaku Zasshi*, **74**, 316 [(1971). *Chem. Abstr.* **74**, 125 566]
342. Bowyer, P. M., Iles, D. H. and Ledwith, A. (1971). *J. Chem. Soc. C*, 2775
343. Lettau, H. (1971). *Z. Chem.*, **11**, 10
344. Alabaster, R. J. and Barry, W. J. (1970). *J. Chem. Soc. C*, 78
345. Nuhn, P. and Wagner, G. (1970). *J. Prakt. Chem.*, **312**, 90
346. Zinner, H. and Peseke, K. (1970). *J. Prakt. Chem.*, **312**, 185
347. Kauffmann, Th. and Wirthwein, R. (1971). *Angew. Chem. Internat. Ed. Engl.*, **10**, 20
348. Markley, L. D. and Dorman, L. C. (1970). *Tetrahedron Lett.*, 1787
349. Guarneri, M., Giori, P. and Benassi, C. A. (1971). *Tetrahedron Lett.*, 665
350. König, W. and Geiger, R. (1970). *Chem. Ber.*, **103**, 788

4
Nitrogen Heterocyclic Molecules, Part 3 The Monocyclic Azines

B. R. T. KEENE
Medway and Maidstone College of Technology, Chatham, Kent

4.1   RING SYNTHESIS

Several interesting syntheses of the pyridine system have appeared, perhaps the simplest stemming from the observation that the attempted alkylation of malononitrile with isopropyl halides and aluminium chloride gives 6-halogeno-3-cyano-2,4-bis(isopropylamino)pyridine[1]. Contrary to an earlier report, malononitrile and hydrogen bromide provide 2-bromo-3-cyano-4,6-diaminopyridine[1] and, in the presence of alkoxide ion, malononitrile condenses with many aldehydes forming 4-substituted-2-amino-3,5-dicyano-6-alkoxypyridines[2]. The reaction between malonyl chloride and nitriles can be made to provide both 2-chloro-4-hydroxypyrid-6-ones and 4-chloro-pyrimidin-6-ones, but its scope appears limited[3]. A detailed examination has been made of the well-known conversion of dehydroacetic acid into 2,6-dimethyl-4-hydroxypyridine; by careful choice of conditions the yield can be raised to 97% (although trace amounts of six other pyridines are formed)[4]. The high-temperature reaction between acetaldehyde and ammonium acetate has long been known to give 5-ethyl-2-methylpyridine, but under milder conditions (a copper(II) acetate catalyst in the presence of oxygen) 2-methylpyridine is obtained in fair yield and n-butanal forms 3,5-diethyl-2-n-propylpyridine when similarly treated[5]. The use of glyoxylic acid in a Hantzsch procedure with, for example, ethyl β-aminocrotonate (or the corresponding aminonitrile) provides pyridine-4-carboxylic acids via the dihydro derivatives[6,7] and an ingenious mechanism of the Hantzsch type accommodates the finding (by ^{14}C labelling) that the γ-carbon atom of the 2,6-diarylpyridines obtained by pyrolysing ketone N,N,N-trimethylhydra-zonium fluoroborates is derived from the trimethylamine moiety[8]. 3-Cyano-4,6-dimethyl-1-hydroxypyrid-2[1H]-one is formed when ethyl cyanoacetate, hydroxylamine and pentane-2,4-dione are heated together in the presence of piperidine[9]. Further examples of pyridine-ring synthesis from 1,3-diketones include the reactions with enamines (of cyclic 1,3-diketones)[16] and base-catalysed condensations with malononitrile and cyanoacetamide[10,11]. A Guareschi-type reaction between ethyl cyanoacetate and α,β-unsaturated ketones provides 4-aryl-3-cyanopyrid-2[1H]-ones[12] and the hetero-ring of pyridinophanes can be constructed similarly by standard methods from ethyl cyanoacetate and, e.g. cyclododecen-2-one[13].

Enamines combine with three-carbon fragments other than 1,3-diketones (see above). 2,3-Cycloalkenopyridines are formed when β-aminoacroleins condense with cyclic ketones in the presence of triethylamine-piperidine[14] and the reaction of β-aminocinnamonitrile with ethyl acetoacetate gives the corresponding pyridone[15]. With benzoyl isothiocyanate, ethyl dimethyl-aminocrotonate forms ethyl 2-benzamido-5-benzoyl-4-dimethylamino-6-thioxonicotinate[17]. Ynamines (e.g. 1-diethylaminoprop-1-yne) with styryl isocyanates give 4-aminopyrid-2[1H]-ones by 1,4-cyclo-addition[18]. The thermal conversion of (1) into (2) provides an interesting example of pyridine-ring formation by the intramolecular addition of a nitrile group to a 1,3-diene; the initial adduct undergoes spontaneous dehydrogenation[19].

Pyrid-2[1H]-ones are formed by heating 1-isocyanatobuta-1,3-dienes[11] and other suitable β-substituted vinyl isocyanates[20, 21]. γ-Cyanocarboxylic acid chlorides form 6-chloropyrid-2[1H]-ones in fair yield on treatment with hydrogen chloride in dibutyl ether[22].

The yield of pyridine obtained in the catalytic ammonolysis of 2-hydroxy-methyltetrahydrofuran is improved by introducing air into the reaction mixture[23]. 2-Chloropyridines can in some cases be isolated along with 3-chloropyridines from the ring expansion (under basic conditions) of 2,5-dialkyl-2-dichloromethyl-2H-pyrroles and an intramolecular carbene process is proposed[24]. 3-Acylpyridines are obtained conveniently from 4-(3-oxoalkyl)isoxazoles by reductive cleavage, followed by dehydration and oxidation of the carbinolamines formed[25]. δ-Aminonitriles are inter-mediates in the formation of 2-aminopyridines from 4-dicyanomethylene-4H-pyrans and secondary amines[10, 11]. 1-Amino- and 1-hydroxy-pyrid-2[1H]-ones have been obtained from pyrones with, respectively, hydrazine and hydroxylamine[26]. Some new pyridthiones have been obtained from pyr-anthiones on treatment with ammonia and amines[26]. Highly-substituted pyrylium salts give good yields of pyridine N-oxides on treatment with hydroxylamine in acid media[27] and [10]-(2,6)-pyridinophane has been obtained in very low yield by ammonolysis of the appropriate pyrylium salt[28]. 2,4,6-Triarylpyrylium perchlorates form 2-amino-3-aroyl-4,6-diaryl-pyridines on treatment with cyanamide and triethylamine[29]. The ring closure of 5-anilino-N-phenylpenta-2,4-dienylidenimine to N-phenylpyridinium ion appears to be an electrocyclic process[30]; rate data are not correlated by the Hammett equation[31].

The thermal ring contraction of 1[H]-1,2-diazepines (3) and their photoproducts (4) to pyridines (6) via ammonium-imines (5) has been further examined[32, 33]. 3-Hydroxypyridine derivatives are formed when 2-tosyl-

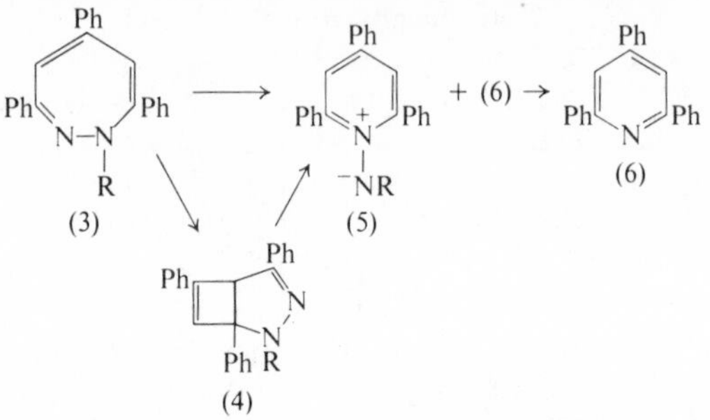

diazepin-4-ones are treated with triethylamine[34] and derivatives of the previously unknown 2,3-pyridocyclobutene system are obtained by the base-catalysed ring contraction of 3,4-dihydro-2-methoxyazocines and of azocinyl dianions[35]. 2-Methoxy-5-arylpyrimidines are cleaved by prolonged heating with ethanolic ammonia to arylacetaldehydes, which form 3,5-diarylpyridines[36].

Aliphatic nitro compounds (and analogous oximes) trimerise to trialkyl-pyridines on treatment with carbon monoxide in the presence of palladium–charcoal and ferric chloride[37].

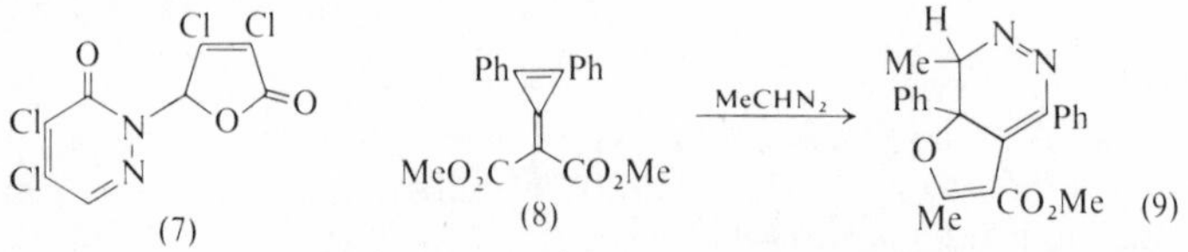

A by-product in the reaction of hydrazine with mucochloric acid has been shown to be (7)[38]. The well-established reactions of hydrazine with γ-ketoacids and dimethoxydihydrofurans have been used to prepare condensed cyclo-alka-[1,2c]-pyridazines[39] and 4,5-bis(hydroxymethyl)pyridazine[40] respectively. The methylenecyclopropene (8) forms (9) with diazoethane; on heating, (9) forms 4-(dimethoxymethyl)-6-methyl-3,5-diphenyl-pyridazine[41].

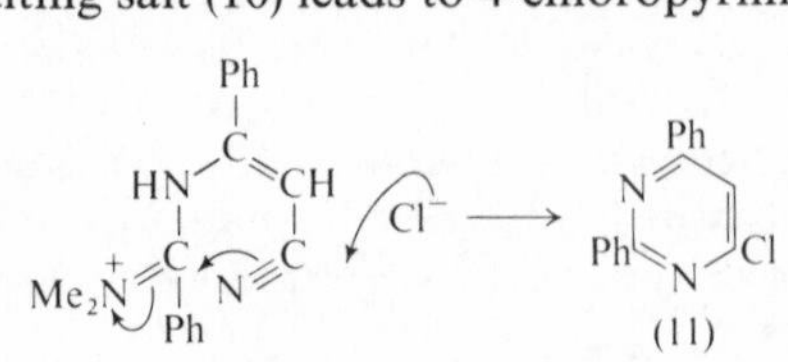

Pyrimidines (including the parent compound) are formed in excellent yield by the piperidinium acetate-catalysed condensation of formamide with β-aminovinyl-aldehydes and -ketones, as well as with the N-alkyl- and N,N-dialkyl-β-enaminocarbonyl compounds reported earlier[42]. 4-Amino-pyrimidines are formed when primary aliphatic amides are heated with formamide and phosphoryl chloride; the low yields (20–30%) are offset by the simplicity of the procedure[43]. The Vilsmeier reagent from N,N-dimethyl-benzamide and phosphoryl chloride acylates enaminonitriles at nitrogen and ring closure of the resulting salt (10) leads to 4-chloropyrimidines (11)[44], and chloropyrimidines are obtained, generally in good yield, when α-dichloro-isocyanide dichlorides are heated with nitriles in the presence of ferric chloride[45]. Aluminium chloride catalyses the addition of nitriles to diketimines

(e.g. 12), giving (13)[46] and lithium (N-phenyl)phenylketenimine is attacked at nitrogen by phenyl isocyanate yielding 6-anilino-1,3,5-triphenyluracil[47].

Imidazolines (rather than pyrimidines[48]) result from the reaction of guanidine (and acetylguanidine) with dimethyl acetylenedicarboxylate[49] but uracil N-1, N-3-bis-sulphofluorides are formed by cyclo-addition of fluoro-sulphonyl isocyanate to alkynes[50].

The principal pyrimidine synthesis (C—C—C+N—C—N) has been widely used[51-59] but failures have been reported[89, 90]. β-Ketoesters and N-hyd-roxyurea give N-3-hydroxyuracils, not the N-1 isomers previously reported[60]. Imino-aldehydes advantageously replace malondialdehydes[61] and, in an interesting variation, a wide range of pyrimidines have been obtained in excellent yields by treating bis(dimethylamino)trimethinium salts with amidines or guanidines[62]. A high yield of 5-acetyl-6-(2-hydroxyphenyl)-2-thiouracil is claimed in the ethoxide-ion catalysed reaction between 3-acetyl-4-hydroxycoumarin and thiourea[63].

Yields in the Whitehead primary synthesis have been improved in some instances by using amidines in excess[54], and the reaction between ethyl ethoxymethylenecyanoacetate and N-methylthiourea provides 5-cyano-2-mercapto-1-methylpyrimidin-6[1H]-one[64] as well as 6-amino-5-carbethoxy-1-methylpyrimidin-2[1H]-thione[65]; the former can be made the major product. Surprisingly, ethoxymethyleneacetonitrile and N-methylthiourea form 4-amino-1-methylpyrimidin-2[1H]-thione rather than the 3-methyl isomer[64]. With trisformylaminomethane, phenylacetamide gives a small amount of 4-hydroxy-5-phenylpyrimidine, but p-chlorobenzyl cyanide and pyridylacetonitriles form the corresponding pyrimidines in fair yields[66, 67]. N-Alkyl-(or N-aryl)-substituted pyrimidinethiones are formed when keto-methylene carbanions react with alkyl (or aryl) isothiocyanates[68].

The pyrolysis of tetrafluoropyridazine near 800 °C gives mainly tetra-fluoropyrimidine, by a mechanism which may involve the formation and sigmatropic rearrangement of diazabenzvalenes[69].

Aminolysis of 1,3-thiazines provides pyrimidines in high yields[70, 71], and the method has been used in a convenient synthesis of 5-deuterated N-sub-stituted-2-thiouracils by direct reaction with the amine in deuterium oxide[70]. A Shaw synthesis has been used to obtain 6-(1-pyrimidinyl)-2-aminocaproic acid; other routes failed[72].

Of considerable theoretical interest are the photoconversions of 4,5-dichloro- and 4,5-di-(perfluoroisopropyl)-3,6-difluoropyridazine (14) into 2,5-dichloro-[73] and 2,5-di(perfluoroisopropyl)-3,6-difluoropyrazine[74] (16) respectively: diazaprismane intermediates are precluded (see p. 122) and (15b) is the first valence isomer of an aromatic diazine to be isolated[74].

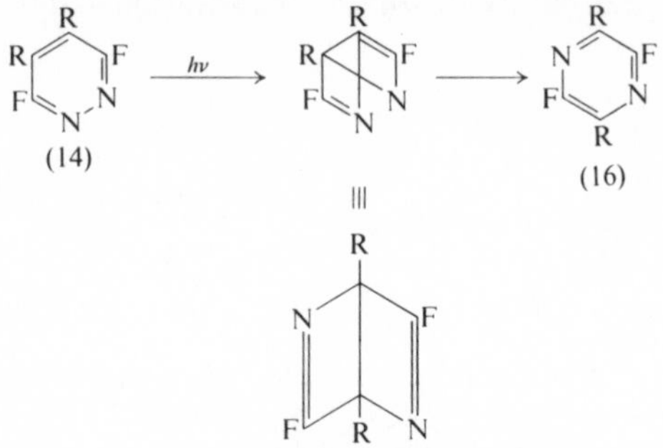

(a) R = Cl
(b) R = C$_3$F$_7$
(15)

Yields in the Sharp and Spring[75] synthesis of 2-aminopyrazine 1-oxides are improved dramatically by using titanium tetrachloride as condensing agent[76].

An improved synthesis of 1,2,4-triazine from S-methylthiosemicarbazide and glyoxal (with subsequent removal of the 3-methylthio group) has been reported[77] and 3,5-diaryl-1,2,4-triazines are prepared simply by heating acyl hydrazides with ω-halogenoacetophenones in ethanol or acetic acid[78]. 1,2,4-Triazinonethiones are formed from glyoxaldoximes and thiosemicarbazide in aqueous alkali; in acid solution triazine-3-thiones are obtained, and 3-(substituted-amino)-1,2,4-triazines can be prepared similarly from phenylglyoxaldoxime and substituted 1-aminoguanidinium bromides[79, 80]. Carboxamide hydrazones condense with 1,2-diketones to form 3-substituted-1,2,4-triazines, but yields are low[81, 82]. The ring closure of 1,2-diketone amidrazones to triazines has been extended to the formation of alkyl- and aryl-1,2,4-triazin-5-ones from α-keto-acid acet- and benz-amidrazone derivatives[83] and an essentially similar method has been used to prepare condensed triazin-5-ones[84]. An indirect preparation of glyoxylic nitrile semicarbazone has permitted a simple synthesis of 6-azacytosine[85].

6-Acylamino- (and thence 6-hydroxy-) triazines are formed in the hydrolysis of s-triazolo-[3,4-f]-1,2,4-triazinium salts[86] and the ring expansion of 2,3-diaryl-1-azirine-3-carboxamides with hydrazine offers a potential route to 1,2,4-triazin-6-ones, via the dihydro compounds described[87]. A related example has been reported[88]. A number of 1,2,4-triazine 4-oxides have been obtained unequivocally by ring-synthetic procedures involving the reactions of 1,2-dicarbonyl compounds with hydrazide oximes, 1,2-dicarbonyl monoximes with amidrazones, and the ring closure of mono-2-ethoxy-methylenehydrazones of 1,2-dicarbonyl compounds[91].

Selective co-trimerisation of acetonitrile and its trichloro derivative (prepared in situ) in the presence of aluminium bromide and hydrogen chloride gives 2-methyl-4,6-bis(trichloromethyl)-1,3,5-triazine in good yield[92]. The anion of urea adds to aryl nitrile in DMSO, giving 4,6-diaryl-2-hydroxy-1,3,5-triazine[93] and a very wide range of triazinethiones has been obtained by the addition of aroyl[94] and ethoxycarbonyl isothiocyanates[95] to, *inter alia*, amidines, isothioureas and guanidines. Benzoyl and thiobenzoyl isocyanates similarly provide triazinones[94] and imidoyl isothiocyanates (17) with amidino and imino compounds form triazinthiones ((18) and (19) respectively) by different routes[96].

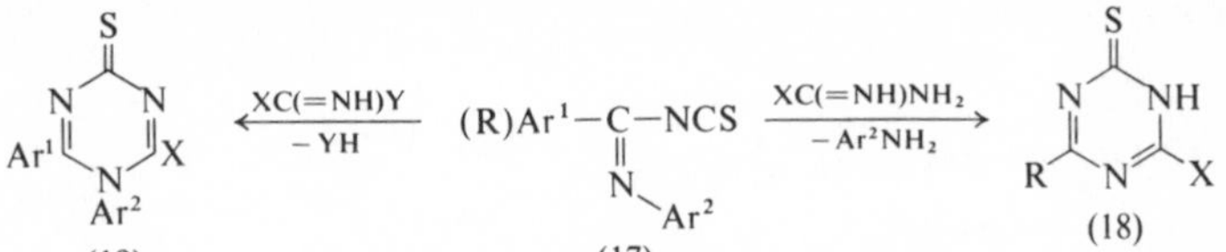

$$\text{(19)} \quad \xleftarrow[-\text{YH}]{\text{XC}(=\text{NH})\text{Y}} \quad \text{(R)Ar}^1-\underset{\underset{\text{Ar}^2}{\overset{\|}{N}}}{\overset{S}{C}}-\text{NCS} \quad \xrightarrow[-\text{Ar}^2\text{NH}_2]{\text{XC}(=\text{NH})\text{NH}_2} \quad \text{(18)}$$

(19)   (17)   (18)

2,4,6-Triaryl-1,3,5-triazines are formed in the chemiluminescence reactions of 2,4,5-triarylimidazoles in the presence of oxygen and a strong base or, better, by the photolysis of the imidazoles in the presence of ammonia and oxygen: the mechanism has been elucidated[97].

4.2 STRUCTURAL AND THEORETICAL STUDIES

4.2.1 Molecular geometry

Accurate crystallographic data are still available for only relatively few simple azine derivatives. 2-Hydroxymethylpyridine N-oxide has a planar ring of the expected geometry, the N—O bond length (0.1322 nm)[98] being shorter than that of pyridine 1-oxide (0.135 nm)[99]. Other pyridine compounds examined include 4-cyanopyridine[100], the 1:1 pyridone:6-chloro-2-hydroxy-pyridine complex[101] and 4-formylpyridine thiosemicarbazone[102].

An accurate analysis of pyrimidin-2[1H]-one has been reported; as in related cases the molecule deviates slightly but significantly from planarity[103]. Various pyrimidines of biological significane have been studied[104-118]. The halogen-bridged structure proposed some time ago[119] for bis(pyrazine) complexes of Co^{II} and Ni^{II} is incorrect: dichlorobis(pyrazine)cobalt(II) contains parallel sheets of cobalt atoms bridged by bidentate pyrazine moieties[120]. Low-temperature x-ray[121] and neutron-diffraction studies[122] of cyanuric acid have given closely similar results, indicating an almost planar trilactam structure.

4.2.2 Dipole moments

A microwave examination of 2-cyanopyridine[123] has been reported, and new microwave data for pyrimidine have both confirmed the general shape and C_{2v} symmetry of the molecule and provided an accurate dipole moment $(2.334 \pm 0.010$ D; $7.785 \pm 0.030 \times 10^{-30}$ Cm)[124]. The dipole moment (microwave) of pyridine N-oxide $(4.13 \pm 0.03$ D; $1.38 \pm 0.01 \times 10^{-29}$ Cm)[125] is marginally lower than values previously measured in benzene solution.

4.2.3 Ionisation potentials

Photo-electron spectroscopy has shed further light on the ionisation potentials of the azines[126, 127] and evidence from related systems[128] supports MINDO calculations that in pyridine the second I.P. corresponds to loss of a lone pair electron[129]. The situation is probably similar in the case of pyrazine but conflicting views are put forward for pyridazine and pyrimidine[127, 129, 130]. Earlier work has been well reviewed[131] and I.P. values have been predicted for 1,3,5-triazine and 1,2,4,5-tetrazine[132]; accurate experimental data are not available.

4.2.4 M.O. calculations

Various MO treatments have been used to calculate the electronic structures of, *inter alia*, pyridine[133], the methylpyridines and their cations[134], cyano-pyridines[135] and 1-methylpyridinium halides[136], and calculated charge distributions in aminopyridines and their mono- and di-cations lend theoretical respectability to the long-accepted initial protonation at ring nitrogen[137]. An SCF MO study of tautomerism in a comprehensive series of heteroaromatic amino and hydroxy compounds provides results in accord with available chemical evidence (e.g. in uracil and cytosine) and the agreement between observed and calculated bond lengths in pyrid-2[1H]-one is reasonable[138]. Related studies on pyridones, pyrimidinones and their thio analogues have been reported[139, 140]; the simple Hückel method is less effective[233].

4.3 SPECTROSCOPY, IONISATION CONSTANTS, POLAROGRAPHY AND ION FRAGMENTATION

4.3.1 N.M.R. spectroscopy

New ¹H n.m.r. data has been reported for many azine derivatives, including amino-[142, 143], alkoxy-[144], fluoro-[145], thioalkyl-[146] and vinyl-pyridines[147], pyrid-2-ones[18, 143, 146, 148] and -thiones[146], α-picolyl-lithium[149], 2-(substituted-alkyl)-4,6-dimethylpyrimidines[150], 4-amino-, 4-hydroxy- and 4-mercapto-pyrimidines (and their derivatives)[56, 151], pyrimidine-2- and pyrimidine-6-sulphonates[152], barbituric acid derivatives[153, 154], 3-substituted-1,2,4-tri-azines[77] and alkoxy-1,3,5-triazines[155]. A range of ring and side-chain quaternary ammonium salts has been examined. The trimethylammonio group shifts ring proton signals to lower field by an extent intermediate between chloro and methylsulphonyl substituents[156]. N-oxides which have received considerable attention include simple pyridine[27] and organosilyl-organogermyl- and organostannyl-pyridine N-oxides[148], pyridazine mono-N-oxides[157] and 1,2-dioxides[158], pyrazine mono-N-oxides (isomeric compounds easily distinguishable by ¹H n.m.r.)[76, 159] and their cations[160], and 1,2,4-triazine N-oxides[91]. Proton shifts and coupling constants in the series pyridone–pyridthione–pyridine confirm that delocalisation (aromaticity) increases in the order shown[146]. The resonance component[161] of the Hammett substituent constant[162] has been used to correlate proton chemical shifts in 3- and 4-substituted pyridines. Simpler LFER plots show that both inductive and conjugative effects are transmitted through the heteroatom of 1-R-pyridinium salts to influence the chemical shifts of 3- and 4-H[163], and N-methyl proton and ^{13}C shifts[164]. Pyridinium ^{14}N shifts in a wide range of 3- and 4-substituted N-methylpyridinium salts have been correlated with σ^+ values[164]. Older values for ring proton couplings in methylpyridines have been corrected, and long-range substituent-ring proton couplings in picolines and pyridine-aldehydes have been measured by multiple resonance techniques[165]. An examination of benzylic coupling in methylpyrazines has shown earlier[166] ring proton assignments to be incorrect[167].

A remarkably simple geometrical model can be used to interpret benzene-induced solvent shifts of the ring protons of pyrazine and pyrimidine and their N-oxides[168]. The effects of chemical shift reagents on the [1]H n.m.r. spectra of pyridine[169, 170] and the diazines[170] has been described; tris(dipivaloyl-methanato)europium (Eu(DPM)$_3$) shifts all the ring proton resonances to lower field, the displacement being unexpectedly large in the case of pyrimidine (especially C-2—H). Loss of fine structure as the amount of reagent is increased limits the applicability of the technique, and signal broadening is even more restrictive on ^{14}N shifts in the presence of Eu(DPM)$_3$ and Yb(DPM)$_3$[171]. [13]C Contact shifts produced by nickel(II)[172, 173] and cobalt(II)[173] acetylacetonates have been observed in pyridine and the methylpyridines.

The effect of protonation on [1]H—[1]H and [13]C—[1]H coupling in pyridine and the diazines has been studied in trifluoroacetic acid solutions[174] and N^+—H—C(2)—H coupling in substituted pyridines has been examined in the same medium[175]. Protonation equilibria of amino- and hydroxy-pyrimidines have been studied in solvents ranging from trifluoroacetic acid to fluorosulphonic acid–antimony pentafluoride–sulphur dioxide mixtures; diprotonation occurs in the latter, the cation structures in both media being derivable from [1]H n.m.r. parameters[176]. Further examples have been reported of characterisable methine (and acetal) σ-complexes in nucleophilic sub-stitutions of nitropyridines[177–180]. Thus, 3,5-dinitropyridine with methoxide ion in DMSO-d_6 rapidly produces (20), the AX$_2$ spectrum of which decays ($t_{\frac{1}{2}} \approx 10$ min) to an AMX spectrum consistent with (21); shift and coupling constant assignments were confirmed by deuteration.

$$\underset{(20)}{\overset{\displaystyle H\ OMe}{O_2N\underset{N}{\bigcirc}NO_2}} \qquad \underset{(21)}{\overset{\displaystyle }{O_2N\underset{MeO\ \ N}{\overset{H}{\bigcirc}}NO_2}}$$

In methanol only (21) could be detected (see p. 100) emphasising again the importance of differential steric and solvation effects in S$_N$Ar reactions[178]. Pyridinium 1-imide cannot be isolated but its existence in solutions of high pH is inferred from the [1]H n.m.r. spectrum (resembling closely that of the isoelectronic N-oxide)[181]. Rate and association constants for pyrid-2-one dimerisation have been measured by several techniques, including [1]H n.m.r.[182].

[19]F N.M.R. parameters have been reported for perfluoropyridines[183] and 5-substituted 2-fluoropyridines; chemical shifts in the latter parallel those in the corresponding p-substituted fluorobenzenes and the same Taft relationship is applicable[184]. [1]H—[19]F Coupling constants in mono- and di-fluorinated pyridines parallel proton–proton values, but with an exagge-

rated range[145]. The structures of some fluoro-pyrimidines[69] and -pyrazines[185] have been deduced by [19]F n.m.r. and fluorine–fluorine coupling constants for di- and tri-fluoropyrazines reported[185].

Natural abundance [13]C n.m.r. spectroscopy has found application in nucleoside[186, 187] and nucleotide chemistry[188] and the technique holds analytical promise[187]. [13]C Shifts in uracil, 5-halogenouracils and thymine have been related to π-charge densities (EHT method) and substituent electronegativity parameters[189]. An almost linear relationship between ^{14}N shifts and π-charge densities (PPP calculations) at ring nitrogen has been claimed for a wide range of N-heteroaromatics including pyridine (and some of its derivatives), the diazines and triazines[190].

[1]H—[31]P Coupling constants (including the first *para* coupling of this type to be reported) have been measured for tri-2-pyridylphosphine and its oxide and sulphide[191, 192].

New ^{14}N quadrupole spectra to be reported and analysed include those of some nitro- and trimethylsilyl-pyridines[193], the pyridine-iodine mono-chloride complex[194], a range of substituted pyrimidines[195] and pyrazine[196].

4.3.2 E.S.R. spectoscopy

Electron spin resonance spectroscopy demonstrates pyridyl radical formation when pyridyl halides are irradiated ($\lambda > 500$ nm) in argon matrices containing sodium[197]. Well-resolved spectra have been obtained from 2- and 4-vinylpyridine radical anions[198]: INDO calculations are in good agreement with proton hyperfine splittings in a number of pyridine anions[199]. The e.s.r. spectrum of the cation (22), from di-(4-pyridyl) ketone and zinc, has been reconstituted fully[200]. The spectrum of 5-methylpyrimidine radical anion

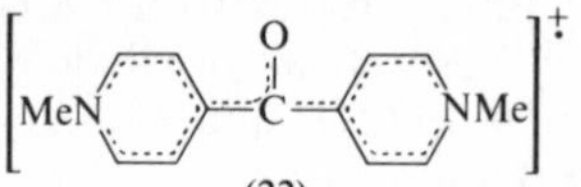

(22)

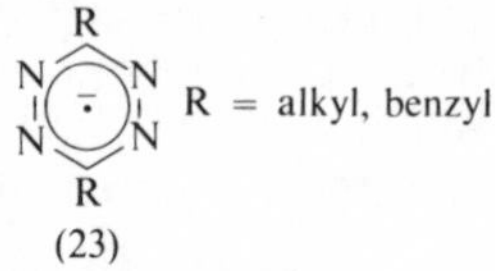

(23)

R = alkyl, benzyl

has been analysed, and more complete assignments made in the case of pyrimidine itself[201]. The spectra of alkyltetrazine radical anions (23) (from the tetrazines and potassium t-butoxide in DMSO) have been outlined[202].

4.3.3 I.R. spectroscopy

All fundamental frequencies in the infrared and Raman spectra of the isomeric dimethylpyridines have been assigned[203]. A re-examination of the vibrational spectrum of pyrimidine has confirmed, substantially, an earlier interpretation[204] but new assignments have been made in the case of 1,2,4,5-tetrazine[205–207]. The vibrational spectrum of pyridine 1-oxide has been analysed with the aid of deuterated analogues[208]. A number of bands in the infrared spectra of N-nitroaminopyridines[209], pyridine carboxylic acid N-oxides[210], 2,2'-bipyridyl 1-oxide cations[211] and anion radicals from 4-nitropyridine and its N-oxide[212] have been identified and both infrared and Raman data have been reported for 2,6-diaminopyridine[213], pentachloropyridine[214], many substituted pyrimidines (200–1700 cm^{-1} region)[215], pyrimidinium and pyrazinium halides[216] and complexes[216, 217] and 2,4,6-trichloro-1,3,5-triazine (for which all fundamental modes were assigned)[218].

Routine infrared spectra have been reported for almost all new compounds described (and many others), but a number of specific applications deserve comment. The appearance of both O—H and N—H stretching bands in the spectra of pyrid-2[1H]-one and -4[1H]-one hexachloro-antimonates and -stannates confirms protonation on oxygen but the solid-phase spectra of these salts with bulky and less nucleophilic anions are sharper and more easily interpreted than those of, for example, the hydrohalides[205] and the technique has been extended to pyrimidines and pyrimidones. (The latter, as expected, undergo protonation on the ring nitrogen not involved in a lactam structure)[219]. Concentrated solutions of 6-chloropyrid-2-ones in chloroform contain mainly dimers, and in the solid phase lactam forms predominate[144] (compare p. 100). A sharp singlet (v_{NH}) at 3387 cm^{-1} in the spectrum of 2-ethoxypyrimid-4-one appears to confirm that, in chloroform, the compound exists in the o-quinonoid form[220], but 3-methylthio-1,2,4-triazin-4[2H]-one provides the first exception to the general rule that o-quinonoid forms have v_{NH} in the range 3360–3420 cm^{-1} whereas in p-quinonoid structures v_{NH} falls between 3415 and 3445 cm^{-1}, since v_{NH} of the triazinone appears at 3401 cm^{-1}, perhaps owing to the proximate lone pair[220]. (However, the rule is obeyed in 3-phenyl-1,2,4-triazin-5[2H]-ones, in which similar p-quinonoid forms predominate[83].) The solid phase spectrum of 2,3,5-trifluoro-6-hydroxypyrazine shows no carbonyl absorption, in sharp contrast to 3-hydroxy-4,5,6-trifluoropyridazine[185].

N—O Stretching frequencies have been reported for a number of N-oxides including those of polysubstituted pyridines[27] and new trialkylsilyl-, trialkyl-germyl- and trialkylstannyl-pyridines[148], the hitherto unaccessible pyridazine dioxides[158] and some 2-substituted pyrazine 1-oxides[159]. The intensity of the v_{N-I} band in complexes of iodine monochloride with a range of methyl-pyridines appears to provide a sensitive measure of base–halogen interaction even though the intensity of the v_{I-Cl} band shows little variation on changing base[221].

4.3.4. U.V. spectroscopy

Ultraviolet absorption spectrophotometry continues to attract attention on both fundamental and diagnostic levels. Singlet and triplet states (370 and 445 nm) have been identified in the high resolution spectrum of pyrid-azine[222] and vibrational assignments made in the phosphorescence spectrum of pyrimidine[223]. Emission, as well as absorption, spectra have been obtained from pyrazine in both a solid matrix and the vapour phase[224, 225]. New vapour-phase absorption and fluorescence spectra have been reported for 1,2,4,5-tetrazine[226, 227]. The polarised single-crystal absorption spectrum of 1-methyluracil (only the second such in the nucleic acid base series) revealed an $n \rightarrow \pi^*$ transition at 264 nm[228]: interesting further applications of the technique can be expected and the spectra of uracil and thymine have recently been described[229]. A new analysis of the spectrum of 2-aminopyridine indicates a $\pi \to \pi^*$ origin for the band at 298 nm, rather than the accepted $n \to \pi^*$ promotion[230] and a rotational analysis of the 342 nm band of pyridine 1-oxide has been interpreted in favour of a $\pi \to \pi^*$ transition[231].

Transition energies and, in some cases, oscillator strengths have been calculated for a wide variety of N-heterocyclic molecules including all the monocyclic azines[132], pyridine[236] and the methylpyridines[232], pyridones[233, 234] pyridonimines[235], the diazines[236, 237], biologically significant pyrimidines[238], aminopyrimidines[239], 5-hydroxypyrimidine[240] and 1,2,4,5-tetrazine[237].

Routine ultraviolet examinations, often coupled with pK_a determinations, have been reported for large numbers of azines including 3-nitropyridines[156], pyrid-2-ones[18, 144] and their annelated derivatives[241–243], aminopyridones[244], dipyridylethylenes[245], stilbazoles[246], 4-aminopyrimidines[151], 2-(substituted-amino)-pyrimidines[247], 2-iminopyrimidines[248], a large number of triamino-pyrimidines[249], 2-hydroxypyrimidines and their O- and N-methyl analogues[61, 250], 4-hydroxypyrimidines[151], aminopyrimidones[244], 2-halogeno-pyrimidines[247], diaminouracil derivatives[251], 4-mercaptopyrimidines[151], pyr-imidine-2- and pyrimidine-6-sulphonic acids[152], alkyl- and cycloalkyl-pyrazines[252], 1,2,4-triazin-5-ones[83], and 1,3,5-triazinthiones and their thio-ethers[95]. The spectra of several silicon-, germanium- and tin-containing pyridines have recently been described[253, 254]. New N-oxides whose spectra have been recorded include several pyridine 1-oxides[27] pyridazine 1,2-dioxides[158] and pyrazine- and amino-pyrazine mono-N-oxides[255].

The rate of formation of (21) (λ_{max} 455 nm) from 3,5-dinitropyridine has been measured by the stopped-flow technique: in solvent mixtures rich in DMSO the thermodynamically less stable (20) λ_{max} 490 nm) is observed and solvation by DMSO stabilises the latter to such an extent that in 92.6% DMSO, $t_{\frac{1}{2}}((20) \to (21)) \approx 5$ s[180] (see p. 97).

Hydroxy tautomers predominate in solutions of 6-chloropyrid-2-ones; solvent effects have been examined[144]. The existence of 3-hydroxy- and 3-mercapto-pyridazine in lactam and thiolactam forms has been confirmed[256]. In aqueous solution 2-ethoxypyrimidin-4-one comprises an equimolar mixture of o- and p-quinonoid forms, but in chloroform 2-ethoxypyrimidin-6[1H]-one predominates[220]. The ultraviolet spectrum of 2,3,5-trifluoro-6-hydroxypyrazine closely resembles that of the 6-methoxy analogue, confirming the hydroxy form[185].

Further studies have been reported on charge-transfer interactions between pyridines and, for example, halogens and interhalogens[257–261], tetracyanoethylene (TCNE)[262] and boron and silicon halides[263]. Earlier spectra are receiving critical attention; thus, for example, care is needed in identifying TCNE complexes since the band observed at 400 nm in the presence of pyridines is due not to complex formation but to the penta-cyanopropenide ion[262]. The pyridine radical anion has a characteristic maximum near 340 nm, the exact position varying with the counter-ion[264].

4.3.5 Ionisation constants

The Hammett (σ_p) constant of the 2-pyrazinyl group has been assessed spectrophotometrically as $c.$ 1.0[265].

Despite the factors which complicate the assessment of their absolute significance, pK_a values provide valid evidence of gross structural features as well as criteria of purity and numerous new data have been reported, including values for amino-, aminonitro- and nitropyridines (both $pK_{a(1)}$ and, where applicable, $pK_{a(2)}$[266]), 6-chloro-2-methoxypyridines[144], amino-[244] and chloro-pyrid-2-ones[144], several annellated pyrid-2-ones[241-243], hydroxy- and mercapto-pyridazines and their N-, O- and S-methyl derivatives[256] and aminopyrimidinones[244]. Revised values (-8.1 and -6.7 respectively) have been obtained for the second protonation of 2- and 4-aminopyridine[267], but spectrophotometric $pK_{a(2)}$ values for a number of 2,4,6-triamino-pyrimidines have been shown invariably to be higher in sulphuric acid than in hydrochloric acid[249], probably owing to differential ion association, so that considerable caution will be required in interpretation. Hammett σ constants for a series of β-substituted-vinyl groups have been obtained from pK_a data for the corresponding 3- and 4-substituted pyridines[268]. Aromatic stabilisation energies, for example, of pyrid-2-one and its analogues, can be calculated conveniently from tautomeric equilibrium constants in dilute aqueous solution (to minimise intermolecular association). The stabilisation energies, relative to pyridine, of pyrid-2-one, -thione and -onimine are 27.2, 22.15 and 35.5 kJ mol^{-1} respectively; the corresponding figure for the methide is 68.4 kJ mol^{-1}, in line with qualitative prediction[269]. The *general* applicability of pK_a values in assessing tautomeric equilibria has been questioned[241].

Thermodynamic pK values and enthalpies and entropies of ionisation for proton loss in uracil, cytosine, thymine and their nucleosides have been reported, correcting earlier values in some cases. ΔH for the protonation, on the other hand, of cytosine and cytidine confirm N-3 as basic centre and the method is probably capable of wide application[270]. The basic strengths of a number of pyridinium, pyrimidinium and pyrazinium ylides (from N-phenacyl quaternary bromides) allow predictions of the stabilisation of the ylides (e.g. (24)) by ring nitrogen atoms[271].

(24)

The first excited state ionisation constants (pK_a^*) of 2-, 3- and 4-amino-pyridine are 9.4, 11.7 and c. 11.9 respectively: as in related cases the mono-protonated species are thus weaker acids than in the ground state[272]. 1,2,4,6-Tetramethyl-3,5-dicyanopyridinium p-tosylate is a relatively strong carbon-acid ($pK_a = 7.8$)[273].

4.3.6 Polarographic behaviour

A much needed and definitive account of the polarographic behaviour of azines shows that meaningful $E_{\frac{1}{2}}$ values (in DMF) must be obtained by cyclic voltammetry rather than simple polarographic reduction. The (reversible) potentials are well correlated by calculated (CNDO) energy differences[274].

4.3.7 Mass spectrometry

Mass spectral data are available for a wide range of pyridine derivatives but there is still a paucity of information on other azines. Fragmentation patterns have been established for alkyl-[275, 276], aryl-[277, 278], carboxyl-[279–281], methoxycarbonyl-[282], fluoro-[283], hydroxy-[284] and nitramino-[285] pyridines and 4-hydroxypyrid-2-ones[286], together with some alkylpyridine N-oxides[275] and N-substituted iminopyridinium betaines[287, 288]. 3,4-Pyridyne (and its dimer) have been identified in the decomposition of pyridine-3-diazonium-4-carboxylate[289]. A detailed examination of the fragmentation of 4-methylpyridine casts doubt on the intermediacy of an azatropylium ion[276] although this is often postulated by analogy with toluene[287]. Fragment ion abundance in the spectra of alkylpyridines and their N-oxides is subject to a marked substituent effect; McLafferty rearrangements are characteristic of ethyl- and butyl-pyridines (and their N-oxides)[275]. p-Fluoro-labelling studies indicate no randomisation prior to molecular ion fragmentation in pentaarylpyridine[278]. Pyridine-2-carboxylic acids undergo initial decarboxylation[284] (in contrast to benzoic acids) but in the case of pyridine-4-carboxylic acid exchange between carboxylic and β-hydrogens precedes elimination of hydroxyl and water: in the 3-isomer similar exchange involves only the α-hydrogen[280]. 3-Hydroxypyridines suffer loss of carbon monoxide[284].

Loss of nitrogen has been confirmed as the major cleavage route in pyridazine and a number of derivatives but an alternative pathway (giving $M-HCN$ or $M-RCN$) has been suggested for some. The representative fragmentation of pyridazine N-oxides involves the formation of $M-30$ $(M-NO)^+$ ions, thus resembling that of 1,2,4-triazine N-oxides; the usual $M-16$ $(M-O)^+$ and $M-17$ $(M-OH)^+$ ions may or may not both be observed[288].

The spectrum of tetrafluoropyridazine has been reported, with metastable ions corresponding to the major fragmentations $(C_4F_4N_2^+ \longrightarrow C_4F_4^+ \longrightarrow C_3F_3^+$ and $C_3F_2^+)^{[69]}$.

Among pyrimidines, the pyrimidones continue to attract most attention. An extensive and detailed study of 6-methylpyrimidin-4-ones[290] has appeared, and N-1- and N-3-substitution in N-alkyluracils can be distinguished conveniently[291]. The mass spectra of barbiturates has found further analytical application[292]. The spectra of some (1-pyrimidinyl)amino acids[72] and thiazolo-[3,2-c-]-pyrimidinium salts[293] have been reported.

Fragmentation patterns have been proposed for several 2-methoxyalkylpyrazines[294, 295] and their methylthio analogues[295]. 2-Methoxy- and 2-methylthio-3-methylpyrazine lose water and hydrogen sulphide from their molecular ions, indicating a prior skeletal rearrangement[295]. A number of naturally-occurring alkyl- and alkoxy-pyrazines have been characterised by their mass spectra[296] and the spectra of some 2-substituted pyrazine N-oxides have been reported[297]. p-Fluoro-labelling establishes no randomisation prior to major fragmentation in 2,5-bis-(p-fluorophenyl)-3,6-diphenylpyrazine[278].

The spectra of 3,6-diaryl-[78] and further 5,6-diphenyl-1,2,4-triazine derivatives[298] have been reported and an earlier interpretation[299] of the spectra of 3-amino-1,2,4-triazine N-oxides (based on wrong structural assignments) has been corrected[300]. 1,3,5-Triazine derivatives have been examined systematically by two groups[301, 302]; the parent compound decomposes by the stepwise loss of two molecules of hydrogen cyanide[301]. Cyanogen chloride is lost similarly in the initial fragmentation of 2,4,6-trichloro-1,3,5-triazine, but aminochlorotriazines instead lose a chlorine atom[302]. Substituent effects are marked and it has been suggested that imino tautomers play a significant role in the breakdown of di- and tri-amino derivatives[301]. More limited studies on 6-substituted 2,4-bis-(m-aminoanilino)-1,3,5-triazines have been reported[303].

A detailed study of the decomposition of metastable ions from the diazines and 1,3,5-triazines has been made with the aid of deuterated analogues[304]. Limited negative-ion studies have been made on, for example, pyridine and pyrazine, their N-oxides, and some substituted derivatives, but insufficient data are available, as yet, for diagnostic purposes[305].

4.4 SUBSTITUTION REACTIONS

4.4.1 Electrophilic, at ring carbon atoms

Reactions of this type remain relatively rare among π-deficient azines, but powerfully electron-releasing substituents permit attack at reasonable rates: thus, in an extreme case, nitration (see below) of 6-hydroxypyrid-2-ones takes place at, or near, the encounter rate.

4.4.1.1 Acylation

Strongly electrophilic acyl halides (e.g. trifluoroacetyl chloride) with 1-lithio-2-phenyl-1,2-dihydropyridine give 5-acyl-2-phenylpyridines with only minor amounts of N-acyl derivatives[306]. Direct acylation of unactivated azines remains unknown but 4,6-dihydroxy-2-methylthiopyrimidine undergoes Vilsmeier formylation[307] and, in a reaction which clearly deserves further investigation, 4-amino-6-anilino-2-methylpyrimidine gives the 5-formyl compound on heating with dimethyl formamide and an equimolar amount of 4-amino-6-chloro-2-methylpyrimidine; the fate of the latter has not been established[308].

Intramolecular acylations, for example, of pyridyl-[309, 310] and pyrimidinyl-aminomethylenemalonates[311], are more common but require vigorous conditions. The report that 2-aminopyridine with ethyl α-methylacetoacetate in PPEt (or PPA) gives 2,3-dimethyl-1,8-naphthyridone[312] is incorrect[313]. Under Kolbé–Schmidt conditions 6-(p-fluorophenyl)-3-hydroxypyridine forms the 2-carboxylic acid in moderate yield[314]. Both 6-aminouracil and its 1,3-dimethyl derivative are attacked at C-5 by ethyl isocyanatoformate, giving 5-[N-(carbethoxy)carboxamide] compounds[315]; uracil and cytosine are not[316].

4.4.1.2 Alkylation and substituted-alkylation

The conversion of (25) into (26) (in good yield) apparently involves a reaction of the Friedel–Crafts type[317]. Several indirect alkylations of pyridines have appeared. In the simplest (and most promising), 3-alkylpyridines are obtained by converting pyridine into lithium tetrakis-(N-dihydropyridyl)aluminate (see p. 112) and treating this with the appropriate alkyl halide. 5-Alkyl-2-phenylpyridines can be obtained similarly from lithiophenyldihydropyridine (above)[318] and the lithio-2-n-butyl analogue, with benzophenone, gives 3-(2-n-butylpyridyl)diphenylcarbinol[319]. The reductive alkylation of (27) with 1-bromo-1-phenylpropane in the presence of sodium in liquid ammonia forms (28), but the reaction appears not to be general[320]. Further examples have been

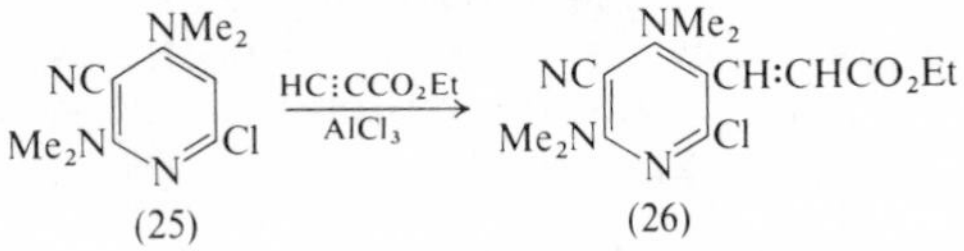

(25)   (26)

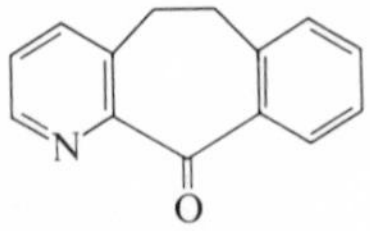

(27) 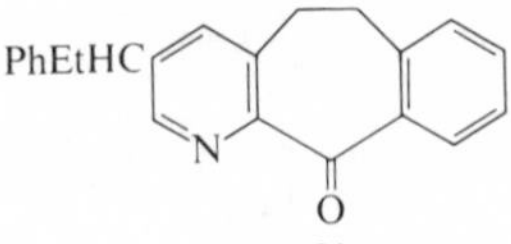   (28)

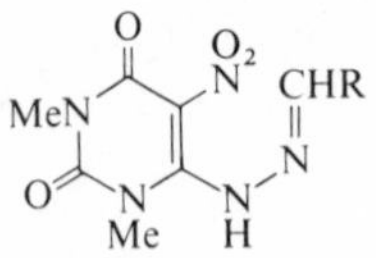

(29) 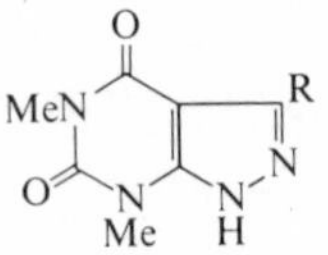   (30)

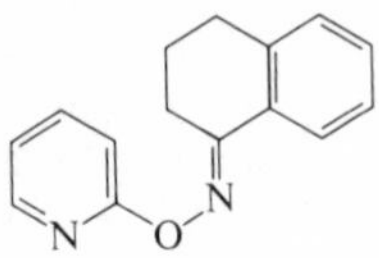

(31) 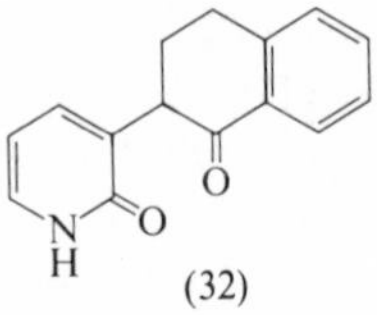   (32) 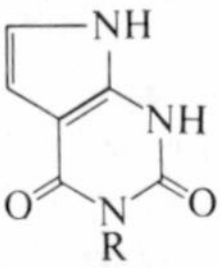   (33)

reported of indolisations of 3- and 4-pyridylhydrazones[568] and, in some cases, their N-oxides[321]. The ring closure of (29) to (30) (see also p. 123) involves an initial electrophilic attack[322]. The thermal rearrangement of the O-(2-pyridyl) oxime (31) to (32) likewise involves initial electrophilic attack on the pyridine ring[323]. The thermal process avoids the deactivation which would accompany acid catalysis.

6-Aryl-3-hydroxypyridines undergo ready hydroxymethylation at C-2[314] and 2-aryl-3-hydroxypyridines at C-6, with subsequent attack at C-4[324]. 1-Methylpyrid-2-one is hydroxymethylated at C-5[325]: 1-methyl-5-hydroxy-uracil at C-6[326].

New examples have been reported of the reaction between aldehydes and barbituric acid[327], and pyrrolopyrimidines (33) are obtained when 3-R-sub-stituted-6-aminopyrimidine-2,4[1H,3H]-diones are heated with chloro-acetaldehyde in alkaline solution[328].

Attempted Mannich reactions on 3-hydroxypyridazine 1-oxide with formaldehyde and amine hydrohalides yield only 6-halogeno-pyridazin-

3[2H]-ones, but substitution of free amine for salt provides 6-(substituted-amino) methyl derivatives[329].

4.4.1.3 Arylation

An interesting example of an electrophilic arylation probably via an ylide, is provided by the base-catalysed rearrangement of 1-aryloxypyridinium salts, e.g. (34) ⟶ (36)[330]. 1-Pyridylbenz[a]anthracenes do not cyclise under Scholl conditions[331].

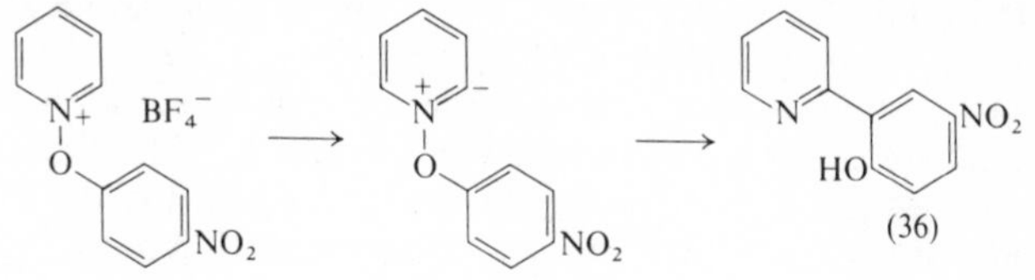

(34)　　　　　　　(35)　　　　　　(36)

4.4.1.4 Deuteration

The known resistance of pyridine to electrophilic deuteration extends, not surprisingly, to the 1-aminopyridinium ion which decomposes at 300 °C in 99% D_2SO_4[181]. Substituent effects in base-catalysed H—D exchanges in pyridinium ions are simply analysed in terms of the Taft σ_1 parameter. Buffer-base catalysis is negligible[332, 333]. Rate profiles for exchange at C-3 in 2-aminopyridines have been obtained[334]. 1-Methylpyrid-4[1H]-one undergoes base-catalysed deuteration by direct proton removal from C-2; 1-methylpyrimidin-4[1H]-one, in contrast, is deuterated initially at N-3, and subsequent deprotonation forms the ylid which then forms 1-methyl-pyrimidin-4[1H]-one-2d[335].

4.4.1.5 Halogenation

Halogenation remains the electrophilic substitution encountered most frequently in the series. Further kinetic data has been assembled on the bromination of 5-substituted 2-aminopyridines, and rate profiles in aqueous sulphuric acid show the free base to be involved: nuclear bromination occurs exclusively at C-3[336].

Pyridine itself can be brominated indirectly by treating lithium tetrakis (N-dihydropyridyl)aluminate (prepared in situ) with bromine; moderate yields of 3-bromopyridine are obtained[318]. 2-Aryl-3-hydroxypyridines are reported to be sufficiently activated to undergo iodination at C-6[337] and the activating influence of the hydroxyl group is apparent also in the ready chlorination in aqueous solution of 4-hydroxy-6-methylpyrid-2[1H]-one at C-3, and the conversion of the 3-nitro analogue into 5-bromo-4-hydroxy-6-methyl-3-nitropyrid-2[1H]-one with bromine in acetic acid[338]. 1-Hydroxy-4,6-dimethylpyrid-2[1H]-one is easily halogenated at C-3 and C-5, the analogous 3-carbonitrile at C-5[9]. 5-Bromonicotinic acid is obtained conveniently by the action of bromine on nicotinyl chloride hydrochloride in refluxing thionyl chloride[339], lending weight to the proposal that this is not a simple electrophilic process[340]. Both tetrachloropyridine-4-sulphonic acid (as the potassium salt) and the sulphonyl chloride give pentachloropyridine on treatment with phosphorus pentachloride[341].

Increased activation is required normally in the diazines, but bromine in acetic acid converts both 4-amino-3,6-dichloro- and 4-amino-6-chloro-3-ethoxy-pyridazine into their 5-bromo derivatives in good yield[342]. Pyrimidines bearing a total of three amino or hydroxy substituents are known to undergo easy bromination, which has been carried out in aqueous alkali, aqueous acetic acid or dimethylformamide[343]. N-Bromosuccinimide in dimethylformamide brominates methyl (2-acetamido-4[3H]-pyrimidon-6-yl)acetate at C-5 only, under suitable conditions; N-chlorosuccinimide is less selective[344]. Direct fluorination at C-5 of uracil and cytosine (and their derivatives) occurs on treatment at low temperatures with trifluoromethyl hypofluorite followed by base[345].

Direct chlorination (with concomitant decarboxylation) of pyrazine-2,3-dicarboxylic acid by heating with phosphorus pentachloride under pressure provides a convenient large scale synthesis of tetrachloropyrazine[185]. Bromine in acetic acid converts 3-aminopyrazine-2-carboxamide into its 6-bromo derivative almost quantitatively[346]. On treatment with sulphuryl chloride in dimethylformamide 2-isopropyl-6-methylpyrazine forms both the 3- and 5-chloro compounds[296].

4.4.1.6   Nitration and nitrosation

Kinetic studies have been reported on the nitration of hydroxypyridines[347], pyridones[347, 348] and hydroxypyridones and their methyl derivatives[347], and pyrimidinones[250]. Rate profiles indicate that 3-hydroxy- and 3-methoxy-pyridine are nitrated as their conjugate acids (at position 2) whereas 6-hydroxypyrid-2[1H]-one and its methyl derivatives react near the encounter rate in free base form (at position 3)[347]. 4-Hydroxy-6-methylpyrid-2[1H]-one is attacked likewise at C-3[338]; 3-hydroxy-2-methoxypyridine at C-4[349]. In a further striking demonstration of the value of rate-acidity studies it has been shown that pyrid-2-one can be nitrated selectively at C-3 or C-5 under conditions chosen with reference to the appropriate rate profiles[348]. Attempts to nitrate pyridinium 1-benzenesulphonylimide (and related compounds) lead only to pyridinium 1-nitroimide[181]. 3-Cyano-4,6-dimethyl-1-hydroxypyrid-2-one is nitrated readily at C-5[9].

In contrast to pyrimidin-2[1H]-one, pyrimidin-4[1H]-one resists nitration[250]; the former, and its methyl derivatives, is attacked in free base form[250] at C-5. Two activating substituents at C-2, -4 or -6 normally guarantee successful nitration at C-5, but although 6-(p-halogenophenyl)uracils are attacked solely at this position by fuming nitric acid in acetic acid–acetic anhydride, concomitant attack on the phenyl ring takes place with fuming nitric acid alone or in fuming sulphuric acid[56]. Solutions of sodium nitrite in sulphuric acid (70%) convert activated pyrimidines, for example the

4-amino-6-hydroxy-2-phenyl compound, into 5-nitro derivatives by initial nitrosation. Oxidation of the 5-nitroso compounds *in situ* was established with authentic nitroso-pyrimidines under similar conditions[350]. Direct nitrosation at C-5 of pyrimidines with two or three activating substituents occurs easily, and a number of new examples of the reaction have appeared[351-354]. The nitrosation of 6-amino-1-methoxyuracil is quantitative[355]. Some 2-amino-methylpyrimidines undergo side-chain nitrosation invariably, with modification of the amino group[351].

4.4.1.7 Miscellaneous electrophilic substitutions

The direct thiocyanation of 6-amino-2,4-dihydroxypyrimidine provides the 5-thiocyanato derivative rather than the thiazolopyrimidine claimed earlier[343]. Pyrimidines bearing at least two strongly electron-releasing substituents give 5-alkylthio derivatives in good yield when treated with n-alkanesulphenyl halides (generated *in situ*)[356]; uracil reacts similarly with trifluoromethyl-sulphenyl chloride[90]. Methanesulphenyl chloride is probably the electrophile involved in the conversion of activated pyrimidines into their 5-methylthio analogues with DMSO in the presence of either acetyl chloride or (better) chloromethyl methyl ether [356, 567].

4.4.2 Electrophilic, at ring nitrogen atoms

4.4.2.1 N-Acylation

1-Acetylpyridinium ions have been detected and characterised in pyridine–acetic anhydride mixtures; their formation is rate-determining in pyridine-catalysed acetylations[357]. Acylpyridinium ions are stabilised extensively by electron-donating substituents[358, 359]. Intramolecular N-acylation has been reported[360].

Quaternisation with alkyl and aralkyl halides (the Menschutkin reaction) continues to attract mechanistic attention, both as a probe in solvent studies[361] and in its own right. The 'methiodide' from iodomethane and 2,6-di-t-butylpyridine is (on ^{1}H n.m.r. evidence) the base hydroiodide, but the normal reactions of less severely hindered pyridines are markedly accelerated by pressure[321]. 3- and 4-Dimethylaminopyridine are attacked by methyl iodide at ring nitrogen, but the 2-isomer reacts at the exocyclic nitrogen[362]. Substituent effects on rates and positions of quaternisation in pyrazine derivatives have been assessed by a competitive method; in some cases rates were determined relative to that of pyridazine, using a ^{1}H n.m.r. method[363]. C–T interactions and photodecomposition interfere with spectrophoto-metric measurements of the absolute rate of pyridazine methiodide formation[564]. Numerous routine quaternisations have been reported, for example of pyridines[156, 364-370] and pyrimidines[156, 248, 371, 570], but electronic factors may sometimes prevent reaction, as in the case of attempted methiodide formation with 4-azido-3-nitro- and 4-azido-3,5-dinitro-pyridine[372]. Further use has been made of trimethyloxonium fluoroborate to overcome resistance to quaternisation[373]. The hindered (and deactivated) 3,5-dicyano-2,4,6-tri-methylpyridine forms the quaternary tosylate on fusion with excess methyl-toluene-*p*-sulphonate[273]. Ring closure by intramolecular attack at ring nitrogen has been used to provide fused ring systems in many instances[57, 374–390]; an interesting example is the conversion of the aldoxime (37) into (39) via the di-anion (38)[376].

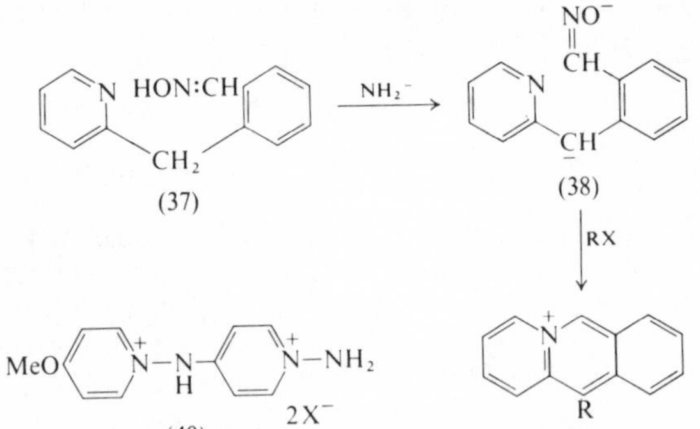

In a reaction related to the Hilbert–Johnson glycosidation pyridine and pyrimidine *O*-trimethylsilyl derivatives form *N*-vinyl lactams on treatment with vinyl acetate, mercuric acetate and sulphuric acid[391], and the methyla-tion of 6-substituted uracil *O*-trimethylsilyl derivatives provides a model for glycosidation[392]. Renewed interest in stable *N*-ylides has yielded an extensive range of phenacylpyridinium bromides[271, 393] and *N*-dicyanomethylides have been obtained from TCNE oxide and pyridazines[394].

1-Aminopyridinium salts are obtained conveniently from the appropriate base and hydroxylamine-*O*-sulphonic acid but 4-methoxypyridine forms (40) by nucleophilic replacement[181]. Both phosphoryl chloride-triethylamine and phosphoramidate ion provide 1-phosphorylpyridinium salts[395].

Acetylation of pyrid-4-one with acetic anhydride–pyridine (or via the thallium salt) gives solid *N*-acetylpyrid-4-one rather than 4-acetoxypyridine although in methylene chloride solution the two are present in roughly equal amounts at equilibrium[396].

4.4.2.2 N-Alkylation

Earlier, illuminating studies on the alkylation of heterocyclic ambident anions have been extended to 5-carbethoxy- and 5-nitro-2-hydroxy-pyridines. The interplay of effects is complicated: although, in general, the electron-withdrawing groups favour *N*-alkylation, relative to pyrid-2-one, the effect is more pronounced with silver salts than sodium salts. Solvent effects are important in the reactions of the former which, under heterogeneous conditions, undergo *O*-alkylation only. Steric influences are important under homogeneous conditions and *O*-alkylation is favoured by bulkier alkylating agents[397]. Alkali metal salts of 4-hydroxypyrimidine are methylated preferen-tially at N-3 in cation solvating media, but the proportion of N-1 substitution increases in hydrogen-bonding solvents. (Silver salts undergo mainly

O-alkylation, but yields are low). Steric effects parallel those observed in the pyridine series but 2-trifluoromethyl-4-hydroxypyrimidine with even methyl halides (in dimethylformamide) gives mainly 4-methoxy-2-trifluoromethyl-pyrimidine; ethyl iodide leads to O-ethylation exclusively[398]. 4-Methylthio-pyrimidin-2[1H]-one (as sodium salt) is alkylated at N-1 by N-(2-bromo-ethoxy)phthalimide[399]. Careful re-examination of the methylation of uracil with diazomethane shows that 1,3-dimethyluracil (73%), 2-methoxy-3-methylpyrimidin-4-one(18%),4-methoxy-1-methylpyrimidin-2-one(4%)and 2,4-dimethoxypyrimidine(5%)are all formed (and thymine behaves similarly). Significantly, the proposed correlation between N:O attack and infrared (amide) frequencies[400] appears untenable[401]. 5-Nitrouracil normally undergoes methylation at N-1, so that the observation that prior conversion into the di-anion allows selective alkylation at N-3 (the more nucleophilic centre) has preparative value[402]. 3-Methyldithiouracil with one equivalent of iodo-methane and alkali gives 1-methyl-2-methylthiopyrimidin-6[1H]-thione[64].

5,6-Diphenyl-1,2,4-triazin-3[2H]-one forms Mannich bases at N-2 and the corresponding 6-phenyl-3,5-dione gives 2,4-bis-(substituted-amino-methyl) derivatives under mild conditions[403]. 5-Bromo-6-azauracil is less reactive (and more selective) than 5-bromouracil towards benzyl chloride, giving 5-bromo-3-benzyl-6-azauracil; in contrast cyanoethylation of both the uracil and its aza-analogue takes place at N-1[404]. Only N-methylation is observed when 6-phenyl-1,2,4-triazin-3,5[2H,4H]-dione is treated with ethereal diazomethane[403], and with the same reagent in ether–dioxane the 1,2,4-triazinonethione (41) gives mainly (42); in ether–ethanol, however, (43) predominates and, when (41) is treated with iodomethane in the presence of methoxide, (43) is the only product[405]. The preference of 1,3,5-triazinethiones

(41)   (42)   (43)

for S-alkylation and -arylalkylation is well known[94, 95]; where, as in the benzylation of triazinone-thiones, competition exists amongst nitrogen, oxygen and sulphur, S-benzylation predominates[95].

2-Hydroxypyridine forms the sulphamate on treatment with sulphur trioxide-dimethylaniline, but 4-hydroxypyridine forms both the sulphamate and 4-pyridyl sulphate[406].

4.4.2.3 N-Oxidation

Pyridines are reported to form their 1-oxides (in yields superior to those by peracid treatment) when treated with t-amyl hydroperoxide–molybdenum pentachloride; tetramethylpyrazine gives its di-N-oxide in excellent yield[407]. Good yields of pyridine and α-picoline N-oxide are also obtained by oxidation with hydrogen peroxide and sodium tungstate in the presence of EDTA[408]. Very weakly basic N-heteroaromatic compounds such as pentachloropyridine (and tetrachloro-pyridazine and -pyrazine) form N-oxides in good yield when treated with mixtures of hydrogen peroxide and concentrated sulphuric acid in acetic or trifluoroacetic acid[409]. Tetrachloropyrazine di-N-oxide has been prepared similarly[410]. Numerous N-oxidations under routine conditions have been reported. Peracetic acid converts 4-trialkylsilyl- and 4-trimethyl-germyl-pyridine into the corresponding oxides and 4-trimethylstannyl-pyridine N-oxide has been obtained using permaleic acid[411]. m-Chloro perbenzoic acid converts a wide range of pyridines into the oxides[27] (and the same reagent has been used to prepare cytosine arabinoside 3-N-oxide[412] and, from 2,4-diamino-5-(3,4,5-trimethoxybenzyl)pyrimidine, a mixture of N-1 and N-3 monoxides[51]). Pyridine-aldehyde N-oxides are obtained conveniently from the acetals with peracetic acid under anhydrous conditions, followed by hydrolysis[413]. 5-Bromonicotinamide forms the 1-oxide in excellent yield under similar conditions[339]. The N-oxidation of 2- and 4-dimethylaminopyridine parallels methiodide formation (see above) but the 3-isomer shows an interesting variation in that oxidation gives the exocyclic oxide in good yield[362]. Careful investigation has shown that 1,2-dioxides are formed in low yields from pyridazines oxidised under a variety of conditions[158] but hydrogen peroxide in polyphosphoric acid smoothly converts pyridazine into the mono-N-oxide, free of dioxide. Simple aminopyridazines form the corresponding nitropyridazine N-oxides in good yields under similar conditions. However, whereas 6-amino-3-substituted pyridazines are oxidised normally at N-1, the newer reagent converts, for example, 3-amino-6-chloropyridazine into 6-chloro-3-nitropyridazine 1-oxide, opening a route to the less easily accessible 3-nitropyridazine 1-oxide series[157]. p-Nitro-perbenzoic acid has been used to prepare other pyridazine mono-N-oxides[414]. Mono- and di-chloropyrazines can be oxidised selectively with peracetic or permaleic acid, but pertrifluoroacetic acid converts 2,5-dichloro-3,6-dimethylpyrazine into the dioxide in good yield. Under similar conditions the further oxidation of 3-chloro-2,5-dimethylpyrazine 1-oxide yields 2-chloro-3,6-dimethylpyrazine as well as the expected dioxide, the former perhaps being formed by the deoxygenation sequence (44) $\longrightarrow$ (46)[415]. The oxidation of other pyrazines has been reported[159].

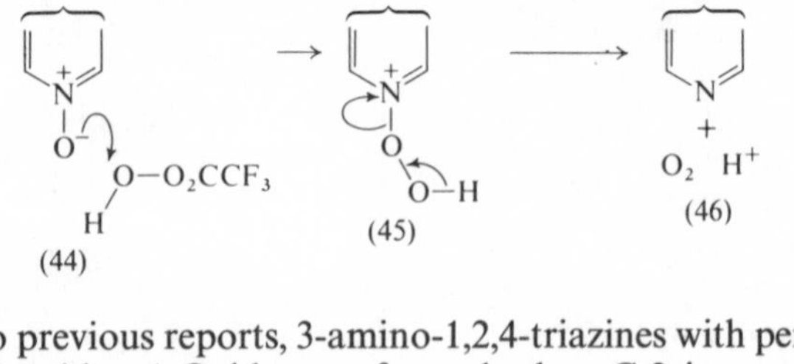

(44) (45) (46)

Contrary to previous reports, 3-amino-1,2,4-triazines with perbenzoic acid give mainly 2-oxides. 1-Oxides are formed when C-3 is unsubstituted, or bears a methoxyl group[416].

4.4.2.4 Complex formation

The structure and spectroscopic properties of halogen and interhalogen complexes of pyridine, the diazines, and their oxides continue to attract attention[417, 418]; chemical interest[419–421] will be further stimulated by reports that the high-melting iodine adduct of pyridazine is a semiconductor[422, 423], and that preparative use can be made of the iodine complex of 2-methyl-pyridines. Heating in DMSO gives pyridine-2-aldehydes in moderate yield[424].

4.4.3 Nucleophilic substitution reactions

No attempt has been made to distinguish between those reactions in which initial products are isolable, and 'direct' nucleophilic replacements.

4.4.3.1 Hydride substitution

Pentachloropyridine forms 2,3,6-trichloropyridine (corresponding to initial 3,4-attack) on reductive dehalogenation with lithium aluminium hydride. The sterically less demanding sodium borohydride leads to 2,3,5,6-tetra-chloropyridine (by 1,4 attack)[378].

4.4.3.2 Alkylation and substituted-alkylation

Renewed interest has been shown in the reactions of azines with Grignard reagents. 2-Allyl-6-chloropyridine and 2,6-diallylpyridine are formed in moderate yield when 2,6-dichloropyridine reacts with appropriate molar quantities of allylmagnesium chloride: in sharp contrast, benzylmagnesium chloride reacts by conjugate addition and subsequent hydrolysis gives 3-benzylglutarimide[425]. Pentachloropyridine is attacked by Grignard reagents at C-4 only[426, 427], but 3,5-dicyanopyridines undergo 1,2- and 1,4-addition with methylmagnesium iodide if C-4 is not blocked[428].

New examples have been reported of the alkylation of pyridazines by Grignard reagents; 3-chloro-6-dimethylaminopyridazine is alkylated at C-5 by reagents with low steric requirements, but at C-4 by t-butylmagnesium chloride[565]. The reaction between 3-chloro-5,6-diphenyl-1,2,4-triazine and alkylmagnesium iodides is anomalous, leading to 5-alkyl-5,6-diphenyl-1,2,4-triazin-3[2H]-ones[403]. Pentachloropyridine 1-oxide gives both 2-methyl- and 2,6-dimethyl-polychloropyridine 1-oxide on treatment with methyl-magnesium iodide[429]. Ring opening may be involved, as in the analogous arylation (below). 1-Methyl (and 1-benzyl)-3-cyanopyridinium salts are alkylated at both C-2 and C-6 by methylmagnesium bromide and, with methylcadmium, 1-methyl (and 1-benzyl)-3-methoxycarbonylpyridinium salts behave similarly[430].

Organo-lithium compounds are more commonly used than Grignard reagents. High yields of 2,6-dialkylpyridines are formed from pyridine and a large excess (10:1) of an alkyl-lithium[431], but t-butyl-lithium is exceptional in that the major product is 2,4,6-tri-t-butylpyridine[431, 432]. The reactions of pyrazine and alkylpyrazines with alkyl-lithiums has been reinvestigated, and 2,3-dialkyl isomers shown to be the major disubstituted products from pyrazine or a monoalkylpyrazine[563]. Although trimethylpyrazine is known to be inert to methyl-lithium it reacts readily with butyl-lithium forming butyl-trimethylpyrazine[426].

2-Lithio-2-phenyl-1,3-dithiane with 2-bromopyridine forms 2-phenyl-2-(2′-pyridyl)-1,3-dithiane, from which phenyl 2-pyridyl ketone is readily obtained; the reaction appears to be of limited scope, however[433]. Propenyl-lithium converts 2,4,6-trifluoro-5-trifluoromethylpyrimidine into (mainly) 4,6-difluoro-2-propenyl-5-trifluoromethylpyrimidine, reaction at C-4 being sterically and electrostatically hindered[434]. Tetrafluoropyrazine forms 2,3,5-trifluoro-6-methylpyrazine on treatment with methyl-lithium (1 mol); with excess n-butyl-lithium a mixture of mono-, di- and (mainly) tri-alkylated derivatives is formed[185].

Carbanion attacks are relatively rare, but interesting. An improved procedure converts 2-chloro-5-nitropyridine into diethyl (5-nitro-2-pyridyl)-methylmalonate in excellent yield[435]. Carbanions from polyfluoroalkenes and fluoride ion are known to displace fluoride ion from heteroaryl fluorides and in this way tetrafluoropyridazine can be converted into 4-perfluoroalkyl- or 4,5-bis-perfluoroalkyl-perfluoropyridazine, depending on conditions. Further substitution is accompanied at higher temperatures by rearrangement[436]. (Similar reactions of pentafluoropyridine provide a striking demonstration of kinetic v. thermodynamic control)[437]. Analogous reactions with tetrafluoropyrimidine and hexafluoropropene give mono-(4-), di-(4,6-) or tri-(2,4,6)-perfluoroisopropyl derivatives directly: the tetraperfluoroisopropyl compound is best obtained from the tri-substituted derivative. The cyano group of 2,5-dicyano-4,6-bisperfluoroisopropylpyrimidine is easily displaced by perfluoroisopropyl carbanion[438]. Tetrafluoropyrazine provides 2,5-difluoro-3,6-bisperfluoroisopropylpyrazine in satisfactory yields[69]. ω-(3-Pyridyl)-1-alkenes cyclize in the presence of alkali metals via picolyl anions; 4-pyridyl isomers do not react[439]. Intramolecular carbanionic attack is involved also in the formation of (48) from (47)[566]. Dimethylsulphoxonium

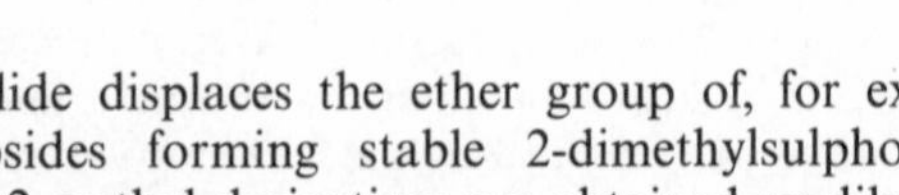

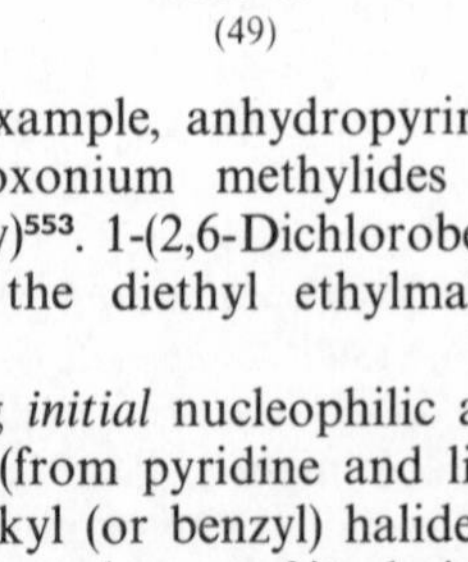

methylide displaces the ether group of, for example, anhydropyrimidine nucleosides forming stable 2-dimethylsulphoxonium methylides (from which 2-methyl derivatives are obtained readily)[553]. 1-(2,6-Dichlorobenzyl)-3-cyanopyridinium chloride is attacked by the diethyl ethylmalonate anion at C-6[430].

In a promising indirect alkylation involving initial nucleophilic attack, lithium tetrakis-(N-dihydropyridyl) aluminate (from pyridine and lithium aluminium hydride, in situ) is converted by alkyl (or benzyl) halides into 3-substituted pyridines in good yield[318]. Pyrazine and some of its derivatives (bearing electron-releasing substituents) are alkylated in good yields by the action of aldehydes or ketones in the presence of (preferably) potassium. The relative positional reactivity is $6 > 3 > 5$: a mechanism has been proposed. Although this interesting reaction bears a similarity to the Emmert alkylation, pyridine itself fails to react[440]; however, 4-methylpyridine forms 2-(diphenyl-hydroxymethyl)-4-methypyridine in moderate yield on treatment with benzophenone and lithium[441].

4.4.3.3 Amination, substituted-amination and related displacements

Examples of simple aminative replacements (S_NAr2) abound throughout the series, but these require little comment. Hetarynes are involved to varying extents in many reactions of strongly basic nucleophiles with unactivated substrates and a further, excellent review of these species has appeared[442]. Kinetic examinations remain infrequent[443, 444]; leaving-group effects are broadly as expected in the piperidinodehalogenation of 2-halogenopyridines but activation by ring nitrogen is less than by a nitro substituent[443]. Most aminative displacements involve halogen atoms but other leaving groups are encountered; the conversion of pyrid-2-one into 2-dimethylaminopyridine (and the analogous reaction of 2,4-dimethylpyrimidin-6[$1H$]-one) by means of tetrakis(dimethylamino)titanium provide an interesting new approach to aminoheterocyclic molecules[445]. Predictably, the hydroxy group at C-4 is replaced preferentially when 2,4-dihydroxy-6-methylpyridine is heated with aqueous ammonia under pressure. Several other 4-hydroxy-pyridines undergo direct amination under similar conditions[4]. Replacements of alkoxy and alkylthio groups by ammonia and amines are well-known[80], particularly in the pyrimidine series and these have found further application in nucleoside synthesis[446, 447]; pyrimidinones[350] and thiones[448] have also been aminated directly in the same connection, although the behaviour of 3-methyldithiouracil with ethanolic ammonia is complicated (and, with aqueous ammonia, anomalous in giving 3-methylisocytosine[64]). In the presence of hydrochloric acid, morpholine replaces the amino group of 4-amino-6-hydroxy-2-phenylpyrimidine on prolonged heating[350]. 2-Methylsulphonylpyrimidines undergo aminolysis smoothly[248, 449] and without the complicating ring cleavage observed with 2-methoxypyridines[449]. (The sulphone substituent appears here to be marginally better as a leaving group than a chlorine atom). Aminative replacements of trimethylsilyloxy groups occur in good yields facilitating, for example, the synthesis of cytidine derivatives (and their 6-aza analogues), from the corresponding uridines (and azauridines)[450]. The method seems capable of wide extension. Most aminolyses in the triazine series involve the displacement of chlorine, but methylthio[80] and phenoxide[94] have also been used as leaving groups. Pyridyne formation by 3-halogenopyridines and sodamide in ammonia has been studied iso-topically[451]. Competition experiments (diethylamine–di-isopropylamine mixtures) show C-3 and C-4 of 3,4-pyridyne to be equally reactive[452]. The aminolysis of perfluoro-(4-phenylpyridine) occurs predominantly α to ring nitrogen, followed by C-4 of the benzene ring rather than C-6 in the pyridine moiety[453]. Polychloropyridazines with trimethylamine give trimethylpyrida-zinylammonium chlorides but these (except 5,6-dichloropyridazin-3-ylammonium chloride) are demethylated to dimethylaminopyridazines in the reaction mixture[454]. The orientation of attack on 4,5-dichloropyridazin-3[2H]-one is dependent on the nucleophile, and morpholino-dechlorination occurs at C-5[455]. Aminolyses (with isopentyl- or 1,4-dimethylpentyl-amine) of 2-halogeno- or 4-halogeno-pyrimidines are slightly more rapid in the case of bromo than of iodo or chloro compounds, but the rate differences are small[247].

Dimethylformamide successfully converts chloropyrimidines into their dimethylamino analogues in satisfactory yields, provided C-5 is not sufficiently activated to undergo simultaneous formylation[350]. The direct amination of 2,4,6-trifluoropyrimidine gives mainly 2-amino-4,6-difluoropyrimidine; the predominance of 2-substitution is attributable to extensive stabilisation of the σ-complex by hydrogen-bonding of the nucleophile to *both* ring nitrogen atoms. 5-Nitro- and 5-cyano-2,4,6-trifluoropyrimidine both undergo disub-stitution, even under mild conditions, but whereas the former gives only 4,6-diamino-2-fluoro-5-nitro-pyrimidine (probably because of hydrogen bonding between nitro group and nucleophile) the cyano compound forms 2,4-diamino-6-fluoro-5-cyanopyrimidine, reflecting the ability of the cyano group to stabilise a carbanionic centre[456]. Advantage has been taken of the easy aminolysis of fluoro compounds in identifying the pyrolysis products of, for example, tetrafluoropyridazine[69]. Further examples have been reported of ready piperidino- and pyrrolidino-dechlorination in 6-chloropyrimi-dines[350, 457] and an interesting transaminative replacement has been observed; 4-amino-6-chloro- (or hydroxy-)2-methylpyrimidines form 6,6'-dichloro-2,2'-dimethyl-4,4'-iminodipyrimidine in good yield on treatment with phosphoryl chloride (which appears to be specific)[457]. Complicating rearrangements are observed in the direct amination of (49) but the 4-amino compound can be obtained satisfactorily by reaction with hydroxylamine followed by hydro-genation[458]. Although substituents at C-5 of pyrimidine are least susceptible to nucleophilic replacement, 5-bromo-1,3-dimethyluracil readily forms its 5-allylamino analogue on heating with the amine[326].

On treatment with potassium amide in liquid ammonia 4-chloropyrimi-dines undergo nucleophilic attack at C-6 followed by ring opening (see below). The 4-aminopyrimidines isolated as minor products appear to arise from recyclisation rather than direct replacement[459] and isotopic studies indicate that the conversion of 4-bromo-6-phenylpyrimidine into the 4-amino compound by potassium amide in liquid ammonia occurs both via an open-chain intermediate (c. 80%) and the hetaryne[460]. Aminodehalogenation via ring-opening takes place only in very strongly basic media but reactions of this type may be more common than suspected hitherto, and the low yield (c. 15%) of 2-aminopyrazine obtained from 2-chloropyrazine probably arises in the same way[461]. Tetrafluoro- and 2,5-difluoro-3,6-bisperfluoroiso-propylpyrazine both form mono-amines on treatment with aqueous ammonia at room temperature[69, 185].

The reactions between chloro-1,3,5-triazines and amines[462], amino-phenols[463] and aminonaphthol derivatives[464] have been studied further: with the last-named, initial O-substitution is followed by rearrangement in alkaline media to N-triazinyl compounds[464]. Both acid- and base-catalysed

Smiles rearrangements have been observed in, for example, 2-(o-amino-phenoxy)-4,6-dimethoxy-1,3,5-triazine[465]. 2-Phenoxy-4-phenyl-1,3,5-triazin-6[1H]-thione forms the 2-dimethylamino compound in excellent yield on heating with dimethylformamide[94].

Substituted-aminodechlorination occurs readily in 2-chloro-5- (and 3-) nitropyridinium salts leading to pyridonimines[466] and the chlorine atom of 6-chloro-3-nitropyridazine 1-oxide is replaced readily by amines[157]; methoxy-dechlorination takes a different course (see below). Aminolysis of 1-methyl-2,6-bismethylthiopyrimidinium iodide occurs at C-6 in near quantitative yield.[64].

The resurgence of interest in organic azides has led to numerous examples of azido-dehalogenation; the reactions are normally rapid and straight-forward[467, 468] even, for example, with 4-chloro-3,5-dinitropyridine[372]. An interesting exception is the reaction between 5-nitropyrimidinones and azide ions, forming 8-azapurines in a single step[469]. Several 3-pyridyl azides have been prepared by the action of azide ion on the corresponding pyridyl-diazonium salts in acetate buffer[470]. An alternative route to azides involves treatment of hydrazino groups with nitrous acid; several hydrazinolyses have been reported in this[372, 414, 467, 468] and other connections[157, 372, 471-474] and the reactions are usually straightforward. 4,6-Dimethoxy-5-nitro-pyrimidine is a notable exception since, with methylhydrazine in pyridine, 4-hydrazino-6-hydroxypyrimidine (and methyl nitrite) are formed. 2,4-Di-methoxypyrimidine gives the 4-hydrazino compound in the normal way[471]. Both 4-chloro-6-hydroxy-5-nitro- and 4,6-dichloro-5-nitro-pyrimidine react normally with methylhydrazine in ethanol[475]. Rather surprisingly, introduc-tion of a hydrazino group proceeds more satisfactorily with simple 3-methoxy-1,2,4-triazines than with their 3-methylthio analogues[77]; a series of hydrazino-1,3,5-triazines has been obtained from both methoxy- and methylthio-triazines, and methoxy-1,3,5-triazin-2[1H]-ones undergo ready substitution under similar conditions[476].

Aldehyde and ketone hydrazones are sufficiently nucleophilic to displace the chlorine atom of 6-chloro-3,4-dimethyl-5-nitrouracil[322].

4.4.3.4 Arylation

With phenyl-lithium, 3-t-butoxypyridine gives 3-t-butoxy-2-phenyl- and 5-t-butoxy-2-phenyl-1,2-dihydropyridine in the ratio 4:1[314]. The same reagent converts 3-methylpyridine into 3-methyl-2-phenyl- and 5-methyl-2-phenyl-pyridine (95:5) after working-up; precedents exist for this lack of steric sensitivity[477]. At higher temperatures an otherwise similar reaction with o-tolyl-lithium gives some 3-methyl-5-o-tolylpyridine at the expense of the 2,3-isomer[477]. 2,5-Dimethyl-4-phenylpyridine undergoes arylation at C-6 with aryl-lithiums[478]. Examples involving substrates other than pyridines are less frequently encountered: however, direct arylation at C-3 accompanies conjugate addition in the reaction between phenylmagnesium bromide and 3-chloro-6-dimethylaminopyridazine, although the overall yield is low[565].

The formation of 2-arylpyridines from pyridine N-oxide and aryl-magnesium halides, followed by treatment with acetic anhydride, is interesting in that the intermediate has been shown to be an acyclic oxime (rather than the presumed dihydropyridine)[479]. In the presence of acylating agents, pyridine N-oxides suffer nucleophilic attack by indoles[480] (or their Grignard reagents)[481] at C-2, with concomitant deoxygenation: in this way indole with the benzoyl chloride adduct of 3-carbethoxypyridine N-oxide forms 2-(3-indolyl)-5-carbethoxypyridine in fair yield. In contrast, N-acylpyridinium salts and indoles form 4-(N-acyl-3-indolyl)pyridinium salts[482] and 4-(3-indolyl)pyridines[483].

4.4.3.5 Cyanide formation

Cuprous cyanide in pyridine (160 °C) converts 4-amino-3-bromo-2,6-dimethylpyridine into the 3-cyano compound in excellent yield[4]. Much less vigorous conditions are required in the case of perfluoro-(4,6-di-isopropylpyrimidine) which forms, rather surprisingly, 2,5-dicyano-4,6-bis(heptafluoroisopropyl)pyrimidine directly[438]. Nitrile formation from quaternary compounds is encountered more commonly. Anhydrous hydrogen cyanide converts 1,2,4,6-tetramethyl-3,5-dicyanopyridinium p-tosylate into 1,2,4,6-tetramethyl-3,4,5-tricyano-1,4-dihydropyridine but in aqueous potassium cyanide the α- and γ-pyridinemethides are major products[273]. Photodimerisation may intervene in the reactions of pyridinium salts with cyanide ion[484] and a thermal isomerisation has been observed in the reaction between 1-methyl-3-substituted-pyridazinium methosulphates and aqueous potassium cyanide[485]. More highly substituted pyridazines undergo reaction without this complication, forming mixtures of cyanopyridazinones[485]. 3,3'-Bipyridazine forms 1,1'-dimethyl-4,4'-dicyano-1,1',4,4'-tetrahydro-3,3'-bipyridazine on treatment with dimethyl sulphate followed by aqueous potassium cyanide; the 6,6'-dimethyl homologue behaves similarly but the 6,6'-dimethoxy compound forms only 1,1'-dimethyl-3,3'-bipyridazine-6,6'[1H,1'H]-dione under Reissert conditions[486]. The quaternary salts of N-oxides undergo cyano substitution quite readily and N-methoxy-4-nitropyridinium methosulphates give 2-cyano-4-nitropyridines in satisfactory yield[487]. When N-alkoxy-3-isobutylpyridinium salts are treated with aqueous potassium cyanide the ratio of C-2:C-4 substitution is dependent on the bulk of the alkoxy group[488]. 5-Methyl-3-phenylpyridazine 1-oxide gives 3-cyano-4-methyl-6-phenylpyridazine in good yield on treatment of the methosulphate with potassium cyanide[414].

4.4.3.6 Halogenation

Many ring syntheses provide hydroxyazines and their conversion into halogeno compounds is frequently the key step in metathesis. Nucleophilic halogenation α or γ to ring nitrogen with, for example, phosphorus halides normally presents little difficulty[18, 22, 59, 244, 247, 248, 307, 329, 338, 372, 455, 460, 489]; leaving groups other than hydroxyl are encountered comparatively rarely although, for example, 4-nitropyridine-2-carboxylic acid is reported to form the 4-chloro compound with hydrochloric acid[487]. Complications occasionally arise, an interesting example being the isolation of bis-(4-chloro-2-methylpyrimidin-6-yl)amine when 4-amino-6-methoxy-2-methylpyrimidine is heated with phosphoryl chloride[457].

Pyridine-4-sulphonyl chloride forms 4-chloropyridine at room temperature by an $S_N i$ process, but the tetrachloro analogue is markedly more stable. The α-chlorine atoms sterically inhibit a similar mechanism, and the peroxide-catalysed formation of pentachloropyridine in boiling acetic acid may be a radical process[490]. 1-Hydroxypyrazin-2-ones with phosphoryl chloride give 2-chloro-6-hydroxypyrazines in moderate to good yields[491].

Halogen exchange reactions are commonly used in the preparation of fluoro compounds and new examples have been reported in the pyridazine[142], pyrimidine[456] and pyrazine[185, 492] series. It is noteworthy that tetrachloro-pyridazine with potassium fluoride (at 200 °C) yields a complex mixture; in contrast, brief treatment with excess lithium chloride in dimethylformamide gives 4,5-dichloro-3,6-difluoropyridazine in high yield[142]. Tetrachloropyrimidine with anhydrous potassium fluoride (530 °C) gives tetrafluoropyrimidine and 5-chloro-2,4,6-trifluoropyrimidine, together with a mixture of isomeric trifluoro-trifluoromethylpyrimidines produced by thermal cleavage. 2,4,6-Trichloro-5-nitropyrimidine with potassium fluoride (250 °C) forms its trifluoro analogue but the less activated 2,4,6-trichloro-5-cyanopyrimidine produces mixtures of mono-, di- and tri-fluoro-5-cyanopyrimidines[456]. Tetrafluoropyrazine is formed when tetrachloropyrazine is heated with potassium fluoride at 310 °C; under milder conditions mixtures of 2,3,5-trichloro-6-fluoropyrazine, 2,6-dichloro-3,5-difluoropyrazine and 2-chloro-3,5,6-trifluoropyrazine are produced[185]. Anhydrous potassium fluoride in N-methylpyrrolidone efficiently displaces chlorine in 2-chloro- and 2,6-dichloro-pyrazines[492]. Interesting studies have recently appeared on the reverse process: when pentafluoropyridine reacts with hydrogen halides in sulpholan replacement of fluorine ortho and para to ring nitrogen occurs: excess hydrogen bromide, for example, produces the 4-bromo- and 2,4,6-tribromo derivatives in 15 and 31% yields respectively. Yields are lower with hydrogen chloride and small amounts of water in the medium inhibit the halogen exchange, which is believed to involve the protonated species[493]. Tetrafluoropyrazine forms mainly tetrabromopyrazine on heating with anhydrous aluminium bromide and hydrogen bromide; monohalogeno-trifluoropyrazines are best prepared by heating monohydrazino-trifluoro-pyrazine with copper(II) or iron(III) halides in the presence of the hydrohalic acid[185].

On treatment with acetyl bromide, 6-chloro-3-nitropyridazine 1-oxide forms the 6-bromo derivative rather than the expected dihalogenopyridazine 1-oxide[157].

4.4.3.7 Hydroxylation and substituted-hydroxylation

Extensive preparative use has been made of reactions of this type, with various leaving groups. Despite the importance of the reaction, however, kinetic data are still sparse. Mention has been made earlier of the characterisation of Meisenheimer intermediates; further examples have been reported in alkoxy-substitutions of pyridines[180] and 1,3,5-triazines[494]. 2-Halogeno-pyridines show the expected values of k_F/k_{Cl} for methoxydehalogenations (in methanol)[443]. The efficiency of the trimethylammonio-group as a leaving group has been assessed in pyridine and the diazines; pyrimidin-2-yltrimethylammonium chloride is c. 700 times more reactive than 2-chloropyrimidine and c. one-fifth as active as 2-methylsulphonylpyrimidine towards hydroxide ion at 20 °C. The reaction with 5-nitropyrimidin-2-yltrimethylammonium chloride shows first-order kinetics[156]. A careful study has been made of the complicated mechanisms of the hydrolysis of 2,4,6-trichloro-1,3,5-triazine in aqueous acetone[494]. A bimolecular process is implicated in the amine-catalysed reaction of 2,4-dichloro-6-phenylamino-1,3,5-triazine with water, alcohol and phenol[495] but an S_N1 reaction may be involved in the hydrolysis of 2-amino-4,6-dihydroxy-1,3,5-triazine in media sufficiently basic to generate the dianion[496].

2- or 4-Chloro-, bromo- and iodo-pyridines give 2- or 4-pyridone when heated with aqueous 4 M potassium hydroxide at 250–350 °C, but 3-halogeno-pyridines form mixtures of 3-hydroxypyridine and 4-pyridone by competing direct substitution and elimination-addition reactions. 4-Aminopyridine is formed as a byproduct in the case of the 4-halopyridines, probably via 1-(4-pyridyl)pyrid-4-one; hydroxydeamination of the 4-amino compound also occurs under similar condition[497]. Examples of direct hydroxylation and substituted hydroxylation are numerous and varied, but positional reactivities are sometimes unpredictable; thus, in the monomethoxydechlorination of pentachloropyridine the yield of 2-methoxy-3,4,5,6-tetrachloropyridine is almost doubled by substituting magnesium methoxide for sodium methoxide[498].

In line with simple electronic prediction, 2-cyano-4-nitropyridine forms 2-cyano-4-alkoxypyridines on treatment with alkoxide[487]. Pentachloropyridine gives 2-hydroxytetrachloropyridine on treatment with a mixture of sulphuric and acetic acids; preferential attack at C-2 suggests that the pentachloropyridinium ion is involved, since aqueous alkali attacks normally at C-4[409]. Displacements of halogen by phenoxide are common[470, 499], and a range of O-(2-pyridyl) oximes have been obtained from 2-fluoropyridine and the appropriate cycloalkanone oximes[323].

Examples of hydroxy- and alkoxy-dechlorination are frequent in pyridazine chemistry[256, 342, 436] and the displacement of trimethylammonio by

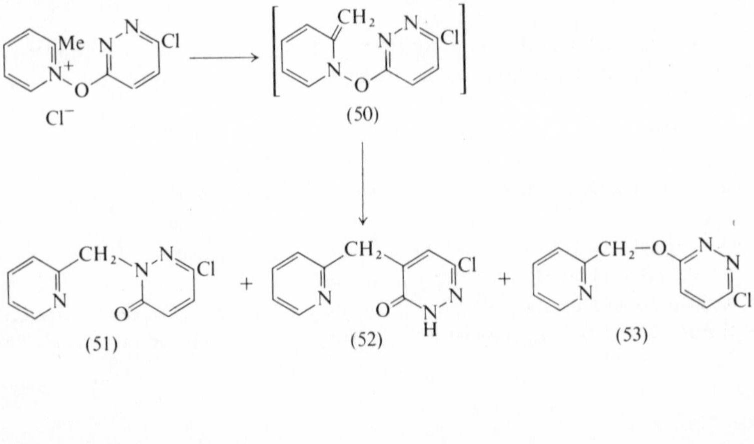

(50)

(51)   (52)   (53)

alkoxy groups has been reported[454]. The reaction between 3,6-dichloro-pyridazine and 2-methylpyridine N-oxide provides a mixture of products (51–53), the formation of which has been rationalised by a sigmatropic rearrangement of (50)[500].

Hydroxy- and alkoxy-dehalogenation reactions are normally easy and rapid in the pyrimidine series, although ring cleavage may sometimes take place (see below). Substituent effects are normally qualitatively predictable[456]. Acid-catalysed hydroxydeamination of simple 4-aminopyrimidines proceeds in fair yield, but reaction times are prolonged[66]. The (specific) deamination of cytosine and its derivatives to the corresponding uracils with aqueous sodium bisulphite provides a further example of an indirect substitution in which transient addition of an external nucleophile occurs[501,502]. With hydroxide, deuteroxide and ethoxide ions N-1 substituted 5-nitrouracils form stable Meisenheimer adducts by nucleophilic attack at C-6[469] and base catalysed proton exchange at C-5 of uracil and its N-methyl derivatives likewise involves initial attack by hydroxyl ion at C-6[503]. Pyrimidine-2- and pyrimidine-6-sulphonates undergo ready hydrolysis over a wide pH range as expected, acid catalysis being the more effective[152]. The cyano group is used relatively rarely as a leaving group, but displacement of a 4-cyano group (by methanolic hydrochloric acid) has been used in nucleoside synthesis and wide applicability is predicted[504]. A novel transetherification procedure (boiling alcohols in the presence of silver oxide) is effective in the case of activated compounds such as 2- and 4-methoxy-5-nitropyrimidine, but fails with simple alkoxy derivatives[471]. 6-Chloro-4-amino- (or substituted amino)-5-nitropyrimidines are hydrolysed to 4-pyrimidinones under near neutral conditions but at low pH ring cleavage occurs, exclusively in the case of tertiary amino compounds[505].

The reactions of 2,6-dichloropyrazines with hydroxide and alkoxide ions have been re-examined and minor products identified[492]. The preparation of 2,5-dimethoxy-3,6-dimethylpyrazine requires less vigorous conditions than reported earlier[492]: methoxydechlorination has been used to locate the halogen atoms in chlorinated pyrazines[296]. Protonation of tetrafluoro-pyrazine renders it susceptible to attack by even weak nucleophiles, so that a solution in concentrated sulphuric acid treated with methanol gives methoxy-trifluoropyrazine in good yield. The orientation of attack by methoxide ion on tri- and tetra-fluoropyrazines has been discussed[185]. 3-Methylthio-1,2,4-triazine forms its 3-methoxy analogue under mild conditions[77].

Predictably, N-oxides undergo hydroxylation easily, as in the attempted N-oxidation of 2,4,6-trifluoro-3,5-dichloropyridine with hydrogen peroxide in polyphosphoric acid; after an aqueous work-up, 2-hydroxy-3,5-dichloro-4,6-difluoropyridine 1-oxide was obtained[409]. 6-Chloro-3-nitropyridazine 1-oxide forms 6-chloro-3-methoxypyridazine 1-oxide on treatment with methoxide, although other nucleophiles displace the chlorine atom[157]. 2,5-Dichloro-3,6-dimethylpyrazine 1,4-dioxide forms the diethoxy and dibenzyloxy compounds on treatment with ethoxide and benzyloxide ion, but aqueous alkali gives only 5-chloro-3,6-dimethyl-1-hydroxypyrazin-2[1H]-one 4-oxide; the hydroxamic acid (as its anion) resists further sub-stitution[415]. The formation of 2-methylaminopyridine-3-aldehyde by the action of alkali on 1-methyl-3-cyanopyridinium iodide involves initial attack by hydroxyl ion at C-2, followed by ring cleavage, recyclisation and finally a Dimroth rearrangement[506]. The effect of 3-substituents on the well-known ferricyanide oxidation of quaternary pyridinium salts has been examined; only in the case of the 3-carboxy and 3-nitro compounds was the 6-pyridone formed exclusively[507]. A careful mechanistic examination has shown that the initial attack by hydroxyl ion is not rate determining and excludes the possibility of ylide intermediates; the mechanism proposed involves the rate-determining formation of a ferricyanide complex[508]. The ferricyanide procedure has been used to obtain N-methylpyridazinones from quaternary methosulphates[414].

The halogen atom of 2-chloro-5-nitro- and 2-chloro-3-nitro-pyridinium salts is replaced rapidly by both neutral and anionic nucleophiles; a second equivalent of, for example, methoxide produces 2,2-dimethoxy-5-nitro-1,2-dihydropyridine[466]. Vinylogous N-acylpyridinium salts are cleaved rapidly at C-2 by aqueous alkali and aliphatic amines[509]. That 2-amino-pyrazine undergoes quaternisation at both ring nitrogen atoms is shown by hydrolysis with aqueous alkali, which gives a small amount of 1-methyl-pyrazin-2-one[363].

Alkaline hydrolysis of 1-methyl-1,2,4-triazinium salts gives the triazin-6-ones as minor products only. Ring contraction (to 1,2,4-triazoles) predominates[510]. An S_Ni mechanism is proposed for the reaction of 2,4,6-trichloro-1,3,5-triazine with alcohols which shows promise as a method of preparing alkyl chlorides[511] and, in the presence of sodium iodide, alkyl iodides[512].

4.4.3.8   Sulphur bond formation

The conversion of N-heterocyclic lactams into their thio analogues is often accomplished by phosphorus pentasulphide in refluxing benzene[256] or, more commonly, pyridine[64, 448, 489, 513] but the substitution of dioxan for the latter is certainly advantageous in nucleoside chemistry[504] and is probably widely satisfactory. 3-Methylcytosine resists thiation under conventional conditions[64] but disodium cytidine 5'-phosphate is converted smoothly into the 4-thiouri-dine analogue by heating at 60 °C under pressure with hydrogen sulphide in pyridine[514]. Substitutions in which halogen is replaced by mercapto[256, 455, 515] and alkylthio[307] groups are commonplace; a further use has been made of the convenient but indirect introduction of the thiol group by means of thiourea, followed by cleavage of the resulting isothiouronium salt[343, 516].

3,4,6-Trichloropyridazine with sodium sulphide gives a compound assumed to be 3,6-dichloro-4-mercaptopyridazine[517]. 2,6-Dichloropyrazine undergoes mono- or di-substitution with the appropriate amounts of sodium benzyl sulphide[492]. Boiling aqueous potassium sulphite converts 2- and 6-chloro-pyrimidines into the corresponding potassium pyrimidine-2- and pyrimidine-6-sulphonates[152], and further examples of thiocyanato-debromination have been reported[343]. In an interesting extension of earlier work with mercaptans, pyridine N-oxides have been shown to undergo deoxidative substitution at C-2 and C-3 on treatment with thiophenols in the presence of sulphonyl chlorides; β-arylthiopyridines constitute a significant proportion (40–60%) of the total product[518].

4.4.3.9 Miscellaneous nucleophilic substitutions

The low temperature reaction between alkali metal diethylphosphonates and N-methoxypyridinium salts constitutes a general method for the preparation of dialkyl pyridine 2-phosphonates[519].

4.4.4 Radical substitution

Radicals ($R_2\dot{N}CO$) generated *in situ* by hydrogen abstraction from formamides attack, for example, pyridine, pyrimidine and pyrazine at positions α- and γ- to ring nitrogen providing carboxamides[520]. In the case of dimethylformamide, N-formyl-N-methylaminomethyl derivatives are also formed from $\dot{C}H_2N(Me)CHO$ radicals[521]; either type of product can be made to predominate by careful choice of oxidant[521] and the reaction has considerable potential. Homolytic methylation of pyridine proceeds more readily (c. 13 fold) in acetic acid than in non-acidic media, and in acetic acid–hydrochloric acid only 2- and 4-substitution occurs[522]. Similarly, 3-substitution is not observed when protonated pyridines react with radicals generated by the silver-catalysed oxidative decarboxylation of acids by peroxydisulphate, although yields of 2- and 4-alkylated products are good[523]. Other radical sources are less effective[141]. Attack at C-2 predominates when pyridinium ion is alkylated in acetic acid–acetic anhydride–hydrochloric acid containing t-butyl peroxide[524]. Competition experiments show that radical cyclohexylation of 4-methylpyridine, mainly at C-2, occurs more readily than methylation or phenylation[525]. Pyridine gives a 4:1 mixture of its 2- and 4-hydroxymethyl derivatives on treatment in aqueous sulphuric acid with methanol and ammonium peroxydisulphate and under similar conditions, but with dioxane as radical source, 4-cyanopyridine and pyrazine form 2-dioxanyl derivatives[526]. Radical benzylation of some substituted pyridinium cations has been reported[527].

Phenylation of pyridine with nitrobenzene at 600 °C gives mainly phenylpyridines and bipyridines. The ratios of isomeric phenylpyridines are similar to those observed with other aryl radical sources in the liquid phase[528]. In competitive arylations at 600 °C the relative (overall) reactivity towards phenyl radicals is benzene, 1; pyridine, 2.3; thiophene, 5[529]. Isomer ratios in the phenylation of pyridine by radicals from the electrolytic reduction of benzenediazonium chloride are generally similar to those observed using benzoyl peroxide or lead tetra-acetate. A new mechanism has been proposed for the non-electrolytic phenylation of pyridine with diazonium salts. Partial rate factors for the phenylation of pyridine N-oxide have been reported: at 0 °C (using the electrolytic procedure) the factors are α, 139; β, 1.5; γ, 31.2 corresponding to the isomer ratios of 89: <1:10[530].

4.5 PHOTOCHEMISTRY

Irradiation of pyridine at 253.7 nm yields Dewar pyridine (54), an intermediate in both photoreduction (NaBH$_4$) to (55) and photohydration to 5-amino-2,4-

pentadienal[531]. Pentakis(fluoroethyl)pyridine gives azaprismane (57) on irradiation, via (56); either can be obtained in essentially quantitative yield[532].

$$\text{(pyridine)} \underset{t_{\frac{1}{2}} \sim 2 \text{ min } (25\,°\text{C})}{\overset{253.7 \text{ nm}}{\rightleftarrows}} \quad (54) \quad \xrightarrow{\text{NaBH}_4} \quad (55)$$

$$(C_2F_5)_5\text{-(pyridine)} \underset{<270 \text{ nm}}{\overset{>200 \text{ nm}}{\rightleftarrows}} (56)\text{-}(C_2F_5)_5 \xrightarrow{>200 \text{ nm}} (57) \, (C_2F_5)_5$$

Perfluoro-2,4,5,6-tetraisopropylpyrimidine remains unchanged under similar conditions[438]. 4,5-Dichloro-3,6-difluoropyrazine forms 2,5-dichloro-3,6-difluoropyrazine, apparently excluding a diazaprismane intermediate[142]; azaprismanes have been implicated[533] (and supporting evidence cited)[534] in the photochemical interconversions of methylpyridines. 2- and 4-Cyclohexyl- and 2,5-dicyclohexyl-pyridine are formed when pyridine is irradiated in oxygen-free cyclohexane. The pyridinyl radical first formed has been characterised by electron spin resonance spectroscopy and a diradical intermediate (perhaps a precursor of (54) above) is proposed to account for 2,5-disubstitution[535]. In methanol–hydrochloric acid mixtures (but not methanol alone) pyridine undergoes photomethylation at C-2 and C-4[536]. One electron-withdrawing (e.g. CN or CO_2Et) group permits hydroxymethylation of the corresponding pyridine at C-2 (and C-4 in 3-cyanopyridine)[537] and the related photopreparation of 3,5-dicarbomethoxy-2,6-dimethyl-2-hydroxymethyl-1,2-dihydropyridine (in good yield from the dicarbomethoxy-dimethylpyridine) is a key step in the synthesis of the corresponding azepine[538]. Pentafluoropyridine undergoes photochemical chlorination to 2,3,4,5-tetrachloropentafluorotetrahydropyridine[539]. The photolysis of 5-bromo-2-methoxypyrimidine in methanol provides only trace amounts of methylated products with larger amounts of hydroxymethylation; overall yields are low[540]. The photolysis of 3,6-dichloropyridazine and related compounds in methanol–hydrochloric acid leads mainly to ring cleavage (forming γ-lactones and diesters)[541]. On the other hand, first reports on the varied photochemistry of 4-halogeno-pyrimidines[542, 543] indicate a promising field. Irradiation of 4-chloro-2,6-dimethoxypyrimidine provides 2,4-dimethoxy-, 2,4,6-trimethoxy- and 4-methyl-2,6-dimethoxypyrimidine (in order of decreasing yield). The addition of triethylamine as hydrogen donor increases the amount of reduction, relative to photoalkoxylation; since 5-bromo-2,4-dimethoxypyrimidine fails to undergo the latter reaction it appears that in this series the order of ground state reactivities is preserved in excited states[542]. In a novel debromination reaction, 5-bromopyrimidine forms 4-hydroxymethylpyrimidine on irradiation in methanol, and a similar intermediate ((58); R = t-butyl) may be involved in the formation of 5-bromo-4-t-butyl-6-methylpyrimidine (by loss of hypobromous, rather than hydrobromic, acid)[543]. The reactions of other halogeno-pyrimidines have been examined briefly[543].

(58)

(59)

Ylide photochemistry continues to attract attention; 1-ethoxycarbonyl-imino-[544] and 1-acetylimino-pyridinium[545] ylides undergo intramolecular, photochemical 1,3-dipolar addition leading to diazepines in good yields, but, in contrast, 1-phenyliminopyridinium ylides form the corresponding pyridines and phenyl nitrene[33, 39].

The photochemical reactions of azine N-oxides have been reviewed[546] and ground- and excited-state properties of pyridine 1-oxide calculated in this connection[547]. Pentaphenyl- and 2,3,5,6-tetraphenyl-pyridine 1-oxides form the corresponding 1,3-oxazepines in good yield, presumably via the oxaziridines[27]. Photochemical rearrangement of pyridine 1-oxide to 2-formylpyrrole proceeds only in low yield[548]; the oxygen transfer previously observed in aromatic solvents (forming, for example, phenol and pyridine from pyridine 1-oxide in benzene) also occurs when pyridazine N-oxides are irradiated in the presence of polymethylbenzenes[549]. The extent of oxygen transfer from 2-cyanopyridine 1-oxide is increased by sensitisation of the N-oxide triplet state[550]. Photolysis of 3,4,5,6-tetraphenylpyridazine 1-oxide in dichloromethane leads mainly to ring opening, and the formation of tetraphenylfuran, isomeric dibenzoylstilbenes and 1-(1,2,3-triphenylcyclo-propenyl)-$\Delta^{3, 6}$-bicyclo[3,2,0]heptadien-2-one can be rationalised in terms of the same intermediate (59)[551]. (In the isoquinolic series, a recent study has failed to find evidence for intermediates of this type).

2- and 4-Hydroxymethylpyridines form the methylpyridines (and 1,2-di-pyridylethanes) on irradiation in isopropanol–aqueous acid solutions[552]. The photolysis of dimethylsulphoxonium 4-hydroxy-2-pyrimidinemethylides in water or methanol leads to 2-hydroxymethyl- or 2-methoxymethyl-4-hydroxypyrimidine; in the presence of sodium borohydride, 2-hydroxy-4-methylpyrimidine is formed[553]. Irradiation of 2-azido-5,6-diphenylpyrazine gives 4,5-diphenylimidazole in low yield, probably by intermolecular cycloaddition of the nitrene followed by valence bond isomerisation[467].

Photochemical Fries reactions have been reported for pyridyl 2- and 3-(but not 4-)benzoates[554]. 2-Aryloxy-1,3,5-triazines undergo photo-Smiles rearrangements if the aryl group bears an ortho-amino substituent; if not, a photo-Fries reaction is preferred[555].

Diphenylacetylene and cycloalkano-2-pyridones undergo photocyclo-addition (initially [4+2] followed by photoreorganisation) giving pentacyclic lactams, although pyridone dimerisation competes[556]. New photoproducts have been characterised from the irradiation of uracil[557], 1,3-dimethyluracil[558] and thymine[559, 560]. New examples of oxidative photocyclisation of 2-stilbazoles have been reported[561], and benzo[c]carbazoles are formed in good yield from 1-(2-indolyl)-2-pyridylethylenes[562]. The photocyclisation of (29) to (30) has been reported without experimental details[322].

References

1. Boldt, P., Thielecke, W. and Oberdörfer, J. (1970). *Angew Chem., Int. Ed. Engl.*, **9**, 377
2. Alvarez-Insua, A., Loro-Tamayo, M. and Soto, J. L. (1970). *J. Heterocycl. Chem.*, **7**, 1305
3. Stensrud, T., Bernatek, E. and Johnsgaard, M. (1971). *Acta Chem. Scand.*, **25**, 523
4. Wang, C. S., Easterly, J. P. and Skelly, N. E. (1971). *Tetrahedron*, **27**, 2581
5. Charman, H. B. and Rowe, J. M. (1971). *Chem. Commun.*, 476
6. Biellmann, J. F. and Callot, H. J. (1970). *Tetrahedron*, **26**, 4799
7. Biellmann, J. F. and Callot, H. J. (1970). *Tetrahedron*, **26**, 4655
8. Newcome, G. R. and Fishel, D. L. (1970). *Chem. Commun.*, 916
9. Rafla, F. K. and Khan, M. A. (1971). *J. Chem. Soc. C*, 2044
10. Van Allen, J. A., Reynolds, G. A., Petropoulos, C. C. and Maier, D. P. (1970). *J. Heterocycl. Chem.*, **7**, 495
11. Reynolds, G. A., Van Allen, J. A. and Petropoulos, C. C. (1970). *J. Heterocycl. Chem.*, **7**, 1061
12. Kametani, T., Kozuka, A. and Tanaka, S. (1970). *J. Pharm. Soc. Japan.*, **90**, 1574
13. Marchesini, A. (1971). *Tetrahedron Letters*, 671
14. Breitmaier, E. and Bayer, E. (1970). *Tetrahedron Letters*, 3291
15. Kato, T., Yamanaka, H. and Horumi, T. (1971). *Yakugaku Zasshi*, **91**, 772
16. Ruangsiyanand, C., Rimek, H. and Zymalkowski, P. (1970). *Chem. Ber.*, **103**, 2403
17. Carney, R. W. J., Wojtkunshi, J., Fechtig, B., Puckett, R. T., Biffar, B. and De Stevens, G. (1971). *J. Org. Chem.*, **36**, 2602
18. Fuks, R. (1970). *Tetrahedron*, **26**, 2161
19. Butsugan, Y., Yoshida, S., Muto, M., Bito, T., Matsuura, T. and Nakashima, R. (1971). *Tetrahedron Letters*, 1129
20. Eloy, F. and Deryckere, A. (1970). *J. Heterocycl. Chem.*, **7**, 1191
21. Eloy, F. and Deryckere, A. (1970). *Helv. Chim. Acta*, **53**, 654
22. Simchen, G. (1970). *Chem. Ber.*, **103**, 389
23. Butler, J. D. and Laundon, R. D. (1970). *J. Chem. Soc. B*, 716
24. Gambacorta, A., Nicoletti, R. and Forcellese, M. L. (1971). *Tetrahedron*, **27**, 985
25. Stork, G., Chashi, M., Kamachi, H. and Kakisawa, H. (1971). *J. Org. Chem.*, **36**, 2784
26. Rafla, F. K. (1971). *J. Chem. Soc. C*, 2048
27. Pedersen, C. L., Harrit, N. and Buchardt, O. (1970). *Acta Chem. Scand.*, **24**, 3435
28. Georgi, U. K. and Retey, J. (1971). *Chem. Commun.*, 32
29. Boyd, G. V. and Dando, S. E. (1971). *J. Chem. Soc. C*, 3873
30. Marvell, E. N., Caple, G. and Shahidi, I. (1970). *J. Amer. Chem. Soc.*, **92**, 5641
31. Marvell, E. N. and Shahidi, I. (1970). *J. Amer. Chem. Soc.*, **92**, 5646
32. Kan, G., Thomas, M. T. and Snieckus, V. (1971). *Chem. Commun.*, 1022
33. Snieckus, V. and Kan, G. (1970). *Chem. Commun.*, 172, 1208
34. Moore, J. A., Volker, E. J. and Kopey, C. M. (1971). *J. Org. Chem.*, **36**, 2676
35. Paquette, L. A. and Kakihara, T. (1971). *J. Amer. Chem. Soc.*, **93**, 174
36. Brown, D. J. and England, B. T. (1970). *Aust. J. Chem.*, **23**, 625
37. Mohan, A. G. (1970). *J. Org. Chem.*, **35**, 3982
38. Landquist, J. K. and Meek, S. E. (1970). *Chem. Ind. (London)*, 688
39. Holava, H. M. and Partyka, R. A. (1971). *J. Med. Chem.*, **14**, 262
40. Novitskii, K. U. and Kasyanova, E. F. (1970). *Khim. Geterotsikl. Soedin*, 1306
41. Eicher, T. and von Angerer, E. (1970). *Chem. Ber.*, **103**, 555
42. Breitmaier, E. (1971). *Angew. Chem. Int. Edn. Engl.*, **10**, 268
43. Morita, K., Kobayashi, S., Shimadzu, H. and Ochiai, M. (1970). *Tetrahedron Letters*, 861
44. Crenshaw, R. R. and Partyka, R. A. (1970). *J. Heterocycl. Chem.*, **7**, 871
45. Braden, R., Findeisen, K. and Holtschmidt, H. (1970). *Angew. Chem. Int. Edn. Engl.*, **9**, 65
46. Hoberg, H. and Barluenga Mur, J. (1970). *Synthesis*, 363
47. Raap, R. (1971). *Can. J. Chem.*, **49**, 1792
48. Keana, J. F. W., Mason, F. P. and Bland, J. S. (1969). *J. Org. Chem.*, **34**, 3705
49. Katner, A. S. and Ziege, E. A. (1971). *Chem. Commun.*, 864
50. Clauss, K. and Jensen, H. (1970). *Tetrahedron Letters*, 119, 123
51. Rey-Bellet, G. and Reiner, R. (1970). *Helv. Chim. Acta*, **53**, 945

52. Nishiwaki, T. (1970). *Bull. Chem. Soc. Japan*, **43**, 937
53. Perini, F. and Tieckelmann, H. (1970). *J. Org. Chem.*, **35**, 812
54. Nishigaki, S., Senga, K., Aida, K., Takabatake, T. and Yondea, F. (1970). *Chem. Pharm. Bull.*, **18**, 1003
55. Nakumizo, N. (1971). *Bull. Chem. Soc. Japan.*, **44**, 2006
56. Clark, J. and Munawar, Z. (1971). *J. Chem. Soc. C*, 1945
57. Kreutzberger, A. and Schucker, R. (1971). *Tetrahedron*, **27**, 3247
58. Jonak, J. P., Zakrewski, S. F., Mead, L. H. and Hakala, M. T. (1970). *J. Med. Chem.*, **13**, 1170
59. Jonak, J. P., Zakrewski, S. F. and Mead, L. H. (1971). *J. Med. Chem.*, **14**, 408
60. Cossey, A. L. and Phillips, J. N. (1970). *Chem. Ind. (London)*, 58
61. Brown, D. J. and Lee, T.-C. (1970). *J. Chem. Soc. C*, 214
62. Wagner, R. M. and Jutz, C. (1971). *Chem. Ber.*, **104**, 2975
63. Péne, C. (1970). *Chim. Ther.*, **5**, 111
64. Brown, D. J. and England, B. T. (1971). *J. Chem. Soc. C*, 2507
65. Whitehead, C. W. and Traverso, J. J. (1956). *J. Amer. Chem. Soc.*, **78**, 5294
66. Tsatsaronis, G. C. and Kehayoglou, A. H. (1970). *J. Org. Chem.*, **35**, 438
67. Tsatsaronis, G. C. and Kehayoglou, A. H. (1971). *Recl. Trav. Chim. Pays-Bas*, **90**, 548
68. Sotiropoulos, J. and Lamazouere, A.-M. (1970). *Compt. Rend. Acad. Sci. Ser. C*, **271**, 1592
69. Chambers, R. D., MacBride, J. A. H. and Musgrave, W. K. R. (1971). *J. Chem. Soc. C*, 3384
70. Mercer, J. F. B. and Warrener, R. N. (1970). *Chem. Ind. (London)*, 927
71. Warrener, R. N. and Cain, E. N. (1971). *Aust. J. Chem.*, **24**, 785
72. de Koning, H. and Pandit, U. K. (1971). *Recl. Trav. Chim. Pays-Bas*, **90**, 874
73. Johnson, D. W., Austel, V., Feld, R. S. and Lemal, D. M. (1970). *J. Amer. Chem. Soc.*, **92**, 7505
74. Chambers, R. D., Musgrave, W. K. R. and Srivastava, K. C. (1971). *Chem. Commun.*, 264
75. Sharp, W. and Spring, F. S. (1951). *J. Chem. Soc.*, 932
76. Karpetsky, T. P. and White, E. H. (1971). *J. Amer. Chem. Soc.*, **93**, 2333
77. Paudler, W. W. and Chen, T.-K. (1970). *J. Heterocycl. Chem.*, **7**, 767
78. Saraswathi, T. V. and Srinivasan, V. R. (1971). *Tetrahedron Letters*, 2315
79. Lalezari, I. and Golgolab, H. (1970). *J. Heterocycl. Chem.*, **7**, 689
80. Lalezari, I., Shafiee, A. and Yalpani, M. (1971). *J. Heterocycl. Chem.*, **8**, 689
81. Case, F. H. (1970). *J. Heterocycl. Chem.*, **7**, 1001
82. Case, F. H. (1971). *J. Heterocycl. Chem.*, **8**, 173
83. Uchytilova, V., Fiedler, P., Prystas, M. and Gut, J. (1971). *Collect. Czech. Chem. Commun.*, **36**, 1955
84. Reimlinger, H., Lingier, W. R. F. and Merenyi, R. (1971). *Chem. Ber.*, **104**, 2793
85. Uchytilova, V. and Gut, J. (1971). *Collect. Czech. Chem. Commun.*, **36**, 2383
86. Becker, H. G. O., Beyer, D., Israel, G., Muller, R., Riediger, W. and Tinipe, H.-J. (1970). *J. Prakt. Chem.*, **312**, 669
87. Nishiwaki, T. and Saito, T. (1970). *Chem. Commun.*, 1479
88. Nishiwaki, T. and Saito, T. (1971). *J. Chem. Soc. C*, 2648
89. Kay, I. T., reported in Reference 210
90. Haas, A. and Hinsch, W. (1971). *Chem. Ber.*, **104**, 1855
91. Neunhoeffer, H., Weischedel, F. and Bohnisch, V. (1971). *Justus Liebigs Ann. Chem.*, **750**, 12
92. Wakabayashi, K., Tsunoda, M. and Yokoo, A. (1971). *Bull. Chem. Soc. Japan*, **44**, 148
93. Singh, B. and Collins, J. C. (1971). *Chem. Commun.*, 498
94. Goerdeler, J. and Neuffer, J. (1971). *Chem. Ber.*, **104**, 1580
95. Goerdeler, J. and Neuffer, J. (1971). *Chem. Ber.*, **104**, 1606
96. Neuffer, J. and Goerdeler, J. (1971). *Chem. Ber.*, **104**, 3498
97. Maeda, K. and Hayashi, T. (1971). *Bull. Chem. Soc. Japan*, **44**, 533
98. Desiderato, R. and Terry, J. C. (1971). *J. Heterocycl. Chem.*, **8**, 617; idem. (1970). *Tetrahedron Letters*, 3203
99. Ulku, D., Huddle, B. P. and Morrow, J. C. (1971). *Acta Crystallogr.* **B27**, 432
100. Laing, M., Sparrow, N. and Sommerville, P. (1971). *Acta Crystallogr.*, **B27**, 1986
101. Almloef, J., Kvick, A. and Olovson, I. (1971). *Acta Crystallogr.*, **B27**, 1201

102. Restivo, R. and Palenik, G. J. (1970). *Acta Crystallogr.* **B26,** 1397
103. Furberg, S. and Solbakk, J. (1970). *Acta Chem. Scand.,* **24,** 3230
104. Oberhänsli, W. E. (1970). *Helv. Chem. Acta,* **53,** 1787
105. Rohrer, D. C. and Sundaralingam, M. (1970). *J. Amer. Chem. Soc.,* **92,** 4950
106. Shefter, E. (1970). *J. Chem. Soc. B,* 903
107. Gatehouse, B. M. and Craven, B. M. (1971). *Acta Crystallogr.,* **B27,** 1337
108. Green, E. A., Shiono, R., Rosenstein, R. D. and Abraham, D. J. (1971). *J. Chem. Soc. D,* 53
109. Hawkinson, S. W. and Coulter, C. L. (1971). *Acta Crystallogr.,* **B27,** 34
110. Lin, G. H. and Sundaralingam, M. (1971). *Acta Crystallogr.,* **B27,** 961
111. Saenger, W. and Suck, D. (1971). *Acta Crystallogr.,* **B27,** 1178
112. Lin, G. H., Sundaralingam, M. and Arora, S. K. (1971). *J. Amer. Chem. Soc.,* **93,** 1235
113. Saenger, W. and Suck, D. (1971). *Acta Crystallogr.,* **B27,** 2105
114. Flippen, J. L. and Karle, I. L. (1971). *J. Amer. Chem. Soc.,* **93,** 2762
115. Gibson, J. W. and Karle, I. L. (1971). *J. Crystallogr. Molec. Struct.,* **1,** 115
116. Konnert, J. and Karle, I. L. (1971). *J. Crystallogr. Molec. Struct.,* **1,** 107
117. Konnert, J., Gibson, J. W., Karle, I. L., Khattak, M. N. and Warg, S. Y. (1970). *Nature (London),* **227,** 953
118. Camerman, N. and Camerman, A. (1970). *J. Amer. Chem. Soc.,* **92,** 2523
119. Lever, A. B. P., Lewis, J. and Nyholm, R. S. (1962). *J. Chem. Soc.,* 1235
120. Carreck, P. W., Goldstein, M., McPartlin, E. M. and Unsworth, W. D. (1971). *Chem. Commun.,* 1634
121. Verschoor, G. C. and Keulen, E. (1971). *Acta Crystallogr.,* **B27,** 134
122. Coppens, P. and Aafje-vos, A. (1971). *Acta Crystallogr.,* **B27,** 146
123. Doraiswamy, S. and Sharma, S. D. (1970). *Curr. Science,* **40,** 398
124. Blackman, G. L., Brown, R. D. and Burden, F. R. (1970). *J. Molec. Spectrosc.,* **35,** 444
125. Brown, R. D., Burden, F. R. and Garland, W. (1970). *Chem. Phys. Letters,* **7,** 461
126. Baker, A. D. and Turner, D. W. (1970). *Phil. Trans. Royal Soc. London,* **268,** 131
127. Gleiter, R., Heilbronner, E. and Hornung, V. (1970). *Angew. Chem. Int. Edn. Engl.,* **9,** 901
128. Baker, A. D., Betteridge, D., Kemp, N. R. and Kirby, R. E. (1970). *Chem. Commun.,* 286
129. Dewar, M. J. S. and Worley, S. D. (1969). *J. Chem. Phys.,* **51,** 263
130. Turner, D. W., Baker, C., Baker, A. D. and Brundle, C. R. (1970). *Molecular Photoelectron Spectroscopy,* 327. (New York: Wiley–Interscience)
131. Worley, S. D. (1971). *Chem. Rev.,* **71,** 295
132. Sundbom, M. (1971). *Acta Chem. Scand.,* **25,** 487
133. Märtensson, O. (1970). *Acta Chem. Scand.,* **24,** 3767
134. Kuthan, J., Koshima, N. V. and Skala, V. (1971). *Collect. Czech. Chem. Commun.,* **36,** 1832
135. Skala, V. and Kuthan, J. (1970). *Collect. Czech. Chem. Commun.,* **35,** 354; **35,** 2378
136. Kuthan, J., Koshima, N. V. and Ferles, M. (1970). *Collect. Czech. Chem. Commun.,* **35,** 3825
137. Konishi, H., Kata, H. and Yonezawa, T. (1970). *Theoret. Chim. Acta,* **19,** 71
138. Bodor, N., Dewar, M. J. S. and Harget, A. J. (1970). *J. Amer. Chem. Soc.,* **92,** 2929
139. Kwiatkowski, J. S. and Olzacka, J. (1970). *Bull. Acad. Pol. Sci. Ser. Sci. Math. Astron. Phys.,* **18,** 215
140. Kwiatkowski, J. S., Berndt, M. and Fabian, J. (1970). *Acta Phys. Pol. A,* **38,** 365
141. Minisci, F., Galli, R., Malatesta, V. and Caronna, T. (1970). *Tetrahedron,* **26,** 4083
142. Johnson, D. W., Austel, V., Feld, R. S. and Lemal, D. M. (1970). *J. Amer. Chem. Soc.,* **92,** 7505
143. Von Ostwalden, P. W. and Roberts, J. D. (1971). *J. Org. Chem.,* **36,** 3792
144. Simchen, G. (1970). *Chem. Ber.,* **103,** 398
145. Thomas, W. A. and Griffin, C. E. (1970). *Org. Magn. Resonance,* **2,** 503
146. Stewart, W. E. and Siddall, T. H. (1970). *J. Phys. Chem.,* **74,** 2027
147. Mielke, I. and Ringsdorf, H. (1970). *J. Polym. Sci. C,* 107
148. Möhrle, H. and Weber, H. (1970). *Tetrahedron,* **26,** 3779
149. Konishi, K., Takahayshi, K. and Asami, R. (1971). *Bull. Chem. Soc. Japan,* **44,** 2281
150. Sutherland, D. R. and Tennant, G. (1971). *J. Chem. Soc. C,* 2156
151. Lardenois, P., Selim, M. and Selim, M. (1971). *Bull. Soc. Chim. France,* 1858
152. Brown, D. J. and Hoskins, J. A. (1971). *J. Chem. Soc. B,* 2214
153. Slesarev, V. I., Ivin, B. A., Smorygo, N. A., Muravich-Alexander, K. L. and Sochilin, E. G. (1970). *Zh. Org. Khim.,* **6,** 1323

154. Ivin, B. A., Slesarev, V. I., Smorygo, N. A. and Sochilin, E. G. (1970). *Zh. Org. Khim.*, **6,** 1326
155. Brandani, V. and Cignitti, M. (1969). *Ric. Sci.*, **39,** 607
156. Barlin, G. B. and Young, A. C. (1971). *J. Chem. Soc. B*, 1675
157. Pollak, A., Stanovnik, B. and Tisler, M. (1970). *J. Org. Chem.*, **35,** 2478
158. Kakadate, M., Sueyoshi, S. and Suzuki, I. (1970). *Chem. Pharm. Bull.*, **18,** 1211
159. Uchimaru, F., Okada, S., Kosasayama, A. and Konno, T. (1971). *Chem. Pharm. Bull.*, **19,** 1337, 1344
160. Ulyanova, T. N., Dvoryantseva, G. G., Alekseeva, L. M. and Sheinker, U. N. (1971). *Khim. Geterot. Soedin.*, 846
161. Swain, C. G. and Lupton, E. C. (1968). *J. Amer. Chem. Soc.*, **90,** 4328
162. Haigh, J. M. and Thornton, D. A. (1970). *Tetrahedron Letters*, 2043
163. Sleinkman, A. K., Kaplan, L. M., Gakh, L. G., Titov, E. V., Baranov, S. N. and Kost, A. N. (1970). *Dokl. Akad. Nauk. SSSR*, **193,** 366
164. Wehrli, F. W., Giger, W. and Simon, W. (1971). *Helv. Chim. Acta*, **54,** 229
165. Bramwell, M. R. and Randall, E. W. (1970). *Spectrochim. Acta*, **26A,** 1877
166. Cox, R. H. and Bothner-By, A. A. (1968). *J. Chem. Phys.*, **72,** 1642, 1646
167. Bramwell, A. F., Riezebos, G. and Wells, R. D. (1971). *Tetrahedron Letters*, 2489
168. Paudler, W. W. and Humphrey, S. A. *Org. Magnetic Res.*, **3,** 217
169. Beauté, C., Wolkowski, Z. W. and Thoai, N. (1971). *Tetrahedron Letters*, 817
170. Armarego, W. L. F., Batterham, T. J. and Kershaw, J. R. (1971). *Org. Magnetic Res.*, **3,** 575
171. Witanowski, M., Stefanik, L., Januszewski, H. and Wolkowski, Z. W. (1971). *Tetrahedron Letters*, 1653
172. Morishima, I., Yonezawa, T. and Goto, K. (1970). *J. Amer. Chem. Soc.*, **92,** 6651
173. Doddrell, D. and Roberts, J. D. (1970). *J. Amer. Chem. Soc.*, **92,** 6839
174. Gil, V. and Pinto, A. J. (1970). *Molec. Phys.*, **19,** 573
175. Giger, W. and Simon, W. (1970). *Helv. Chim. Acta*, **53,** 1609
176. Wagner, R. and von Philipsborn, W. (1970). *Helv. Chim. Acta*, **53,** 299
177. Biffin, M. E. C., Miller, J., Moritz, A. G. and Paul, D. B. (1970). *Aust. J. Chem.*, **23,** 957
178. Biffin, M. E. C., Miller, J., Moritz, A. G. and Paul, D. B. (1970). *Aust. J. Chem.*, **23,** 963
179. Chatrousse, A. P., Terrir, F. and Schaal, R. (1970). *Compt. Rend. Acad. Sci. Ser. C*, **271,** 1477
180. Schaal, R., Terrier, F., Halle, J. C. and Chatrousse, A. P. (1970). *Tetrahedron Letters*, 1393
181. Epsztayn, J., Lunt, E. and Katritzky, A. R. (1970). *Tetrahedron*, **26,** 1665
182. Hammes, G. G. and Lilliford, P. J. (1970). *J. Amer. Chem. Soc.*, **92,** 7578
183. Chambers, R. D., Jackson, J. A., Musgrave, W. K. R., Sutcliffe, L. H. and Tiddy, G. J. T. (1970). *Tetrahedron*, **26,** 71
184. Giam, C. S. and Lyle, J. L. (1970). *J. Chem. Soc. B*, 1516
185. Allison, C. G., Chambers, R. D., MacBride, J. A. H. and Musgrave, W. K. R. (1970). *J. Chem. Soc. C*, 1023
186. Jones, A. J., Grant, D. M., Winkley, M. W. and Robins, R. K. (1970). *J. Amer. Chem. Soc.*, **92,** 4079
187. Jones, A. J., Grant, D. M., Winkley, M. W. and Robins, R. K. (1970). *J. Phys. Chem.*, **74,** 2684
188. Dorman, D. E. and Roberts, J. D. (1970). *Proc. Natl. Acad. Sci. U.S.*, **65,** 27
189. Tarpley, A. R. and Goldstein, J. H. (1971). *J. Amer. Chem. Soc.*, **93,** 3573
190. Witanowski, M., Januszewski, H. and Webb, G. A. (1971). *Tetrahedron*, **27,** 3129
191. Jakobsen, H. J. (1970). *J. Molec. Spectrosc.*, **34,** 245
192. Griffin, G. E. and Thomas, W. A. (1970). *J. Chem. Soc. B*, 477
193. Guibe, L., Linscheid, P. and Lucken, E. A. C. (1970). *Molec. Phys.*, **19,** 317
194. Fleming, H. C. and Hanna, M. W. (1971). *J. Amer. Chem. Soc.*, **93,** 5030
195. Schempp, E. and Bray, P. J. (1971). *J. Magnetic Res.*, **5,** 78
196. Eletr, S. H. (1970). *Molec. Phys.*, **18,** 119
197. Kasai, P. H. and McLeod, D. (1970). *J. Amer. Chem. Soc.*, **92,** 6085
198. Buick, A. R., Kemp, T. J. and Stone, T. J. (1970). *J. Phys. Chem.*, **74,** 3439
199. Hirst, D. M. (1971). *Theoret. Chim. Acta*, **20,** 292
200. Filipesui, N., Geiger, F. E., Trichilo, C. L. and Minn, F. L. (1970). *J. Phys. Chem.*, **74,** 4311

201. Sevilla, M. D. (1970). *J. Phys. Chem.*, **74**, 805
202. Malkus, H., Battiste, M. A. and White, R. M. (1970). *Chem. Commun.*, 479
203. Green, J. H. S., Harrison, D. J., Kynaston, W. and Paisley, H. M. (1970). *Spectrochim. Acta*, **26A**, 2139
204. Kanaskova, Y., Pentin, Y. A. and Sukhorukov, B. I. (1970). *Opt. Spektrosk*, **29**, 679
205. Shoffner, J. P., Bauer, L. and Bell, C. L. (1970). *J. Heterocycl. Chem.*, **7**, 479
206. Sigworth, W. D. and Pace, E. L. (1971). *J. Chem. Phys.*, **54**, 5379
207. Sigworth, W. D. and Pace, E. L. (1971). *Spectrochim. Acta*, **27A**, 747
208. Varsanyi, G., Szoke, S., Keresztury, G. and Gelleri, A. (1970). *Acta Chim.*, **65**, 73
209. Czuba, W. and Poradowska, H. (1970). *Roczniki. Chem.*, **44**, 1447
210. Szafran, M. and Brzezinski, B. (1970). *Bull. Akad. Pol. Sci., Ser. Sci. Chim.*, **18**, 247
211. Dega-Szafran, Z. (1970). *Roczniki. Chem.*, **44**, 2371
212. Ezumi, K., Miyazaki, H. and Kubota, T. (1970). *J. Phys. Chem.*, **74**, 2397
213. Baruah, G. D., Amma, R. A., Dube, P. S. and Rai, S. N. (1970). *Indian J. Pure Appl. Phys.*, **8**, 761
214. Bailey, R. T. and Strachan, G. P. (1970). *Spectrochim. Acta*, **26A**, 1129
215. Lafaix, A. J. and Lebas, J. M. (1970). *Spectrochim. Acta*, **26A**, 1243
216. Foglizzo, R. and Novak, A. (1970). *Appl. Spectrosc.*, **24**, 601
217. Foglizzo, R. and Novak, A. (1970). *Spectrochim. Acta*, **26A**, 2281
218. Thomas, D. M., Bates, J. B., Bandy, A. and Lippincott, E. R. (1970). *J. Chem. Phys.*, **53**, 3698
219. Shoffner, J. P., Bauer, L. and Bell, C. L. (1970). *J. Heterocycl. Chem.*, **7**, 487
220. Pitha, J. (1970). *J. Org. Chem.*, **35**, 903
221. Yarwood, J. (1970). *Spectrochim. Acta*, **26A**, 2099
222. Tincher, W. C. and Pearson, E. F. (1970). *J. Molec. Spectrosc.*, **36**, 114
223. Nishi, N., Shimada, R. and Kanda, Y. (1970). *Bull. Chem. Soc. Japan*, **43**, 41
224. Nakamura, K. (1971). *J. Amer. Chem. Soc.*, **93**, 3138
225. Ishi, H., Koyanagi, M. and Kanda, Y. (1971). *Bull. Soc. Chem. Japan*, **44**, 1205
226. Kimble, C. D. (1970). *U.S. Govt. Res. Develop. Rep.*, **70**, 78
227. Livak, D. T. and Innes, K. K. (1971). *J. Molec. Spectrosc.*, **39**, 115
228. Eaton, W. A. and Lewis, T. P. (1970). *J. Chem. Phys.*, **53**, 2164
229. Tanaka, M. and Tanaka, J. (1971). *Bull. Chem. Soc. Japan*, **44**, 938
230. Hollas, J. M., Kirby, G. H. and Wright, R. A. (1970). *Molec. Phys.*, **18**, 327
231. Brand, J. C. D. and Tang, K. T. (1971). *J. Molec. Spectrosc.*, **39**, 171
232. Kuthan, J., Koshima, N. V. and Skala, V. (1971). *Collect. Czech. Chem. Commun.*, **36**, 1832
233. Kuthan, J. and Ichova, M. (1971). *Collect. Czech. Chem. Commun.*, **36**, 1413
234. Kwiatkonski, J. S. (1971). *Acta Phys. Polon.*, **A39**, 587
235. Yoshida, Z. and Tsuetoshi, T. (1970). *Theoret. Chim. Acta*, **19**, 377
236. Bustard, T. M. and Jaffé, H. H. (1970). *J. Chem. Phys.*, **53**, 534
237. Gropen, O. and Skancke, P. N. (1970). *Acta Chem. Scand.*, **24**, 1768
238. Bailey, M. L. (1970). *Theoret. Chim. Acta*, **16**, 309
239. Carper, W. R. and Stengl, J. (1970). *Spectrochim. Acta*, **26A**, 307
240. Kwiatkowski, J. S. (1970). *Theoret. Chim. Acta*, **16**, 243
241. Spinner, E. and Yeoh, G. B. (1971). *J. Chem. Soc. B*, 279
242. Spinner, E. and Yeoh, G. B. (1971). *J. Chem. Soc. B*, 289
243. Spinner, E. and Yeoh, G. B. (1971). *J. Chem. Soc. B*, 296
244. Barlin, G. B. and Pfleiderer, W. (1971). *J. Chem. Soc. B*, 1425
245. Beggiato, G., Favaro, G. and Mazzucato, U. (1970). *J. Heterocycl. Chem.*, **7**, 583
246. Tandon, S. P. and Bhutra, M. P. (1970). *Z. Naturforsch. B*, **25**, 343
247. Arantz, B. W. and Brown, D. J. (1971). *J. Chem. Soc. C*, 1889
248. Brown, D. J. and England, B. T. (1971). *J. Chem. Soc. C*, 250
249. Roth, B. and Strelitz, J. Z. (1970). *J. Org. Chem.*, **35**, 2696
250. Johnson, C. D., Katritzky, A. R., Kingsland, M. and Scriven, E. F. V. (1971). *J. Chem. Soc. B*, 1
251. Pfleiderer, W. (1971). *Justus Liebigs Ann. Chem.*, **747**, 111
252. Smith, H. E. and Hicks, A. A. (1970). *Chem. Commun.*, 1112
253. Musker, W. K. and Scholl, R. L. (1971). *J. Organometallic Chem.*, **27**, 37
254. Dua, S. S., Edmondson, R. C. and Gilman, H. (1971). *J. Organometallic Chem.*, **27**, 33
255. Kaganskii, M. M., Dvoryantseva, G. G. and Elina, A. S. (1971). *Dokl. Akad. Nauk. SSSR*, **197**, 832

256. Barlin, G. B. and Young, A. C. (1971). *J. Chem. Soc. B*, 1261
257. Mackay, R. A. and Poziomek, E. J. (1970). *J. Amer. Chem. Soc.*, **92**, 2432
258. Aloisi, G., Beggiata, G. and Mazzucato, U. (1970). *Trans. Faraday Soc.*, **66**, 3075
259. Vernulapalli, G. K. (1970). *J. Amer. Chem. Soc.*, **92**, 7589
260. Mackay, R. A., Landolph, J. R. and Poziomek, E. J. (1971). *J. Amer. Chem. Soc.*, **93**, 5026
261. Carter, S., Gray, N. A. B. and Wood, J. L. (1971). *J. Molec. Struct.*, **7**, 481
262. Shine, H. J. and Goodin, R. D. (1970). *J. Org. Chem.*, **35**, 949
263. Hensen, K., Messer, K. P. and Pickel, P. (1970). *Chem. Ber.*, **103**, 2091
264. Kalyanaraman, V., Rao, C. N. R. and George, M. V. (1971). *J. Chem. Soc. B*, 2406
265. Reichardt, C. and Halbritter, K. (1971). *Chem. Ber.*, **104**, 822
266. Bellobono, I. R. and Favini, G. (1971). *J. Chem. Soc. B*, 2034
267. Brown, E. V. and Plasz, A. C. (1970). *J. Heterocycl. Chem.*, **7**, 335
268. Ellam, G. B. and Johnson, C. D. (1971). *J. Org. Chem.*, **36**, 2284
269. Cook, M. J., Katritzky, A. R., Linda, P. and Tack, R. D. (1971). *Chem. Commun.*, 510
270. Christensen, J. J., Rytting, J. H. and Izatt, R. M. (1970). *J. Chem. Soc. B*, 1643
271. Phillips, W. G. and Ratts, K. W. (1970). *J. Org. Chem.*, **35**, 3144
272. Schulman, S. G., Capomachia, A. C. and Rietta, M. S. (1971). *Analyt. Chim. Acta*, **56**, 91
273. Wallenfels, K. and Hanstein, W. (1970). *Justus Liebigs Ann. Chem.*, **732**, 139
274. Wiberg, K. B. and Lewis, T. P. (1970). *J. Amer. Chem. Soc.*, **22**, 7154
275. Lightner, D. A., Nicoletti, R., Quistad, G. B. and Irwin, E. (1970). *Org. Mass Spectrom.*, **4**, 571
276. Neeter, R., Binnering, N. M. M. and de Boer, Th. J. (1970). *Org. Mass Spectrometry*, **3**, 597
277. Terent'ev, P. B., Khmel'nitskii, R. A., Khromov, I. S., Kost, A. N., Gloriozov, I. P. and Islam, M. (1970). *Zh. Org. Khim.*, **6**, 606
278. Bursey, M. M. and Elwood, T. A. (1970). *J. Org. Chem.*, **35**, 793
279. Moser, R. J. and Brown, E. V. (1970). *Org. Mass Spectrometry*, **4**, 555
280. Neeter, R. and Nibbering, N. M. M. (1971). *Org. Mass Spectrom.*, **5**, 735
281. Brown, E. V. and Moser, R. T. (1971). *J. Heterocycl. Chem.*, **8**, 189
282. McCloskey, J. A., Leemans, R. A. and Prochaska, P. O. (1970). *Arch. Mass. Spectral. Data*, **1**, 40
283. Smithson, L. D., Bhattacharya, A. K. and Tamborski, C. (1970). *Org. Mass Spectrom.* **4**, 1
284. Filyugina, A. D., Dyumaev, K. M., Dubrovin, A. A. and Rotermel, I. A. (1970). *Zhur. Org. Khim.*, **6**, 2131
285. Wilson, J. G., Barnes, C. S. and Godsack, R. J. (1970). *Org. Mass Spectrom.*, **4**, 365
286. Wulfson, N. S., Zaikin, V. G., Ziyavidinova, Z. S., Burikov, V. M. and Mukerjee, S. K. (1971). *Org. Mass Spectrom.*, **5**, 743
287. Ikeda, M., Tsijimoto, N. and Tamura, Y. (1971). *Org. Mass Spectrom.*, **5**, 61, 389, 935
288. Ikeda, M., Kato, S., Sumida, Y. and Tamura, Y. (1971). *Org. Mass Spectrom.*, **5**, 1383
289. Kramer, J. M. and Berry, R. S. (1971). *J. Amer. Chem. Soc.*, **93**, 1305
290. Kato, T., Yamanaka, H., Ichikawa, H., Chiba, T., Abe, H. and Sasaki, S. (1970). *Org. Mass Spectrom.*, **4**, 181
291. Falch, E. (1970). *Acta Chem. Scand.*, **24**, 137
292. Fales, H. M., Milne, G. W. A. and Azenrod, T. (1970). *Analyt. Chem.*, **42**, 1432
293. Undheim, K. and Hvistendahl, G. (1971). *Org. Mass Spectrom.*, **5**, 325
294. Seifert, R. M., Buttery, R. G., Guadagni, D. G., Black, D. R. and Harris, J. G. (1970). *J. Agr. Food Chem.*, **18**, 246
295. Kolor, M. G. and Rizzo, D. J. (1971). *Org. Mass Spectrom.*, **5**, 959
296. Burrell, J. W. K., Lucas, R. A., Michalkiewicz, D. M. and Riezebos, G. (1970). *Chem. Ind. (London)*, 1409
297. Uchimaru, F., Okada, S., Kosasayama, A. and Konno, T. (1971). *J. Heterocycl. Chem.*, **8**, 99
298. Palmer, M. H., Preston, P. N. and Stevens, M. F. G. (1971). *Org. Mass Spectrom.*, **5**, 1085
299. Sasaki, T., Minamoto, K., Nishikawa, N. and Shima, T. (1969). *Tetrahedron*, **25**, 1021
300. Paudler, W. W. and Chen, T. (1971). *J. Heterocycl. Chem.*, **8**, 317
301. Preston, P. N., Steedman, W., Palmer, M. H., Mackenzie, S. M. and Stevens, M. F. G. (1970). *Org. Mass Spectrom.*, **3**, 863

302. Ross, J. A. and Tweedy, B. G. (1970). *Org. Mass Spectrom.*, **3**, 219
303. Karliner, J. and Seltzer, R. (1971). *J. Heterocycl. Chem.*, **8**, 629
304. Beynon, J. H., Caprioli, R. M. and Ast, T. (1971). *Org. Mass Spectrom.*, **5**, 229
305. Paudler, W. W. and Humphrey, S. A. (1970). *Org. Mass Spectrom.*, **4**, 513
306. Giam, C. G. and Knaus, E. E. (1971). *Tetrahedron Letters*, 4961
307. Santilli, A., Kim, D. H. and Wanser, S. V. (1971). *J. Heterocycl. Chem.*, **8**, 445
308. Nishigaki, S., Senga, K. and Yoneda, F. (1971). *Chem. Pharm. Bull.*, **19**, 1526
309. Carboni, S., Da Settimo, A. and Tonetti, I. (1970). *J. Heterocycl. Chem.*, **7**, 875
310. McCausland, D. J. and Cheng, C. C. (1970). *J. Heterocycl. Chem.*, **7**, 467
311. Minami, S., Shono, T. and Matsumoto, M. (1971). *Chem. Pharm. Bull.*, **19**, 1482, 1426
312. Mullock, E. B., Searby, R. and Suschitzky, H. (1970). *J. Chem. Soc. C*, 829
313. Bowden, K. and Brown, T. H. (1971). *J. Chem. Soc. C*, 2163
314. Walford, G. H., Jones, H. and Shen, T. Y. (1971). *J. Med. Chem.*, **14**, 339
315. Niess, R. and Robins, R. K. (1970). *J. Heterocycl. Chem.*, **7**, 243
316. Jones, A. S. and Warren, J. H. (1970). *Tetrahedron*, **26**, 791
317. Sasaki, T. and Kojima, A. (1971). *Tetrahedron Letters*, 4593
318. Giam, C. S. and Abbott, S. D. (1971). *J. Amer. Chem. Soc.*, **93**, 1294
319. Levine, R. and Kadunce, W. M. (1970). *Chem. Commun.*, 921
320. Villani, F. J., Ellis, C. A., Yudis, M. D. and Morton, J. B. (1971). *J. Org. Chem.*, **36**, 1709
321. Le Noble, W. J. and Ogo, Y. (1970). *Tetrahedron*, **26**, 4119
322. Maki, Y., Izuta, K. and Suzuki, M. (1971). *Chem. Commun.*, 1442
323. Sheradsky, T. and Salemnick, G. (1971). *J. Org. Chem.*, **36**, 1061
324. Smirnov, L. D., Kuzimin, V. I., Lezina, V. P. and Dyumaev, K. M. (1970). *Izv. Akad. Nauk. SSSR. Ser. Khim.*, 2784
325. Tomisawa, H., Kobayashi, Y., Hongo, H. and Fujita, R. (1970). *Chem. Pharm. Bull.*, **18**, 932
326. Otter, B. A., Taube, A. and Fox, J. J. (1971). *J. Org. Chem.*, **36**, 1251
327. Schulte, K. E., von Weissenborn, V. and Tittel, G. L. (1970). *Chem. Ber.*, **103**, 1250
328. Winkley, M. W. (1970). *J. Chem. Soc. C*, 1869
329. Kamiya, S., Okusa, G. and Hirakawa, H. (1970). *Chem. Pharm. Bull.*, **18**, 632
330. Abramovitch, R. A., Kato, S. and Singer, G. M. (1971). *J. Amer. Chem. Soc.*, **93**, 3075
331. Vingiello, F. A., Yanez, J. and Campbell, J. A. (1971). *J. Org. Chem.*, **36**, 2053
332. Zoltewicz, J. A. and Helmick, L. S. (1970). *J. Amer. Chem. Soc.*, **92**, 7547
333. Zoltewicz, J. A., Smith, C. L. and Kauffmann, G. M. (1971). *J. Heterocycl. Chem.*, **8**, 337
334. El-Anani, A., Jones, P. E. and Katritzky, A. R. (1971). *J. Chem. Soc. B*, 2363
335. Beak, P. and Watson, R. N. (1971). *Tetrahedron Letters*, 953
336. Brignell, P. J., Jones, P. E. and Katritzky, A. R. (1970). *J. Chem. Soc. B*, 117
337. Smirnov, L. D., Kuz'min, V. I., Lezina, V. P. and Dymaev, K. M. (1970). *Izv. Akad. Nauk. SSSR. Ser. Khim.*, 2400
338. Wang, C (1970). *J. Heterocycl. Chem.*, **7**, 389
339. Greco, C. V. and Hunsberger, I. M. (1970). *J. Heterocycl. Chem.*, **7**, 761
340. Schofield, K. (1967). *Heteroaromatic Nitrogen Compounds*, 228. (London: Butterworths)
341. Ager, E., Iddon, B. and Suschitzky, H. (1970). *J. Chem. Soc. C*, 1530
342. Yanai, M., Kinoshita, T., Takeda, S., Sadaki, H. and Watanabe, H. (1970). *Chem. Pharm. Bull.*, **18**, 1680
343. Baker, J. A. and Chatfield, P. V. (1970). *J. Chem. Soc. C*, 2478
344. Keana, J. F. W. and Mason, F. P. (1970). *J. Org. Chem.*, **35**, 838
345. Robins, M. J. and Naik, S. R. (1971). *J. Org. Chem.*, **93**, 5277; (1972). *Chem. Commun.*, 18
346. Albert, A. and Ohta, K. (1971). *J. Chem. Soc. C*, 3727
347. Katritzky, A. R., Tarhan, H. O. and Tarhan, S. (1970). *J. Chem. Soc. B*, 114
348. Burton, A. G., Halls, P. J. and Katritzky, A. R. (1971). *Tetrahedron Letters*, 2211
349. Smirnov, L. D., Avezov, M. R., Lezina, V. P., Zaitsev, B. E. and Dyumaev, K. M. (1971). *Izv. Akad. Nauk. SSSR. Ser. Khim.*, 845
350. Nishigaki, S., Ogiwara, K. and Yoneda, F. (1971). *Chem. Pharm. Bull.*, **19**, 418
351. Hurst, D. (1970). *Tetrahedron Letters*, 979
352. Yoneda, F., Shinomura, K. and Nishigaki, S. (1971). *Tetrahedron Letters*, 851
353. Rogers, G. T. and Ulbricht, T. L. V. (1971). *J. Chem. Soc. C*, 2364

354. Chatterjee, S. S., Garrison, D. R., Kaprove, R., Moran, J. F., Triggle, A. M., Triggle, D. J. and Wayne, A. (1971). *J. Med. Chem.*, **14**, 499
355. Birdsall, N. J. M., Lee, T., Delia, T. J. and Parham, J. C. (1971). *J. Org. Chem.*, **36**, 2635
356. Gray, E. A., Hulley, R. M. and Snell, B. K. (1970). *J. Chem. Soc. C*, 986
357. Fersht, A. R. and Jencks, W. P. (1970). *J. Amer. Chem. Soc.*, **92**, 5432
358. Wakselmann, M. and Guibé-Jampel, E. (1970). *Tetrahedron Letters*, 1521
359. Guibé-Jampel, E. and Wakselmann, M. (1971). *Chem. Commun.*, 267
360. Kappé, T. and Lube, W. (1971). *Monatsch*, **102**, 781
361. Haberfield, P., Budelman, A., Bloom, A., Romm, R. and Ginsberg, H. (1971). *J. Org. Chem.*, **36**, 1792
362. Frampton, R., Johnson, C. and Katritzky, A. R. (1971). *Justus Liebigs Ann. Chem.*, **749**, 12
363. Deady, L. W. and Zoltewicz, J. A. (1971). *J. Amer. Chem. Soc.*, **93**, 5475
364. Dickeson, J. E. and Summers, L. A. (1970). *J. Heterocycl. Chem.*, **7**, 719
365. Adamson, J. and Glover, E. E. (1971). *J. Chem. Soc. C*, 861
366. Black, A. L. and Dummers, L. A. (1971). *J. Chem. Soc. C*, 2271
367. Hammer, C. F. and Craig, J. H. (1971). *J. Heterocycl. Chem.*, **8**, 411
368. Ried, W., Schmidt, A. H. and Gildmeister, H. (1971). *Synthesis*, 256
369. Joshi, R. K., Krasnec, L. and Laks, I. (1971). *Helv. Chim. Acta*, **54**, 112
370. Paris, G. Y., Garmaise, D. L. and Komlossy, J. (1971). *J. Heterocycl. Chem.*, **8**, 169
371. Hickmott, P. W. (1971). *J. Chem. Soc. C*, 1675
372. Bailey, A. S., Heaton, M. W. and Murphy, J. I. (1971). *J. Chem. Soc. C*, 1211
373. Quast, H. and Schmitt, E. (1970). *Justus Liebigs Ann. Chem.*, **732**, 64
374. Reimlinger, H., Vandewalle, J. J., King, G. S., Lingier, W. R. and Merenyi, R. (1970). *Chem. Ber.*, **103**, 1918
375. Reimlinger, H., Vandewalle, J. J. and Lingier, W. R. F. (1970). *Chem. Ber.*, **103**, 1960
376. Crowther, C., Johnston, M. Y. and Bradsher, C. K. (1971). *J. Heterocycl. Chem.*, **8**, 157
377. Bradsher, C. K., Burnham, W. S. and Zinn, M. F. (1970). *J. Heterocycl. Chem.*, **7**, 779
378. Binns, F., Roberts, S. M. and Suschitzky, H. (1970). *J. Chem. Soc. C*, 1375
379. Pollak, A., Stanovnik, B. and Tisler, M. (1971). *J. Org. Chem.*, **36**, 2457
380. Potts, K. T. and Armbruster, R. (1971). *J. Org. Chem.*, **36**, 1846
381. Goody, R. S. and Walker, R. T. (1971). *J. Org. Chem.*, **36**, 727
382. Greco, C. V. and Warchol, J. F. (1971). *J. Org. Chem.*, **36**, 604
383. Calligaris, M., Fabrissen, S., de Nardo, M. and Nisi, C. (1971). *J. Org. Chem.*, **36**, 602
384. Brown, G. R. and Dyson, W. R. (1971). *J. Chem. Soc. C*, 1527
385. Clark, J. and Ramsden, T. (1971). *J. Chem. Soc. C*, 1942
386. Reimlinger, H., Jacquier, R. and Daunis, J. (1971). *Chem. Ber.*, **104**, 2702
387. Guesret, P., Jacquier, R. and Maury, G. (1971). *J. Heterocycl. Chem.*, **8**, 643
388. Kröck, F. W. and Kröhnke, F. (1971). *Chem. Ber.*, **104**, 1629
389. Undheim, K. and Tvesta, B. O. (1971). *Acta Chem. Scand.*, **25**, 5
390. Okamoto, Y., Takada, A. and Ueda, T. (1971). *Chem. Pharm. Bull.*, **19**, 764
391. Kaye, H. and Chang, S. (1970). *Tetrahedron*, **26**, 1369
392. Wittenberg, E. (1971). *Collect. Czech. Chem. Commun.*, **36**, 246
393. Sasaki, T., Kanematsu, K., Yukimoto, Y. and Ochiai, S. (1971). *J. Org. Chem.*, **36**, 813
394. Ogura, H., Sugimoto, S., Igeta, H. and Tsuchiya, T. (1971). *J. Heterocycl. Chem.*, **8**, 391
395. Jameson, G. W. and Lawlor, J. M. (1970). *J. Chem. Soc. B*, 53
396. Fleming, I. and Philippides, D. (1970). *J. Chem. Soc. C*, 2426
397. Chung, N. M. and Tieckelmann, H. (1970). *J. Org. Chem.*, **35**, 2517
398. Jonak, J. F., Hopkins, G. C., Minnemeyer, H. J. and Tieckelmann, H. (1970). *J. Org. Chem.*, **35**, 2512
399. Brown, D. M., Coe, P. F. and Green, D. P. L. (1971). *J. Chem. Soc. C*, 867
400. Gompper, R. (1962). *Advan. Heterocycl. Chem.*, **2**, 245
401. Wong, J. T. and Fuchs, D. S. (1971). *J. Org. Chem.*, **36**, 848
402. Blank, H. U. and Fox, J. J. (1970). *J. Heterocycl. Chem.*, **7**, 735
403. Mustafa, A., Mansour, A. K. and Zaher, H. A. (1970). *Justus Liebigs Ann. Chim.*, **733**, 177
404. Novacek, A. (1971). *Collect. Czech. Chem. Commun.*, **36**, 1964
405. Daunis, J., Jacquier, R. and Viallefont, P. (1971). *Bull. Soc. Chim. France*, 3658
406. Jerfy, A. and Roy, A. B. (1970). *Aust. J. Chem.*, **23**, 847

407. Tolstikov, G. A., Jemilev, U. M., Jurjev, V. P., Gershanov, F. B. and Rafikov, S. R. (1970). *Tetrahedron Letters*, 2807
408. Hamana, M., Nomura, S. and Kawakita, T. (1971). *J. Pharm. Soc. Japan*, **19**, 134
409. Chivers, G. E. and Suschitzky, H. (1971). *J. Chem. Soc. C*, 2867
410. Kyriacou, D. (1971). *J. Heterocycl. Chem.*, **8**, 697
411. Weiner, M. A. (1970). *J. Organometallic Chem.*, **23**, C20
412. Panzica, R. P., Robins, R. K. and Townsend, L. B. (1971). *J. Med. Chem.*, **14**, 259
413. Krüger, W. and Krüger, G. (1970). *Z. Chem.*, **10**, 184
414. Leclerc, G. and Wermuth, C. G. (1971). *Bull. Soc. Chim. France*, 1752
415. Blake, K. W. and Sammes, P. G. (1970). *J. Chem. Soc. C*, 1070
416. Paudler, W. W. and Chen, T. K. (1971). *J. Org. Chem.*, **36**, 787
417. Kuleosky, N. and Severson, R. G. (1970). *Spectrochim. Acta*, **26A**, 2227
418. Vemulapalli, G. K. (1970). *J. Amer. Chem. Soc.*, **92**, 7589
419. Dratler, R. and Laszlo, P. (1970). *Chem. Commun.*, 180
420. Hoppé, J. I. and Keene, B. R. T. (1970). *Chem. Commun.*, 188
421. Trofimchuk, A. K., Govenbein, E. Y. and Abarbarchuk, I. L. (1970). *Zh. Obshch. Khim.*, **40**, 1435
422. Hoare, R. J. and Pratt, J. M. (1969). *Chem. Commun.*, 1320
423. Falla, L. (1971). *Bull. Soc. Roy. Sci. Liege*, **40**, 37
424. Markovae, A., Stevens, C. L., Ash, A. B. and Hackley, B. E. (1970). *J. Org. Chem.*, **35**, 841
425. Koppang, R., Ranade, A. C. and Gilman, H. (1970). *J. Organometallic Chem.*, **22**, 1
426. Ivashchenko, Y. N., Moshchitskii, S. D. and Yeliseyeva, A. K. (1970). *Khim. Geterotsikl. Soedinenii*, **58**
427. Haiduc, I. and Gilman, H. (1971). *Rev. Roum. Chim.*, **16**, 597
428. Kuthan, J., Kohoutova, A. and Helesic, L. (1970). *Collect. Czech. Chem. Commun.*, **35**, 2776
429a. Binns, F. and Suschitzky, H. (1970). *Chem. Commun.*, 750
429b. idem. (1971). *J. Chem. Soc. C*, 1223
430. Lyle, R. E. and White, E. (1971). *J. Org. Chem.*, **36**, 772
431. Francis, R. F., Wisener, J. T. and Paul, J. M. (1971). *Chem. Commun.*, 1420
432. Scalzi, F. V. and Golob, N. F. (1971). *J. Org. Chem.*, **36**, 2541
433. Baarschers, W. H. and Loh, T. L. (1971). *Tetrahedron Letters*, 3483
434. Banks, R. E., Field, D. S. and Haszeldine, R. N. (1970). *J. Chem. Soc. C*, 1280
435. Cooper, G. H. and Rickard, R. L. (1971). *J. Chem. Soc. C*, 197
436. Chambers, R. D., Cheburkov, Yu. A., MacBride, J. A. H. and Musgrave, W. K. R. (1971). *J. Chem Soc. C*, 532
437. Chambers, R. D., Corbally, R. P., Gribble, M. Y. and Musgrave, W. K. R. (1971). *Chem. Commun.*, 1345
438. Drayton, C. J., Flowers, W. T. and Haszeldine, R. N. (1971). *J. Chem. Soc. C*, 2750
439. Pines, H., Kannan, S. V. and Stalick, W. M. (1971). *J. Org. Chem.*, **36**, 2308
440. Bramwell, A. F., Payne, L. S., Riezebos, G., Ward, P. and Wells, R. D. (1971). *J. Chem. Soc. C*, 1627
441. Crawforth, C. E., Russell, C. A. and Meth-Cohn, O. (1970). *Chem. Commun.*, 1406
442. Kauffmann, T. and Wirthwein, R. (1971). *Angew. Chem. Int. Ed. Engl.*, **10**, 20
443. Bressan, G. B., Giardi, I., Illuminati, G., Linda, P. and Sleiter, G. (1971). *J. Chem. Soc. B*, 225
444. Zaguleva, O. A., Shein, S. M., Shvets, A. Y., Mamaev, V. P. and Krivopalov, V. P. (1971). *Reakts. Sposobnost. Org. Soedin.*, **7**, 1133
445. Wilson, J. D., Wager, J. S. and Weingarten, H. (1971). *J. Org. Chem.*, **36**, 1613
446. Rokos, H. and Pfleiderer, W. (1971). *Chem. Ber.*, **104**, 739
447. Lichtenthaler, F. W., Trummlitz, G., Banbach, G. and Rychlik, I. (1971). *Angew. Chem. Int. Edn. Engl.*, **10**, 334
448. Verhayden, J. P. H., Wagner, D. and Moffatt, J. C. (1971). *J. Org. Chem.*, **36**, 250
449. Brown, D. J. and England, B. T. (1971). *J. Chem. Soc. C*, 425
450. Vorbrüggen, H. and Niedballa, U. (1971). *Angew. Chem. Int. Edn. Engl.*, **10**, 657
451. Zoltewicz. J. A., Smith, C. L., Grahe, G. and Kauffman, G. M. (1969). *Amer. Chem. Soc. (Div. Petrol Chem.) Reprints*, **14**, C64–C67; *Chem. Abstr.*, **74**, 3238
452. Kauffmann, T., Fischer, H., Nürnberg, R. and Wirthwein, R. (1970). *Justus Liebigs Ann. Chem.*, **731**, 23

453. Banks, R. E., Barlow, M. G., Haszeldine, R. N. and Phillips, E. (1971). *J. Chem. Soc. C,* 1957
454. Fenton, R. S., Landquist, J. K. and Meek, S. E. (1971). *J. Chem. Soc. C,* 1536
455. Kuzuya, M. and Kaji, K. (1970). *Chem. Pharm. Bull.,* **18,** 2420
456. Banks, R. E., Field, D. S. and Haszeldine, R. N. (1970). *J. Chem. Soc. C,* 1280
457. Nishigaki, S., Senga, K., Ogiwara, K. and Yoneda, F. (1970). *Chem. Pharm. Bull.,* **18,** 997
458. Falco, E. A., Otter, B. A. and Fox, J. J. (1970). *J. Org. Chem.,* **35,** 2326
459. van Meeteren, K. W. and van der Plas, H. C. (1971). *Recl. Trav. Chim. Pays-Bas,* **90,** 105
460. de Valk, J. and van der Plas, H. C. (1971). *Recl. Trav. Chim. Pays-Bas,* **90,** 1239
461. Lont, P. J., van der Plas, H. C. and Koudijs, A. (1971). *Recl. Trav. Chim. Pays-Bas,* **90,** 207
462. Nohara, N., Sekiguchi, S. and Matsui, K. (1970). *J. Heterocycl. Chem.,* **7,** 519
463. Harayama, T., Sekiguchi, S. and Matsui, K. (1970). *J. Heterocycl. Chem.,* **7,** 975
464. Budziarek, R. (1971). *J. Chem. Soc. C,* 74
465. Harayama, H., Okada, K., Sekiguchi, S. and Matsui, K. (1970). *J. Heterocycl. Chem.,* **7,** 981
466. Severin, T., Batz, D. and Lerche, H. (1970). *Chem. Ber.,* **103,** 1
467. Sasaki, T., Kanematsu, K. and Murata, M. (1971). *J. Org. Chem.,* **36,** 446
468. Sasaki, T., Kanematsu, K. and Murata, M. (1971). *Tetrahedron,* **27,** 5121
469. Blank, H. U., Wempen, I. and Fox, J. J. (1970). *J. Org. Chem.,* **35,** 1131
470. Eatough, J. J., Fuller, L. S., Good, R. H. and Smalley, R. K. (1970). *J. Chem. Soc. C,* 1874
471. Brown, D. J. and Sugimoto, T. (1971). *J. Chem. Soc. C,* 2616
472. Clark, J. and Yates, F. S. (1971). *J. Chem. Soc. C,* 2475
473. Collins, I., Roberts, S. M. and Suchitsky, H. (1971). *J. Chem. Soc. C,* 167
474. Ivin, B. A., Glushkov, R. K. and Sochilin, E. G. (1970). *J. Gen. Chem. USSR,* **40,** 184
475. Stahl, Q., Lehmkuhl, F. and Christensen, B. E. (1971). *J. Org. Chem.,* **36,** 2462
476. Kobe, J., Stanovnik, B. and Tisler, M. (1970). *Tetrahedron,* **26,** 3357
477. Abramovitch, R. A., Giam, C. S. and Poulton, G. A. (1970). *J. Chem. Soc. C,* 128
478. Prostakov, N. S., Mikhailova, N. M. and Talanov, U. M. (1970). *Khim. Geterotsikl. Soedin.,* 1359
479. van Bergen, T. J. and Kellogg, R. M. (1971). *J. Org. Chem.,* **36,** 1705
480. Kametani, T. and Suzuki, T. (1971). *J. Chem. Soc. C,* 1053
481. Hamana, M. and Kumadaki, I. (1971). *Chem. Pharm. Bull.,* **19,** 1669
482. Bergman, J. (1970). *J. Heterocycl. Chem.,* **7,** 1071
483. Deubel, H., Wolkenstein, D., Jokisch, H., Messerschmitt, T., Brodka, S. and von Dobeneck, H. (1971). *Chem. Ber.,* **104,** 705
484. Winters, L. J., Smith, N. G. and Cohen, M. I. (1970). *Chem. Commun.,* 642
485. Igeta, H., Tsuchuja, T. and Kaneko, C. (1971). *Tetrahedron Letters,* 2883
486. Igeta, H., Tsuchuja, T., Okuda, C. and Yokogawa, H. (1971). *Chem. Pharm. Bull.,* **19,** 1297
487. Matsumura, E., Ariga, M. and Ohfuji, T. (1970). *Bull. Chem. Soc. Japan.,* **43,** 3210
488. Jankovsky, M. and Ferles, M. (1970). *Collect. Czech. Chem. Commun.,* **35,** 2797
489. Senga, K., Yoneda, F. and Nishigaki, S. (1971). *J. Org. Chem.,* **36,** 1829
490. Ager, E. and Iddon, B. (1970). *Chem. Commun.,* 118
491. Cheeseman, G. W. H. and Godwin, R. A. (1971). *J. Chem. Soc. C,* 2977
492. Cheeseman, G. W. H. and Godwin, R. A. (1971). *Chem. Commun.,* 2973
493. Chambers, R. D., Hole, M., Musgrave, W. K. R. and Thorpe, J. G. (1971). *J. Chem. Soc. C,* 61
494. Rys, P., Schmitz, A. and Zollinger, H. (1971). *Helv. Chim. Acta,* **54,** 163
495. Ostrogovich, G., Fliegl, E. and Bacaloglu, R. (1971). *Tetrahedron,* **27,** 2885
496. Ostrogovich, G., Fliegl, E. and Bacaloglu, R. (1971). *Tetrahedron,* **27,** 3869
497. Zoltewicz, J. A. and Sale, A. A. (1971). *J. Org. Chem.,* **36,** 1455
498. Berry, D. J., Wakefield, B. J. and Cook, J. D. (1971). *J. Chem. Soc. C,* 1227
499. Temple, C., Laseter, A. G. and Montgomery, J. A. (1970). *J. Heterocycl. Chem.,* **7,** 1219
500. Deegan, A. and Rose, F. L. (1971). *J. Chem. Soc. C,* 2756
501. Hayatsu, H., Wataya, Y. and Kai, K. (1970). *J. Amer. Chem. Soc.,* **92,** 742
502. Shapiro, R., Servis, R. E. and Welcher, M. (1970). *J. Amer. Chem. Soc.,* **92,** 422

503. Santi, D. V., Brewer, C. F. and Farber, D. (1970). *J. Heterocycl. Chem.*, **7**, 903
504. Klein, R. S., Wempen, I., Watanabe, K. A. and Fox, J. J. (1970). *J. Org. Chem.*, **35**, 2330
505. Clark, J., Gelling, I., Southon, I. W. and Morton, M. S. (1970). *J. Chem. Soc. C*, 494
506. Blanch, J. H. and Fretheim, K. (1971). *J. Chem. Soc. C*, 1892
507. Möhrle, H. and Weber, H. (1971). *Chem. Ber.*, **104**, 1478
508. Abramovitch, R. A. and Vinutha, A. R. (1971). *J. Chem. Soc. B*, 131
509. Fischer, G. W. (1970). *Chem. Ber.*, **103**, 3489
510. Lee, J. and Paudler, W. W. (1971). *Chem. Commun.*, 1636
511. Sandler, S. R. (1970). *J. Org. Chem.*, **35**, 3967
512. Sandler, S. R. (1971). *Chem. Ind. (London)*, 1416
513. Saneyoshi, M. (1971). *Chem. Pharm. Bull.*, **19**, 493
514. Ueda, T., Imazura, M., Muira, K., Iwata, R. and Odajima, K. (1971). *Tetrahedron Letters*, 2507
515. Clark, J. and Yates, F. S. (1971). *J. Chem. Soc. C*, 2278
516. Yamazaki, T., Nagata, M., Nohara, F. and Urano, S. (1971). *Chem. Pharm. Bull.*, **19**, 159
517. Poole, A. J. and Rose, F. L. (1971). *J. Chem. Soc. C*, 1285
518. King, K. F. and Bauer, L. (1971). *J. Org. Chem.*, **36**, 1641
519. Redmore, D. (1970). *J. Org. Chem.*, **35**, 4114
520. Minisci, F., Gardini, G. P., Galli, R. and Bertini, F. (1970). *Tetrahedron Letters*, 15
521. Gardini, G. P. (1971). *Tetrahedron Letters*, 59
522. Bass, K. C. and Nababsing, P. (1970). *J. Chem. Soc. C*, 2169
523. Minisci, F., Bernardi, R., Bertini, F., Galli, R. and Perchinunno, M. (1971). *Tetrahedron*, **27**, 3575
524. Dou, H. J., Vernin, G. and Metzer, J. (1971). *Bull. Soc. Chim. France*, 1021
525. Dou, H. J., Vernin, G. and Metzer, J. (1971). *Bull. Soc. Chim. France*, 3553
526. Buratti, W., Gardini, G. P., Minisci, F., Bertini, F., Galli, R., Perchinunno, M. (1971). *Tetrahedron*, **27**, 3655
527. Dou, H., Vernin, G., Dufour, M. and Metzer, J. (1971). *Bull. Soc. Chim. France*, 111
528. Fields, E. K. and Meyerson, S. (1970). *J. Org. Chem.*, **35**, 62
529. Fields, E. K. and Meyerson, S. (1970). *J. Org. Chem.*, **35**, 67
530. Elofson, R. M., Gadallah, F. F. and Schulz, K. F. (1971). *J. Org. Chem.*, **36**, 1526
531. Wilzbach, K. E. and Rausch, D. J. (1970). *J. Amer. Chem. Soc.*, **92**, 2178
532. Barlow, M. G., Dingwall, J. G. and Haszeldine, R. N. (1971). *Chem. Commun.*, 1580
533. Caplain, S. and Lablache-Combier, A. (1970). *Chem. Commun.*, 1247
534. Caplain, S., Catteau, J. P. and Lablache-Combier, A. (1970). *Chem. Commun.*, 1475
535. Caplain, S., Castellano, A., Catteau, J. P. and Lablache-Combier, A. (1971). *Tetrahedron*, **27**, 3541
536. Travecedo, E. F. and Stenberg, V. I. (1970). *J. Chem. Soc. D*, 609
537. Natsume, M. and Wada, M. (1971). *Tetrahedron Letters*, 4503
538. Van Bergen, T. J. and Kellogg, R. M. (1971). *J. Org. Chem.*, **36**, 978
539. Banks, R. E., Chang, W. M., Haszeldine, R. N. and Shaw, G. (1970). *J. Chem. Soc. C*, 55
540. Nasielski, J., Kirsch-Demesmaeker, A., Kirsch, P. and Nasielski-Hinkins, R. (1970). *Chem. Commun.*, 302
541. Tsuchiya, T., Arai, H. and Igeta, H. (1971). *Chem. Pharm. Bull.*, **19**, 1108
542. Hamilton, L. R. and Kropp, P. J. (1971). *Tetrahedron Letters*, 1625
543. Deeleman, R. A. F., van der Plas, H. C., Koudijs, A. and Darwinkel-Risseeuw, P. S. (1971). *Tetrahedron Letters*, 4159
544. Sasaki, T., Kanematsu, K., Kakehi, A., Ichikawa, I. and Hayakawa, K. (1970). *J. Org. Chem.*, **35**, 426
545. Balasubramanian, A., McIntosh, J. M. and Snieckus, V. (1970). *J. Org. Chem.*, **35**, 433
546. Spence, G. G., Taylor, E. C. and Burchardt, O. (1970). *Chem. Rev.*, **70**, 230
547. Leibovici, C. and Streith, J. (1971). *Tetrahedron Letters*, 387
548. Streith, J. and Sigwalt, C. (1970). *Bull. Soc. Chim. France*, 1157
549. Tsuchiya, T., Arai, H. and Igeta, H. (1970). *Tetrahedron Letters*, 2213
550. Bellamy, F., Barragan, L. G. and Streith, J. (1971). *Chem. Commun.*, 456
551. Tsuchiya, T., Arai, H. and Igeta, H. (1971). *Tetrahedron Letters*, 2579
552. Stenberg, V. I. and Travecedo, E. F. (1971). *Tetrahedron*, **27**, 513

553. Kuneida, T. and Witkop, B. (1971). *J. Amer. Chem. Soc.*, **93**, 3487
554. Le Goff, M. and Bengelmans, M. R. (1970). *Tetrahedron Letters*, 1355
555. Shizuka, H., Kanai, T., Morita, T., Ohota, Y. and Matsui, K. (1971). *Tetrahedron*, **27**, 4021
556. Meyers, A. I. and Singh, P. (1970). *J. Org. Chem.*, **35**, 3022
557. Wagner, P. J. and Bucheck, D. J. (1971). *J. Amer. Chem. Soc.*, **93**, 181
558. Elad, D., Rosenthal, I. and Sasson, S. (1971). *J. Chem. Soc. C*, 2053
559. Wang, S. Y. and Rhoades, D. F. (1971). *J. Amer. Chem. Soc.*, **93**, 2554
560. Wang, S. Y. (1971). *J. Amer. Chem. Soc.*, **93**, 2768
561. Kumber, P. L. and Dybas, R. A. (1970). *J. Org. Chem.*, **35**, 125
562. De Silva, O. and Snieckus, V. (1971). *Synthesis*, 254
563. Schwaiger, W. and Ward, J. P. (1971). *Recl. Trav. Chim. Pays-Bas*, **90**, 513
564. Atkinson, P. and Keene, B. R. T., unpublished work
565. Crossland, I. and Kofed, H. (1970). *Acta Chem. Scand.*, **24**, 751
566. Wilson, R. M. and DiNinno, F. (1970). *Tetrahedron Letters*, 289
567. Anzai, K. and Suzuki, S. (1966). *Agric. Biol. Chem.*, **30**, 597
568. Kelly, A. H. and Pasrick, J. (1970). *J. Chem. Soc. C*, 303
569. Hamana, M. and Kumadaki, I. (1970). *Chem. Pharm. Bull.*, **18**, 1742
570. Schmidt, R. R., Schwille, D. and Wolf, H. (1970). *Chem. Ber.*, **103**, 2760

5

Nitrogen Heterocyclic Molecules Part 4 Benzazines (Quinolines, Isoquinolines, Quinolizinium compounds, Cinnolines, Quinazolines, Quinoxalines, Phthalazines, Benzotriazines)

W. L. F. ARMAREGO
Australian National University, Canberra

5.1 THEORETICAL STUDIES

Among the studies made in 1970 and 1971, SC_β and $SC_{\alpha,\beta}$-MO calculations were applied to quinoline[1], and electronic indices were obtained for quinoline, quinoxaline and several 6-substituted derivatives[2]. The latter work confirmed the reliability of the simple sum rule of molecular indices. MO calculations on quinoline and isoquinoline in connection with tautomerism in the hydroxy derivatives gave good correlations with data derived from pK_a values and spectra[3]. Also MO calculations with regard to coupling constants from the e.s.r. spectra of radical ions of quinolines and isoquinolines were discussed and criticised[4]. Bond-order and bond-length relations for 2,3-di-t-butylquinoxaline were derived[5].

5.2 SPECTRA

5.2.1 Electronic spectra

The ultraviolet spectra of several reduced quinolones and quinolizidinones were studied with regard to 'Woodward's rule' and used to distinguish between *cis* and *trans* vinylogous amides and imides[6]. The spectra of many nuclear *N*-methyl derivatives of dimethylamino-quinolines, isoquinolines, cinnolines, quinoxalines and phthalazines and their cations in aqueous solution were measured and discussed[7]. Electronic relaxation was shown to be the cause of diffuseness in the electronic spectra (vapour) of quinoline, isoquinoline, quinazoline and quinoxaline[8]. The spectra of benzazines were included in a review on the ultraviolet spectra of heterocyclic molecules[9]. Phosphorescence polarisation studies of phthalazine in ether–isopentane–ethanol (5:5:2) at 77 K led to the conclusion that the phosphorescent triplet state was the $^{3}\beta_{2}(\pi,\pi^{*})$ state[10]. Photoelectron spectroscopy of benzazines was briefly reviewed[11]. The singlet–triplet absorption spectra of quinoline and 2-, 4-, 6-, 7- and 8-methylquinoline were measured in dimethylmercury which enhanced the efficiency of forbidden radiative and non-radiative transitions. The relative efficiency of dimethylmercury, ethyl iodide and tetraethyl-lead in the enhancement was compared and the effects of substituents on the $^{1}L_{a} \leftarrow ^{1}A$, $^{1}L_{b} \leftarrow ^{1}A$ and $^{3}L_{a} \leftarrow ^{1}A$ transitions of quinoline were discussed[12].

5.2.2 Mass spectra

The mass spectra of amino-[13] and methyl- quinolines[14], several quinazolines[15], quinoxalines[16], and 1,2,3-benzotriazines[17] were measured and, typically, hydrogen cyanide was an important fragment. A monograph on mass spectrometry of heterocyclic compounds appeared in 1971[18].

5.2.3 Nuclear magnetic resonance spectra

There has been much interest recently in the effect of complexes of the lanthanide rare earth metals on the induced chemical shifts of ring protons in heterocyclic molecules. In particular, tris-(dipivaloylmethanato)europium was used to study the induced shifts in quinoline[19–22] isoquinoline[19–21] and other benzazines[19]. In a detailed study, which included benzazines, it was shown that the angular dependence term must be included in the calculations of pseudo-contact shifts in order to arrive at reasonable values for the Eu—N distances. Also, the induced shifts of proton signals in azanaphthalenes can be predicted from this work[19]. Other studies included solvent and concentration effects on the chemical shifts in N-methylquinolinium salts[23], correlation of substituent effects on chemical shifts in benzazines[24], ^{14}N chemical shifts in cinnoline, quinazoline, quinoxaline and phthalazine[25], dependence of ^{15}N—H coupling constants on the hybridisation of the nitrogen atom in quinoline and 6-nitroquinoline[26], and the coupling constant between the ^{15}N nucleus of the nitro group and the aromatic protons in several 4-nitroquinoline 1-oxides[27]. Studies of the conformations of 3,4-dihydro-1-phenoxymethylquinolines[28], N-acyl-1,2,3,4-tetrahydroquinolines[29] and 3-aryl-1,2,3,4-tetrahydro-4-methoxycarbonylisoquinolines were also reported[30]. Nitrogen inversion in aziridinecarboxylic esters was slowed down (on the n.m.r. time scale) by having 1,2-dihydroquinol-2-on-1-yl and 3,4-dihydro-2-methyl-quinazol-4-on-3-yl substituents on the nitrogen atom. It was thus possible to study the effect of the alkyl substituent in the ester group on nitrogen inversion in the aziridines[31].

5.3 ELECTROCHEMISTRY OF BENZAZINES

The electrolysis of N-heterocyclic compounds was reviewed in 1971[32]. Rate constants of protonation of benzodiazines in water were obtained from electrochemical reduction experiments[33]. 1,2,3,4-Tetrahydro-1,4-diphenyl-cinnoline was formed by the electrochemical reduction of 1,2-diphenyl-hydrazine in the presence of styrene, and it was oxidised electrochemically to the 1,4-dihydro derivative and then to the 1,4-diphenylcinnolinium cation[34]. Polarographic reduction of alkyl- and aryl-quinoxalines in neutral or alkaline solution led to 1,2-, 1,4- or 3,4-dihydro, or 1,2,3,4-tetrahydro derivatives[35]. The initial product from the polarographic (two step) reduction of 2-phenylquinoxaline was the 1,4-dihydro compound but this rearranged to the thermodynamically more stable 1,2-dihydro derivative. The rearrangement was slow, even at pH=0, because it required the transfer of a proton from a nitrogen to a carbon atom[36]. Similarly, reduction of phthalazine and its alkyl derivatives in alkaline medium gave 1,2-dihydro- or 1,2,3,4-tetra-hydrophthalazines, or dimeric products, depending on the cathode potential. In acid solution a mixture of o-bis-(aminomethyl)benzene and isoindoline was formed[37]. The observed energy differences between benzazines and their radical anions obtained from polarographic work were in good agreement with results from CNDO and self-consistent field π-electron calculations[38].

5.4 QUINOLINES

5.4.1 Ring synthesis

A convenient preparation of 2,3- and 1,2,3-substituted quinolines from α-methylene ketones via β-chlorovinylaldehydes with aryl amines was reported[39]. The formation of 4-benzoyl-2,3-diphenylquinoline and 3,4-disubstituted 2-phenylquinolines from trans-dibenzoylstilbene (1) and cis-

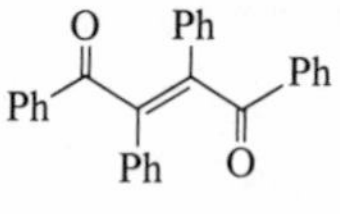

(1) 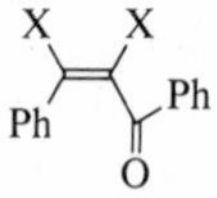 (2) X = Ph, Br 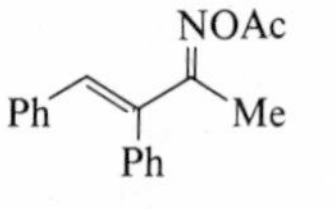 (3)

disubstituted ω-benzoylstyrenes (2) respectively, with hydrazoic acid in sulphuric acid, should be a useful method for making highly substituted derivatives[40]. In a direct cyclisation, 3,4-diphenylbut-3-en-2-one O-acetyl-oxime (3) gave 2-methyl-3-phenylquinoline[41]. The acid-catalysed reaction of o-nitrostyrene oxides to form 3,4-dihydro-1,3-dihydroxy-4-oxoquinolines[42, 43] was essentially an extension of the cyclisation reaction of o-nitrostyrenes to quinolines[44, 45].

Of the recent procedures for making quinolines by ring enlargement of indoles[46, 47], the most ingenious was the reaction of 2,3-cycloalkylindoles (4) with dichlorocarbene to generate 3-chlorobenzopyridino-2,4-cyclophanes (5). The limiting value of m was 6; thereafter a rearrangement occurred to form the less strained 4-chloro-2,3-cycloalkylquinolines (6)[48]. Thermolysis of methanetricarboxylate derivatives (7) under reduced pressure gave good

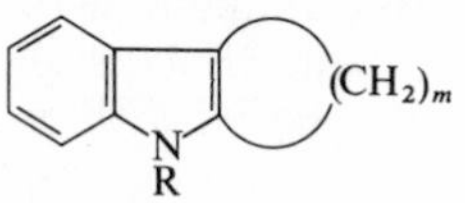

(4) R = H, Ac;
m = 10, 8, 6, 5

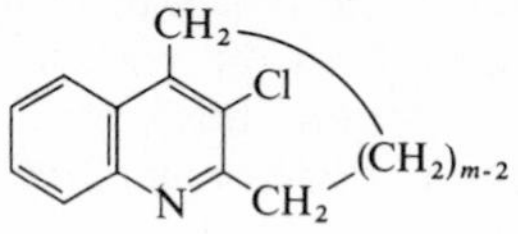

(5)

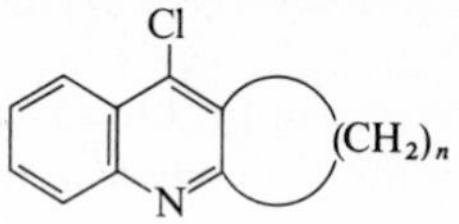

(6)

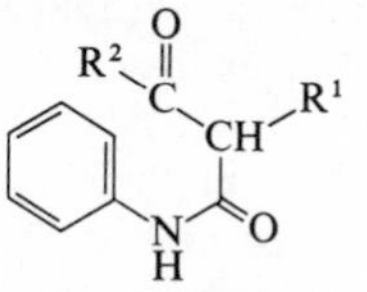

(7) R[1] = CN, COMe, CONHPh
R[2] = OEt, NHPh

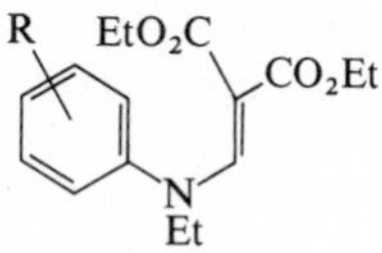

(8)

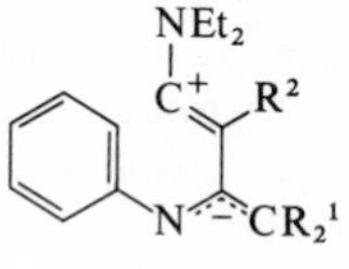

(9)

yields of 3-substituted 4-hydroxycarbostyrils[49, 50]. 4-Hydroxycarbostyrils with alkyl and aryl substituents on C-3 were formed by cyclisation of N-acyl anthranilic esters[51]. The malonates (8) were successfully converted in one step into 3-ethoxycarbonyl-1-ethyl-4-chloroquinolinium chlorides with phosphoryl chloride[52]. In the reactions between ketenimines derived from aniline and ynamines derived from diethylamine, the dipolar intermediates (9) formed had cyclised to 4-diethylaminoquinolines[53].

The synthesis of quinoline and its methyl derivatives from aniline and toluidines and aldehydes (e.g. acetaldehyde, acrolein) in the gas phase and catalysed by a fluorine-containing aluminium silicate, has possibilities for industrial preparations[54]. The pyrolysis of nitrobenzene with pyridine at 600 °C to give low yields of quinoline is of theoretical interest[55]. Ultraviolet irradiation of alcoholic solutions of N-allylaniline and cinnamylidene aniline gave quinoline and 2-phenyl- (but not 4-phenyl-) quinoline respectively[56]. Substituted carbostyrils were prepared in 50–90% yields by a photoinduced synthesis from 3-(o-aminobenzylidene)-N-arylsuccinimides (10)[57].

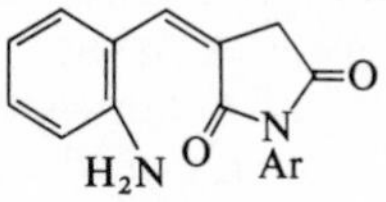

(10)

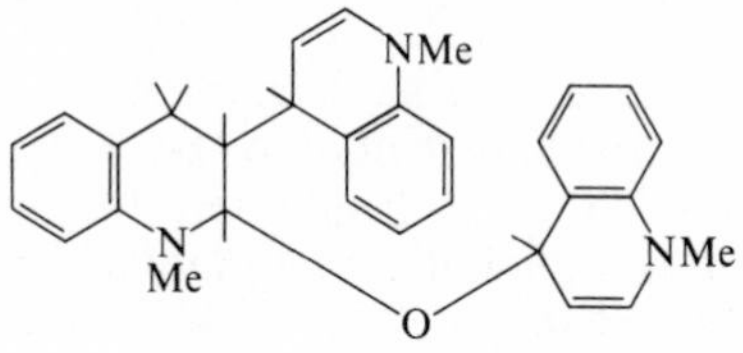

(11)

5.4.2 Substitution reactions

5.4.2.1 Electrophilic

Chlorination of quinoline in the absence of solvent at 160–190 °C gave 3,4-dichloro-, 3,4,6-trichloro-, 3,4,8-trichloro-, 3,4,6,8-tetrachloro- and

3,4,6,7,8-pentachloro-quinoline as major components[58]. A detailed study of the rates of halogenation of 8-hydroxyquinoline with N-halogenosuccinimides showed that although the orientation (towards C-5 and C-7) was unchanged, the rates depended on the reagent; N-iodosuccinimide, which gave mainly 8-hydroxy-5-iodoquinoline, reacted faster than N-chlorosuccinimide which gave a mixture of 5-chloro- and 7-chloro-, and 5,7-dichloro-8-hydroxyquinolines[59].

Nitration of the quinolinium cation in 80.05% sulphuric acid at 25 °C gave 3-(0.001%), 5-(50.1%), 6-(1.6%), 7-(0.01%) and 8-(42.3%) nitroquinolines, and the partial rate factors were determined[60]. 1-Methylquinol-4-one cation was also involved in nitration, and as in 4-methoxyquinoline substitution occurred at C-6, and to a lesser extent C-8 [61]. Nitration of 2-anilino-4-methylquinoline in 50% sulphuric acid took place in the anilino group, and in both the anilino group and at C-6 at higher acidity. With acetyl nitrate, however, substitution occurred only in the anilino group, and the authors postulated that 2-anilino-1-nitroquinolinium cation was involved in contrast with the quinolinium cation in sulphuric acid[62].

Tritiated 4-nitro- and 2-methyl-4-nitro-quinoline 1-oxides were prepared for the study of their carcinogenic properties[63]. The kinetics of electrophilic deuterium exchange in quinoline and its N-oxide in deuteriosulphuric acid showed that substitution in the carbocyclic ring proceeded via the conjugate acids. Exchange in the heterocyclic ring most probably occurred via deprotonation at C-2, and via a covalent hydrated species at C-3 [64].

Hydroxymethylation (HCHO–HCl) [65] and acetylation (Friedel–Crafts)[66] of 1-methylquinol-2(1H)-one took place at C-6, and alkylation of 4-hydroxyquinol-2(1H)-one occurred at C-3 [67, 68]. Methylation of 4-acylamino- and 4-tosylamino-quinoline in neutral media gave the 1-methyl derivatives, but in the presence of sodamide these were accompanied by substantial amounts of the isomers methylated on the extra-cyclic nitrogen atom[69]. 2-Chloro- and 2-azido-quinolines were successfully alkylated on N-1 with triethyloxonium borofluoride[70].

5.4.2.2 Nucleophilic

1-methylquinolinium methosulphate in dilute sodium hydroxide deposited di-(1,2-dihydro-1-methylquinol-2-yl)ether which gave a trimer (11) by boiling in ethanol[71, 72]. In an interesting amination 2- and 4-chloroquinolines gave 2- and 4-dialkylaminoquinolines respectively when boiled in dialkyl formamide[73]. Alcoholysis of 4-chloroquinolines to quinol-4(1H)-ones in the

$$\text{(structure)} \xrightarrow[\text{H}^+]{\text{R''OH}} \text{(structure)} + R_2''O + R''Cl + H^+ \qquad (5.1)$$

presence of acid proceeded as in equation (5.1)[74]. Rate constants and activation parameters for nucleophilic displacement in 1-chloro- and 4-chloro-, and 1-fluoro- and 4-fluoro-quinolines, and 2-halogenoquinoxalines by methoxide or piperidine did not show new features which could differentiate halogen activation by a ring-nitrogen atom in heterocyclic molecules, from that by a nitro group in carbocyclic aromatic systems[75].

Displacement of fluorine in perfluoroquinoline with hydrogen halides took place readily in the 2- and 4- positions[76]. The 3-CF_3 group in quinoline and its N-oxide was converted into an ethoxycarbonyl group by boiling with sodium ethoxide, but the 2- and 4-CF_3 groups required more severe conditions[77]. However, 2-trifluoromethylquinoline reacted with sodamide in liquid ammonia to form 2-aminoquinoline in good yield; the 3-isomer gave a mixture of 3-cyano- and 4-amino-3- cyano-quinoline, and the 4-isomer gave 2-amino-4-trifluoromethylquinoline in poor yields[78]. 3-Nitroquinoline gave a variety of methoxy-bromo- and methoxy-bromo-3-nitroquinolines in low yields when it was reacted with sodium hypobromite in alcoholic alkali[79]. The displacement of the nitro group in 4-nitroquinoline-1-oxide by methionine or homocystein formed the respective sulphides without loss of the N-oxide function, and its activity in biological *trans*-methylation was implied[80]. Similarly, a reaction with potassium cyanide in methanol gave 4-methoxyquinoline 1-oxide; but 3-nitroquinoline 1-oxide under these conditions gave a mixture of 4-cyano-3-methoxyquinoline 1-oxide and 9-carbamoyl-3-methoxy-1H-pyrazolo[4,3-b]quinoline (12); 5-nitroquinoline 1-oxide gave a mixture of 6-cyano-5-methoxyquinoline 1-oxide and 7-aminoisoxazolo[3,4-f]quinoline 1-oxide (13); and 6-nitroquinoline 1-oxide gave 5-cyano-6-methoxyquinoline 1-oxide, 2-carbamoyl-5-cyano-6-methoxyquinoline 1-oxide and 1-hydroxy-6-nitrocarbostyril[81]. The nitro group in 2-substituted 4-nitroquinoline 1-oxides was also easily displaced by malonate (or cyanoacetate) in the presence of potassium cyanide, but the corresponding 3-cyanoquinol-4-yl malonates (or cyanoacetates) were formed, i.e. oxygen was eliminated. In the absence of these esters, 2-substituted 3,4-dicyanoquinolines were produced[82]. 1-Methyloxindole reacted with quinoline 1-oxides in the presence of acetic anhydride or benzoyl chloride to give 1-methyl-3-(quinol-2-yl)oxindoles. 2- and 4-Chloroquinoline 1-oxides formed the quinol-4-yl and quinol-2-yl oxindoles respectively, and it was also shown that indole, 1-methyl- and 2-phenyl-indole but not skatole behaved similarly[83]. The condensations of 2-unsubstituted quinoline 1-oxides with enol–ethers[84] and with acylmethylpyridinium salts[85] to give the respective 2-substituted quinolines were also reported.

From the kinetics of hydrolysis of 2-, 5-, 6-, 7- and 8-quinolyl ketoximes in sulphuric acid, it was shown that the rate determining step was the base-catalysed loss of hydroxylamine from the tetrahedral intermediate, which was formed by attack of water on the di-cation derived from protonation of the heterocyclic ring and the oximino group[86]. By generating hetarynes from variously substituted chloro (or bromo) quinolines followed by reaction with dialkylamines, Kauffmann and co-workers obtained the relative reactivities at all positions (other than the C-2), and compared them with those of benzyne, naphthalyne, phenanthryne and pyridyne. The conclusions were: replacement of C—H by a nitrogen atom lowered the reactivity of the aryne bond, i.e. quinolyne was less reactive than naphthalyne; the α-carbon atoms (naphthalene nomenclature), i.e. C-4 and C-5, were more reactive than the β-carbon atoms, i.e. C-3 and C-6, except for 7,8-dehydroquinoline in which this was reversed probably because of the *peri*-nitrogen atom and/or the electronic factors at C-7 [87]. 2-Lithiothiophene reacted with quinoline at 20 °C to give *only* 2-(2-thienyl)quinoline (38 %), but the yield was improved (75 %) by raising the temperature to 45 °C. Absence of the 4-isomer was attributed to steric hindrance by the *peri*-hydrogen atom at C-5 [88].

The insertion of dimethylcarbamoyl groups into the 2- and 4-positions of quinolines by reaction with dimethyl formamide in the presence of a variety of oxidants involved free radicals[89].

5.4.3 Ring-modifying processes

Thermal interconversion of quinoline and isoquinoline was shown to occur at 850 °C. Quinoline gave 0.7 % of isoquinoline, whereas the reverse yield was 2.3 %; the only other products were hydrogen cyanide and acetylene. Caution was therefore advised in the interpretation of mass spectra[90]. The heterocyclic ring can be modified by ylide formation, thus the ylide from benzylquinoline condensed exothermically with cyanoacetylene to form benzindolizines[91]. Ylides formed during the decarboxylation of 1-methyl-quinolinium 2- and 4-carboxylates were demonstrated by trapping with electrophilic reagents, e.g. diazonium ions, azides[92]. Methyl bromocyano-acetate did not give ylides with quinolines in contrast to its behaviour with isoquinoline, but two molecules added across the 1,2- and 3,4-double bonds to form the azirino[1,2-*a*]cyclopropa[*c*]quinoline (14)[93]. Quinoline and

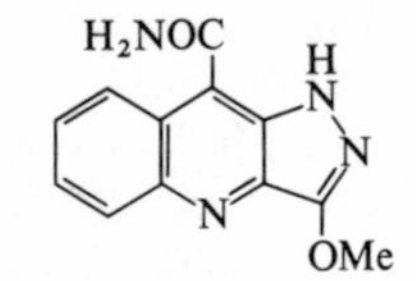

(12) 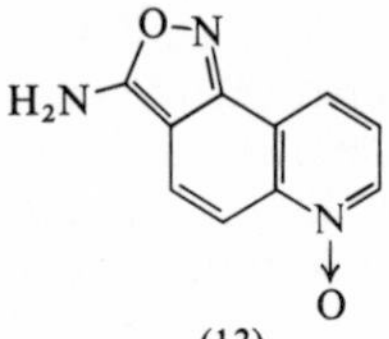   (13) 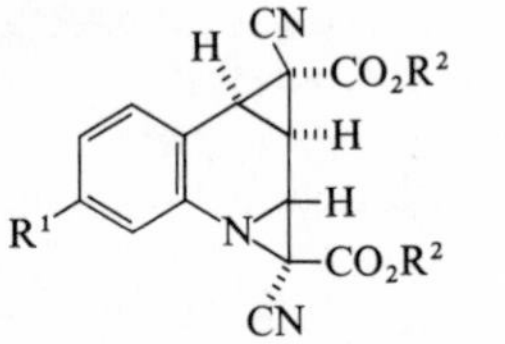   (14)

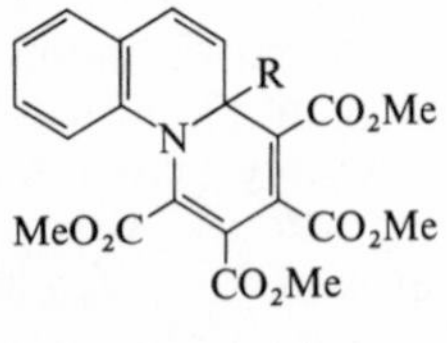

(15) 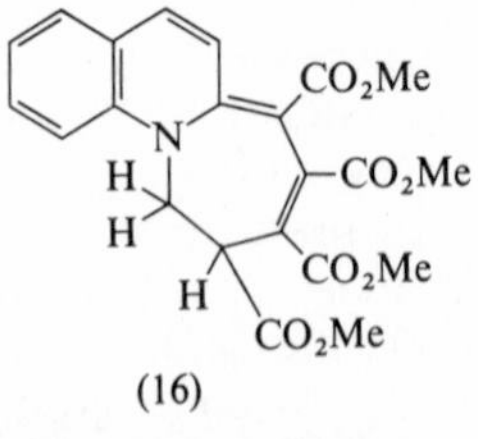   (16) 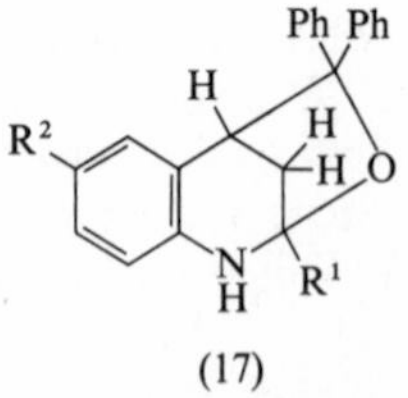   (17)

2-alkylquinolines reacted with two molecules of dimethyl acetylenedicarboxy-late to form 4*a*-alkyl-4*aH*-benzo[*c*]quinolizines (15) together with 10,11-dihydroazepin[1,2-*a*]quinolines (16)[94, 95]. Ethyl 3-(quinol-2-yl)-*trans*-acrylate reacted in a similar manner, but required only one molecule of the acetylene ester to yield benzo[*c*]quinolizines[96]. In a lithium analogue of the Emmert reaction, quinoline and 4-methylquinoline reacted with benzophenone in ether to provide a convenient route to 2,2-diphenyl-2(quinol-2-yl)methanols, but with 2-methyl-, 2,6-dimethyl- and 2-phenyl-quinoline, 4,4-diphenyl-

1,2,4,5-tetrahydro-2,5-methano-3,1-benzoxazepines (17) were formed[97]. The rate of racemisation of optically active 4,4'-biquinolyl altered slightly from one solvent to another but varied considerably in acid solutions, the higher the acid concentration the slower the rate. This effect was attributed mainly to changes in the molecular dimensions on protonation[98].

5.4.4 Photochemistry

Photolysis of quinolines in alcohols produced quinol-2-ylcarbinols, but in the presence of acid the 4-isomers were also formed[99]. Further photoreduction of the side chain was achieved in degassed methanolic solutions[100]. In ether or cyclohexane the respective 2- and 4-alkoxymethyl- or 2- and 4-cyclohexyl-quinolines were formed only when the C-2 and C-4 positions in the quinolines were unsubstituted[101]. The cyano group in 2-cyanoquinoline, however, was replaced by alcohol on irradiation in the presence of air[102].

Photocycloaddition of olefins to carbostyrils occurred readily together with dimerisation across the 2,3-positions[103, 104]. The stereochemistry of the adducts was elucidated spectroscopically[103, 105] and the dimerisation was shown to proceed through a bimolecular interaction between the triplet state and the unexcited molecules[106]. 5,6,7,8-Tetrahydrocarbostyril, on the other hand, gave the 3,8a- dimer (18)[107]. Photo-addition of methanol to 1,4-dimethyl-2,3,5,6,7,8-hexahydroquinolinium perchlorate, in a chloride dependent reaction, gave 7,10-dimethyl-10-aza-11-oxatricyclo[5,3,2,0^{1,6}]-dodecane (19)[108]. 1-Benzoyliminoquinoline ylide rearranged to 2-benzamido-quinoline[109].

5.4.5 Reduced quinolines

These were prepared by ring syntheses and by reduction of quinolines. The formation of several hydroquinolines[110-112] from enamines was another example of the versatility of enamines in organic synthesis. 1-Azidoindene, or indan-1-one and sodium azide, in sulphuric acid and benzene gave good yields of 3,4-dihydrocarbostyril together with small amounts of the isomeric 3,4-dihydroisocarbostyril[113]. 1-Methylquinolinium iodide gave 1,2-dihydro-1-methylquinoline on reduction with lithium aluminium hydride, but with 1.4% sodium amalgam the 1,4-dihydro isomer was formed[114]. Reduction of quinoline and its 6-methoxy derivative with lithium in liquid ammonia occurred preferentially in the benzene or heterocyclic ring depending on whether methanol was present or absent in the medium[115]. With lithium in tetrahydrofuran, 2-methyl- and 2,6-dimethyl-quinoline gave dimeric reduced products whose structures were deduced spectroscopically[116].

The formation of 2-phenyl-, 2-methyl-4-phenyl- and 4-methyl-2-phenyl-quinoline, when 4-methylquinoline reacted with phenyl-lithium, occurred by an initial 1,4-addition reaction followed by elimination. Evidence for this came from the nature of the products and from the formation of 1-ethoxy-carbonyl-1,4-dihydro-4-methyl-4-phenylquinoline in the presence of ethyl chloroformate[117]. 3-Cyanoquinoline also gave the 1,4-adduct, 3-cyano-4-

ethyl-1,4-dihydroquinoline, with ethyl magnesium bromide[118]. Lithium carboranes (20), however, gave 1,2-adducts with 1-methylquinolinium iodide[119].

5.4.6 Properties of functional groups

The reduction of 8-nitroquinolines to 8-aminoquinolines with Raney nickel W-7 and hydrazine hydrate was an improvement on earlier methods[120]. 2-Aminoquinoline was successfully converted into its N-oxide in 66% yield with hydrogen peroxide in the presence of sodium tungstate and EDTA[121], but was accompanied by a small amount of 2-nitroquinoline.

Oxidation of 1-aminocarbostyril with lead tetra-acetate gave 1-carbostyrilnitrene which reacted with olefins to produce 1,2-dihydro-2-oxoquinolin-1-ylaziridines[122]. 6-Hydroxyquinoline was oxidised, in methanol containing secondary amines, with cupric ions or chlorine in acetic acid to yield 8-dialkylaminoquinolin-5,6-diones or 5,6,7,8-tetrahydro-5,5,7,7,8-pentachloro-6-oxoquinoline[123]. Oxidation of 7-amino-8-hydroxyquinoline-6-sulphonic acid

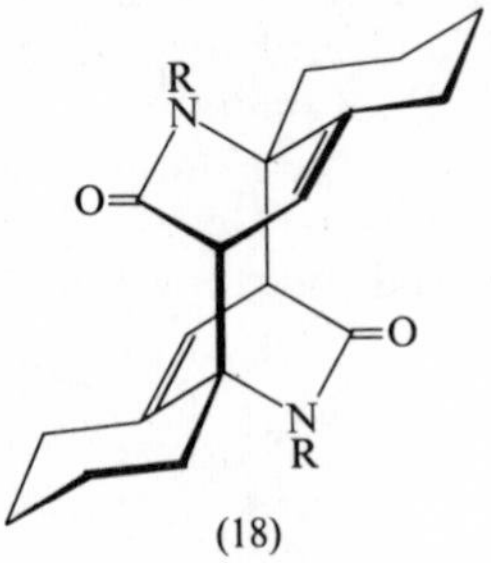

(18)

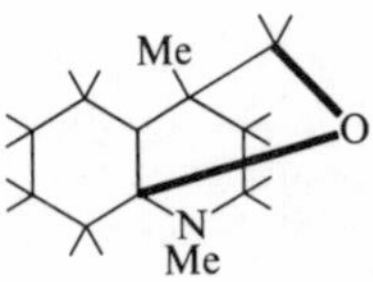

(19)

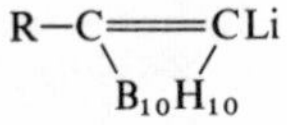

(20)

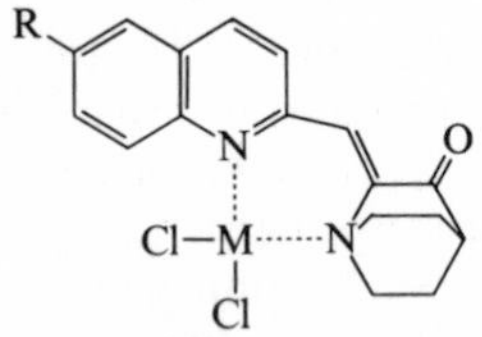

(21)

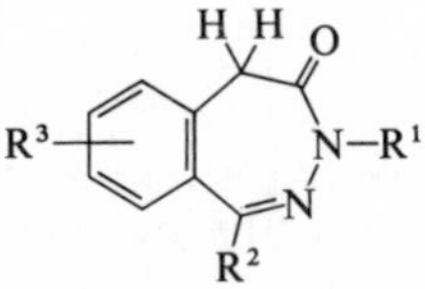

(26)

with silver oxide gave 7,8-dihydro-7-imino-8-oxoquinoline-6-sulphonic acid which reacted with aniline to form the quinone-imine, 7-amino-5,8-dihydro-8-oxo-5-phenyliminoquinoline[124].

Reduction of 2,3-bisethoxycarbonylquinoline with di-isopropylaluminium hydride or lithium aluminium hydride in toluene at −70 °C gave the corresponding dialdehyde[125].

Substituents, e.g. chloromethyl[126] and amino or nitro[127], in the benzene ring of quinolines reacted in much the same way as they did in benzene. However, the O-acetyl derivative of 5-chloro-8-hydroxyquinoline, because of the proximity of the oxygen atom to the ring nitrogen atom, behaved differently and was a good acetylating agent. It was used to acetylate the free amino groups in peptidyl-nucleosides, nucleotides and oligonucleotides[128].

5.4.7 Complexes of quinoline

Several papers have appeared recently on the properties and stabilities of complexes of 8-hydroxyquinoline and related compounds with a variety of metals[129]. The stability of complexes of 8-hydroxyquinoline with 23 metal ions were shown to be higher than those of the corresponding N-oxide[130]. A most interesting report was on the interconversion of the nickel complexes (21) of trans-2(quinol-2-ylmethylene)-3-quinuclidinones between the square planar and tetrahedral forms[131]. Quinoline formed 1:1 and 1:2 complexes with iodine bromide which were more stable than those formed by pyridine[132]. Quinoline successfully displaced ether from its magnesium iodide complex with the formation of the complex: $MgI_2 \cdot 2C_9H_7N$[133].

5.5 ISOQUINOLINES

5.5.1 Synthesis

Isoquinolines were recently synthesised from the four basic structures (22–25). The Pomeranz–Fritsch cyclisation (i.e. 22) involving alkyl- and alkoxy-benzylidene 1,1-diethoxyethylamines was improved by using boron trifluoride in trifluoroacetic anhydride[134]. In a variant of this reaction N-1,1-diethoxyethyl-N-tosyl-(2-benzyloxy-3-methoxybenzyl)amine was cyclised to the corresponding N-tosyl-1,2-dihydroisoquinoline with hydrochloric acid in dioxan, and converted into 8-benzyloxy-7-methoxyquinoline by elimination of the tosyl group promoted by potassium t-butoxide[135]. 7,8-Methylenedioxyisoquinol-4-yl acetic acid was prepared by cyclisation of N-1',1'-diethoxyethyl-2,3-methylenedioxybenzylamine in the presence of glyoxylic acid[136]. N-Cyanomethyl benzylamines were cyclised with sulphuric acid to 1,2-dihydro-4-hydroxyisoquinolines which gave isoquinolines by aerial oxidation[137]. Formamide was incorporated into C-1 and N-2 of 3-substituted

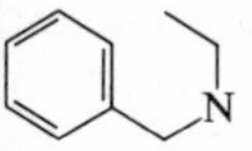

$^-NC_6H_{11}$

(22) 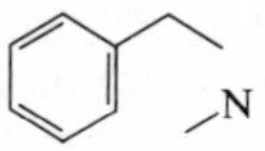 '(23) 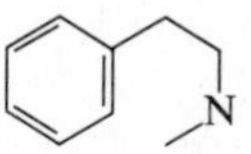 (24) 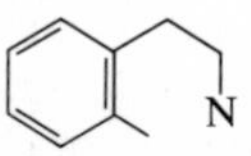 (25)

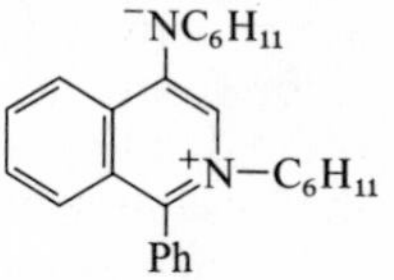

$^+N-C_6H_{11}$
Ph
(27)

isoquinolines (i.e. 23) by reaction with o-unsubstituted acetonyl- and benzoyl-methyl-benzenes in the presence of phosphoryl chloride[138]. A Friedel–Crafts reaction on o-chlorophenylacetyl isothiocyanate (i.e. 24) gave 5-chloro-3-hydroxy-1-mercaptoisoquinoline[139]. The synthesis of chlorinated isoquinolines from α-substituted o-carboxybenzylnitriles with phosphorus pentachloride (i.e. 25) involved reduced isoquinolines as intermediates[140].

The following syntheses did not arise from the basic structures (22–25). The cyclisation of *o*-carboxymethylbenzoyl compounds with hydrazines, or their hydrazones directly, gave *N*-aminoisoquinol-3-ones together with diazepinones (26)[141]. *N*-Aminoisoquinolines were also prepared from 2-thiocoumarins by reaction with hydrazine[142], and from *N*-2',4'- dinitrophenylisoquinolinium chloride by ring cleavage with hydrazines followed by cyclisation and elimination of 2,4-dinitroaniline (N^+—N^- ylides were formed)[143]. When hydroxylamine replaced hydrazine in this reaction, a 52% yield of isoquinoline *N*-oxide was obtained[144]. 6-Substituted 2,3-dichloro-pyrid-4,5-ynes condensed with furan, and the Diels–Alder adducts were then reduced to 5,8-endoxo-5,6,7,8-tetrahydroisoquinolines and deoxygenated to the respective isoquinolines with bromine in acetic acid[145]. In a formally disallowed thermal valence tautomerism 1-cyclohexyl-6'-(cyclohexylimino)-1*a*-phenylindano[1,2-*b*]aziridine gave the red isoquinolinium imine (27)[146].

5.5.2   Reactions

The kinetics of acid-catalysed hydrogen exchange by deuterium in isoquinoline and its *N*-oxide have been studied (cf. Section 5.4.2.1)[64]. Bromination, acylation, nitration and acid-catalysed condensation with formaldehyde of *N*-alkylisocarbostyrils occurred exclusively at C-4 under mild conditions[147]. Arylisocyanates reacted with 4-hydroxy-2-methylisocarbostyril to form the 3-*N*-arylcarbamoyl derivatives[148].

Nucleophilic displacements of the halogen in 1-chloroisoquinoline by carbanions produced a variety of 1-acylmethylisoquinolines[149]. The displacement of fluorine in perfluoroisoquinoline by hydrogen halide was slower than in perfluoroquinoline (cf. Section 5.4.2.2)[76], and the 1-trifluoromethyl group in isoquinoline can be replaced by amino with sodamide[78]. The displacement of the 1-methylsulphonyl group by a variety of nucleophiles (e.g. OEt^-, PhS^-, $CH_2^- COR$) makes this a very useful substituent for synthetic purposes[150]. Isoquinoline 2-oxide reacted with nucleophiles in the presence of acylating agents with loss of the oxygen atom and with substitution at C-1[84, 151]. 1-Methylisoquinoline 2-oxide, however, gave a mixture of 1-hydroxymethyl- and 4-hydroxy-1-methyl-isoquinoline when boiled with acetic anhydride followed by acid hydrolysis; but phosphoryl chloride in chloroform gave only 1-chloromethylisoquinoline[152]. Ethyl magnesium bromide added across the 1,2-double bond of 4-cyanoisoquinoline[118], whereas the 1-cyano isomer under a similar treatment gave 2-propionylisoquinoline[153].

N-Ylides of isoquinoline, derived from the reactions with phenacyl halides[91, 93] and ethoxycarbonylcarbene[154], gave 1,3-dipolar cycloaddition products with acetylenes. Isoquinoline underwent ring enlargement when treated with the ynamine, $PhC{\equiv}CNMe_2$, in the presence of boron trifluoride, but this was followed by extrusion of hydrogen cyanide with the formation of 2-dimethylamino-3-phenylnaphthalene[155]. *N*-Benzylisoquinolinium halides reacted with carbon disulphide in aqueous dioxan containing sodium hydroxide and gave two products: 3-arylthiazolo[2,3-*a*]isoquinolin-2-thiones (mesoionic betains) (28) and *N*-benzylisoquinolinium-4-dithiocarboxylates

(29). The molar ratio of these two products varied with the substituents in the phenyl ring of the benzyl group, and by using ^{35}S isotopic dilution and quantitative product isolation techniques, the Hammett plot correlation [betain/dithiocarboxylate] = 2.32σ -0.12 ($R = 0.971$) was obtained[156].

5.5.3 Photochemistry

The photochemistry of isoquinoline in alcohols and ethers[101], of N-acylimino-isoquinoline betaines[109] and of isocarbostyril in the presence of olefins[104] was similar to that of the corresponding quinolines. The main products of the photolysis of 1,3,4-substituted isoquinoline 2-oxides in acetone were benzo[ƒ][1,3]oxazepines (30). The structure of one of these oxazepines was confirmed by an x-ray analysis[157]. Diphenylcyclopropenethione (31) added across the 1,2-positions of isoquinoline and benzodiazines in chloroform containing methanol to yield 1-methoxy-1,2-diphenyl-3-thionpyrrolidino-[1,2-a]isoquinoline (32)[158]. 2,3-Dihydro-2-methyl-3-oxoisoquinoline furnished Diels–Alder adducts, with maleic anhydride and tetracyanoethylene,

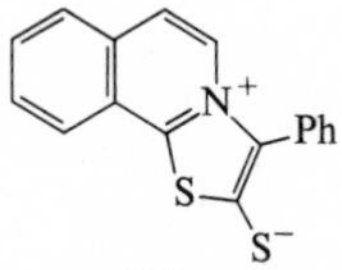

(28) 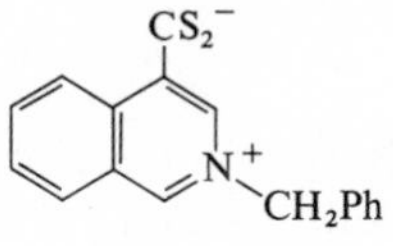   (29) 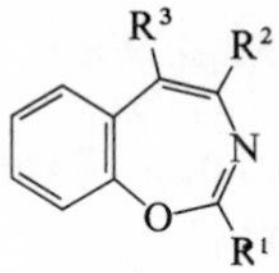   (30)

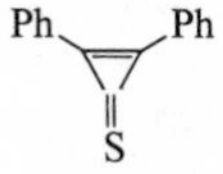

(31) 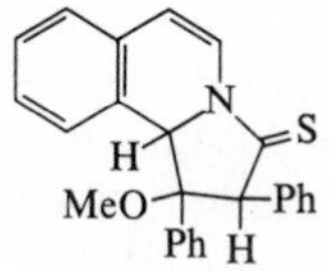   (32) 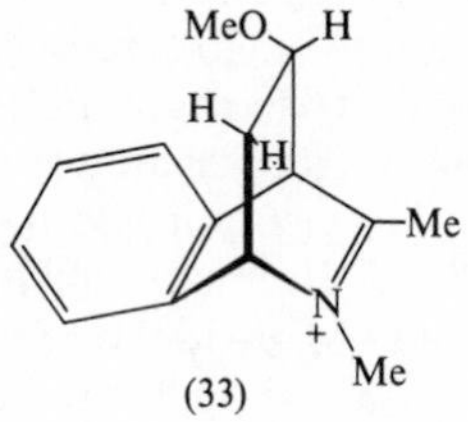   (33)

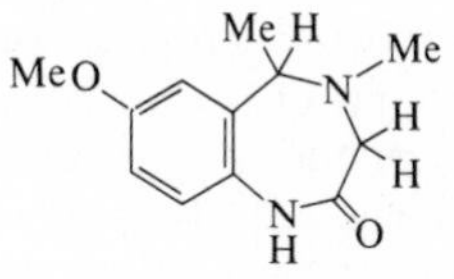

(34) 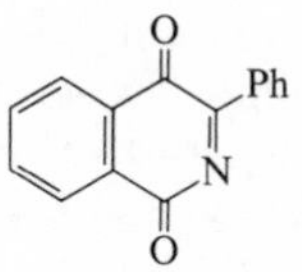   (35)

in which addition had occurred across C-1 and C-4[159]. Methyl vinyl ether added across C-1 and C-4 of 2,3-dimethylisoquinolinium iodide in 97% yield, and was stereospecific. An x-ray analysis showed that the methoxy group was on the carbon atom adjacent to C-4 and was in the direction of the benzene ring, i.e. on the opposite side of the 3-methyl group (33)[160].

5.5.4   Reduced isoquinolines

5.5.4.1   Synthesis

Reduction of isoquinoline with zinc in acetic anhydride yielded *meso*- and *dl*-N,N-diacetyl-1,1'-dihydro-1,1'-bi-isoquinolinyl[161]. Several 3,4-dihydroiso-quinolines were prepared by the classical Bischler–Napieralski reaction[162, 163]. 1-Spiro-1,2,3,4-tetrahydroisoquinolines were made from phenethylamine and carbonyl compounds[164]. Under the conditions of the Graf–Ritter reaction a number of 1,2,3,4-tetrahydroisoquinolines were obtained from benzylcyclopropanes and alkylnitriles[165]. In a new synthesis, *o*-vinylbenzyl-amines cyclised to 1,2,3,4-tetrahydroisoquinolines under basic conditions[166]. 1,2,3,4-Tetrahydroisoquinoline was obtained in a two step synthesis by reaction of benzylamine with 2-bromoethanol followed by hydrobromic acid[167]. 1,2,3,4-Tetrahydro-4-phenylisoquinolines were formed from N-1,1-diethoxyethylbenzylamines and phenols[168], and N-benzyl-1,2,3,4-tetrahydro-3,5-dimethoxy-4-oxoisoquinoline was made from N-benzyl-N-ethoxycar-bonylmethyl-3,5-dimethoxybenzylamine in the presence of acid[169].

5.5.4.2   Reactions

2-Acyl-1-cyano-1,2-dihydroisoquinolines (Reissert compounds) were useful for the preparation of 1-substituted isoquinolines by reaction with olefins[170] and aldehydes[171]. 2-Benzoyl-1-cyano-1,2-dihydroisoquinoline formed a chlorohydrin which reacted with a base to eliminate hydrogen chloride and, depending on the conditions, yielded 1-benzoylamino-1-cyanoisochromene, 1-benzoyliminoisochromene, 1-ethoxy-3-phenylisoquinoline and 1-ethoxy-4-formyl-3-phenylisoquinoline[172].

Aziridinium intermediates were formed when 3,4-dihydro-2-methyl-6,7-methylenedioxyisoquinolinium perchlorate reacted with diazomethane[173]. 3,4-Dihydro-6,7-dimethoxyisoquinolinium chloride condensed with methyl vinyl ketone or 1-dimethylaminopentan-3-one and 2-methyl-1-dimethyl-aminobutan-3-one to form benzo[*a*]hexahydroquinolizin-2-ones. The last two dimethyl compounds most probably eliminated dimethylamine prior to condensing[174].

Electrolytic oxidation of 1,2,3,4-tetrahydro-7-hydroxy-6-methoxy-1,2-dimethylisoquinoline was stereoselective in that only molecules of identical configuration coupled to form the corresponding 8,8'-bitetrahydroiso-quinolyl, i.e. S gave S,S and R gave R,R [175]. The oxidative (phenolic) coupling using a platinum catalyst, on the other hand, was not stereoselective[175, 176]. Dehydration of the oxime of 1,2,3,4-tetrahydro-7-methoxy-2-methyl-4-oxo-1-phenylisoquinoline with polyphosphoric acid gave 1,3,4,5-tetrahydro-7-methoxy-4-methyl-5-phenyl-2H,1,4-benzodiazepin-2-one (34) and a 4-amino-7-methoxy-2-methyl-1-phenylisoquinolinium salt[177]. 1-*p*-Chlorophenoxy-methyl-3,4-dihydroisoquinoline 2-oxide also dehydrated to 1-*p*-chloro-phenoxymethylisoquinoline in an unusual fashion[178].

The addition of nucleophiles and dienes across the C-3 and N-2 double bond in 1,4-dihydro-1,4-dioxo-3-phenylisoquinoline (35) was facile[179].

1-n-Butyl-3,4-dihydroisoquinolines, and other isoquinolines in which a hydrogen atom was present on the γ-atom of the 1-side chain, underwent a Norrish type II photo-elimination to give the 1-methylisoquinoline derivative (and propene in the case of 1-n-butyl)[180].

5.6 QUINOLIZINIUM COMPOUNDS

These compounds have attracted little attention recently but some interest has come from alkaloids bearing this ring system. 2,3,4-Trisubstituted-4-oxo-quinolizines were prepared from ethyl pyrid-2-ylacetate and methyl 2,2-bismethylthioacrylate[181]. 3-Methoxycarbonyl and 3,4-bismethoxycarbonyl 1-ethoxycarbonyl-4H-quinolizines were formed when ethyl 3-(pyrid-2-yl)-trans-acrylate reacted with one molecule of dimethyl acetylenedicarboxylate. In acid solution the tricarboxylic ester gave mostly one cation which was protonated on C-3, whereas the dicarboxylic ester gave a 9:1 mixture of the C-3 (36) and C-1 (37) protonated species[96]. The synthesis of benzo[c]quino-lizines from quinolines was mentioned earlier[94, 95, 96].

5-Substituted 2-bromo-1,2,3,4-tetrahydro-4-oxoquinolizines were obtained by a base-catalysed condensation of 6-substituted 2-picolinic esters with butyrolactone to 2-(6-substituted pyrid-2-ylcarbonyl)butyrolactones followed by treatment with hydrobromic acid. The bromo-oxoquinolizines that

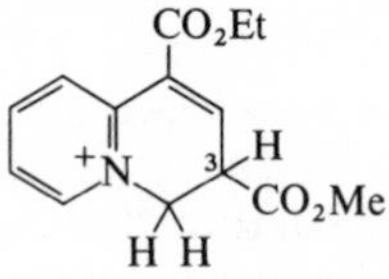

(36) 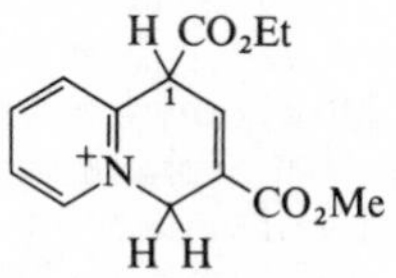   (37) 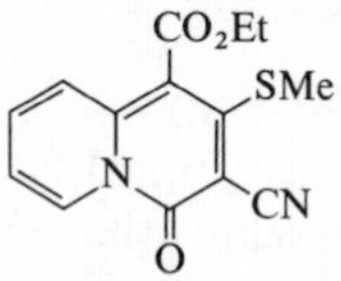   (38)

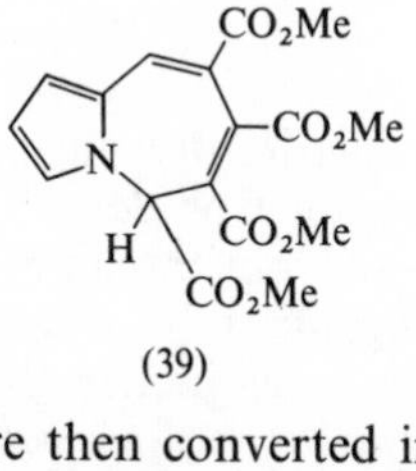

(39) 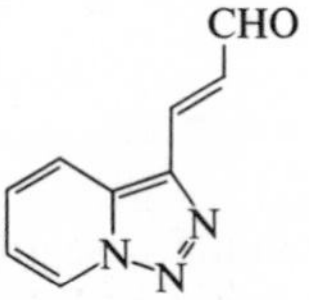   (40)

were formed were then converted into 2-(substituted-amino)quinolizinium compounds by reaction with bromine–acetic acid followed by amines, giving a new class of anthelmintic agents[182].

In a synthetic approach to the Lythraceae alkaloids an α,β-unsaturated ketone was condensed with 3,4,5,6-tetrahydropyridine to form the quinolizi-dine nucleus[183]. Two independent syntheses of the alkaloid (±) lamprolobine were reported. The first used an acid-catalysed intramolecular cyclisation of 3-cyano-1,4,5,6-tetrahydro-1-(2,2-ethylenedioxybut-4-yl)pyridine to 1-cyano-7,7-ethylenedioxyquinolizidine[184], and the second started from 1,10-de-hydroquinolizidine which gave the 1-ethoxycarbonyl derivative with ethylchloroformate[185].

The mechanism of thermal rearrangement of tetramethyl 7,9-dimethyl-9aH-quinolizin-1,2,3,4-tetracarboxylate to the 4-H tautomer was studied using deuterium which indicated that an intramolecular shift took place[186]. The methylthio group in 3-cyano-1-ethoxycarbonyl-2-methylthio-4H-quinolizin-4-one (38) can be displaced by benzylamine and the 3-cyano group can be hydrolysed to a carbamoyl group. Reduction of the 3-benzylamino derivative gave first 2-benzylamino-3-cyano-1-ethoxycarbonyl-6,7,8,9-tetrahydro-4H-quinolizin-4-one and then the 1,6,7,8,9,9a-hexahydro derivative[187]. Under the influence of ultraviolet light several 1,2,3,4-tetramethoxycarbonyl-9aH-quinolizines isomerised to pyrrolo[1,2-a]azepines (39), i.e. nitrogen analogues of azulene. No reaction occurred when the bridgehead carbon atom C-9 carried a phenyl substituent[95].

A number of 1-aminoquinolizinium salts rearranged when heated with nitrous acid and gave first the cis-3-(v-triazolo[1,5a]pyridyl) acraldehydes which rearranged further to the trans isomers (40)[188]. In a new ring cleavage reaction quinolizinium salts reacted with aliphatic amines to give (vinylogous) amidines of 2-substituted pyridines with trans–trans conjugated tetramethine chains[189].

5.7 CINNOLINES

5.7.1 Synthesis

In a new synthesis, 1,4,5,6,7,8-hexahydro-3-phenyl-1-ureidocinnoline was conveniently prepared from 1-morpholinocyclohex-1-ene and phenacylbromide semicarbazone, and converted into 5,6,7,8-tetrahydro-3-phenylcinnoline with alcoholic alkali[190]. 5,6,7,8-Tetrahydrocinnolines were dehydrogenated to cinnolines with palladium on charcoal[190, 191].

5.7.2 Reactions

Semi-empirical calculations of electron distribution in cinnolines showed that there was little difference between the electron density at N-1 and N-2, and preferential protonation on N-2 arose from the more favourable electronic energy for N-1 protonation being offset by a larger energy from nuclear repulsion[192]. It was also shown that the greater reactivity of N-2 in the alkylation of 3- and 3,4-alkyl substituted cinnolines resulted from steric effects. Product ratios in sterically balanced cinnolines, e.g. 3-ethyl-8-methylcinnoline, showed that the two nitrogen atoms had nearly identical properties[193].

In contrast with the situation in the quinolines series, the free base of 4-hydroxycinnoline may be involved on nitration in sulphuric acid[61]. Cinnoline 2-oxide and its 4-methyl derivative reacted with phenyl magnesium bromide to give the corresponding 1,2-dihydro-2,3-diphenyl derivative among other products[194].

Ring contraction occurred when cinnolin-3(2H)-one reacted with chloramine or hydroxylamine-O-sulphonic acid to yield 3-carbamoylindazole and oxindole respectively[195]. Reductive formylation (formic acid and formamide) of cinnoline and its 3- and 4-alkyl derivatives produced N-formylindoles[196]. Photochemical isomerisation of hexafluorocinnoline to hexafluoroquinazoline has been demonstrated[197]. Dimethylketene added across the 1,2-double bond of cinnolines to form oxadiazino[4,3-a]-(41) or oxadiazino-[3,4-a]-(42) 1,2-dihydrocinnolines depending on whether C-3 was substituted or not[198]. Diphenylcyclopropenethione added in a similar manner to give 6,7-diphenyl-8-thionpyrazolo[1,2-a]cinnolines (43)[158].

The well known reactivity of a γ-methyl group in nitrogen heterocyclic molecules was demonstrated in 4-methylcinnoline by conversion into 4-hydroxyiminomethylcinnoline with ethyl nitrite in the presence of excess hydrogen chloride. The formation of a quasi-p-quinonoid intermediate was thought to be a necessary step in the course of this nitrosation reaction[199].

5.8 QUINAZOLINES

5.8.1 Synthesis

Aryl pentachlorophenyl-ketone imines, obtained from pentachlorophenyllithium and aryl nitriles, reacted further with nitriles to form 2,4-diaryl-5,6,7,8-tetrachloroquinazolines[200]. 2-Oxo- (or thio-) 4-chloro (or bromo)-quinazolines were prepared in high yields by heating o-isocyano- (or isothiocyano-) benzonitriles with hydrogen chloride (or bromide) in dibutyl ether[201]. The Niementowski reaction is still one of the best methods for preparing quinazolin-4(3H)ones. N-Acetyl (or benzoyl) anthranilic acids

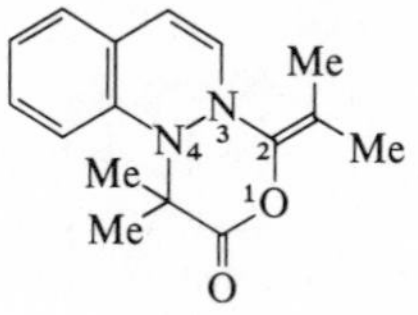

(41) 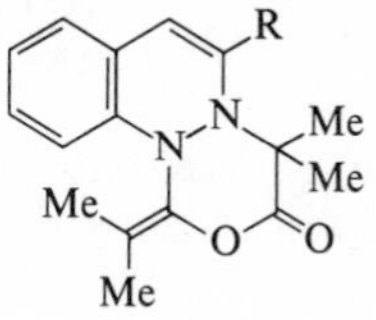   (42) 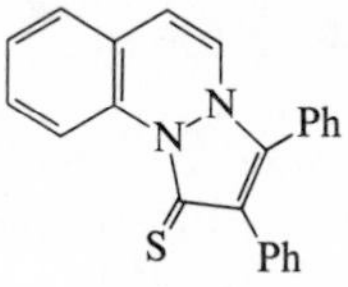   (43)

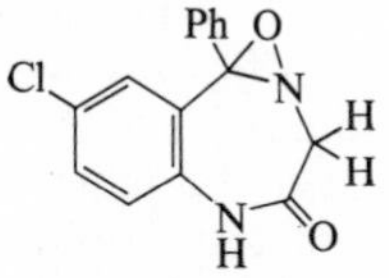

(44) 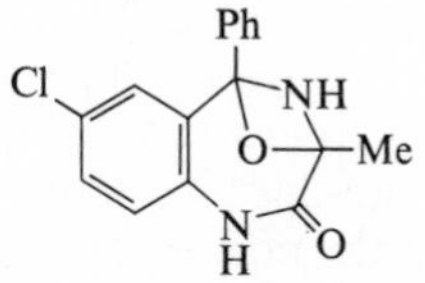   (45) 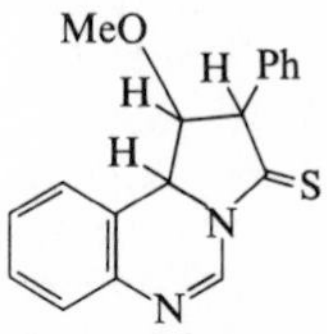   (46)

gave the respective 2-methyl (or phenyl) quinazolin-4(3H)-ones with formamide[202]. The reaction of anthranilamide with phenylacetic acid in xylene containing phosphorus pentoxide gave a mixture consisting of 2-benzyl-quinazolin-4(3H)-one, o-aminobenzonitrile, tricycloquinazoline, 2-(o-aminophenyl)quinazolin-4(3H)-one, 6,12-diaminophenhomiazine and a tricyclic compound with the composition $C_{29}H_{21}N_5O$ [203].

Ring contraction of 7-chloro-4,5-epoxy-1,3,4,5-tetrahydro-5-phenyl-2H,1,4-benzodiazepin-2-one (44) by water in tetrahydrofuran, and of 7-chloro-2,5-endoxo-1,2,4,5-tetrahydro-2-methyl-5-phenyl-3H,1,4-benzodiazepin-3-one (45) by methanolic hydrogen chloride afforded 6-chloro-4-phenylquinazolin-2(1H)-one[204] and 2-acetyl-6-chloro-4-phenylquinazoline[205] respectively. The conversion ·of 1,4-benzodiazipines into quinazolines and *vice versa* was reviewed by Sternbach in 1971[206].

A rearrangement and oxidation occurred to yield 4-amino-2-phenyl-quinazoline in an attempt to prepare 3-amino-1-benzylindazole by cyclisation of *o*-(1-benzylhydrazino)benzonitrile[207]. A detailed kinetic study of the intramolecular nucleophilic attack of the ureido group upon carbonyl groups in *o*-ureidobenzoyl compounds to form quinazolin-2,4(1H,3H)-dione was made[208]. Quinazolin-2,4(1H,3H)-diones were prepared by the standard reaction of anthranilamides with phosgene[209]. Ethyl *o*-hydroxyamino-benzoate was the source for 3-methyl-1-N-methylcarbamoyloxyquinazolin-2,4-dione (by reaction with methylisocyanate) and several 1,3-disubstituted quinazolin-4-one-2-thiones (by reaction with isothiocyanates)[210]. Acetylation of *o*-aminobenzamidoximes and hydrolysis of *O,N*-diacetyl-*o*-aminobenzami-doximes gave derivatives of 4-aminoquinazoline 3-oxide. The isomeric 4-hydroxyimino-3,4-dihydroquinazolines were obtained from 4-methoxy-quinazolines and hydroxylamine[211]. 5,8-Dihydro-6-methoxy-4-methyl-2-methylthio-5,8-dioxoquinazoline was readily formed from 2-acetyl-3-amino-5-methoxy-1,4-benzoquinone and *S*-methylisothiouronium sulphate[212]. α-(Quinazolin-4-yl)glycine was identified among the photo-oxidation products of tryptophan and its ethyl ester, but not its N-acetyl derivative[213].

5.8.2 Reactions

Quaternising methylation of quinazoline, previously considered to yield only the 3-methyl derivative, was recently shown to give a mixture of the 1- and 3-methyl derivatives in the ratio 1:5[214]. Trimethylsilylation of quinazolin-2,4(1H,3H)-dione gave 2,4-bistrimethylsilyloxyquinazoline which reacted with methyl iodide to provide 1-methylquinazolin-2,4(3H)-dione, and with α-bromo-acetylglucose to yield the 1-glucopyranosyl derivative[215]. Phosphoryl chloride or phosphorus pentachloride converted quinazolin-2,4(1H,3H)-diones into 2-chloroquinazolin-4(3H)-ones in which the halogen can be displaced by a variety of nucleophiles. Quinazolindiones also reacted with aryl magnesium halides to give 2,4-disubstituted quinazolines[216].

Several ketones, in the presence of sodamide, replaced the halogen in 4-chloroquinazoline to give the corresponding 4-acylmethylquinazolines in variable yields[149]. In a detailed study of the displacement of the trimethyl-ammonio group by hydroxide in nitrogen heterocyclic molecules, quinazolin-4-yltrimethylammonium chloride was shown to react *c.* 700 times faster than 4-chloroquinazoline at 20 °C[7].

Oxidation of 5,8-dihydroxyquinazoline to the quinone, 5,8-dihydro-5,8-dioxoquinazoline, was achieved without oxidation at C-4[217]. The 3,4-adduct, 1-methoxy-1,2-diphenyl-3-thionpyrrolidino[1,2-*c*]quinazoline (46) was obtained from quinazoline and diphenylcyclopropenethione in chloroform-methanol solution[158]. Quinazolines, unlike cinnolines and phthalazine, reacted with diketene to give 1:2 adducts; i.e. across the 1,2- and 3,4-double bonds[218].

The nitrene generated from 3-aminoquinazolin-4-one with lead tetraacetate (cf. Section 5.4.6), reacted stereospecifically with olefins to form the 3-ylaziridines[122]. 1-(3,4-dihydroquinazolin-4-on-3-yl)-2-vinylaziridines undergo thermal rearrangement in the aziridino group, but at high temperatures cleavage of the N—N bond occurred with the liberation of the 3-unsubstituted quinazolinone[219].

7-Chloro-2-chloromethyl-4-phenylquinazoline 3-oxide reacted with hydroxylamine and was modified to 8-chloro-1,2-dihydro-2-hydroxyimino-6-phenyl-3H,4,1,5-benzoxadiazocine (47). Reduction of the diazocine gave back a 2-hydroxymethyl-4-phenylquinazoline[220]. 3,4-Dihydro-3-propargylquinazolin-4-one, with various basic reagents, afforded the isomeric 3-propadienyl (allene) derivative. On prolonged treatment with base both these compounds gave 2-(2-formamidophenyl)-5-methylisoxazole and 1,2,3,4-tetrahydro-2-vinylquinazolin-4(3H)-one. Formation of the latter necessitated ring opening between C-2 and N-3, and cyclisation of N-1 to the α-carbon atom of the allene side chain with elimination of C-2 as formate[221].

Photolysis of quinazolin-4-on-3-ylaziridines gave the 3-ylnitrene as a reactive intermediate[222]. The photostability of 2- and 4-(o-hydroxyphenyl)-quinazolines was attributed to intramolecular hydrogen bonding[223].

The reactions of 2-benzylthio-3-phenyl-4-thioquinazoline with hydrazine under a variety of conditions has been described[224], and the nitrosation of 4-methylquinazoline to 4-hydroxyiminomethylquinazoline was similar to that of 4-methylcinnoline (cf. Section 5.7)[199].

The biological activity of quinazolones was reviewed in 1970[225].

5.8.3 Reduced quinazolines

The structure of 3-amino-6-chloro-3,4-dihydro-4-hydroxy-4-phenylquinazoline, previously regarded as a 1,3,4-benzotriazepine, was determined by its chemical reactions and by an x-ray analysis[226]. Several 4-(substituted-amino)- and 4-hydroxy-1,2,3,4-tetrahydro-4-phenylquinazolin-2(1H)-ones were synthesised from N-ethoxycarbonyl-2-benzoyl-4-chloroaniline and aliphatic amines[227]. Condensation of anthranilamide with dimethyl acetylenedicarboxylate gave dimethyl o-carbamoylanilinofumarate which reacted with phosphorus pentasulphide to yield 1,2-dihydro-2-methoxycarbonyl-2-methoxycarbonylmethylquinazolin-4(3H)-thione[228]. Ring enlargement of 1-(N-substituted carbamoyl)isatins with ethanolic triethylamine gave 3-substituted 4-ethoxycarbonyl-3,4-dihydro-4-hydroxyquinazolin-2(1H)-ones[229].

N-Alkyl cyclohexen-1-ylcarbamoylisothiocyanates (48) cyclised readily to 1-substituted 5,6,7,8-tetrahydroquinozolin-2-one-4(3H)-thiones, and the various transformations of these products, e.g. alkylation, reduction, were carried out by standard procedures[230]. A number of hexahydroquinazolin-2(1H)-ones and -thiones were prepared from 2-hydroxymethylcyclohexanone and ureas or thioureas[231]. Cis- and trans-octahydroquinazolin-2(1H)-ones and -thiones, among other hexa- and octa-hydroquinazolines, were obtained from *cis*- and *trans*-2-aminomethylcyclohexylamines, and their spectral properties were compared[232]. Satisfactory yields of 3-substituted 1,2,3,4,5,6,7,8-octahydro-5-oxoquinazolines were obtained by condensation of 3-aminocyclohex-2,3-en-1-one with formaldehyde and primary amines[233].

All four (+) and (−), *cis*- and *trans*-decahydroquinazolines and 2-amino-octahydroquinazolines were synthesised by methods which defined their absolute stereochemistry[234]. 8*a*-Aminomethyl-3-phenyl-*cis*-perhydroquinazolin-2,4-dione was obtained by an acid-catalysed rearrangement of 3*a*-(*N'*-phenylureido)-*trans*-perhydroisoindol-1-one[235]. 8*a*-Nitromethyl-*cis*-perhydroquinazolin-2-one (49) was formed in high yield by a stereospecific

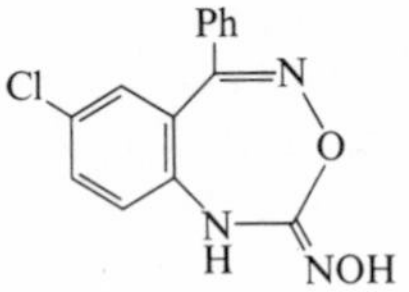

(47) 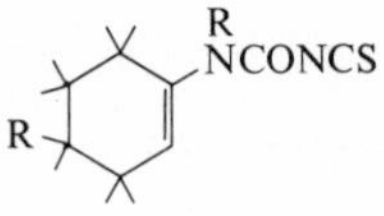　　　　　　　　(48) 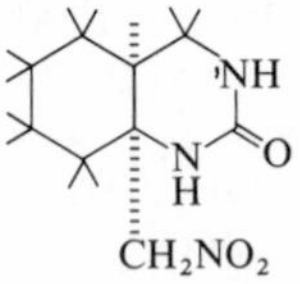　　　　　　　　(49)

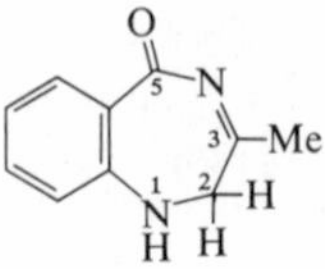

(50) 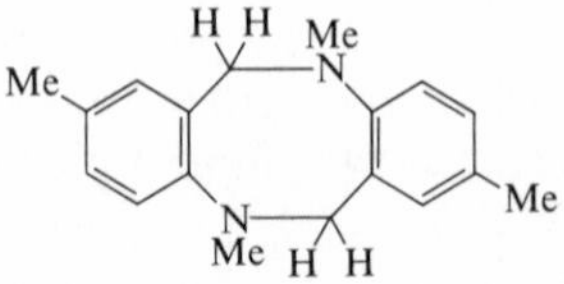　　　　　　　　(51)

cis-addition of the elements of nitromethane when 3,4,5,6,7,8-hexahydro-quinazolin-2(1*H*)-one was fused with nitroacetic acid. The stereochemistry was deduced by transforming the 8*a* side chain to an 8*a*-methyl group and an independent synthesis of this product from *cis*-2-aminomethyl-1-methyl-cyclohexylamine of known stereochemistry[236].

Ring enlargement of 2-chloromethyl-1,2,3,4-tetrahydro-2-methylquina-zolin-4(3*H*)-ones with potassium t-butoxide gave 1,2-dihydro-3-methyl-5*H*,1,4-benzodiazepin-5-ones (50), whereas the 1-methylquinazoline derivative gave 1,2-dihydro-1-methyl-2-methylene-5*H*,1,4-benzodiazepin-5-one[237]. 1,2,3,4-Tetrahydro-1,6-dimethyl-3-*p*-tolylquinazoline was converted to 5,6,11,12-tetrahydro-2,5,8,11-tetramethylphenhomiazine (51) in chloroform only when one molecular equivalent of methyl bromide was added[238].

5.9　QUINOXALINES

5.9.1　Synthesis

The conventional method for preparing quinoxalin-2(1*H*)-ones from *o*-phenylene diamine and α-keto-esters[239] was extended to dimethyl acetylenedicarboxylate which furnished 1,4-dihydro-2-methoxycarbonyl-methylenequinoxalin-2-(1*H*)-one[240].

Several quinoxaline 1,4-dioxides were prepared from 1,2-bishydroxyimino-cyclohexa-3,5-diene, generated from benzofurazan 1-oxide and 2,5-di-t-butylhydroquinone, and α-diketones or α-keto-alcohol[241]. Benzofurazan 1-oxide (52) also reacted directly with enamines to form quinoxaline-1,4-dioxides with elimination of an amine. This was not possible with N,N-dimethylisobutylenylamine, and 2,3-dihydro-2,2-dimethyl-3-dimethyl aminoquinoxaline 1,4-dioxide (53) was formed[242]. 1,2,3,4-Tetrahydro-2-hydroxy-1,4-dimethyl-6-nitro-2-phenylquinoxalin-3-one was made by ring expansion from diethyl α-(1,3-dimethyl-5-nitrobenzimidazol-2-yl)benzyl-phosphate iodide (54) in dimethyl sulphoxide[243].

Ring contraction of certain 1,4-benzodiazepine oxides with phosphoryl chloride[244], or ultraviolet light[245], gave quinoxalines. An unequivocal synthesis of (S)-1,2,3,4-tetrahydro-2-methylquinoxaline (laevo-rotatory in CHCl$_3$ and 65% EtOH, but dextro- rotatory in tetrahydrofuran) was accomplished from 2,4-dibromonitrobenzene and (S)-α-alanine[246].

5.9.2 Reactions

2,3-Diphenyl-1-thiopyrrolo[1,2-a]quinoxaline (55) was formed from quinoxaline and diphenylcyclopropenethione (compare with other benzazines)[158].

The 2-methyl group in 1,2,3-trimethylquinoxalinium salts was particularly reactive towards electrophiles. It reacted with the p-chlorophenyldiazonium ion and p-nitroso-N,N-dimethylaniline to form the 2-p-chlorophenylazo-methyl and 2-p-N,N-dimethylaminophenyliminomethyl derivatives. 1,3-Dimethyl-2-methylenequinoxaline was also isolated from the trimethyl salt by alkaline treatment[247]. 2-Methylquinoxaline and two equivalents of tetrachloro-1,2-benzoquinone gave two products: a coloured substance related to quinoxaline orange and 2-(2-quinoxalino)-4,5,6,7-tetrachloro-benzo-1,3-dioxole (56)[248].

1-Ribofuranosylquinoxalin-2(1H)-ones were formed from 3-substituted 2-trimethylsilyloxyquinoxalines[249] (compare with quinazolines, Section 5.8). In boiling aqueous solution, hydrazine added across the 3,4-double bond of quinoxalin-2(1H)-one; cleavage of the 3,4-bond, cyclisation and elimination of nitrogen followed to give 2-methylbenzimidazole. The 1-methyl derivative gave o-methylaminoaniline, the 3-methyl derivative failed to react, and quinoxalin-2,3(1H,4H)-dione and its 1-methyl derivative gave 3-hydrazino-quinoxalin-2(1H)-ones[250].

Reduction of 2,3-bis-2'-diethylaminomethyl-6,7-dimethoxyquinoxaline with lithium aluminium hydride gave the 1,2,3,4-tetrahydro derivative but the stereochemistry at C-2 and C-3 was assumed to be cis[251]. Aerial oxidation of 3,4-dihydro-7-methoxy-1,3-(diphenyl- or di-p-methoxyphenyl-)quinoxalin-2(1H)-ones proceeded in the presence of acid to yield the corresponding 3,4-dehydro compounds. A similar oxidation of a 1-methyl derivative, however, gave 5-methoxy-3-(p-methoxyphenyl)-1-methylbenzimidazol-2-one. The former oxidation was also achieved photochemically in the presence of 4,4'-dimethoxyazobenzene which was reduced to p-anisidine[252]. Quinoxaline 1,4-dioxides also gave benzimidazol-2-ones on photolysis[253].

Homolytic benzylation of quinoxaline in acetic acid using dibenzylmercury as the radical source gave only 2-benzylquinoxaline[254]. Selective homolytic α-oxyalkylation of quinoxalines (also quinolines and isoquinolines) was achieved with several oxidising agents, e.g. H_2O_2, t-BuOOH, $(NH_4)_2S_2O_8$. Thus in dioxane, tetrahydrofuran and 1,3-dioxalane the products were 2-(dioxan-2-yl)-, 2-(tetrahydrofuran-2-yl)- and 2-(dioxalan-2(and 4)-yl)-quinoxaline respectively[255].

Substituents in the heterocyclic ring and in side chains react in the usual manner, e.g. 3-substituted 2-chloroquinoxalines yielded ω-hydroxyalkyl-amino compounds with ω-aminoalcohols[256], 2-(2-chloro-1-hydroxyethyl)-quinoxaline was dehydrochlorinated with ethanolic alkali, and the resulting

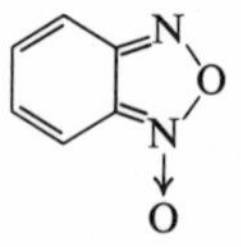

(52) 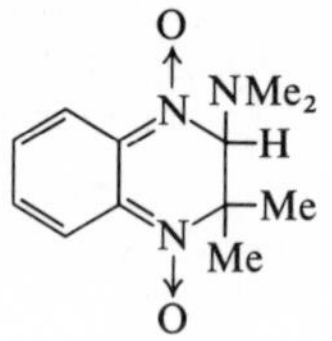   (53) 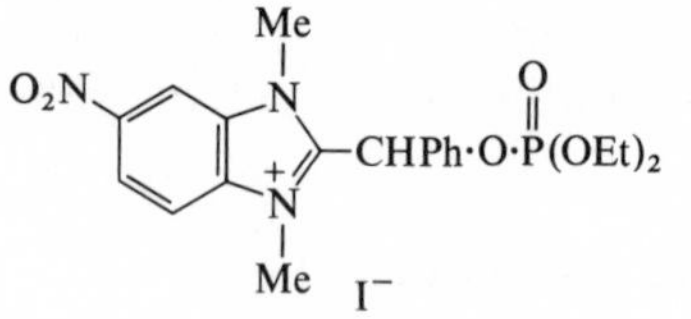   (54)

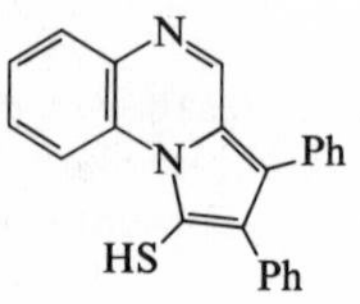

(55) 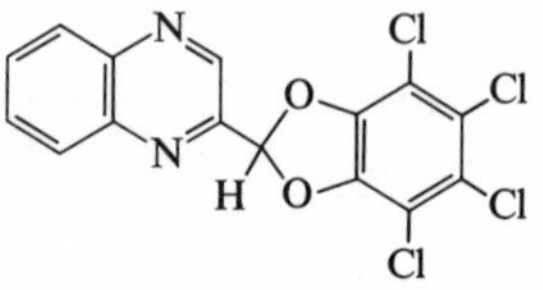   (56)

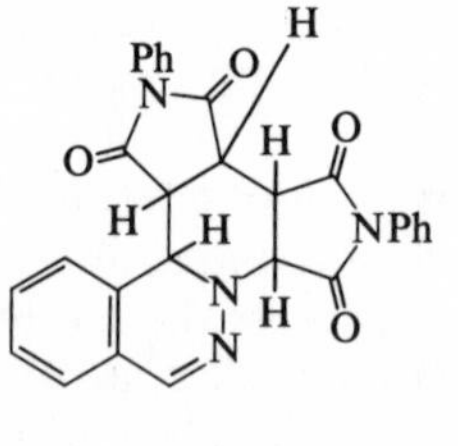

(57) 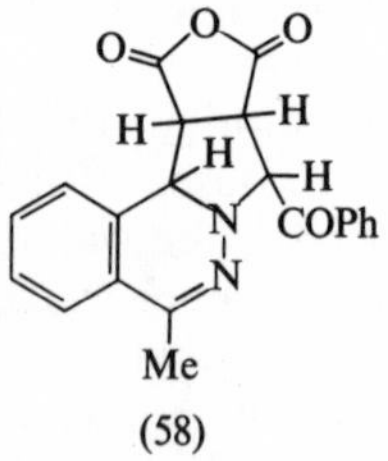   (58)

epoxide reacted with a variety of alkylamines to give 2-(2-alkylamino-1-hydroxyethyl)quinoxalines[257], and 2-formylquinoxaline formed a Schiff's base with methyl p-aminobenzoate which was used for making a tetrahydro-folic acid analogue[258].

Similarly, substituents in the benzene ring of quinoxaline reacted in the usual manner. Thus 5-aminoquinoxalines were diazotised and coupled with β-naphthol[259], and 2,3-disubstituted 6-nitroquinoxalines reacted with alco-holic potassium cyanide and gave the respective 5-cyano-6-nitroquinoxa-lines[260].

Acetophenone and 3-substituted quinoxaline 1-oxides in the presence of sodamide in dry benzene afforded 3-substituted 2-phenacylquinoxaline 1-oxides in good yields, i.e. no deoxygenation took place[261].

5.10 PHTHALAZINES

One new synthesis was reported, and it required 3-arylphthalimidine as starting material. This was alkylated to 1-ethoxy-3-aryl-3H-isoindole, with triethyloxonium borofluoride, and converted into 1-amino-6-arylphthalazines in high yields with 97% hydrazine[262]. In a standard synthesis, 2-acetyl-6-nitrobenzoic acid and methyl hydrazine gave only one of the two possible isomers: 1,3-dimethyl-5-nitrophthalazin-4-one[263].

Phthalazine, like other benzodiazines, reacted with one molecule of diphenylcyclopropenethione to form 1-methoxy-1,2-diphenyl-3-thionpyrrolidino[1,2-a]phthalazine[158]; whereas two molecules of N-phenylmaleimide added across the C-1, N-2 double bond to yield a pyrido[1,2-a]phthalazine (57)[264]. The ylide derived from 3-benzoylmethyl-1-methylphthalazinium bromide probably reacted as a 1,3-dipolar reagent with maleic anhydride to form a pyrrolidino[2,1-a]phthalazine (58); and as a carbene (after a prototropic rearrangement and leaving two electrons on C-4) to give the 4-benzoyl derivative with benzaldehyde[265].

A detailed kinetic study of methoxydechlorination of 4-substituted 1-chlorophthalazines showed that the usual S_NAr2 mechanism operated. By comparison with the respective pyridazines, annelation was shown to enhance the rates 50–100-fold, and further annelation, i.e. to benzo[g]-phthalazines, increased the rates further but to a lesser extent[266]. 1,4-Dihalogenophthalazines with chloro, fluoro or trifluoromethyl groups in the benzene ring reacted as usual towards nucleophiles, and chlorine was readily displaced from both rings by reaction with potassium fluoride[267, 268].

Phthalazin-1,4-diones underwent ring contraction to phthalimides on ultraviolet irradiation in the presence of oxygen[269]. The mechanism of the ferricyanide-catalysed chemiluminescence of luminol,5-aminophthalazin-1,4-(2H,3H)-dione, was studied. It proceeded by a one-electron oxidation of the luminol dianion to 5-aminophthalazin-1,4-semidione. The semidione then reacted with oxygen to produce electronically-excited 3-aminophthalic acid and nitrogen. Alternatively, the semidione was oxidised further by ferricyanide in a non-luminescent reaction[270, 271].

Aryl and alkylsulphonyl chlorides reacted with phthalazin-1,4(2H,3H)-dione at the oxygen atom to form the O-sulphonyl esters, but 2-chloroethanesulphonic acid reacted at the nitrogen atom and gave the N-2-ethylsulphonic acid[272].

5.11 BENZOTRIAZINES

5.11.1 1,2,3-Benzotriazines

Although 1,2,3-benzotriazines have been known for a long time, the parent base was synthesised for the first time by Rees and his co-workers in 1971.

1,2,3-Benzotriazine and 4-substituted derivatives were prepared in high yields by oxidation of 1- and 2-aminoindazoles and their 3-methyl, methoxy or phenyl derivatives with lead tetra-acetate in methylene chloride. 1,2,3-Benzotriazine had the expected high reactivity towards nucleophiles and their exclusion during the preparation was imperative[273]. 4-Alkyl- and 4-aryl-1,2,3-benzotriazines were also obtained by a similar oxidation of the hydrazones of o-aminophenyl alkyl or aryl ketones, and thermolysis of 1-(o-azidophenyl)diazoethane gave high yields of 4-methyl-1,2,3-benzotriazine[274]. A series of 3-aryl-3,4-dihydro-4-imino-1,2,3-benzotriazines were made by cyclisation of the appropriate o-cyanophenyl*triazenes* which rearranged to 4-anilino-1,2,3-benzotriazines[275].

The reaction of nucleophiles with 3-substituted 3,4-dihydro-4-oxo-[276] or 3,4-dihydro-4-imino-1,2,3-benzotriazines[277] resulted in ring cleavage with the formation of the respective o-carbonyl- or o-cyano-anilines. Ring cleavage of 4-amino- and 4-oxo-1,2,3-benzotriazines was also effected by Raney nickel and hydrazine[278]. Oxidation of 3-amino-4-oxo-1,2,3-benzotriazines with lead tetra-acetate produced nitrenes which reacted in two ways: by loss of a molecule of nitrogen to yield indazolones, and by loss of two nitrogen molecules to form the unstable benzocyclopropanones[279]. The use of 3,4-dihydro-3-hydroxy-4-oxo-1,2,3-benzotriazine in peptide synthesis by the dicyclohexylcarbodi-imide method minimised racemisation[280].

5.11.2 1,2,4-Benzotriazines

1-Aminoquinoxalin-2-one and its 3,6- and/or 7-methyl or phenyl derivatives, obtained by the reaction of 1-unsubstituted quinoxalin-2(1H)-ones with hydroxylamine-O-sulphonic acid, were oxidised with lead tetra-acetate to N-nitrenes which rearranged to 3,6 and/or 7 substituted 1,2,4-benzotriazines[281]. Several 3-amino-1,2,4-benzotriazine 1-oxides were synthesised by condensation of o-nitroanilines with cyanamide. Oxidation of 3-amino-1,2,4-benzotriazines with peracetic acid, on the other hand, gave at room temperature the 2-oxides, and at 50 °C the 1,4-dioxides[282].

References

1. Heidrich, D. and Scholz, M. (1970). *Monatsh. Chem.*, **101,** 1394
2. Sasaki, Y. and Suzuki, M. (1970). *Chem. Pharm. Bull.* (Tokyo), **18,** 1774
3. Bodor, N. Dewar, M. J. S. and Harget, A. J. (1970). *J. Amer. Chem. Soc.*, **92,** 2929
4. Lunazzi, L., Mangini, A., Pedulli, G. F. and Taddei, F. (1970). *J. Chem. Soc. B,* 163
5. Häfelinger, G. (1970). *Chem. Ber.*, **103,** 2902
6. Ostercamp, D. L. (1970). *J. Org. Chem.*, **35,** 1632
7. Barlin, G. B. and Young, A. C. (1971). *J. Chem. Soc. B,* 2323
8. Byrne, J. P. and Ross, I. G. (1971). *Aust. J. Chem.*, **24,** 1107
9. Armarego, W. L. F. (1971). *Physical Methods in Heterocyclic Chemistry, 3,* 67. (New York: Academic Press)
10. Baba, H., Yamazaki, I. and Takemura, T. (1971). *Spectrochim. Acta,* **27A,** 1271
11. Worley, S. D. (1971). *Chem. Rev.,* **71,** 295
12. Vander Donckt, E. and Vogels, C. (1971). *Spectrochim. Acta,* **27A,** 2157
13. Brown, E. V., Plasz, A. C. and Mitchell, S. R. (1970). *J. Heterocycl. Chem.,* **7,** 661
14. Draper, P. M. and MacLean, D. B. (1970). *Can. J. Chem.,* **48,** 746

15. Bogentoft, C., Kronberg, L. and Danielsson, B. (1970). *Acta Chem. Scand.*, **24**, 2244
16. Koch, C. W. and Markgraf, J. H. (1970). *J. Heterocycl. Chem.*, **7**, 235
17. Johnstone, R. A. W., Payling, D. W., Preston, P. N., Stevens, H. N. E. and Stevens, M. F. G. (1970). *J. Chem. Soc. C*, 1238
18. Porter, Q. N. and Baldas, J. (1971). *Mass Spectra of Heterocyclic Compounds*, (Wiley-Interscience: New York), p. 564
19. Armarego, W. L. F., Batterham, T. J. and Kershaw, J. R. (1971). *Org. Mag. Res.*, **3**, 575
20. Sanders, J. K. M. and Williams, D. H. (1971). *J. Amer. Chem. Soc.*, **93**, 641
21. Huber, H. and Pascual, C. (1971). *Helv. Chim. Acta*, **54**, 913
22. Beauté, C., Wolkowski, Z. W. and Thoai, N. (1971). *Tetrahedron Lett.*, 817
23. Beaumont, T. G. and Davis, K. M. C. (1970). *J. Chem. Soc. B*, 592
24. Katritzky, A. R., Takeuchi, Y., Ternai, B. and Tiddy, G. J. T. (1970). *Org. Mag. Res.*, **2**, 357
25. Witanowski, M., Stefaniak, L., Januszewski, H. and Webb, G. A. (1971). *Tetrahedron*, **27**, 3129
26. Axenrod, T., Wieder, M. J., Berti, G. and Barili, P. L. (1970). *J. Amer. Chem. Soc.*, **92**, 6066
27. Kawazoe, Y., Araki, M., Sawaki, S. and Ohnishi, M. (1970). *Chem. Pharm. Bull.* (Tokyo), **18**, 381
28. Jones, R. A. Y., Katritzky, A. R., Shapiro, B. B., Tute, M. S., Gadsby, B. and Broadbent, R. W. (1971). *J. Chem. Soc. B*, 1325
29. Monro, A. M. and Sewell, M. J. (1971). *J. Chem. Soc. B*, 1227
30. Haimova, M. A., Spassov, S. L., Novkova, S. I., Palamareva, M. D. and Kurtev, B. J. (1971). *Chem. Ber.*, **104**, 2601
31. Anderson, D. J., Horwell, D. C. and Atkinson, R. S. (1971). *J. Chem. Soc. C*, 624
32. Lund, H. (1970). *Advan. Heterocycl. Chem.*, **12**, 213
33. van der Meer, D. (1970). *Rec. Trav. Chim. Pays-Bas*, **89**, 51
34. Cauquis, G. and Genies, M. (1970). *Tetrahedron Lett.*, 3403
35. Pinson, J. and Armand, J. (1971). *Coll. Czech. Chem. Commun.*, **36**, 585
36. Schellenberg, M. (1970). *Helv. Chim. Acta*, **53**, 1151
37. Lund, H. and Jensen, E. Th. (1970). *Acta Chem. Scand.*, **24**, 1867
38. Wiberg, K. B. and Lewis, T. P. (1970). *J. Amer. Chem. Soc.*, **92**, 7154
39. Gagan, J. M. F. and Lloyd, D. (1970). *J. Chem. Soc. C*, 2488
40. Pratt, R. E., Welstead, Jr., W. J. and Lutz, R. E. (1970). *J. Heterocycl. Chem.*, **7**, 1051
41. Goszczyński, S. and Salwinska, E. (1971). *Tetrahedron Lett.*, 3027
42. Spence, T. W. M. and Tennant, G. (1970). *Chem. Commun.*, 1100
43. Sword, I. P. (1971). *J. Chem. Soc. C*, 820
44. Sword, I. P. (1970). *J. Chem. Soc. C*, 1916
45. Pettit, G. R., Stevenson, B. and Gever, G. (1971). *Organic Preparation Procedures Int.*, **3**, 93
46. Eistert, B., Kurze, W. and Müller, G. W. (1970). *Justus Liebigs Ann. Chem.*, **732**, 1
47. Foltz, C. M. and Kondo, Y. (1970). *Tetrahedron Lett.*, 3163
48. Parham, W. E., Davenport, R. W. and Biasotti, J. B. (1970). *J. Org. Chem.*, **35**, 3775
49. Ziegler, E., Junek, H. and Metallidis, A. (1970). *Monatsh. Chem.*, **101**, 92
50. Ziegler, E., Sterk, H. and Steiger, W. (1970). *Monatsh. Chem.*, **101**, 762
51. Hörlein, U. and Geiger, W. (1971). *Arch. Pharm. (Weinheim)*, **304**, 130
52. Agui, H., Mitani, T., Nakashita, M. and Nakagome, T. (1971). *J. Heterocycl. Chem.*, **8**, 357
53. Ghosez, L. and de Perez, C. (1971). *Angew. Chem., Internat. Edn.*, **10**, 184
54. Uebel, H. J., Moll, K. K. and Mühlstädt, M. (1970). *J. Prakt. Chem.*, **312**, 263
55. Fields, E. K. and Meyerson, S. (1970). *J. Org. Chem.*, **35**, 62
56. Ogata, Y. and Takagi, K. (1971). *Tetrahedron*, **27**, 1573
57. Gailey, R. G. and Zimmer, H. (1970). *Tetrahedron Lett.*, 2839
58. Tong, Y. C. (1970). *J. Heterocycl. Chem.*, **7**, 171
59. Gershon, H. and McNeil, M. W. (1970). *J. Org. Chem.*, **35**, 3993; Gershon, H., Mc.Neil, M. W. and Schulman, S. G. (1971). *J. Org. Chem.*, **36**, 1616
60. Crout, D. H. G., Penton, J. R. and Schofield, K. (1971). *J. Chem. Soc. B*, 1254
61. Moodie, R. B., Penton, J. R. and Schofield, K. (1971). *J. Chem. Soc. B*, 1493
62. Hamada, Y., Ito, Y. and Hirota, M. (1970). *Chem. Pharm. Bull.* (Tokyo), **18**, 2094

63. Uehara, N. and Kawazoe, Y. (1970). *Chem. Pharm. Bull.* (Tokyo), **18**, 203
64. Bressel, U., Katritzky, A. R. and Lea, J. R. (1971). *J. Chem. Soc. B*, 11
65. Tomisawa, H., Kobayashi, Y., Hongo, H. and Fujita, R. (1970). *Chem. Pharm. Bull.* (Tokyo), **18**, 932
66. Tomisawa, H., Watanabe, M., Fujita, R. and Hongo, H. (1970). *Chem. Pharm. Bull.* (Tokyo), **18**, 919
67. Nishimura, H., Nagai, Y., Suzuki, T. and Sawayama, T. (1970). *Yakugaku Zasshi,* **90**, 818
68. Chamberlain, T. R. and Grundon, M. F. (1971). *J. Chem. Soc. C*, 910
69. Renault, J. and Cartron, J.-C. (1971). *Bull. Soc. Chim. Fr.,* 888
70: Balli, H. and Schelz, D. (1970). *Helv. Chim. Acta,* **53**, 1903
71. Vorsanger, H. and Vorsanger, J.-J. (1970). *Bull. Soc. Chim. Fr.,* 589
72. Vorsanger, H., Ferrand, R., Mazza, M. and Vorsanger, J.-J. (1970). *Bull. Soc. Chim. Fr.,* 593
73. Heindel, N. D. and Kennewell, P. D. (1970). *J. Med. Chem.,* **13**, 166
74. Heindel, N. D. and Fine, S. A. (1970). *J. Org. Chem.,* **35**, 796
75. Bressan, G. B., Giardi, I., Illuminati, G., Linda, P. and Sleiter, G. (1971). *J. Chem. Soc. B*, 225
76. Chambers, R. D., Hole, M., Musgrave, W. K. R. and Thorpe, J. G. (1971). *J. Chem. Soc. C*, 61
77. Kobayashi, Y., Kumadaki, I. and Taguchi, S. (1971). *Chem. Pharm. Bull. (Tokyo),* **19**, 624
78. Kobayashi, Y., Kumadaki, I., Taguchi, S. and Hanzawa, Y. (1970). *Tetrahedron Lett.,* 3901
79. Okamoto, T. and Nagakura, M. (1971). *Chem. Pharm. Bull. (Tokyo),* **19**, 1745
80. Okano, T. and Kano, T. (1971). *Chem. Pharm. Bull. (Tokyo),* **19**, 1293
81. Okamoto, T. and Takahashi, H. (1971). *Chem. Pharm. Bull. (Tokyo),* **19**, 1809
82. Himeno, J., Noda, K. and Yamazaki, M. (1970). *Chem. Pharm. Bull. (Toyko),* **18**, 2138
83. Hamana, M. and Kumadaki, I. (1970). *Chem. Pharm. Bull. (Tokyo),* **18**, 1822, 1742
84. Hamana, M. and Noda, H. (1970). *Chem. Pharm. Bull. (Tokyo),* **18**, 26
85. Yamazaki, M., Noda, K. and Hamana, M. (1970). *Chem. Pharm. Bull. (Tokyo),* **18**, 901
86. Gregory, B. J., Moodie, R. B. and Schofield, K. (1970). *J. Chem. Soc. B*, 1687
87. Kauffmann, T., Fischer, H., Nürnberg, R. and Wirthwein, R. (1970). *Justus Liebigs Ann. Chem.,* **731**, 23; Kauffmann, T. and Wirthwein, R. (1971). *Angew. Chem. Internat. Edn.,* **10**, 20
88. Kauffmann, T., Jackisch, J., Streitberger, H.-J. and Wienhöfer, E. (1971). *Angew. Chem. Internat. Edn.,* **10**, 744
89. Gardini, G. P., Minisci, F., Palla, G., Arnone, A. and Galli, R. (1971). *Tetrahedron Lett.,* 59
90. Patterson, J. M., Issidorides, C. H., Papadopoulos, E. P. and Smith, Jr., W. T. (1970). *Tetrahedron Lett.,* 1247
91. Sasaki, T., Kanematsu, K. and Yukimoto, Y. (1970). *J. Chem. Soc. C*, 481
92. Quast, H. and Schmitt, E. (1970). *Justus Liebigs Ann. Chem.,* **732**, 43
93. Kobayashi, Y., Kutsuma, T., Morinaga, K., Fujita, M. and Hanzawa, Y. (1970). *Chem. Pharm. Bull. (Tokyo),* **18**, 2489
94. Acheson, R. M. and Nisbet, D. F. (1971). *J. Chem. Soc. C*, 3291
95. Acheson, R. M. and Stubbs, J. K. (1971). *J. Chem. Soc. C*, 3285
96. Acheson, R. M. and Woollard, J. McK. (1971). *J. Chem. Soc. C*, 3296
97. Crawforth, C. E., Russell, C. A. and Meth-Cohn, O. (1970). *Chem. Commun.,* 1406
98. Crawford, M. and Ingle, R. B. (1971). *J. Chem. Soc. B*, 1907
99. Stremitz, F. R., Wei, C. C. and O'Donnell, C. M. (1970). *J. Amer. Chem. Soc.,* **92**, 2745
100. Rubin, M. B. and Fink, C. (1970). *Tetrahedron Lett.,* 2749
101. Castellano, A. and Lablache-Combier, A. (1971). *Tetrahedron,* **27**, 2303
102. Hata, N., Ono, I. and Ogawa, S. (1971). *Bull. Chem. Soc. Jap.,* **44**, 2286
103. Evanega, G. R. and Fabiny, D. L. (1970). *J. Org. Chem.,* **35**, 1757
104. Evanega, G. R. and Fabiny, D. L. (1971). *Tetrahedron Lett.,* 1749
105. Paolillo, L., Ziffer, H. and Buchardt, O. (1970). *J. Org. Chem.,* **35**, 39
106. Yamamuro, T., Tanaka, I. and Hata, N. (1971). *Bull. Chem. Soc. Jap.,* **44**, 667
107. Meyers, A. I. and Singh, P. (1970). *J. Org. Chem.,* **35**, 3022
108. Gault, R. and Meyers, A. I. (1971). *Chem. Commun.,* 778

109. Tamura, Y., Ishibashi, H., Tsujimoto, N. and Ikeda, M. (1971). *Chem. Pharm. Bull. (Tokyo)*, **19**, 1285
110. Evans, D. A. (1970). *J. Amer. Chem. Soc.*, **92**, 7593
111. Ninomiya, I., Naito, T., Higuchi, S. and Mori, T. (1971). *Chem. Commun.*, 457
112. Stevens, R. V., Fitzpatrick, J. M., Kaplan, M. and Zimmerman, R. L. (1971). *Chem. Commun.*, 857
113. Hassner, A., Ferdinandi, E. S. and Isbister, R. J. (1970). *J. Amer. Chem. Soc.*, **92**, 1672
114. Bunting, J. W. and Meathrel, W. G. (1971). *Tetrahedron Lett.*, 133
115. Remers, W. A., Gibs, G. J. Pidacks, C. and Weiss, M. J. (1971). *J. Org. Chem.*, **36**, 279
116. Jones, A. M., Russell, C. A. and Meth-Cohn, O. (1971). *J. Chem. Soc. C*, 2453
117. Otsuji, Y., Yutani, K. and Imoto, E. (1971). *Bull. Chem. Soc. Jap.*, **44**, 520
118. Matsumori, K., Ide, A. and Watanabe. H. (1971). *J. Chem. Soc. Jap. (Pure Chem. Sect.)*, **92**, 80
119. Zakharkin, L. I., Litovchenko, L. E. and Kazantsev, A. V. (1970). *J. Gen. Chem. U.S.S.R.*, **40**, 113
120. Ainscough, E. W. and Plowman, R. A. (1970). *Aust. J. Chem.*, **23**, 403
121. Hamana, M., Nomura, S. and Kawakita, T. (1971). *Yakugaku Zasshi*, **91**, 134
122. Anderson, D. J., Gilchrist, T. L., Horwell, D. C. and Rees, C. W. (1970). *J. Chem. Soc. C*, 576
123. Bullock, F. J. and Tweedie, J. F. (1970). *J. Med. Chem.*, **13**, 261
124. Bullock, F. J. and Tweedie, J. F. (1970). *J. Heterocycl. Chem.*, **7**, 1125
125. Godard, A., Queguiner, G. and Pastour, P. (1971). *Bull. Soc. Chim. Fr.*, 906
126. Sparrow, J. T. (1971). *J. Heterocycl. Chem.*, **8**, 477
127. Bailey, D. M., Archer, S., Wood, D., Rosi, D. and Yarinsky, A. (1970). *J. Med. Chem.*, **13**, 598
128. Gut, V., Chládek, S. and Žemlička, J. (1970). *Collect. Czech. Chem. Commun.*, **35**, 2398
129. Avinashi, B. K. and Banerji, S. K. (1971). *J. Indian Chem. Soc.*, **48**, 174 (and other similar papers in vols. **47** and **48**); Riolo, C. B., Fullesoldi, T. and Spini, G. (1970). *Ann. Chim. (Rome)*, **60**, 836; Akimov, V. K., Busev, A. I. and Bragina, S. I. (1970). *J. Gen. Chem. U.S.S.R.*, **40**, 1320
130. Gupta, R. D., Manku, G. S., Bhat, A. N. and Jain, B. D. (1970). *Aust. J. Chem.*, **23**, 1387
131. Coffen, D. L. and McEntee, Jr., T. E. (1970). *J. Org. Chem.*, **35**, 503
132. Trofimchuk, A. K., Gorenbein, E. Ya. and Abarbarchuk, I. L. (1970). *J. Gen. Chem. U.S.S.R.*, **40**, 1422
133. Esafov, V. I. and Shlyapnikov, D. S. (1970). *J. Gen. Chem. U.S.S.R.*, **40**, 2259
134. Bevis, M. J., Forbes, E. J., Naik, N. N. and Uff, B. C. (1971). *Tetrahedron*, **27**, 1253
135. Jackson, A. H. and Stewart, G. W. (1971). *Chem. Commun.*, 149
136. Sainsbury, M., Dyke, S. F. and Moon, B. J. (1970). *J. Chem. Soc. C*, 1797
137. Harcourt, D. N. and Waigh, R. D. (1971). *J. Chem. Soc. C*, 967
138. Koyama, T., Toda, M., Hirota, T., Katsuse, Y. and Yamato, M. (1970). *Yakugaku Zasshi*, **90**, 11 and 1207
139. Dey, A. S., Rosowsky, A. and Modest. E. J. (1970). *J. Org. Chem.*, **35**, 536
140. Pangon, G. (1970). *Bull. Soc. Chim. Fr.*, 1997
141. Wermuth, C. G. and Flammang, M. (1971). *Tetrahedron Lett.*, 4293
142. Legrand, L. and Lozach, N. (1970). *Bull. Soc. Chim. Fr.*, 2237
143. Tamura, Y., Tsujimoto, N. and Uchimura, M. (1971). *Yakugaku Zasshi*, **91**, 72
144. Tamura, Y. and Tsujimoto, N. (1970). *Chem. Ind. (London)*, 926
145. Berry, D. J., Wakefield, B. J. and Cook, J. D. (1971). *J. Chem. Soc. C*, 1227
146. Lown, J. W. and Matsumoto, K. (1971). *J. Org. Chem.*, **36**, 1405
147. Horning, D. E., Lacasse, G. and Muchowski, J. M. (1971). *Can. J. Chem.*, **49**, 2777 and 2797
148. Lombardino, J. G. (1970). *J. Heterocycl. Chem.*, **7**, 1057
149. Higashino, T., Tamura, Y., Nakayama, K. and Hayashi, E. (1970). *Chem. Pharm. Bull. (Tokyo)*, **18**, 1262
150. Hayashi, E. and Tamura, Y. (1970). *Yakugaku Zasshi*, **90**, 594
151. Hamana, M. and Matsumoto, T. (1971). *Yakugaku Zasshi*, **91**, 269
152. Agrawal, K. C., Cushley, R. J., McMurray, W. J. and Sartorelli, A. C. (1970). *J. Med. Chem.*, **13**, 431
153. Ide, A., Matsumori, K., Ishizu, K. and Watanabe, H. (1971). *J. Chem. Soc. Jap. (Pure Chem. Sect.)*, **92**, 83

154. Zugrăvescu, I., Rucinschi, E. and Surpăteanu, G. (1970). *Tetrahedron Lett.*, 941
155. Fuks, R. and Viehe, H. G. (1970). *Chem. Ber.*, **103**, 573
156. Baldwin, J. E. and Duncan, J. A. (1971). *J. Org. Chem.*, **36**, 3156
157. Simonsen, O., Lohse, C. and Buchardt, O. (1970). *Acta Chem. Scand.*, **24**, 268
158. Lown, J. W. and Matsumoto, K. (1971). *Can. J. Chem.*, **49**, 3119
159. Mruk, N. J. and Tieckelmann, H. (1970). *Tetrahedron Lett.*, 1209
160. Bradsher, C. K., Day, F. H., McPhail, A. T. and Wong, P.-S. (1971). *Tetrahedron Lett.*, 4205
161. Nielsen, A. T. (1970). *J. Org. Chem.*, **35**, 2498
162. Harmon, R. E., Jensen, B. L., Gupta, S. K. and Nelson, J. D. (1970). *J. Org. Chem.*, **35**, 825
163. Merchant, J. R., Mhatre, R. R. and Patell, J. R. (1971). *J. Indian Chem. Soc.*, **48**, 427
164. Kametani, T., Kigasawa, K., Hiiragi, H. and Ishimaru, H. (1970). *J. Heterocycl. Chem.*, **7**, 51, and (1971). *J. Chem. Soc. C*, 2632
165. Engel, W., Seeger, E., Teufel, H. and Machleidt, H. (1971). *Chem. Ber.*, **104**, 248
166. Freter, K., Dubois, E. and Thomas, A. (1970). *J. Heterocycl. Chem.*, **7**, 159
167. Deady, L. W., Pirzada, N. and Topsom, R. D. (1971). *Chem. Commun.*, 799
168. Bobbitt, J. M.'and Shibuya, S. (1970). *J. Org. Chem.*, **35**, 1181
169. Teitel, S. and Brossi, A. (1970). *J. Heterocycl. Chem.*, **7**, 1401
170. McEwen, W. E., Berkebile, D. H., Liao, T.-K. and Lin, Y.-S. (1971). *J. Org. Chem.*, **36**, 1459
171. Popp, F. D., Klinowski, C. W., Piccirilli, R., Purcell, Jr., D. H. and Watts, R. F. (1971). *J. Heterocycl. Chem.*, **8**, 313
172. Kirby, G. W., Tan, S. L. and Uff, B. C. (1970). *Chem. Commun.*, 1138
173. Bernhard, H. O. and Shieckus, V. (1971). *Tetrahedron*, **27**, 2091
174. Buzas, A., Cossais, F. and Jacquet, J.-P. (1971). *Bull. Soc. Chim. Fr.*, 1701
175. Bobbitt, J. M., Noguchi, I., Yagi, H. and Weisgraber, K. H. (1971). *J. Amer. Chem. Soc.*, **93**, 3551
176. Bobbitt, J. M., Weisgraber, K. H., Steinfeld, A. S. and Weiss, S. G. (1970). *J. Org. Chem.*, **35**, 2884
177. Fryer, R. I., Earley, J. V., Evans, E., Schneider, J. and Sternbach, L. H. (1970). *J. Org. Chem.*, **35**, 2455
178. Tute, M. S., Brammer, K. W., Kaye, B. and Broadbent, R. W. (1970). *J. Med. Chem.*, **13**, 44
179. Ben-Ishai, D., Inbal, Z. and Warshawsky, A. (1970). *J. Heterocycl. Chem.*, **7**, 615
180. Ogata, Y. and Takagi, K. (1971). *Tetrahedron*, **27**, 2785
181. Kobayashi, G., Furukawa, S., Matsuda, Y., Natsuki, R. and Matsunaga, S. (1970). *Chem. Pharm. Bull. (Tokyo)*, **18**, 124
182. Alaimo, R. J., Hatton, C. J. and Eckman, M. K. (1970). *J. Med. Chem.*, **13**, 554
183. Rosazza, J. P., Bobbitt, J. M. and Schwarting, A. E. (1970). *J. Org. Chem.*, **35**, 2564
184. Wenkert, E. and Jeffcoat, A. R. (1970). *J. Org. Chem.*, **35**, 515
185. Goldberg, S. I. and Lipkin, A. H. (1970). *J. Org. Chem.*, **35**, 242
186. Acheson, R. M. and Jones, B. J. (1970). *J. Chem. Soc. C*, 1301
187. Kobayashi, G., Furukawa, S., Matsuda, Y., Natsuki, R. and Matsunaga, S. (1970). *J. Chem. Soc. Jap.*, **90**, 127
188. Davies, L. S. and Jones, G. (1970). *J. Chem. Soc. C*, 688
189. Mörler, D. and Kröhnke, F. (1971). *Justus Liebigs Ann. Chem.*, **744**, 65
190. Sprio, V., Maccioni, A., Marongiu, E. and Plescia, S. (1970). *Ann. Chim. (Rome)*, **60**, 168
191. Sprio, V., Plescia, S. and Migliara, O. (1971). *Ann. Chim. (Rome)*, **61**, 271
192. Palmer, M. H., Gaskell, A. J., McIntyre, P. S. and Anderson, D. W. W. (1971). *Tetrahedron*, **27**, 2921
193. Palmer, M. H. and McIntyre, P. S. (1971). *Tetrahedron*, **27**, 2913; see also Ames, D. E., Ansari, H. R., France, A. D. G., Lovesay, A. C., Novitt, B. and Simpson, R. (1971). *J. Chem. Soc. C*, 3088
194. Igeta, H., Tsuchiya, T., Nakai, T., Okusa, G., Kumagai, M., Miyoshi, J. and Itai, T. (1970). *Chem. Pharm. Bull. (Tokyo)*, **18**, 1497
195. Rees, C. W. and Sale, A. A. (1971). *Chem. Commun.*, 531 and 532
196. Ames, D. E. and Novitt, B. (1970). *J. Chem. Soc. C*, 1700
197. Chambers, R. D., McBride, J. A. H. and Musgrave, W. K. R. (1970). *Chem. Commun.*, 739

198. Shah, M. A. and Taylor, G. A. (1970). *J. Chem. Soc. C,* 1642
199. Bredereck, H., Simchen, G. and Speh, P. (1970). *Justus Liebigs Ann. Chem.,* **737,** 39
200. Berry, D. J. and Wakefield, B. J. (1971). *J. Chem. Soc. C,* 642
201. Simchen, G., Entenmann, G. and Zondler, R. (1970). *Angew. Chem.* (English Ed.), **9,** 523
202. Mehta, H. J., Patel, V. S. and Patel, S. R. (1970). *J. Indian Chem. Soc.,* **47,** 124
203. Pakarshi, S. C. (1971). *J. Org. Chem.,* **36,** 642
204. Ning, R. Y., Douvan, I. and Sternbach, L. H. (1970). *J. Org. Chem.,* **35,** 2243
205. Walser, A., Silverman, G., Blount, J., Fryer, R. I. and Sternbach, L. H. (1971). *J. Org. Chem.,* **36,** 1465
206. Sternbach, L. H. (1971). *Angew. Chem. Internat. Edn.,* **10,** 34
207. Finch, N. and Gschwend, H. W. (1971). *J. Org. Chem.,* **36,** 1463
208. Hegarty, A. F. and Bruice, T. C. (1970). *J. Amer. Chem. Soc.,* **92,** 6561, 6575
209. Jacobs, R. L. (1970). *J. Heterocycl. Chem.,* **7,** 1337
210. Capuano, L., Ebner, W. and Schrepfer, J. (1970). *Chem. Ber.,* **103,** 82
211. Gonçalves, H., Mathis, F. and Foulcher, C. (1970). *Bull. Soc. Chim. Fr.,* 2599 and 2615
212. Schäfer, W., Aguado, A. and Sezer, U. (1971). *Angew. Chem. Internat. Edn.,* **10,** 406
213. Savige, W. E. (1971). *Aust. J. Chem.,* **24,** 1285
214. Bunting, J. W. and Meathrel, W. G. (1970). *Can. J. Chem.,* **48,** 3449
215. Wittenburg, E. (1971). *Collect. Czech. Chem. Commun.,* **36,** 246
216. Abdel-Megeid, F. M. E., Elkaschef, M. A-F., Mokhtar, K-E. M. and Zaki, K-E. M. (1971). *J. Chem. Soc. C,* 1055
217. Malesani, G., Marcolin, F. and Rodighiero, G. (1970). *J. Med. Chem.,* **13,** 161
218. Shah, M. A. and Taylor, G. A. (1970). *J. Chem. Soc. C,* 1651
219. Gilchrist, T. L., Rees, C. W. and Stanton, E. (1971). *J. Chem. Soc. C,* 3036
220. Giraldi, P. N., Fojanesi, A., Tosolini, G. P., Dradi, E. and Logemann, W. (1970). *J. Heterocycl. Chem.,* **7,** 1429
221. Bogentoft, C., Ericsson, Ö. and Danielsson, B. (1971). *Acta Chem. Scand.,* **25,** 551
222. Gilchrist, T. L. Rees, C. W. and Stanton, E. (1971). *J. Chem. Soc. C,* 988
223. Pater, R. (1970). *J. Heterocycl. Chem.,* **7,** 1113
224. Gupta, C. M., Bhaduri, A. P. and Khanna, N. M. (1970). *Indian J. Chem.,* **8,** 1055
225. Amin, A. H., Mehta, D. R. and Samarth, S. S. (1970). *Progress in Drug Research,* (E. Jucker, editor), **14,** 218–268. (Basel: Birkhäuser)
226. Derieg, M. E., Blount, J. F., Fryer, R. I. and Hillery, S. S. (1970). *Tetrahedron Lett.,* 3869; see also (1971). *J. Org. Chem.,* **36,** 782
227. Sato, Y., Tanaka, T. and Nagasaki, T. (1970). *Yakugaku Zasshi,* **90,** 629
228. Heindel, N. D. and Chun, M. C. (1971). *J. Heterocycl. Chem.,* **8,** 685
229. Capuano, L. Welter, M. and Zander, R. (1970). *Chem. Ber.,* **103,** 2394
230. Chupp, J. P. (1971). *J. Heterocycl. Chem.,* **8,** 565
231. Zigeuner, G., Eisenreich, V. and Immel, W. (1970). *Monatsh. Chem.,* **101,** 1745, see also pp. 1731 and 1686
232. Armarego, W. L. F. and Kobayashi, T. (1971). *J. Chem. Soc. C,* 238
233. Roth, H. J. and Hagen, H-E. (1971). *Arch. Pharm. (Weinheim),* **304,** 331
234. Armarego, W. L. F. and Kobayashi, T. (1970). *J. Chem. Soc. C,* 1597
235. Armarego, W. L. F. and Kobayashi, T. (1971). *J. Chem. Soc. C,* 3222
236. Armarego, W. L. F. (1971). *J. Chem. Soc. C,* 1812
237. Field, G. F., Zally, W. J. and Sternbach, L. H. (1971). *J. Org. Chem.,* **36,** 777
238. Swan, G. A. (1971). *J. Chem. Soc. C,* 2880
239. Iwanami, Y., Seki, T. and Inagaki, T. (1971). *Bull. Chem. Soc. Jap.,* **44,** 1316
240. Iwanami, Y. (1971). *Bull. Chem. Soc. Jap.,* **44,** 1311
241. Abushanab, E. (1970). *J. Org. Chem.,* **35,** 4279
242. McFarland, J. W. (1971). *J. Org. Chem.,* **36,** 1842
243. Takamizawa, A., Hamashima, Y., Sato, H. and Matsumoto, Y. (1970). *Chem. Pharm. Bull. (Tokyo),* **18,** 1576
244. Walser, A., Silverman, G., Fryer, R. I., Sternbach, L. H. and Hellerbach, J. (1971). *J. Org. Chem.,* **36** 1248
245. Ning, R. Y., Field, G. F. and Sternbach, L. H. (1970). *J. Heterocycl. Chem.,* **7,** 475
246. Fisher, G. H., Whitman, P. J. and Schultz, H. P. (1970). *J. Org. Chem.,* **35,** 2240
247. Le Bris, M.-T. (1970). *Bull. Soc. Chim. Fr.,* 563 and 2277
248. Lown, J. W., Westwood, R. and Aidoo, A. S. K. (1970). *Can. J. Chem.,* **48,** 327
249. Pfleiderer, W. and Schanner, M. (1971). *Chem. Ber.,* **104,** 1915

250. Cheeseman, G. W. H. and Rafiq, M. (1971). *J. Chem. Soc. C*, 452
251. Stogryn, E. L. (1971). *J. Med. Chem.*, **14**, 171
252. Morrow, D. F. and Regan, L. A. (1971). *J. Org. Chem.*, **36**, 27
253. Haddadin, M. J., Agopian, G. and Issidorides, C. H. (1971). *J. Org. Chem.*, **36**, 514
254. Bass, K. C. and Nababsing, P. (1971). *Organic Preparations and Procedure Int.*, **3**, 45
255. Buratti, W., Gardini, G. P., Minisci, F., Bertini, F., Galli, R. and Perchinunno, M. (1971). *Tetrahedron*, **27**, 3655
256. Otomasu, H., Yoshida, K. and Takahashi, H. (1970). *Yakugaku Zasshi*, **90**, 1391
257. Moreno, H. R. and Schultz, H. P. (1970). *J. Med. Chem.*, **13**, 119
258. Mertes, M. P. and Lin, A. J. (1970). *J. Med. Chem.*, **13**, 77
259. Sherif, S., Ekladious, L. and Abd Elmalek, G. (1970). *J. Prakt. Chem.*, **312**, 759
260. Takahashi, H. and Otomasu, H. (1970). *Chem. Pharm. Bull. (Tokyo)*, **18**, 22
261. Iijima, C. and Hayashi, E. (1971). *Yakugaku Zasshi*, **91**, 721
262. Eberle, M. and Houlihan, W. J. (1970). *Tetrahedron Lett.*, 3167
263. Ghelardoni, M. and Pestellini, V. (1970). *Ann. Chim. (Rome)*, **60**, 775
264. Zirngibl, L., Kunz, G. and Pretsch, E. (1971). *Tetrahedron Lett.*, 4189
265. Caprosu, M., Petrovanu, M., Druță, I. and Zugrăvescu, I. (1971). *Bull. Soc. Chim. Fr.*, 1834
266. Hill, J. H. M. and Ehrlich, J. H. (1971). *J. Org. Chem.*, **36**, 3248
267. Vigevani, A., Cavalleri, B. and Gallo, G. G. (1970). *J. Heterocycl. Chem.*, **7**, 677
268. Chambers, R. D., MacBride, J. A. H., Musgrave, W. K. R. and Reilly, I. S. (1970). *Tetrahedron Lett.*, 57
269. Omote, Y., Yamamoto, H. and Sugiyama, N. (1970). *Chem. Commun.*, 914
270. Shelvin, P. B. and Neufeld, H. A. (1970). *J. Org. Chem.*, **35**, 2178
271. Gundermann, K-D., Fiege, H. and Klockenbring, G. (1970). *Justus Liebigs Ann. Chem.*, **738**, 140
272. Le Berre, A., Dumaitre, B. and Petit, J. (1970). *Bull. Soc. Chim. Fr.*, 4376
273. Adams, D. J. C., Bradbury, S., Horwell, D. C., Keating, M., Rees, C. W. and Storr, R. C. (1971). *Chem. Commun.*, 828
274. Bradbury, S., Keating, M., Rees, C. W. and Storr, R. C. (1971). *Chem. Commun.*, 827
275. Stevens, H. N. E. and Stevens, M. F. G. (1970). *J. Chem. Soc. C*, 765 and 2289
276. Murray, A. W. and Vaughan, K. (1970). *J. Chem. Soc. C*, 2070
277. Stevens, H. N. E. and Stevens, M. F. G. (1970). *J. Chem. Soc. C*, 2284
278. Stevens, H. N. E. and Stevens, M. F. G. (1970). *J. Chem. Soc. C*, 2308
279. Adamson, J., Forster, D. L., Gilchrist, T. L. and Rees, C. W. (1971). *J. Chem. Soc. C*, 981
280. König, W. and Geiger, R. (1970). *Chem. Ber.*, **103**, 2024
281. Adger, B., Rees, C. W., Sale, A. A. and Storr, R. C. (1971). *Chem. Commun.*, 695
282. Mason, J. C. and Tennant, G. (1970). *J. Chem. Soc. B*, 911

6
Nitrogen Heterocyclic Molecules Part 5. Naphthyridines and Polyazaheteroaromatic Compounds

D. G. WIBBERLEY
University of Aston in Birmingham

6.1 INTRODUCTION

The literature of 1970 and 1971 contains almost 800 references to poly-azaheteroaromatic compounds and naphthyridines. Of these publications, more than 20% describe work on purines and about 10% work on pteridines. Many other references to nucleosides, nucleotides and nucleic acids were considered to be out of the scope of this particular review, but the total volume of work related to these two ring systems is undoubtedly a reflection of their unique biological significance.

The importance of the purines and pteridines is also seen in the many closely related compounds which have been prepared for an evaluation of their chemotherapeutic or physiological properties, or for comparison of chemical and physico-chemical properties. Only six other ring systems receive mention in more than ten separate publications in the literature of these two years and of these, five are structurally related compounds. Thus the pyrimido[5,4-d]pyrimidines[1] are isomeric with the pteridines and are referred to in most publications (18 refs.). Pyrimido[5,4-e]triazines[2] are 7-azapteridines and pyrido[2,3-d]pyrimidines[3] are 5-de-azapteridines. Similarly, pyrazolo[3,4-d]pyrimidines[4] are isomers of purines and pyrrolo[2,3-d]pyrimidines[5] are 7-de-azapurines. 1,8-Naphthyridines[6] are mentioned in ten publications.

A further significant factor in this field is that 160 different ring systems have been described in the two years under review, many of them for the first time and in only one publication. No attempt can therefore be made in the space of this chapter to cover comprehensively all the work on the purines or all the work on the preparation of individual ring systems. Instead, the various ring systems are considered as one group with respect to methods of synthesis, chemical and physical properties, etc., and any subdivisions of the relevant sections are made as required by the nature of the topic under consideration rather than on the basis of separate ring systems.

During the course of these two years review articles and books have appeared on naphthyridines[7], 1,5-naphthyridines[8], azapteridines[9], purines[10], n.m.r. studies of purines, nucleosides, nucleotides and nucleic acids[11], proton and metal ion interaction with nucleic acids and their constituent bases[12], electrolysis of N-heterocyclic compounds[13], and 1,3-dipolar additions to nitrile oxides[14].

6.2 SYNTHESES

The syntheses are considered in the first place on the basis of the size of the ring formed in the final stage of preparation. Since there are no such syntheses of polyclic N-heteroaromatic compounds containing other than 5- or 6-membered rings the two main categories discussed are 6-membered ring systems (Section 6.2.1) and 5-membered ring systems (Section 6.2.2). In those cases where multiple ring closures occur, e.g. in syntheses of 5,6 fused-ring systems from purely aliphatic starting materials, an arbitrary placing has been made into one or other of these major groups. Within these two groups the syntheses are further subdivided according to the type of bond formation involved in the final ring-closure step and wherever known the mechanism of this final stage is classified. An attempt has been made to ensure that the review is comprehensive in respect of the methods and mechanisms of syntheses of these polycyclic N-heteroaromatic compounds, and as comprehensive as space allows in the description of new ring systems. Papers which describe the use of known routes for the synthesis of well-established ring systems, even though many new compounds may have been described, will probably receive no mention.

6.2.1 Syntheses which terminate in the formation of a 6-membered ring

These ring-closures are further subdivided according to the position of cyclisation, the nature, and the mechanism of bond formation. This is not always known and certain preparations may equally be considered in one of two different groups, so that several arbitrary selections have been made.

6.2.1.1 Electrophilic attack at a ring C atom

This is the route most commonly used for the synthesis of naphthalenes, quinolines, isoquinolines and extended in the past to the various naphthyridines[7] and other ring systems with b- or c-fused pyridine rings. It has

continued to be the method of choice, because of the ease of preparation of
starting materials for the 1,8-[15] and 1,5-naphthyridines[16] and their benzo-
analogues[17]. Acids[18] or Lewis acids[19] may be used to enhance the electrophilicity of the attacking carbonyl group or the reactions may be carried out
in a solvent of high boiling point. Where there is a choice of positional attack
by the electrophilic side chain then the direction is controlled by electronic
and steric factors. Thus 3-aminoquinoline (1) condenses with ethyl ethoxy-
methylenemalonate (2) (EMME) to give in high yield the benzo[*g*][1,5]
naphthyridine (4)[17] and not the benzo[*g*][1,6]naphthyridine, without

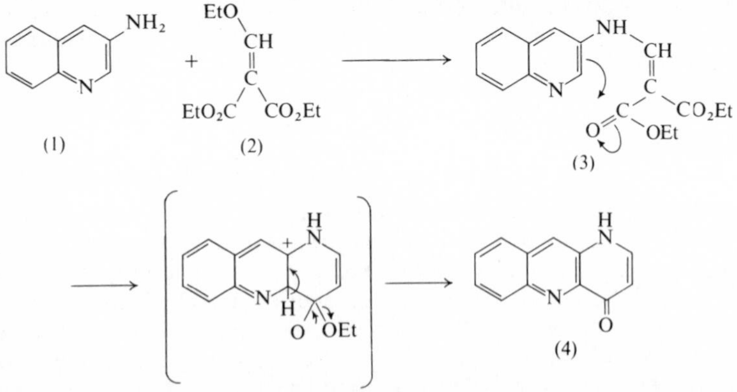

the necessity for the isolation of the intermediate quinol-3-ylaminomethyl-
enemalonate (3), and 1,5-naphthyridines are formed in preference to 1,6-
from 3-aminopyridines[16]. In the derivation of 1,8-naphthyridines from 2-
aminopyridines or benzo[*h*][1,6]naphthyridines from 4-aminoquinolines
the intermediates once formed will cyclise easily into the 3-position of the
pyridine ring since this is the most electron rich, but the initial condensation
is hindered because of the low basicity of the exocyclic 2- or 4-amino groups.
Thus 4-aminoquinoline did not undergo reaction with aceto-acetates in
acid-catalysed Conrad–Limpach reactions or with aldehydes in a Doebner
cyclisation, but did react with the more electrophilic EMME[20]. Where the
intramolecular electrophilic cyclisation involves a ring of more π-excessive
character then yields are good and conditions are mild and in this manner
4-chloropyrazolo[3,4-*b*]quinolines (5)[18], benzo[*b*][1,8]naphthyridin-5-ones
(6)[22], pyrazolo[3,4-*b*]pyridin-4-ones (7)[23] and pyrrolo[1,2-*b*]pyrazines (8)[24]

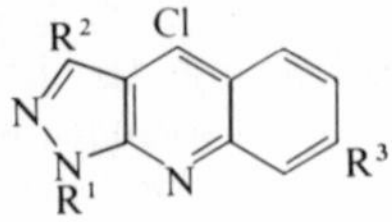

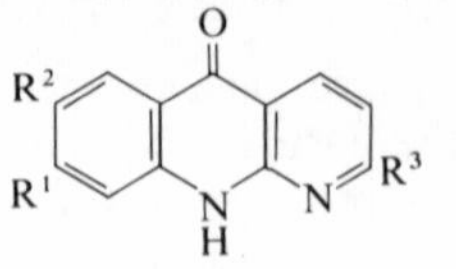

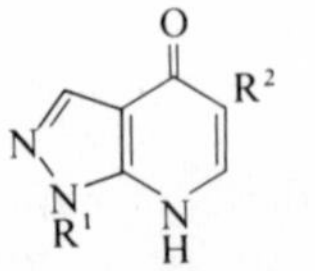

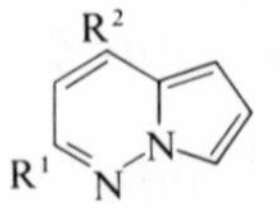

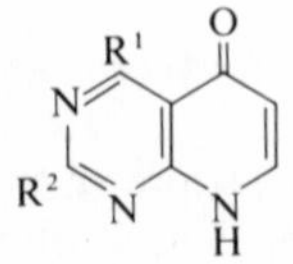

have been prepared from the appropriately substituted pyrazine, benzene or pyrrole. Where ring closure occurs into a more π-deficient ring system then electron-donating groups are required in positions where they can effect a delocalisation of the charge in the intermediate complex, and in such a way anthyridines[25] and pyrido[2,3-d]pyrimidines (9)[3] have been synthesised from the appropriate ethoxycarbonylvinylamino derivatives of naphthyridi-nones and pyrimidines respectively. Careful control of the catalyst in Skraup syntheses[26] and the use of esters derived by treatment of polyphos-phoric acid with ethanol (PPEt)[27] in EMME-type syntheses holds promise for increased scope of this type of cyclisation.

6.2.1.2   Electrophilic attack at a ring N atom

Fifteen different N-bridgehead ring systems have been prepared in 1970 and 1971 by this type of method. If a heterocyclic compound containing an amino group *ortho* to a ring N atom is converted into a suitable derivative such as the ethoxycarbonylvinylamino compound (11) cyclisation can take place either at the adjacent C-atom if this is unsubstituted, or at the ring N-atom. Generally speaking an sp^2-hybridised ring-nitrogen atom will be the more nucleophilic centre and cyclisation will give the N-bridgehead compound but such compounds are more susceptible to hydrolysis or thermal ring-opening than their isomers and so the catalyst and temperature can vary

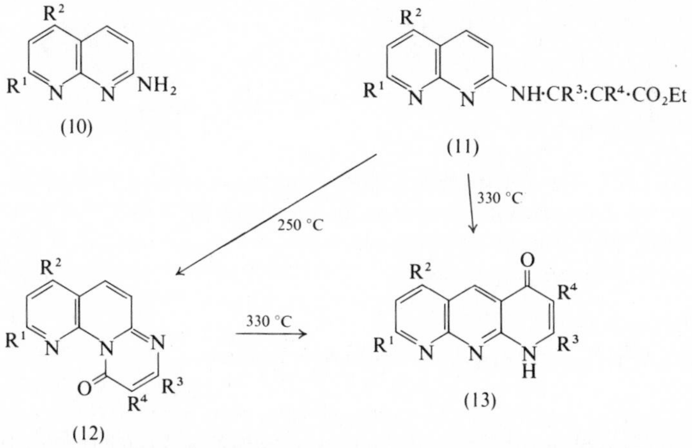

the direction of cyclisation. Thus anthyridinones (13)[28] or pyrimido[4,5-b] [1,8]naphthyridinones (12)[25] may be prepared from 2-amino-1,8-naphthyri-dines (10) and the latter (12) may be obtained directly in certain cases from suitable 2,6-diethoxycarbonylvinylaminopyridines[28]. In a similar manner the condensation of 3-aminopyrazoles with the acetals of 1,3-dicarbonyl compounds yields either pyrazolo[3,4-b]pyridines or the isomeric N-bridgehead pyrazolo[1,5-a]pyrimidines[19].

Pyrazolo[2,3-a]pyrimidines[30], triazolo[1,5-a]pyrimidines[31], pyrido[1,2-a] pyrimidines[32], and pyrimido[1,2-b]pyrazines[33] have all been prepared from the appropriate *o*-aminoheterocyclic compounds by treatment with β-keto-

or β-acetylenic-esters, ethoxymethylene-malonates or -malononitriles at a high temperature and often in the presence of an acid catalyst. The attacking electrophilic group may be, as well as an ester, an acetal[34], aldehyde[35], ketone[36], amide carbonyl[37], thioamide[38], acetylene[31] or dichloromethylene[41] and other new ring systems synthesised by this route include the *as*-triazino [3,4-*a*]isoquinolines[39], triazino[4,5-*a*]benzimidazoles[34], pyridazino[6,1-*c*]-*as*-triazines[36], pyrazolo[1,5-*c*]pyrido[3,2-*c*]pyrimidines[40], and a series of fused *s*-triazines[41].

6.2.1.3 Nucleophilic attack at a ring C-atom

The only examples of this type of ring cyclisation in 1970 and 1971 all involved an exocyclic amino group and a ring C atom which was present as a carbonyl group. For example, thiosemicarbazones (14) derived from pyrazole-4,5-diones cyclise in the presence of base to yield pyrazolo[3,4-*e*]

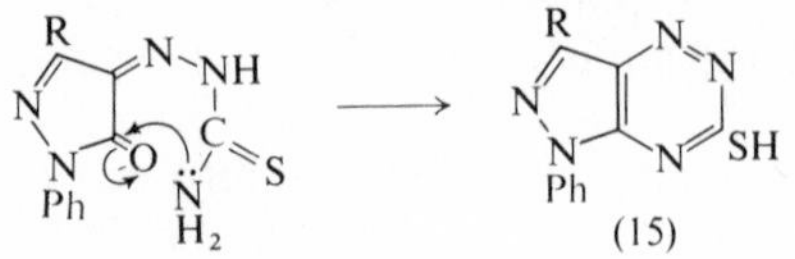

(14)   (15)

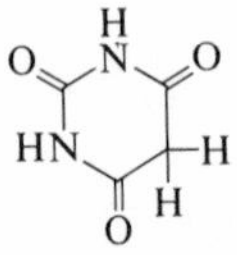

(16) 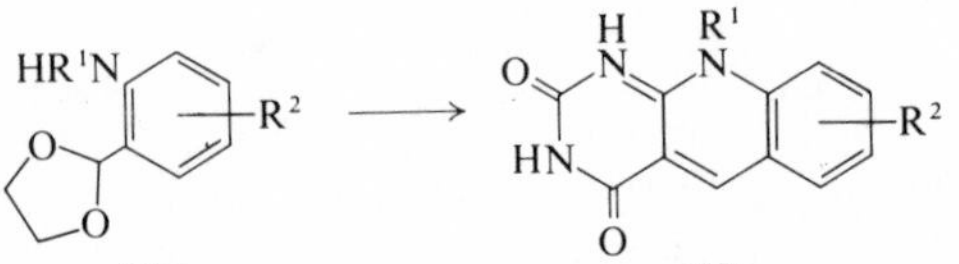   (17)   (18)

triazines (15)[43]. De-azariboflavin and related pyrimido[4,5-*b*]quinolines[44] (18) may be conveniently prepared in one step from barbituric acid (16) and acetals. Substituted anthranilaldehyde (17) and *o*-diamines have been used for the similar preparation of other tri-[45] and tetra-cyclic[45] systems.

6.2.1.4 Amine-carbonyl condensation reactions

The commonest route for the synthesis of 6-membered N-heteroaromatic compounds involves, at the final stage, an addition of a nucleophilic primary or secondary amino group to a carbonyl group with subsequent elimination of water or another small molecule. This is the type of route used for the synthesis of quinoxalines, quinazolines and benzotriazines from suitable *o*-substituted benzenes and which has been extended with considerable success over many years to the preparation of similar fused N-heteroaromatic compounds from other ring systems. Thus the preparation of quinoxalines from *o*-phenylene diamine has been extended to give the most popular route for the synthesis of pteridines from 4,5-diaminopyrimidines which has been used in very many publications[46-48] in 1970 and 1971. It has been confirmed that when 4,5,6-triaminopyrimidines are condensed with unsymmetrical 1,2-dicarbonyl compounds the more reactive carbonyl group reacts with the most nucleophilic amino group. Pyrido[1,2-*b*]-*as*-triazinium salts (19)[49] and imidazo[2,1-*b*]pteridines (20)[50] have been formed similarly from the appropriate diamines.

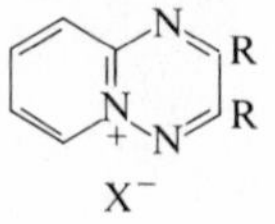

(19) 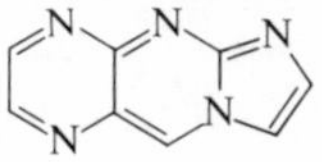   (20)

Pyrimido[4,5-*d*]pyridazines[51], pyrido[3,4-*d*]pyridazines[52], and pyrazolo [3,4-*d*]pyridazines[53] have all been prepared by reaction of the appropriate *o*-di-ester or acid with hydrazine.

o-Amino-esters, -nitriles, and -amides have all been used as precursors of fused pyrimidines but fairly common intermediates are the *o*-acylamino-amides (e.g. 21) which cyclise on heating in the presence of base or Lewis acid catalysts[54]. Formamide is useful in these reactions for its dual capacity

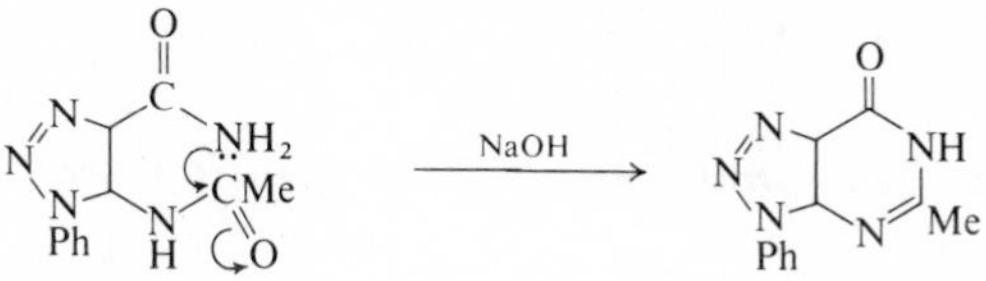

(21)

of high-boiling solvent and formylating agent[55] and the stronger nucleo-philicity of the amino group in a thio-amide group has been adopted with advantage[56].

The powerful poison toxoflavin has stimulated some research into its parent ring system, the pyrimido[5,4-*e*]-*as*-triazines (24), for which the most convenient route involves the treatment of 5-amino-4-hydrazino-pyrimidines (22) with 1-carbon containing electrophiles, such as the ortho-formates, to yield 1,2-dihydro-derivatives (23) which are readily oxidised to the fully aromatic system (24)[57]. A route to other 1,2-dihydro-derivatives of

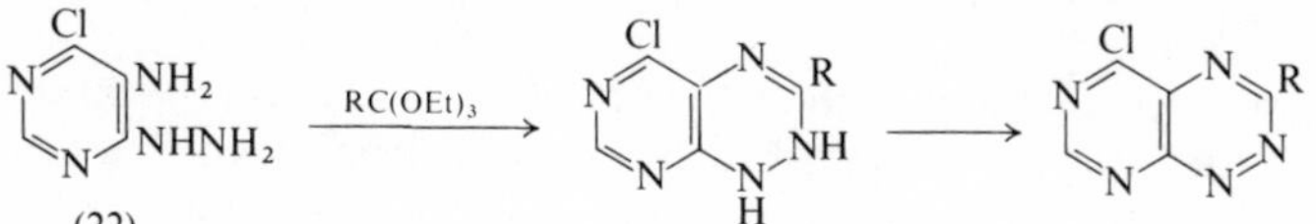

(22)   (23)   (24)

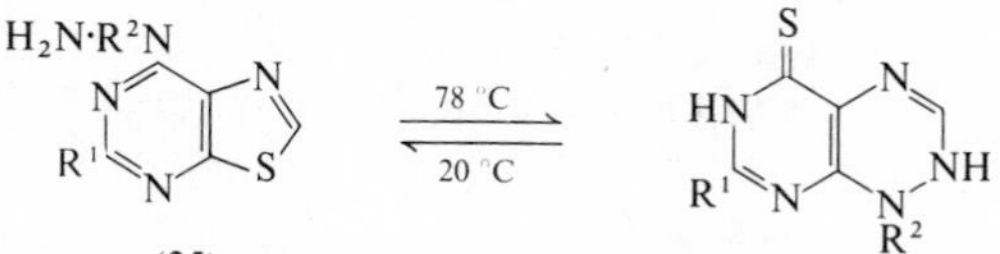

(25)

the same ring-system (24) has been developed which involves the reversible rearrangement of thiazolo[5,4-*d*]pyrimidines containing hydrazino groups (25)[58].

N-Bridgehead 5,6-membered ring-systems in which the 6-membered ring

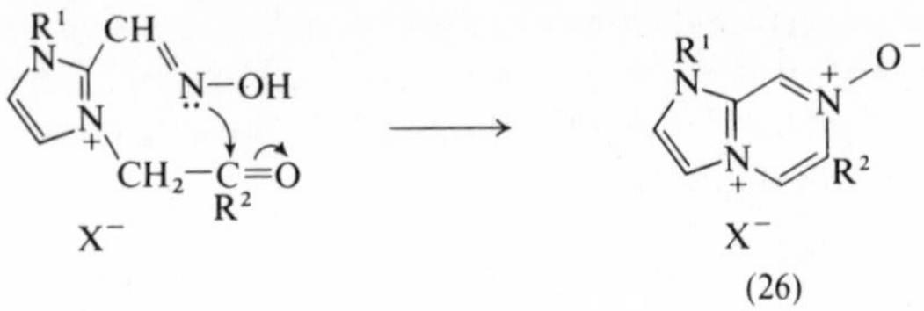

(26)

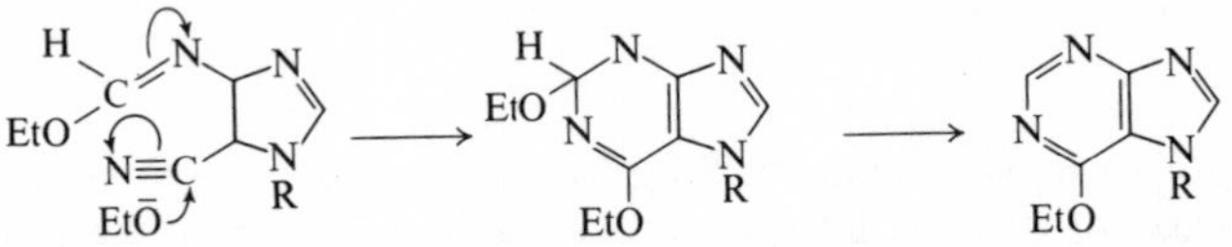

(27)

is finally closed by cyclisation at the non-bridgehead N atom are the pyrrolo [1,2-a]-pyrazines[59] and -quinoxalines[59] and imidazo[1,2-a]pyrazinium salts[60]. In the latter case cyclisation by nucleophilic attack of an oxime N atom yields the N-oxide (26).

6.2.1.5   Amine–imine and amine–nitrile addition reactions

Syntheses of this type may have the same initial mechanisms as those of the amine-carbonyl condensations but the imines formed initially may tautomerise to amines as in the case of the pyrrolo[4,3,2-de]pyrimido[4,5-c] pyridazines[61], or be stable as such as in the case of the pyrido[1,2-a]pyrimido [4,5-b]pyridines[62]. An interesting purine synthesis which has some parallels in the synthesis of quinoxalines is initiated by nucleophilic attack of a nitrile group at a C=N (27)[63].

6.2.1.6   Carbanion–nitroso reactions

A new unequivocal route to the pteridines (30) which appears to offer scope for other ring systems is described by workers from both Princeton[64] and Jerusalem[65]. This involves the reaction of phosphonate anions (29) with 4-amino-5-nitrosopyrimidines (28). The phosphonate intermediate is a doubly-activated methylene compound which loses one of its activating groups, the phosphonate, during the course of the condensation.

A similar double activation of a methylene group has been used in a syn-

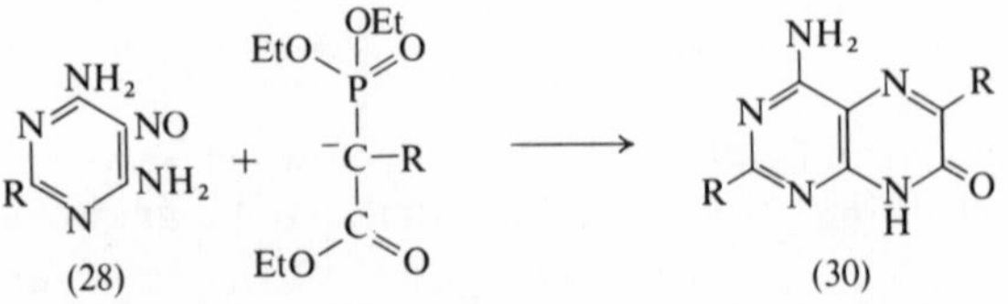

(28)   (29)   (30)

thesis of pteridines from 4-amino-5-nitrosopyrimidines and N-acylmethyl-pyridinium salts[66].

6.2.1.7   Nitrogen–nitrogen bond formation

Nitrogen–nitrogen bond formation has been accomplished by two different methods. Thus the interaction of an amine and a diazo group has been used for the first synthesis of pyrazolo[4,3-d]-vic-triazines (31)[67] and similar attacks of diazo groups at ring N atoms yield N-bridgehead compounds (32 and 33)[68, 69]. In the second type of reaction suitably positioned amino

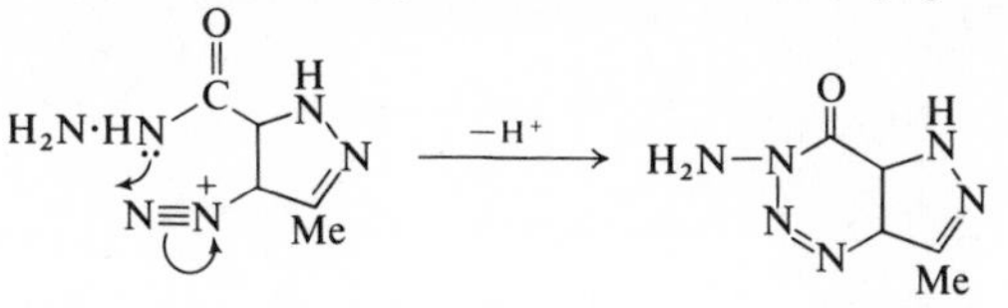

(31)

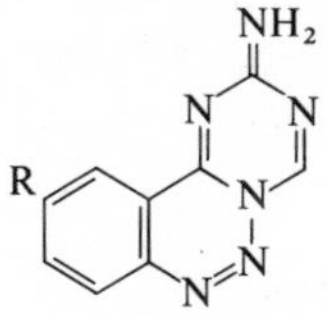

(32)

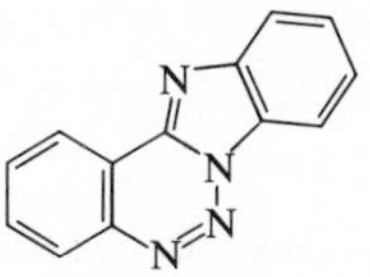

(33)

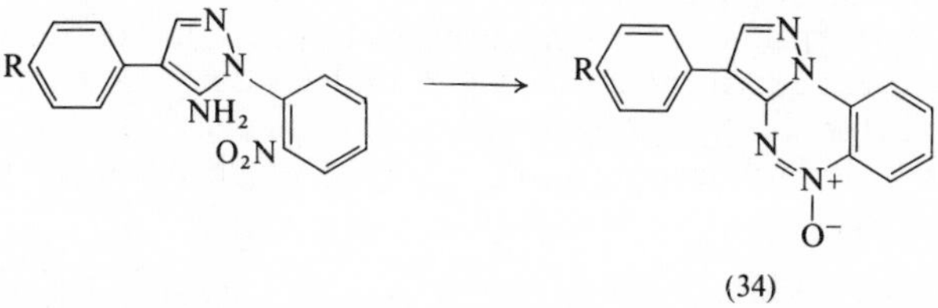

(34)

groups may be induced to react with nitroso or nitro groups. The expected interference if nitroso nitrene intermediates are formed could not be detected in the synthesis of the pyrazolotriazines (34) but the mechanism of these reactions has not been established[70]. Pyrido[4,3-e]-as-triazines[71] and pyrazolo[5,1-c][1,2,4] benzotriazines have been similarly prepared[72].

6.2.1.8   Oxidation of di- or tetra-hydro compounds

Several useful synthetic routes to the pteridines[73], pyrido-[74] and pyrimido-triazines[75] yield dihydro-derivatives in the cyclisation step. Such compounds are often readily oxidised under mild conditions using $KMnO_4$, H_2O_2, $K_3Fe(CN)_6$ and oxygen from the air. The high sensitivity of such ring systems to covalent hydration has often resulted in further oxidation, e.g. to pteridin-4-ones[73], and the best of the mild oxidising agents normally appears to be

activated MnO_2[73, 76]. 1-Chlorobenzotriazole has been successfully used for
the preparation of a pyrimido[5,4-*e*]-*as*-triazine when Ag_2O, MnO_2,
$KMnO_4$ and $K_3Fe(CN)_6$ had all failed[2]. D. J. Brown and his co-workers
have shown that 7-chloro-1,2-dihydropyrimido[5,4-*e*]-*as*-triazines can be
transformed into the fully aromatic di-substituted derivatives (35) which

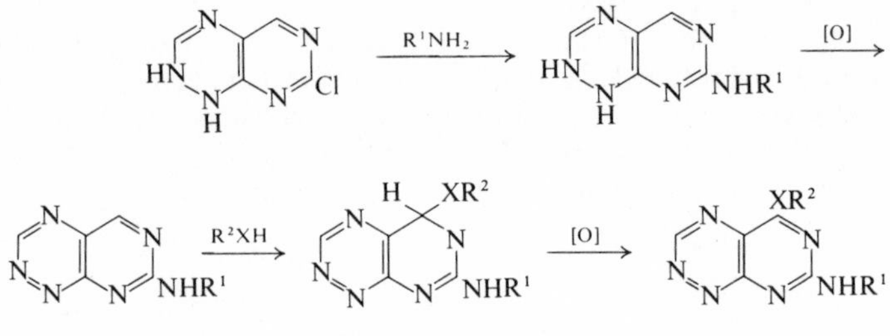

(35) X = O or NH

occurs in one stage. Covalent addition and two aerial oxidation steps are
postulated[78]. A new sensitive test for aldehydes has been developed which
depends on the rapid oxidation at the liquid-air interface of an intermediate
dihydro derivative of 6-mercapto-3-substituted *sym*-triazolo[4,3-*b*]-*sym*-
tetrazines[79].

6.2.1.9 Miscellaneous cyclisation reactions

Syntheses of bi- or tri-cyclic systems which occur in one stage from purely
aliphatic starting materials are not easily classified in mechanistic terms.
Imidazo[1,2-*a*]pyridines have been prepared by reaction of propargylamine
with 1-cyano-2-methylpent-1-en-3-yne[80], and tetra-acetylethylene is a new
intermediate for both pyridazino[4,5-*d*]pyridazines and pyrrolo[3,4-*d*]
pyridazines[81]. Several reactions are reported which occur by mechanistic

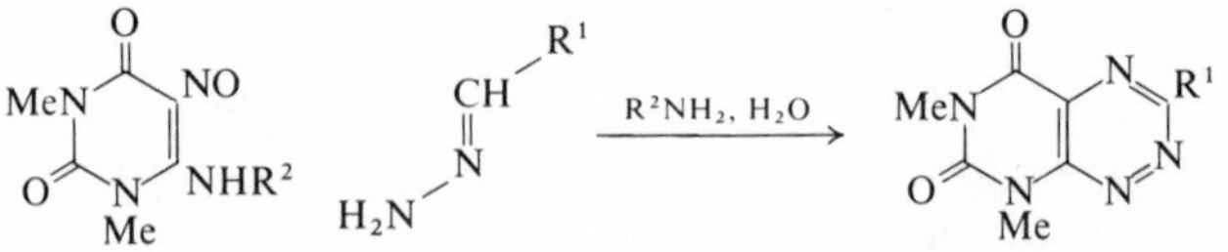

(36)

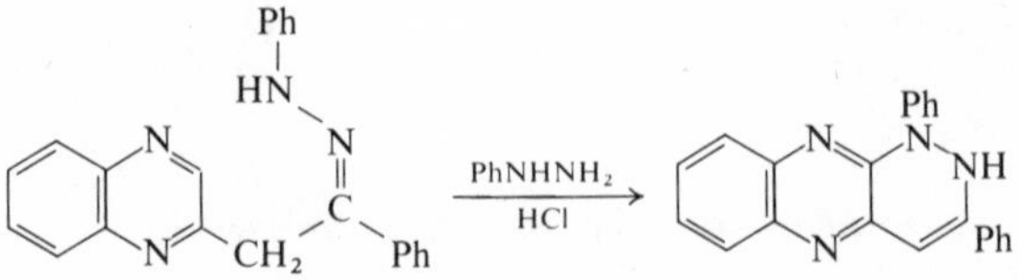

(37)

processes which have not been classified in the above scheme or by mechanisms which are not obvious. Thus a new preparation of fervenulin[82] (36,
R^1 = H) appears to involve a nitroso–imine interaction, and osazones are
suggested as intermediates in the preparation of pyrazino[3,4-*b*]quinoxa-
lines[83] (37). Pyrazolo[1,5-*a*]quinolines[84] have been prepared from 5-(2'-*m*-

chlorophenylethyl) pyrazoles by a treatment with potassamide which has been shown to involve aryne intermediates.

6.2.2 Syntheses which terminate in the formation of a 5-membered ring

Subdivision has again been made, as far as possible on the basis of position and nature of the bond-formation in the cyclisation stage.

6.2.2.1 Electrophilic attack at a ring C atom

Cyclisation reactions of this type are rare for 5-membered ring systems. A 6-amino-3-ribofuranosylpyrimidin-2,4-dione has been converted into a 3-ribofuranosyl-7H-pyrrolo[2,3-d]pyrimidin-2,4-dione by treatment with chloroacetaldehyde[85]. The Fischer–Indole synthesis is normally initiated by an acid-catalysed intramolecular electrophilic cyclisation[86] of a phenyl-hydrazone but in the extension of the reaction to pyridylhydrazones acid catalysis is often ineffective and cyclisations of this type possibly operate by a concerted mechanism[87] which is discussed further in Section 6.2.2.6.

6.2.2.2 Electrophilic attack at a ring N atom

Nineteen different ring systems have been formed either by direct electro-philic cyclisation on to a ring N atom or by isomerisation under mild con-ditions of the products formed by such a route. Where the electrophile is suitably orientated as part of a substituent group on a position *ortho* to the ring N atom the mechanism of the cyclisation step is clear. For example pyrazin-2-ylamino-acetates and -acetamides cyclise to imidazo[1,2-a] pyrazines (38)[88] and imidazo[1,2-a]-fused pyridines[89] and purines[90] have

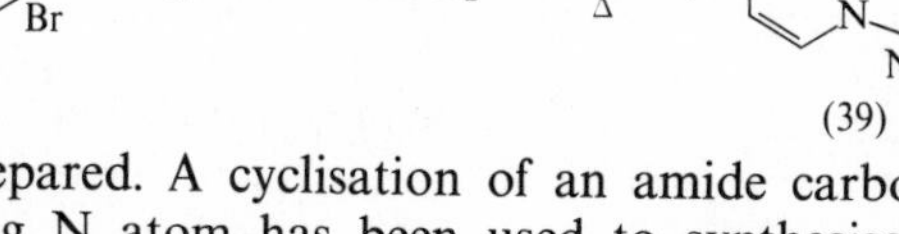

(38)

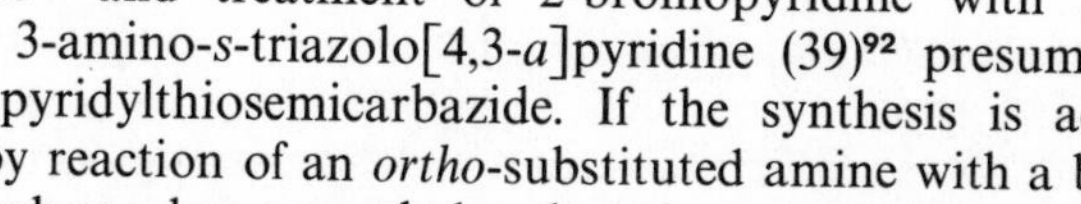

(39)

been similarly prepared. A cyclisation of an amide carbonyl group at a benzimidazole ring N atom has been used to synthesise imidazo[5,1-b] benzimidazoles[91] and treatment of 2-bromopyridine with thiosemicar-bazide yields 3-amino-s-triazolo[4,3-a]pyridine (39)[92] presumably via an intermediate pyridylthiosemicarbazide. If the synthesis is accomplished in one stage by reaction of an *ortho*-substituted amine with a bi-functional electrophile such as a bromomethyl carbonyl compound or a 1,2-dicarbonyl compound, or by the reaction of an *ortho*-substituted hydrazine with a 1-carbon reagent such as formic acid, then the mechanism of the final stage is often unknown and could involve either electrophilic cyclisation at the ring N atom or an amine–carbonyl addition. In two syntheses of imidazo [1,2-*a*]pyridines (41)[89] the problem does not arise because of the identical R^2 and R^3 groups (40 and 41), but in a reaction of 2-aminopyrimidines with *p*-substituted phenacyl bromides, the position of the *p*-substituted phenyl group in the product (42) indicates that the initial step has been the quaternisation of the more basic ring N atom[93]. Fused *sym*-triazolo-pyridines (44)[94], -pyrimidines[95], -pyridazines[96], -isoquinolines[97], -quinazolines[98], -furopyridazines[99], and -benzothienopyridazines[100] have all been prepared by reaction

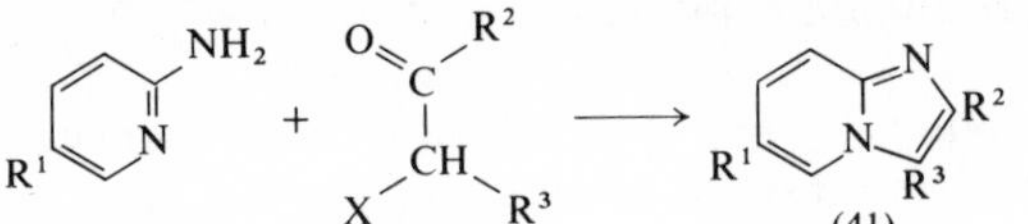

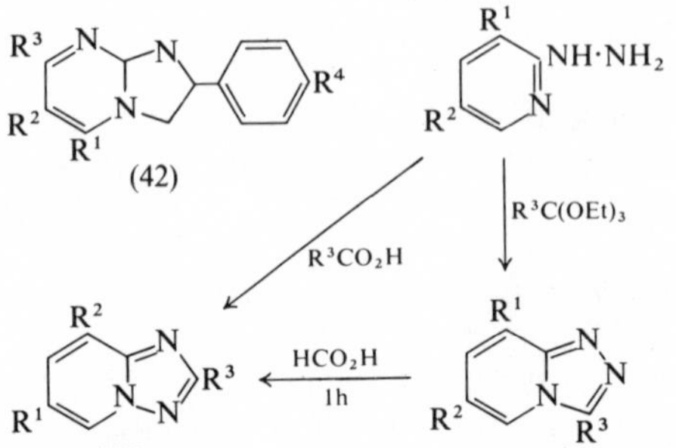

of the appropriate hydrazino-bases with formic acid, trialkyl *ortho*-esters, methyl isothiourea, phosgene or cyanogen bromide. If the *sym*-triazolo-ring is fused to a diazine, triazine or pyridine ring with suitably orientated electron-withdrawing substituents then ring-opening and recyclisation can occur (cf. Section 6.6.2). Thus *sym*-triazolo[1,5-*a*]pyridines (43) are the final products from the reaction of a 2-hydrazinopyridine with formic acid providing an electron-attracting group is present in the pyridine ring ($R^1 = NO_2$)[101].

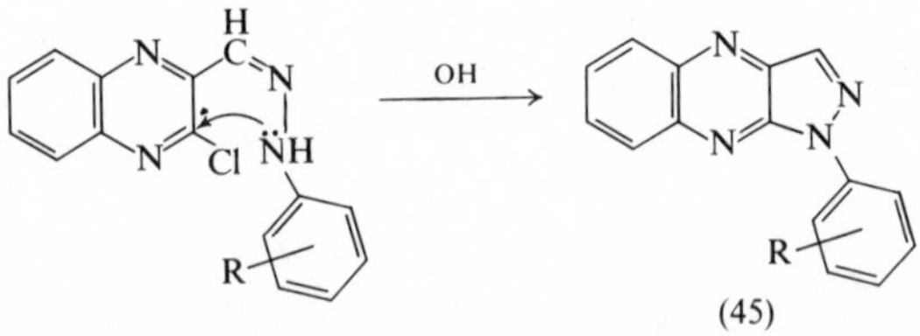

6.2.2.3 Nucleophilic attack at a ring C atom

Intramolecular nucleophilic attack at a π-deficient ring occurs more easily when the displaced group is one which is easily lost with its attacked electron pair. Thus pyrazolo[3,4-*b*]quinoxalines (45) are formed from the phenyl-hydrazones of 2-chloro-3-formylquinoxaline[102] and OH⁻ is eliminated in the formation of an *as*-triazino[5,6-*b*]indole[103]. One interesting synthesis of imidazo[1,2-*a*]pyridines involves the treatment of a pyridinium dicyano-methylide with sodium methoxide[104].

6.2.2.4 Amine addition to C=O, C≡N, C=N, C=S groups

This is the commonest route employed for the ring closure of 5-membered ring systems. The long-established Traube synthesis of purines is accomplished by the reaction of a 4,5-diaminopyrimidine with a carboxylic acid. Acids continue to be used as the reagents for inserting the required single carbon atom in the synthesis of purines but esters[105], nitriles[105], amidines[106], imino-ethers[107], carbon disulphide[108], thiourea[112], anhydrides[105], orthoformates[109], and diethoxymethyl acetate[110] have all been used. Initial acylation invariably occurs at the more nucleophilic 5-amino group and under mild conditions

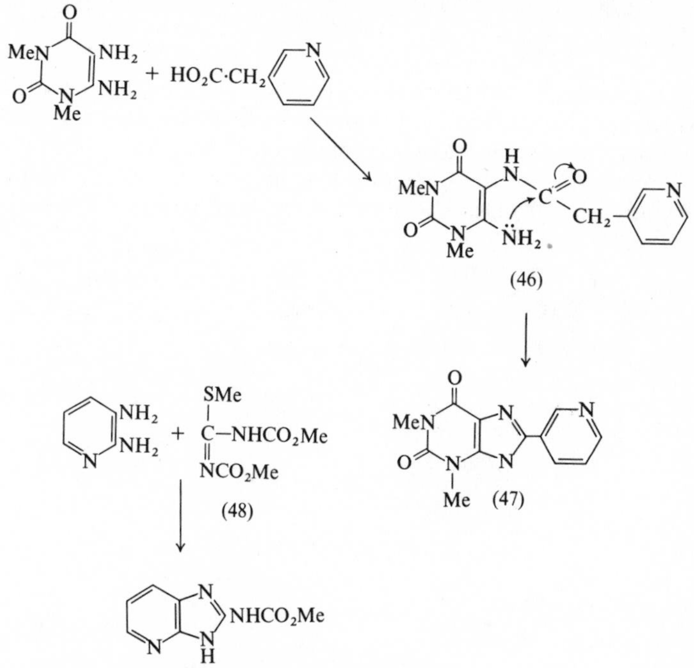

(46)

(48)

(47)

(49)

the expected 5-acylamido-derivatives (e.g. 46) may be isolated and cyclised to the purine (e.g. 47) by treatment with base or acid or simply on heating[111]. Imidazo[4,5-*b*]pyridines (49) have been prepared for biological evaluation by the reaction of analogous 2,3-diaminopyridines with the unusual amidine (48) derived from *S*-methylisothiourea and methyl chloroformate[112].

4-Amino-5-nitrosopyrimidines (50) are common precursors for the 4,5-diaminopyrimidines and Pfleiderer has shown that milder syntheses of purines can be accomplished by acylation (of (50) followed by reduction to 5-amino-4-acylamidopyrimidines (51)). These isomers, because of the more basic 5-amino group, may be cyclised to the purines (54) under much milder conditions[113] than the isomeric 4-amino-5-acylamidopyrimidines (e.g. 46). Pyrimidin-2,4-diones yield 4-acylimino-5-acyloxyimino-pyrimidines (52)

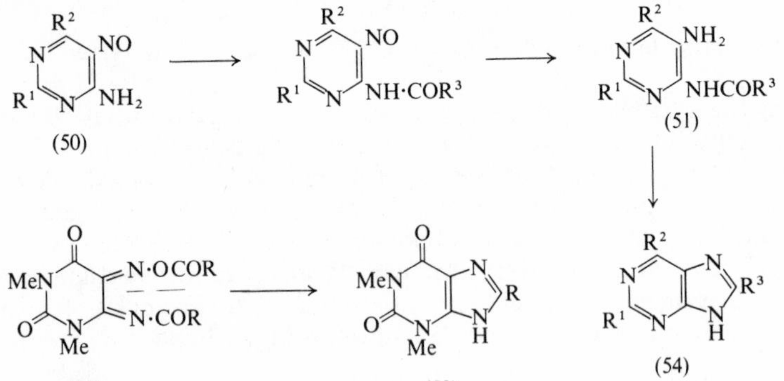

on acylation which transfer both acyl groups very readily to nucleophiles and which are also converted in to purines (53) under mild conditions[114].

Ten other ring systems have been prepared by reactions of this type. Thus a new preparation of the pyrazolo[3,4-d]pyrimidine, allopurinol (55), has been accomplished from 4,6-dichloro-5-formylpyrimidine[115], and the sydnone,

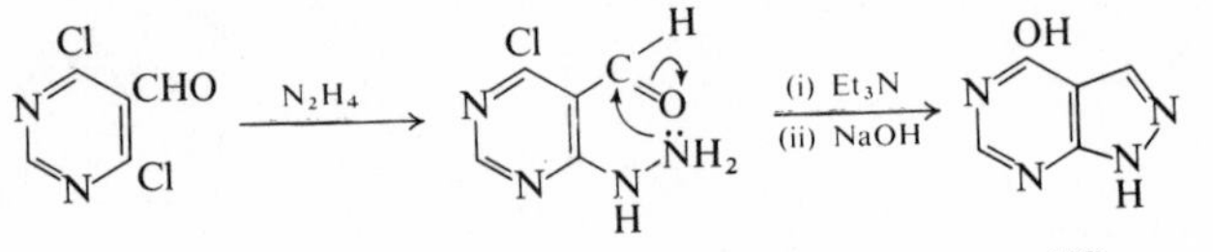

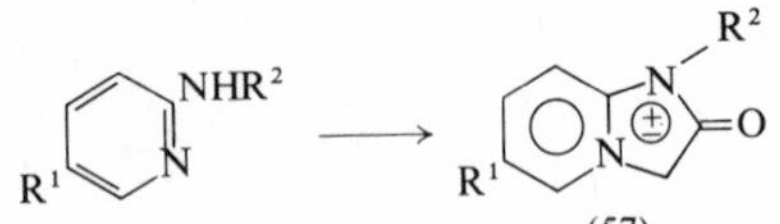

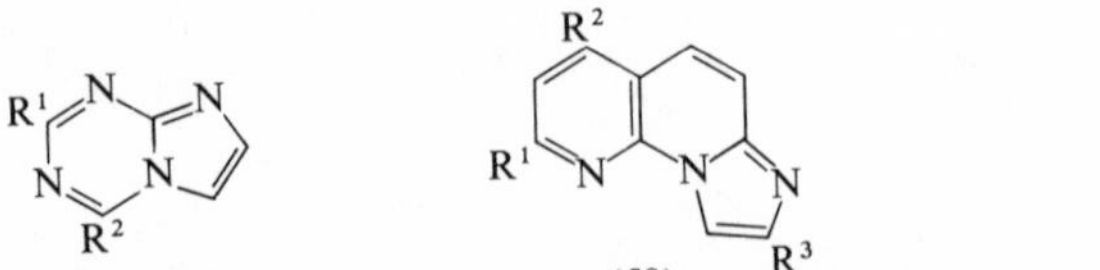

imidazo[1,2-a]pyridine (57), has been prepared by reaction of 2-amino-pyridines (56) with chloroacetic anhydride[116]. Imidazo[1,2-a]-sym-tri-azines[117] (58) and -[1,8]naphthyridines (59)[118] have been prepared by the type of reaction with α-bromomethyl carbonyl compounds described in Section 6.2.2.3, and the position of the substituent in the imidazole ring in the latter ring system confirms that quaternisation of the ring N atom is again the first step.

6.2.2.5   Nitrogen–nitrogen bond formation involving amine–diazonium group interactions

as-Triazolo-pyridines[119], -alloxazines[120], and -pyrimidines[121] have been prepared by diazotisation of *o*-diamines when *o*-aminodiazonium compounds are formed which undergo spontaneous cyclisation (60).

6.2.2.6   1,5-Dipolar cyclisation reactions

There is an increasing interest in this type of concerted ring closure which obeys the rules of orbital symmetry. Reimlinger has pointed out that the azide–tetrazole tautomerism commonly seen in several *o*-azido ring systems falls into this class of electrocyclic process and has demonstrated criteria necessary for effected ring-closures of the type (61) ⟶ (62)[122]. Cyclisation is generally effected by treatment of an ortho hydrazino derivative with nitrous acid and has been successful for the synthesis of pyrazolo[1,5-*a*] tetrazolo[1,5-*c*]pyrimidines[123] and tetrazolo[1,5-*a*][1,8]naphthyridines[124]. Reimlinger has also shown that conjugated nitrilamines are intermediates in a thermolytic synthesis of *s*-triazolo[4,3-*a*]pyridines (63)[125].

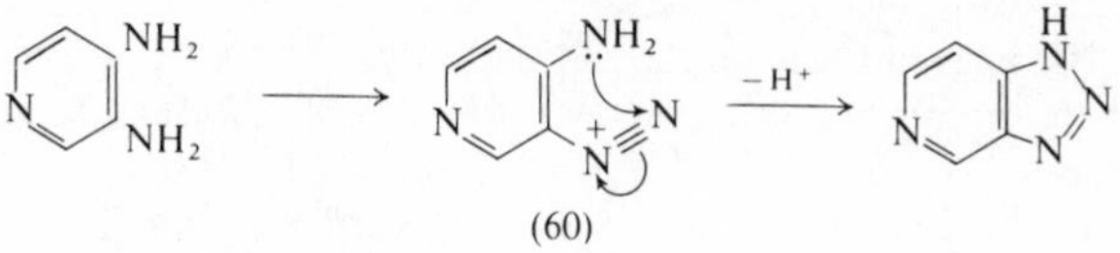

(60)

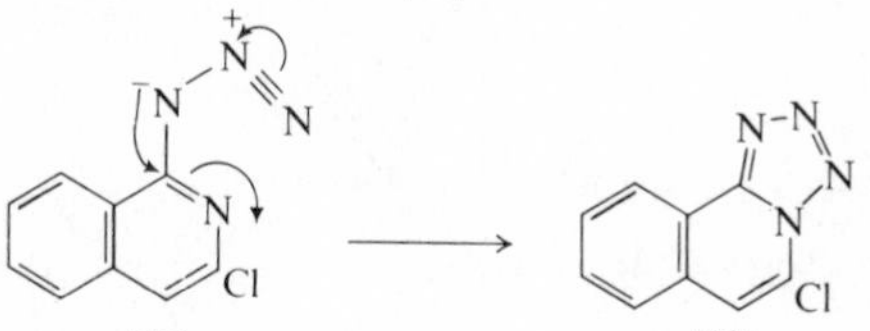

(61)   (62)

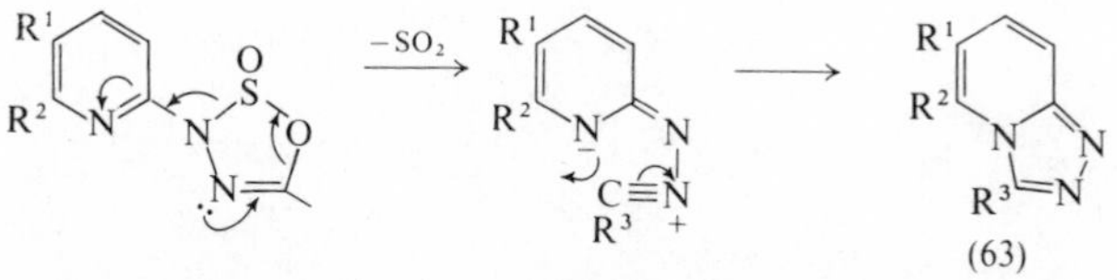

(63)

Norman and his co-workers have recently shown[126] that the unstable nitrilimines $(R^1C\colon N\cdot N^- R^2 \leftrightarrow R^1\overset{+}{C}\colon N\cdot \bar{N}R^2 \leftrightarrow$ etc.) are the products of the oxidation of aryl aldehyde hydrazones. It therefore appears very probable that dipolar compounds (65)[127] are involved in the cyclisation of various heterocyclic hydrazones (e.g. 64) with lead tetra-acetate or manganese dioxide. This is a method which has been exploited since its first use for the synthesis of *s*-triazolo[1,5-*a*]pyridines in 1957[128] and which has been used in the period under review for the preparation of two new ring systems, s-triazolo[4,3-a]-s-triazines (66)[127] and s-triazolo[4,5-b]-s-triazoles[128].

The author suggests that 1,5-dipolar intermediates could similarly be involved in the oxidative cyclisation of N-isoquinolylamidines to s-triazolo

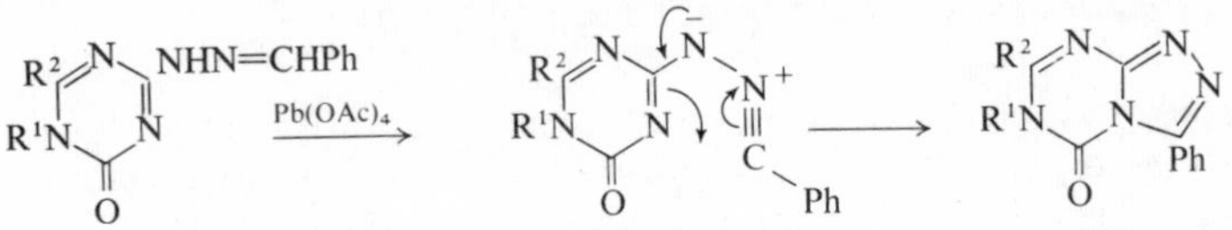

(64)   (65)   (66)

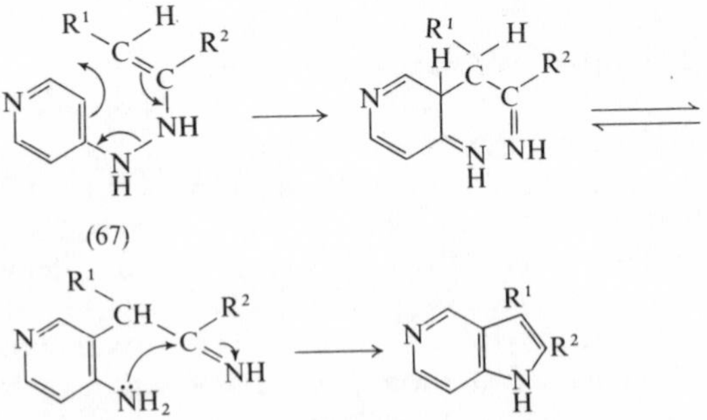

(67)

(68)

[5,1-a]isoquinolines[129] and of pyrid-2-yl aldehyde hydrazones to vic-triazolo[5,1-a]pyridines.

It has been proposed that the mechanism of a thermally-induced indolisation reaction is, in part, analogous to that of a Claisen rearrangement. A concerted reaction of this type (67) ⟶ (68) would explain why attack at the 3-position of a pyridine ring is possible in the cyclisation of 2- and 4-pyridylhydrazones.

6.2.2.7 1,3-Dipolar cyclisation reactions

Pyrazolo[2,3-a]pyridines (70) and pyrazolo[2,3-a]isoquinolines have been prepared by the 1,3-dipolar cyclo-addition reaction of N-iminopyridines (69) or N-iminoisoquinolines with cyanoacetylenes or diethylacetylene dicarboxylate.

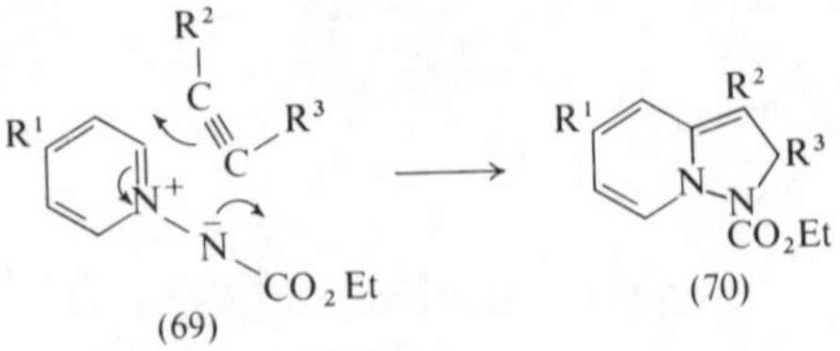

(69)   (70)

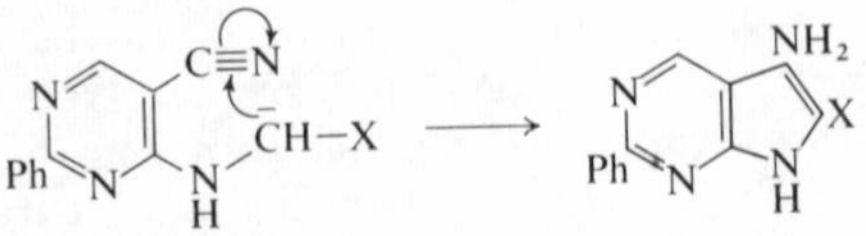

(71)   (72)

6.2.2.8   Exocyclic carbon–carbon bond formation

A new route to pyrrolo[2,3-*d*]pyrimidines (72) involves an intramolecular Dieckmann-type condensation of anions derived from 4-methylamino-5-nitriles (71) or -esters.

6.2.2.9   Photochemical cyclisation reactions

A series of 2,2-pyridylamines (73) and 6-anilinopyrimidines have been irradiated to yield pyrido[2,3-*b*]pyro-olo[2,3-*b*]pyridines (74) and pyrimido-[4,5-*b*]indoles.

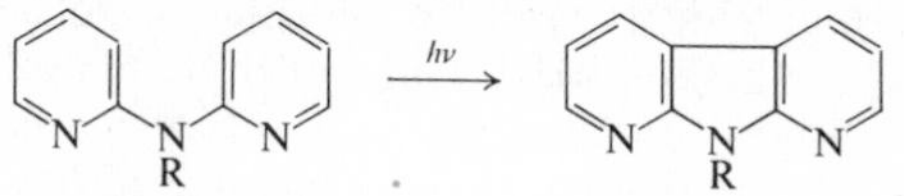

(73)   (74)

6.2.2.10   Miscellaneous reactions

A one-step preparation of 5-aminotriazolo[1,5-*a*]pyrimidines (75) has been shown to be terminated by an intramolecular attack of an imino group at a ring N atom[130]. Less obvious is the mechanism involved in the one-step synthesis of allopurinol (pyrazol[3,4-*d*]pyrimidin-4-one)[131]. An unusual synthesis of an imidazo[5,1-*b*]-*s*-triazole (77) from trifluoro-acetonitrile and sodium cyanide involves the formation of a nitrogen to nitrogen bond

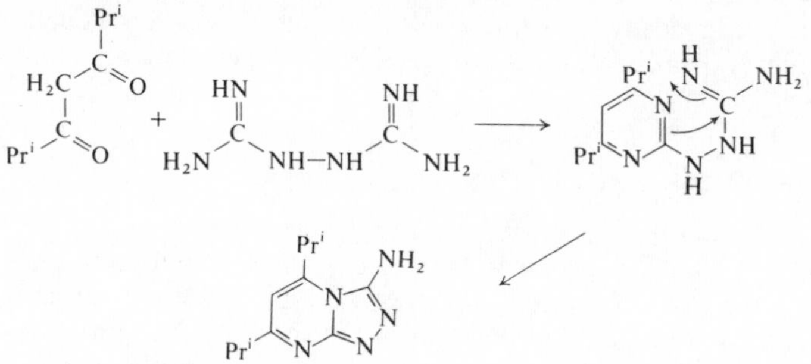

(75)

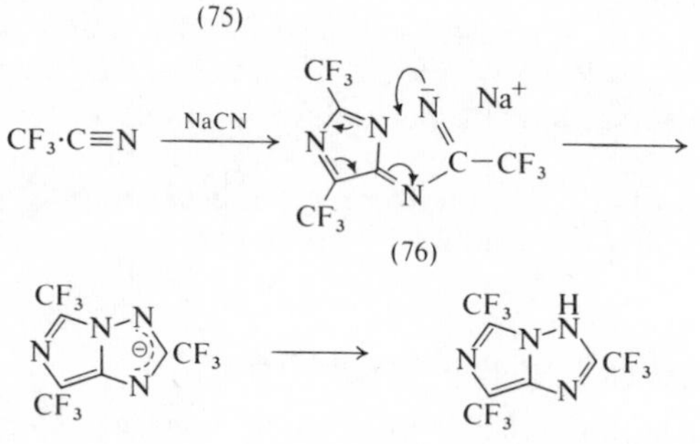

(76)

(77)

by the addition of a carbon–nitrogen double bond in a direction opposite to that normally observed (76)[132].

6.3 THEORETICAL STUDIES

There have been relatively few accounts of the use of molecular orbital calculations on polyaza compounds. The problems in selection of the correct parameters for the ring N atoms and their α-C atoms are even more serious than the simpler N-heteroaromatic systems. Furthermore, the correlation between the various reactivity indices and aromatic substitution reactions is complicated by the fact that exact mechanisms of substitution or the nature of the substituted species are often unknown. Attempts have, however, been made, e.g. to correlate frontier electron densities, which were calculated by simple LCAO MO methods, with the susceptibility of each of the ring nitrogen atoms in the purines to electrophilic attack[133]. Similar calculations for 4H-imidazo[5,1-b]benzimidazoles have yielded electron structure indices which have been correlated both with ^{1}H n.m.r. chemical shifts and chemical properties[134]. Molecular orbital calculations on pteridin-2,4(1H,3H)-diones (78) and pyrimido[5,4-e]-as-triazin-2,4(1H,3H)-diones (79) have shown that the C-6 position in the pteridine (78) carries a smaller positive charge and

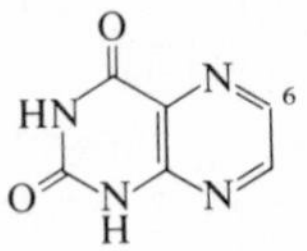

(78) 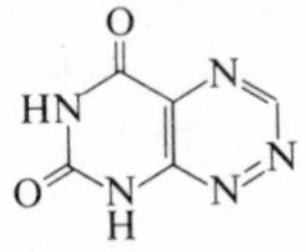   (79)

exhibits a lower super-delocalisability for nucleophilic attack than the corresponding position in the 7-aza analogue (79). The calculations were borne out by the relative rates and positions of attack of mammalian xanthine oxidase[135].

6.4 PHYSICAL PROPERTIES

It has now become the accepted norm in the chemistry of polyazaheterocyclic compounds that the proof of structure of new compounds rests largely, and in some cases completely, on spectroscopic evidence. Infrared and ^{1}H n.m.r. spectra are offered in the majority of correct publications as evidence for structural assignments and, increasingly, mass-spectral data are presented, either in the form of electron-induced fragmentation pathways which are correlated with molecular formulae, or in the form of accurate mass measurements of molecular ions in place of, or in addition to, analytical elemental determinations.

In the present chapter, therefore, references are only made to those publications in which the amount of spectroscopic data presented exceeds this norm.

6.4.1 Ultraviolet and visible spectra

The ultraviolet and visible absorption spectra of many purines[113] and pteridines[136] have been recorded and u.v. spectra and mass spectra used together in the assignment of the structures of four closely similar ribo-nucleosides which were responsible for cytokinin activity in wheat germ tRNA[137]. A number of spectroscopic procedures, including u.v. and visible were used in the establishment of the stacking pattern of a dinucleoside monophosphate derived from adenine-8-cyclonucleosides[138].

6.4.2 N.M.R. spectra

The use of [1]H n.m.r. spectroscopy to distinguish between two theoretically possible isomers, e.g. s-triazolo-[4,3-a]pyrimidines and -[1,5-a]pyrimidines[139], between positions of protonation[140], and between keto and enol tautomers[141] are typical of the type of application which still continues. More recently, however, an increasing use has been made of [1]H n.m.r. and particularly high frequency (220 MHz) spectra of fairly complex molecules of biological importance. For example, the M and P helices of reduced pyridine di-nucleotides have been studied[142] and the nature of the configuration and con-formation in the complexes formed between N-methyl-N-ethylnicotinamide and adenine dinucleotide investigated[143]. Tertiary complexes have been shown to be formed of the types adenosine–Co–guanosine and guanosine–Zn–imidazole[144], and the interaction of mercury(II) complexes with cytidine, adenosine and guanosine has been demonstrated[145]. [1]H N.M.R., fluores-cence and phosphorescence spectra have all been used to investigate the stacking interactions of N^2-dimethylguanosine with adenosine and cytidine by the use of model compounds in which a trimethylene bridge was sub-stituted for the ribose–phosphate–ribose linkage[146]. In the pteridine field [1]H n.m.r. spectroscopy has been used to examine deuterium exchange of both methyl[147] and ring[148] protons in lumazine derivatives.

6.4.3 Mass spectra

In a series of papers[149], Clark has proposed fragmentation pathways for a large number of pteridines and correlated these with earlier work both in the same system and with other N-heteroaromatic analogues, and the use of trimethylsilyl derivatives in the determination of the mass spectra of poorly volatile hydroxypteridines has been demonstrated[150]. Similar derivatisation procedures have enabled the mass spectra of very many derivatives of adeno-sine[151] and of uridine, adenosine, cytidine, and guanosine[152], to be determined. The mass spectra of simpler purines[153], of pyrazolo[3,4-b]quinoxalines and the isomeric 5-triazolo-[4,3-a]pyridines and -[1,5-a]pyridines[154] have also been described and fragmentation pathways postulated.

6.4.4 x-Ray diffraction methods

The value of x-ray diffraction methods in the determination of structures of relatively complex and biologically important compounds has been recog-nised for a number of years and has recently been more widely exploited.

In one important paper[155] the three-dimensional structure of adenosine triphosphate (ATP) in the form of its hydrated disodium salt is reported to a resolution of 0.9 Å. The shape of the molecule, the co-ordination about the Na ions and the molecular packing have all been fully established. The cell constants indicated that there were a total of eight molecules of ATP, 16 Na ions and 24 H_2O molecules in the unit cell. There was a stacking of the aromatic rings on top of each other and a bending back of the phosphate chain to the aromatic ring. X-Ray studies of 1:1 purine:pyrimidine complexes have shown that either a Watson–Crick type base-pairing configuration is formed or in other cases that there is a 'reversed' Watson–Crick type of hydrogen-bonded configuration[156]. The nature of the bonding in the 2:1 complex of barbitone with caffeine has been determined by x-ray diffraction methods and the cell has been shown to be comprised of four barbitone and two caffeine molecules[157].

6.4.5 Phase-solubility methods

Suggestions in the last few years, that base-stacking interactions play a more important role in stabilising single- and double-stranded helical structures of nucleic acids, than hydrogen-bonded interaction, has stimulated current interest in the structure and conformation of polynucleotides in water. An examination of the interactions of a large number of purines, pyrimidines and other heteroaromatic compounds in aqueous media by a phase-solubility method, which employed radioactive solutes, has demonstrated that in these circumstances base-stacking is primarily responsible for interactions[158].

6.4.6 Optical rotatory dispersion

Optical rotatory dispersion methods find little use in polyazaheterocyclic compounds because of the infrequent occurrence of optical isomerism. *Anti-* and *syn*-conformational equilibria have, however, been demonstrated in solutions of oligoinosinic acids[159].

6.4.7 Electron spin resonance spectra

The formation of 1:1 radical chelates between isolloxazine anionic radicals and Zn^{2+} as well as Cd^{2+} has been shown by electron spin resonance and isotopic substitution methods[160].

6.5 SUBSTITUTION REACTIONS

The susceptibility of polyazaheterocyclic compounds to substitution reactions has long been accepted as a measure of their aromaticity and there are very many isolated reports of substitution reactions by electrophilic, nucleophilic and free radical reagents.

6.5.1 Substitution by electrophilic reagents

In general terms electrophilic substitution at ring C atoms of π-excessive ring systems occurs more readily, and of π-deficient ring systems occurs less readily, than in benzene or analogous polyhomocyclic compounds. Electrophilic substitution either of a 'lone-pair' of electrons or of a H atom can also occur at a ring N atom in the former ring systems more easily and in the latter less easily than at the ring C atoms. In all cases the nature of existing substituent groups and the presence of positive or negative charges can profoundly affect the extent and position of a subsequent substitution reaction. Three groups of electrophilic substitution reaction are attracting current interest and are mentioned in several separate publications. In the first place several workers have investigated N-oxide formation. Paudler has synthesised all of the mono-N-oxides of 1,5-, 1,6-, 1,7-, 1,8- and 2,7-naphthyridines and of di-N-oxides of all but the 1,8-naphthyridine by the action of perbenzoic acid in chloroform either at room temperature (for the mono N-oxides) or by heating under reflux for 1 h [161]. The position of substitution has been shown by ^{1}H n.m.r. spectroscopy. The nature of a 6-substituent is shown to be one factor determining whether substitution of 6-substituted purines (80) occurs at the 1- (81) or 3- (82) position [162]. A 6-CN

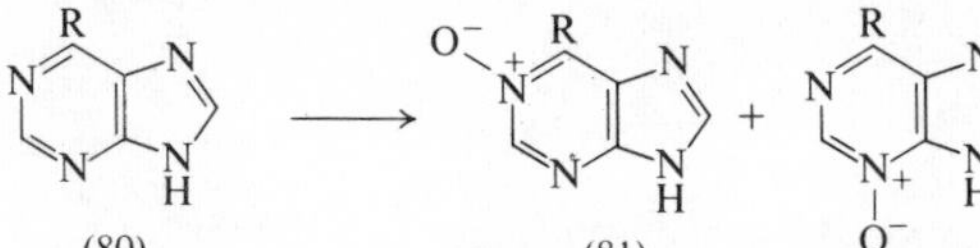

(80) (81) (82)

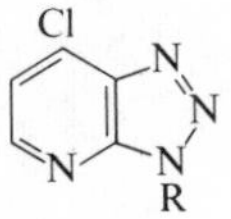

(83) 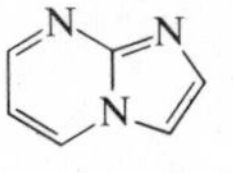 (86) 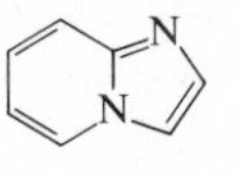 (85)

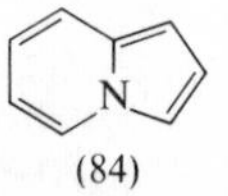

(84)

substituent favours 3-N-oxide formation and a 6-Me or 6-H yields both 1- and 3-N-oxides. More of the 1-oxide is formed with peroxyacetic acid as the electrophile and more of the 3-oxide with m-chloroperoxybenzoic acid [162]. Guanines give good yields of 3-N-oxides [163] whereas adenines yield the 1-N-oxides [164] and both xanthine and guanine N-oxide are shown to exist preferentially as the cyclic hydroxamic acid tautomer [165].

Another electrophilic substitution reaction which continues to attract considerable attention is N-alkylation. Many authors have investigated methods for the synthesis of nucleosides or nucleoside analogues which are dependent on the alkylation of a purine or similar ring system. Thus Robins and his co-workers have demonstrated the first example of the use of a 1-O-methyl-2-deoxy sugar in the direct fusion synthesis of 2′-deoxy-L-adenosine and 2′-deoxy-L-guanosine and their α-anomers[166]. Simpler alkylations of purines have been effected with ethylchloroacetate in DMSO[167], and chloromethyl ethyl ethers[167a]. Ribosylation of silylated pyrrolo[2,3-d] pyrimidines yields[168] mixtures of 1-, 3- and 7-ribosyl derivatives in high yield and N-alkylation of s-triazolo[4,5-d]pyrimidines is accomplished by treatment with chloroacetamide in the presence of NaOH and Na[169]. Critics of the aza- and deaza-nomenclature will note the description, albeit in brackets, of the ribosyl derivatives of s-triazolo[4,5-b]pyridines (83) as 1-deaza-8-azapurines[170]. The third major investigation has been of the electrophilic substitution of various N-bridgehead compounds. Three groups of workers have investigated the electrophilic substitution reactions of imidazo[1,2-a]pyrimidines (86)[93, 171, 172] and in one publication comparisons are made of the ease of substitution of this ring system compared with imidazo[1,2-a]pyridines (85) and indolizines (84)[171].

Bromination of 3,6-disubstituted pyrrolo[1,2-c]pyrimidines with either N-bromosuccinimide or bromine in chloroform yielded the expected 5- and 7-monobromo and 5,7-dibromo (87) derivatives[173]. In addition, per-bromides (88) were formed which could be converted into 4-alkoxy-5,7-dibromo (89) or 4,5,7-tribromo compounds on treatment with an alcohol[173].

(87) (88) (89)

Halogenation of imidazo[4,5-b]pyridines[174] and pyrrolo[1,2-b]pyridazines[175] also results in the insertion of halogen atoms into the 6-membered rings. Six different electrophilic reagents have all been shown to attack pyrrolo-[3,2-b]pyridine at the 3-position[176].

6.5.2 Substitution by nucleophilic reagents

It has recently been demonstrated that nucleophilic substitution reactions in the simpler π-deficient heterocyclic compounds may occur by a variety of different mechanistic pathways. In the case of the polyazaheterocyclic compounds containing more than one ring, however, investigators have been content to use nucleophilic substitution reactions to effect particular syntheses or to investigate the scope of such reactions. Nucleophilic substitution reactions can of course involve the replacement of either a ring H atom or a substituent with its bonding pair of electrons and those substituents which easily carry a negative charge are the more easily replaced. Several authors have investigated the replacement of the chlorine in 6-chloropurine[177] or in 6-chloropurine 3-N-oxide[178], and the 6-chloro group is shown to be the most reactive towards nucleophilic displacement in 2,6- and 6,8-dichloro- and in 2,6,8-trichloro-purines. Trimethylammonium groups introduced by such reactions with trimethylamine underwent thermal rearrangements to 6-dimethylamino-3- and 6-dimethylamino-6-methyl purines[179]. The greater the number of ring N-atoms in polyazaheterocyclic compounds the more susceptible they become to nucleophilic attack at ring C atoms and 7-chloroimidazo[4,5-*b*]pyridine (90) is less reactive than 7-chloro-*s*-tri-azolo[4,5-*b*]pyridines (91) towards amines, thiourea or mercaptans[180].

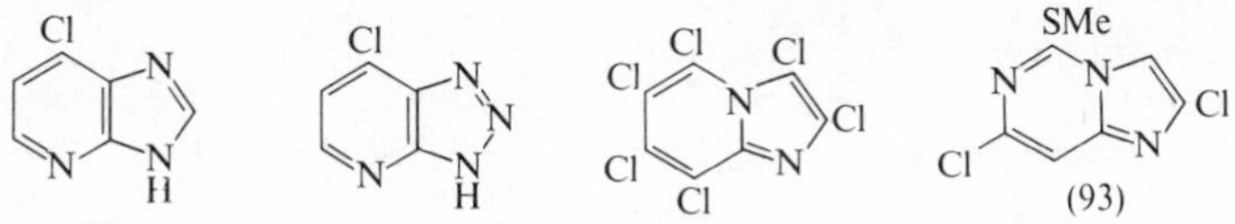

(90)   (91)   (92)   (93)

In separate investigations of two N-bridgehead systems a chloro group in the 6-membered ring at the position adjacent to the bridgehead N atom was found to be the most reactive. Thus nucleophilic substitution takes place preferentially at C-5 in the perchloroimidazo[1,2-*a*]pyrimidine (92)[181] and the 5-methylmercaptoimidazo[1,2-*c*]pyrimidine (93) undergoes reaction with water, ammonia or methanol at the 5-position despite the fact that a chloro group is a better leaving group than a methylmercaptan[182]. The conversion of 'hydroxy' N-heteroaromatic compounds into amino-derivatives in the laboratory has usually been accomplished in two stages. An interesting use of phosphoramines allows one-stage reactions to be accomplished which mimic the biochemical processes for the conversion of hypoxanthines into adenines ((95) ⟶ (94), (95) ⟶ (96)) [183].

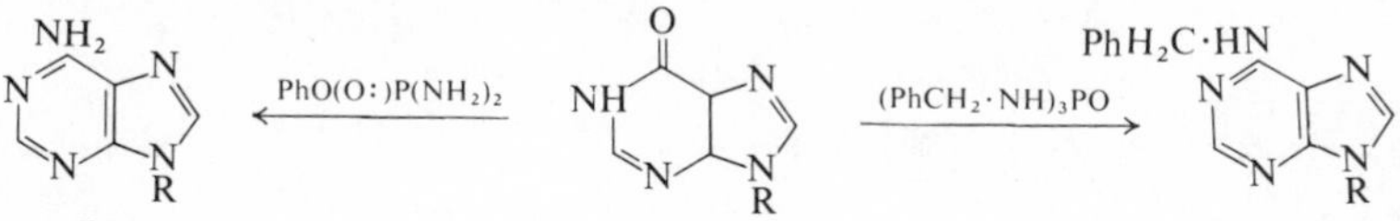

(94)   (95)   (96)

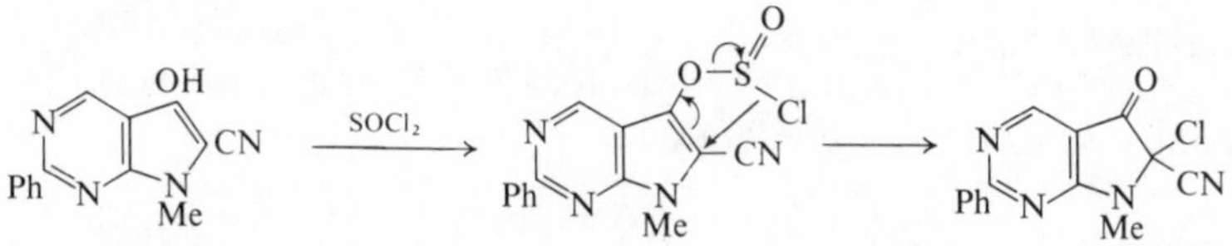

(97)   (98)   (99)

Naphthyridines are particularly susceptible to nucleophilic substitution at positions *ortho*- or *para*- to the ring N atom whether or not a substituent is present and calculations have been made by the Hückel molecular orbital method of various reactivity indicies for 1,5-, 1,6- and 1,8-naphthyridine[184]. Methylation in dimethyl sulphoxide has recently[185a] been shown to occur by nucleophilic attack of the $CH_3 \cdot SO_2 \cdot CH_2^-$ ion and the position and extent of the substitution was shown to be in agreement with π-electron and frontier electron densities but not with super-delocalisability calculations in the case of 1,6-naphthyridines[184]. Ethylation at the 2-position is effected in ethyl 7-methyl-1,8-naphthyridin-4(1*H*)-one-3-carboxylate[185] but with a vacant 4-position, 1,5-naphthyridines yield 4-amino- and not 2-amino-derivatives as previously claimed[186]. A novel intramolecular nucleophilic attack followed by an allylic-type rearrangement (98) is proposed to explain the product (99) which results from the treatment of 6-cyano-5-hydroxy-7-methylpyrrolo[2,3-*d*]pyrimidine (97) with thionyl chloride[187].

6.5.3 Cyclo-addition reactions

The meso-ionic pyrido[1,2-*a*]pyrimidine (100) which is readily prepared in one stage from 2-methylaminopyridine undergoes cyclo-addition reactions with acetylenic dipolarophiles to yield 1,2-disubstituted quinolizin-4-

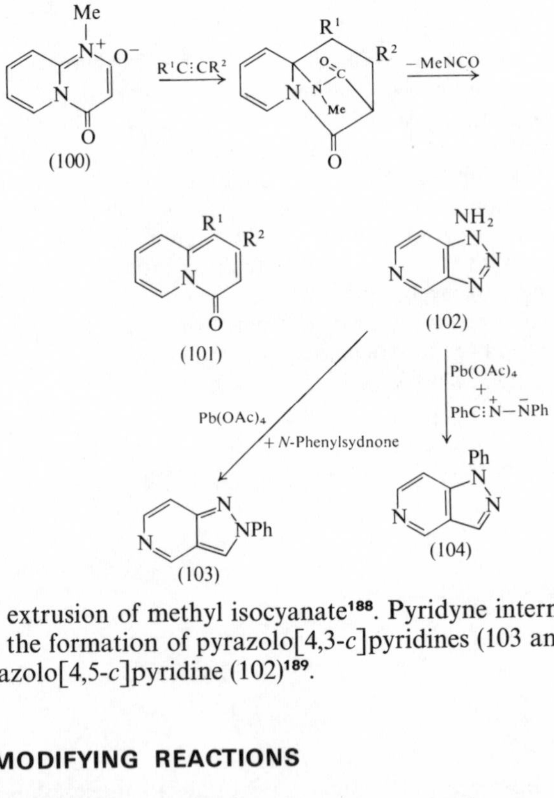

(100)

(101)

(102)

(103)

(104)

ones (101) by extrusion of methyl isocyanate[188]. Pyridyne intermediates are implicated in the formation of pyrazolo[4,3-*c*]pyridines (103 and 104) from 1-amino-*s*-triazolo[4,5-*c*]pyridine (102)[189].

6.6. RING-MODIFYING REACTIONS

N-Heteroaromatic rings may be structurally modified either by addition reactions which yield hydrogenated — or derivatives of hydrogenated — rings, or, more drastically, by reactions which result in ring fission.

6.6.1 Addition reactions which yield hydrogenated ring systems

N-Heteroaromatic compounds are reduced more easily than their aromatic counterparts and nucleophilic hydrogenating agents such as sodium borohydride are particularly effective in the hydrogenation of π-deficient rings with initial attack occurring at the positions of lowest electron density. The two pteridine research schools of E. C. Taylor at Princeton University and W. Pfleiderer at Konstanz University have combined again to publish a paper which shows conclusively that borohydride reduction of 8-alkyl-pteridin-7(8H)-one-6-carboxylic acid derivatives (106) yields 3,4-dihydro derivatives (105) and not 4,8- or 5,8- as previously suggested, whereas catalytic reduction gives 5,6-dihydro compounds (107) of very different chemical and physical properties[190].

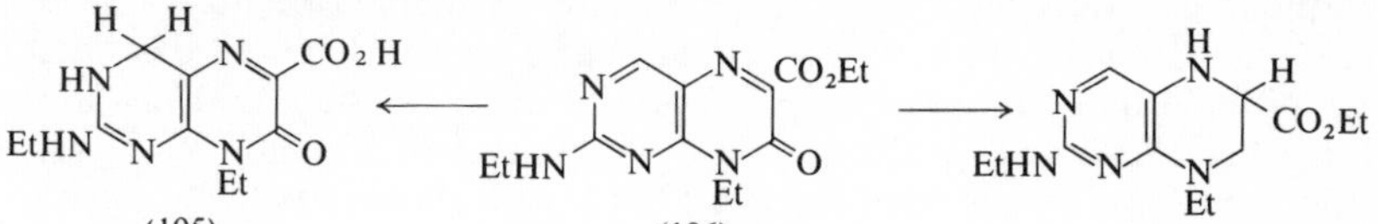

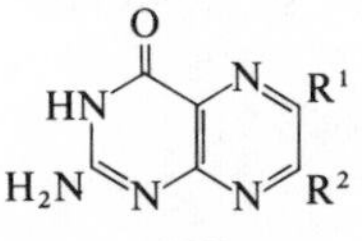

Where two electron-donating groups are present in the pyrimidine ring, however, pteridines are attacked by both borohydride or catalytically-produced hydrogen in the pyrazine ring. The pterins (108 and 109)[191], and lumazines (110)[192], for example yield 7,8-di- or easily oxidisable 5,6,7,8-tetrahydro derivatives. Quaternary salt formation in both the pteridines[193] and the purines[194] renders the C atom adjacent to a quaternary N atom susceptible to nucleophilic hydrogenation and certain pyrido[3,4-d] pyrimidin-4(3H)-ones are reduced by lithium aluminium hydride at both the exocyclic C=O group and the ring 1,2 C=N group to yield 1,2,3,4-tetrahydroderivatives[195].

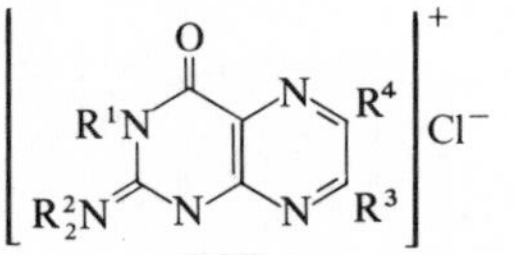

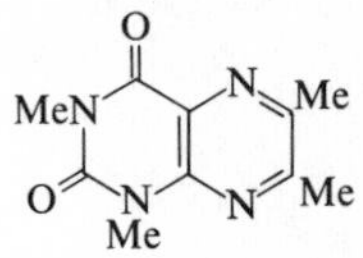

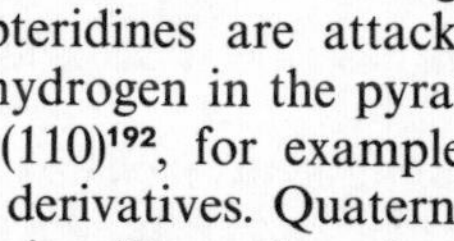

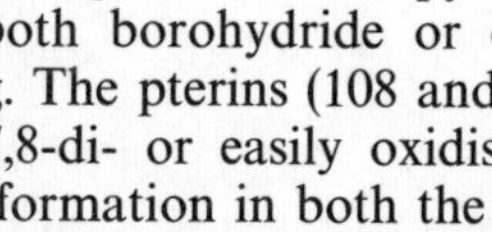

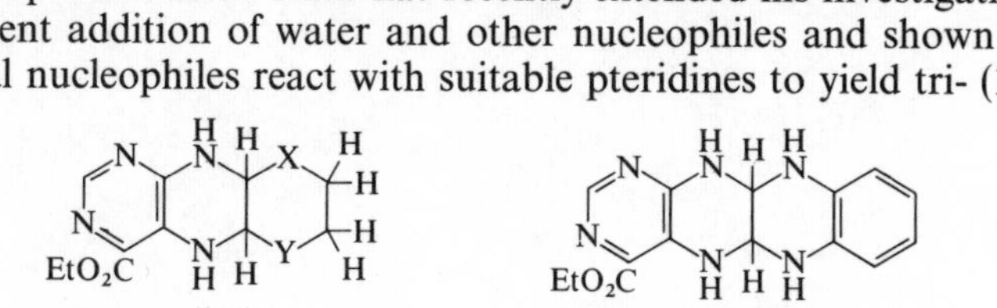

The painstaking and yet exciting work of Adrien Albert in the field of the pteridines and related ring systems over the last 20 years has resulted in the present detailed chemical knowledge. Two of Albert's former co-workers Clark and Brown are now themselves independent authorities in this extremely important area. Clark has recently extended his investigations into the covalent addition of water and other nucleophiles and shown that bifunctional nucleophiles react with suitable pteridines to yield tri- (111) and tetra-cyclic (112) ring systems[196]. The nature of the 2-substituent in the 4-trifluoromethylpteridines[197] and the 7-substituent in their aza-analogues, the 5-trifluoromethylpyrimido[5,4-e]-as-triazines[2], considerably influences the position and stability of covalent hydrates. Pteridin-2-thione, for example, irreversibly yields a 3,4-dihydrate, the stability of which is attributed to favourable resonance factors (113) ↔ (114) ↔ (115)[197].

The involvement of covalent addition reactions[78] in the synthesis of pyrimido[5,4-e]-as-triazines (35) has already been referred to in Section 6.2.1.8 and Brown and his co-workers have also shown that 5-alkoxy derivatives of the same ring system (117) may be prepared by oxidation of the appropriate 5,6-covalent alcoholates (116). Transetherification of 5-alkoxy derivatives by treatment with AgO in a boiling alcohol is similarly suggested to involve covalent adducts[198].

The detection and degree of covalent hydration and the covalent addition of other nucleophiles has continued to be evaluated by physico-chemical methods, in particular ^1H n.m.r. and u.v. spectroscopy and pK_a determinations[2, 78, 197].

6.6.2 Ring-fission reactions

Three distinct types of ring-fission reaction have been reported in the 2 years under review.

In the first place ring fission, either by spontaneous isomerisation or by the application of a reagent, results in the formation of a product which can be recyclised, either spontaneously, or by treatment with another reagent, to re-form the original starting material. Such a process occurs in the tetrazolo–azido isomerisations which have been reported for a variety of ring systems. Thus the equilibrium between the tetrazolonaphthyridine (118) and the azidonaphthyridine (119) is dependent on solvent and pH[199], and the

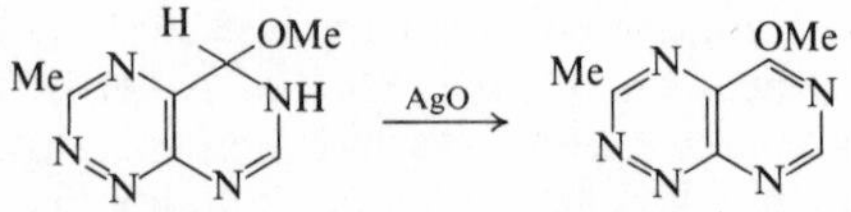

(116) → AgO → (117)

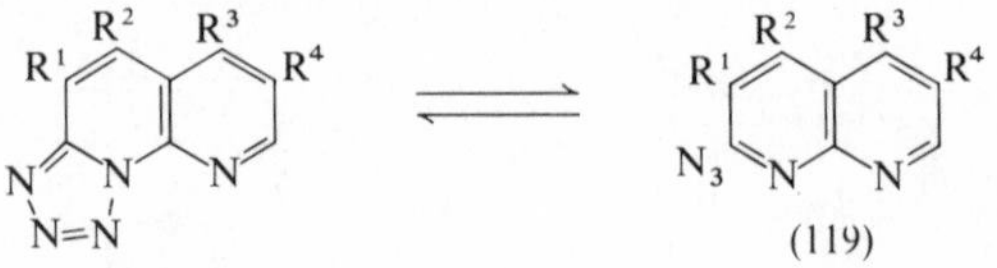

(118) ⇌ (119)

azides are the more preferred the more electron-attracting substituents or ring N atoms there are present at positions ortho- or para- to the azido group.

In the second place the first ring-fission step can result in the formation of an intermediate compound which then undergoes a cyclisation to yield a different ring-system. Thus n.m.r. evidence shows the presence of 67% of 6-azidopyrido[4,3-d]tetrazolo[1,5-b]pyridazine (120) and 33% of the isomeric 6-azidopyrido[3,4-d]tetrazolo[1,5-b]pyridazine (121) in solution in dimethyl sulphoxide at room temperature[200].

It is not essential for the two open chain groups to be the same; the hydrazone (122) may be converted into the azide (123) by treatment with polyphosphoric acid and the reaction reversed with hydrochloric acid[201].

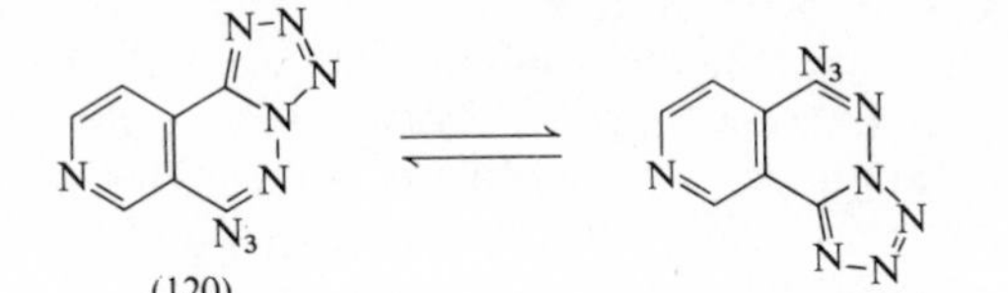

(120)   (121)

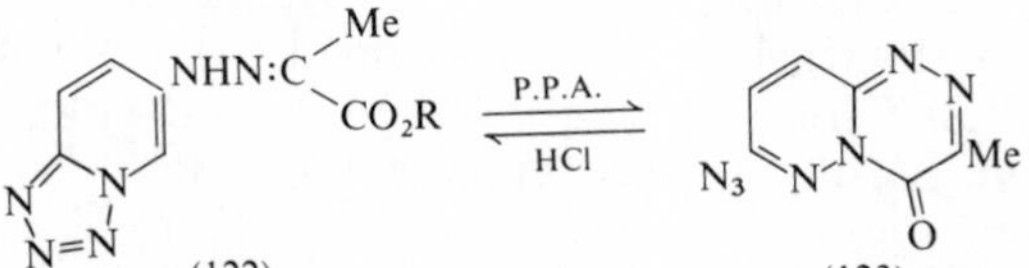

(122)   (123)

A number of workers[202-204] have investigated Dimroth rearrangements in polyazaheteroaromatic systems and one careful investigation has been made of the various electronic and steric factors which influence these reactions. It has been shown for example that imidazo[1,2-a]pyridines (124) cannot be rearranged but that imidazodiazines (125)–(127) can, and that the most susceptible were the imidazo[1,2-c]pyrimidines (127)[202]. Overall pseudo-first-order kinetics have been established by kinetic determinations followed by n.m.r. studies. It has been suggested that the rearrangement

(124)   (125)   (126)   (127)

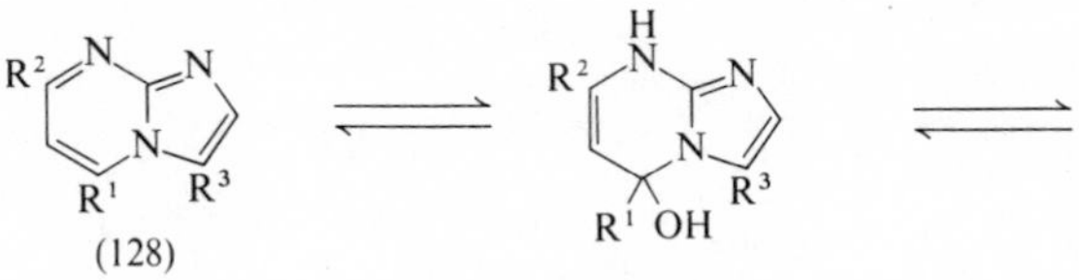

(128)

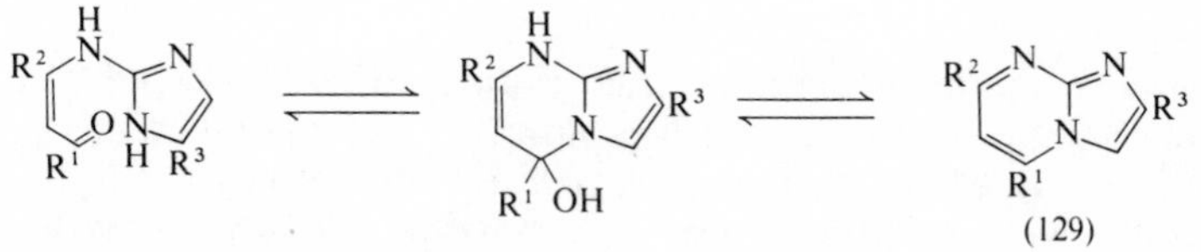

(129)

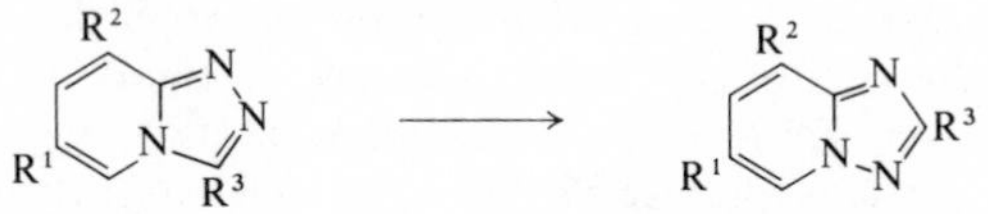

(130)   (131)

is a rather complex sequence of several steps (128) ——→ (129) involving two fundamental phenomena: covalent hydration at the position adjacent to the N-bridgehead atom and ring-chain tautomerism. Tautomeric structures can be drawn for each of the proposed intermediates[202]. Two independent schools have given examples of the first isomerisations of s-triazolo[3,4-a]-isoquinolines to s-triazolo[5,1-a]isoquinolines[203, 204] and the presence of a suitably-orientated acraldehyde grouping has been shown to facilitate the thermally-induced ring opening and rearrangement of vic-triazolo[1,5-a]-pyridines[205]. The importance of the initial nucleophilic attack in many of these reactions is neatly demonstrated by the successful ring opening and rearrangement of the s-triazolo[4,3-a]pyridine (130) to the s-triazolo[1,5-a]-pyridine (131) when the group R^1, para to the site of ring opening, was nitro and the failure of the reaction when this same R^1 group was amino[101].

The third group of ring-opening reactions comprises those in which an external reagent, usually a nucleophile, initiates an irreversible reaction. Such reagents are sodium hydroxide[206], ammonia[207], hydrazine[208], aniline[209], lithium aluminium hydride[195], sodium borohydride[193], resorcinol[210] and 2-naphthol[68]. Benzotriazines and related compounds can act as 'masked' diazonium compounds, thus the conversion of the s-triazino[1,2-c] [1,2,3]-benzotriazine (132) into the azonaphthol (133) proceeds via a diazo compound and the conversion of the s-triazolo[5,1-a]pyridine (134) into the pyridyl-triazole (135) involves a similar N—N cleavage initiated by the external nucleophile[68].

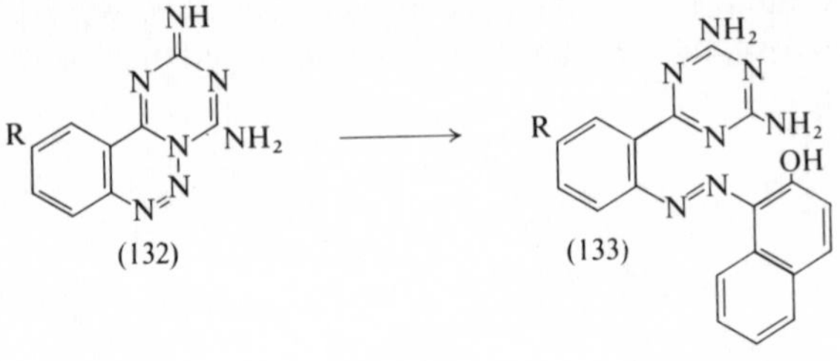

(132)   (133)

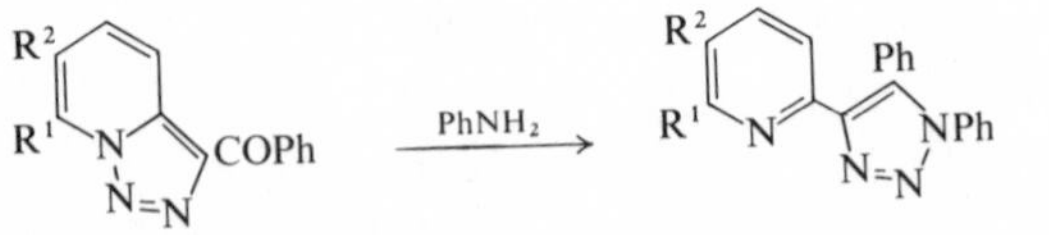

(134)   PhNH₂   (135)

Pyridopyrimidines, pteridines and purines are all susceptible to attack by nucleophiles at the 2- and 4-positions of the pyrimidine ring. Consequently, many of these compounds yield di- and tetra-hydro derivatives on treatment with metal hydrides. 3-Arylpyrido[3,4-d]pyrimidin-4(3H)-ones (136) are converted into 3-aminopyridines (138) under extremely mild conditions on treatment with lithium aluminium hydride and it is suggested that the delocalisation of the negative charge on the N atom (137) either to the aryl group or additionally to an adjacent C=O group ($R^1R^2 = O$) accelerates the cleavage process[195]. Adenine (139) has been shown to undergo ring opening to the ethoxycarbonylamidine (140) on treatment with diethyl

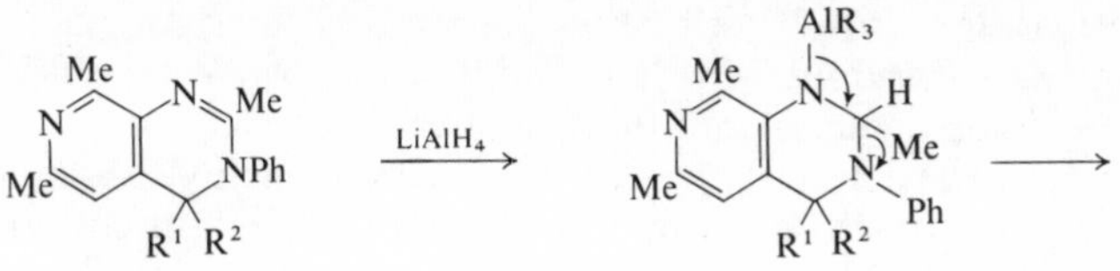

(136)

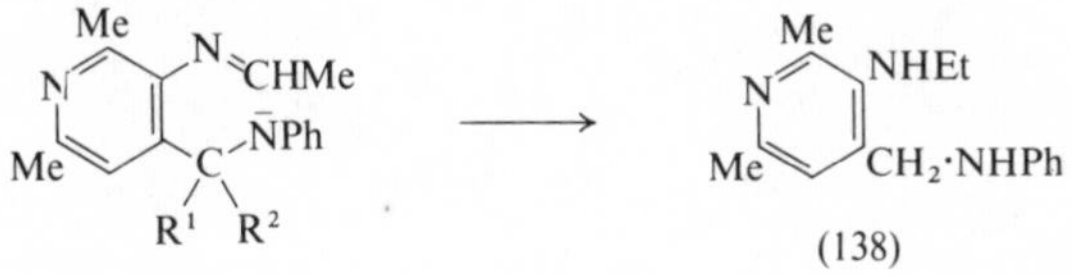

(137)

(138)

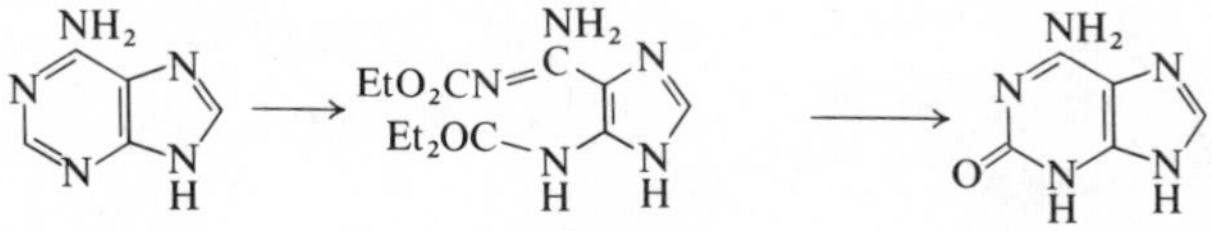

(139)   (140)   (141)

pyrocarbonate and treatment of the latter with ammonia yields isoguanine (141)[211].

6.7 PHOTOCHEMISTRY

Photochemically-induced reactions have rarely been used in polyazahetero-aromatic compounds. Photodealkylation of 8-hydroxyalkylpurines by irradiation at > 290 nm has been shown to yield caffeine, adenines and guanosines[212]. This type of 8-hydroxyalkylpurine has been shown to possess similar chromatographic mobility to the products obtained by sensitised or direct ultraviolet light irradiation of DNA in the presence of 2-propanol.

Some heterocyclic acrylaldehydes have been converted into the corresponding propionic acids by irradiation of their solutions in methanol[213], and photolysis of 1-benzotriazolylbenziminazoles yielded benzimidazo-[1,2-a]benzimidazoles[214].

6.8 BIOLOGICAL PROPERTIES

The stimulus for much of the work on polyazaheteroaromatic compounds has been the search for new compounds of chemotherapeutic or physiological value, and the investigation of the mode of action of drugs at a molecular level.

The ideal anti-malarial drug has yet to be found and a considerable amount of work has been reported in this field in the 2 years under review. The naphthyridines in particular have been investigated by several groups of workers and 7-chloro-4-(4-diethylamino-1-methylbutlyamino)-1,5-naphthy-ridine (142)[215], 8-chloro-4-(2'-N,N-dibutylamino-1'-hydroxyethyl)benzo [h]-1,6-naphthyridine (143)[20], and 4-(3-diethylamino-1-methylpropylamino)-

1,6-naphthyridine (144)[216] all show interesting activity in preliminary screening tests. Pyrido[2,3-b]-[218] and pyrido[3,4-b]-pyrazines[219] have also been prepared as potential anti-malarial agents.

The high antibacterial activity of 1-ethyl-7-methyl-1,8-naphthyridin-4(1H)-one-3-carboxylic acid (nalidixic acid) (145) continues to stimulate

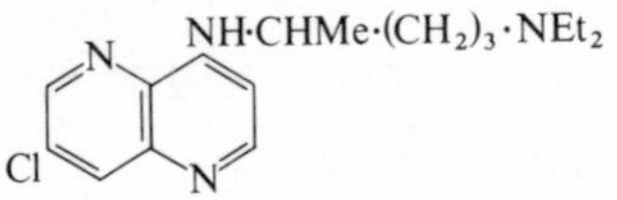

(142)

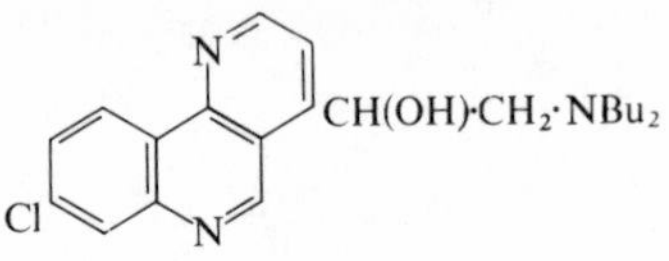

(143)

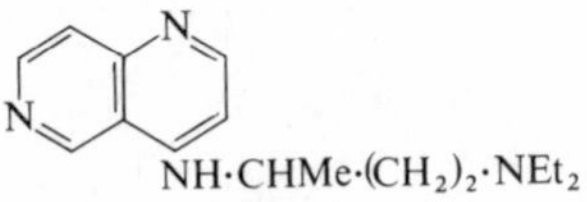

(144)

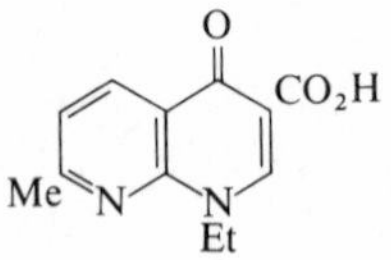

(145)

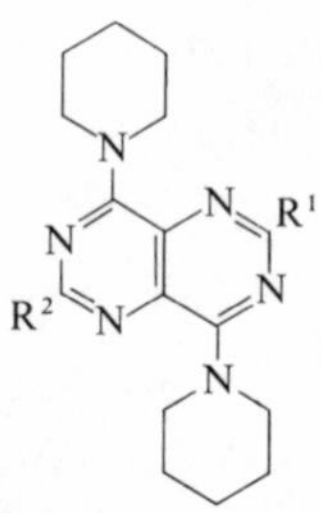

(146)

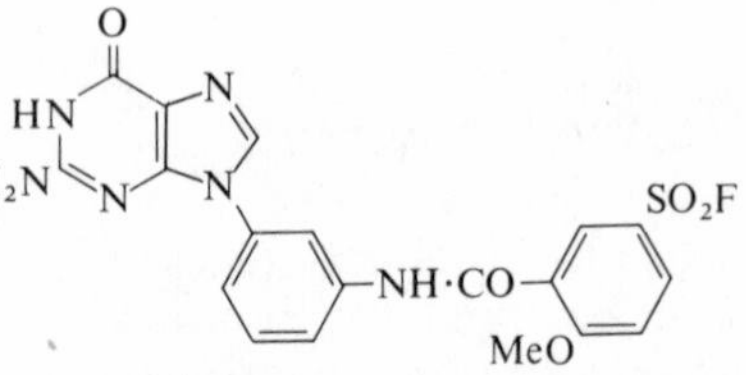

(147)

the preparation of structurally related compounds and derivatives many of which are recorded in the patent literature[219]. Pyrimido[5,4-d]pyrimidines are known inhibitors of platelet aggregation[220] and a large number of structurally similar 2,6-disubstituted derivatives of 4,8-diperidino-pyrimido-[5,4-d]pyrimidine (146) have been prepared, several of which with sulphur-containing 2- and 6-substituents are claimed to be coronary vasodilator's[221]. The search for enzyme inhibitors which structurally resemble the known purine metabolites still continues to attract considerable attention[222]. In his 185th paper on *Irreversible Enzyme Inhibitors*, B. R. Baker has demonstrated the irreversible inhibition of Walker 256 guanine deaminase by a number of 9-phenylguanines bearing terminal sulphonylfluoride groups (e.g. 147)[223].

The literature continues to contain very many reports of biological activity which may or may not be sustained on further evaluation. Of particular interest and worthy of further study, however, are claims of antihypertensive

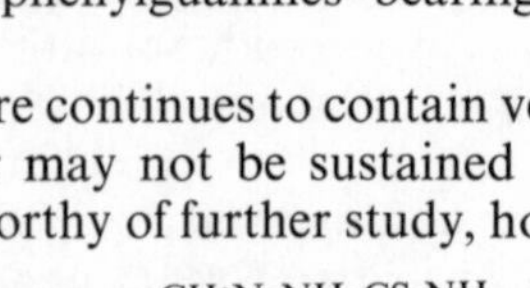

(148)

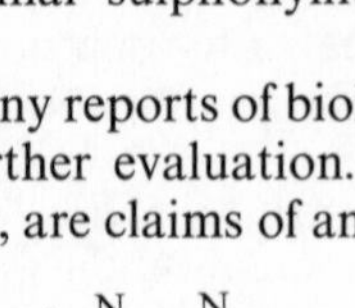

(149)

activity in the thiosemicarbazone (148)[224], diuretic activity in pyrido[2,3-d]
pyrimidines[225] and pyrimido[4,5-d]pyridazines[226] and hypoglycemic activity
in pyrido[1,2-a]pyrimidines[227]. As the knowledge of the mode of action of
existing drugs increases, the search for new compounds becomes less
random and Weinstock and his co-workers have evaluated the factors
necessary for binding to receptors in a series of natriarctic 2,4,7-triamino-
pteridines (149)[228].

References

1. Kurakami, M., Kawahara, S., Inukai, N., Ishida, S., Imai, K. and Ozasa, T. (1971).
   Japan Pat., 71 12, 735
2. Clark, J. and Yates, F. S. (1971). *J. Chem. Soc. C*, 2475
3. Shono, T. and Matsumoto, J. (1971). *Chem. and Pharm. Bull. Japan*, **19**, 1426
4. Scheiffele, E. (1971). *Ger. Offen.*, 1 950 076
5. Hinshaw, B. C., Gerster, J. F., Robins, R. K. and Townsend, L. B. (1970). *J. Org. Chem.*,
   35, 236
6. Yoneda, F., Nishigaki, S., Mizushima, N. and Takahashi, H. (1971). *J. Med. Chem.*, **14**,
   638
7 Paudler, W. W. and Kress, T. J. (1970). *Advan. Heterocycl. Chem.*, **11**, 123
8. Pomorski, J. (1970). *Wind. Chem.*, **24**, 773
9. Yoneda, F. (1970). *Kagaku No Ryoiki*, **24**, 1077
10. Lister, J. H. (1971). *Purines*, (A. Weissberger, editor). Vol. 24, 1 (New York: Inter-
    science Publishers)
11. Rowe, J. J. M., Hinton, J. and Rowe, K. L. (1970). *Chem. Rev.*, **70**, 1
12. Izatt, R. M., Christensen, J. J. and Rytteng, J. H. (1971). *Chem. Rev.*, **71**, 439
13. Lund, H. (1970). *Advan. Heterocycl. Chem.*, **12**, 213
14. Grundman, C. (1970). *Synthesis*, **7**, 344
15. Landquist, J. K. and Jones, W. G. M. (1970). *Brit. Pat.*, 1 208 279
16. Temple, C., Laseter, A. G. and Montgomery, J. A. (1970). *J. Heterocycl. Chem.*, **7**, 1219
17. Heindel, N. D. and Fine, S. A. (1970). *J. Med. Chem.*, **13**, 760
18. Stein, R. G., Biel, J. H. and Singh, T. (1970). *J. Med. Chem.*, **13**, 153
19. Khan, M. A. and Lynch, B. M. (1970). *J. Heterocycl. Chem.*, **7**, 247
20. Roseman, K. A., Gould, M. M., Linfield, W. M. and Edwards, B. E. (1970). *J. Med.
    Chem.*, **13**, 230
22. Swirska, A., Piechaczek, T. and Nantka-Namirski, P. (1970). *Pol. Pat.*, **60**, 795
23. Hoehn, H. and Chasin, M. (1970). *Ger. Offen.*, 2 028 869
24. Flitsch, W. and Kraemer, U. (1970). *Liebigs Ann. Chem.*, **735**, 35
25. Harper, J. F. and Wibberley, D. G. (1971). *J. Chem. Soc. C*, 2991
26. Hamada, Y. and Takeuchi, I. (1971). *Chem. and Pharm. Bull. Japan*, **19**, 1857
27. Mullock, E. B., Searby, R. and Suschitzky, H. (1970). *J. Chem. Soc. C*, 829
28. Carboni, S., Da Settimo, A., Ferrarini, P. L. and Tonetti, I. (1971). *Gazz. Chim. Ital.*, **101**,
    129
30. Checchi, S., Auzzi, G. (1970). *Ann. Chim. (Rome)*, **60**, 225
31. Reimlinger, H., Jacquier, R. and Daunis, J. (1971). *Chem. Ber.*, **104**, 2702
32. Chinoin Gyogyszer es Vegyeszeti Termekek Gyara Rt (1970). *Brit. Pat.*, 1 209 946
33. Pollak, A., Stanovnik, B. and Tisla, M. (1970). *Chimia*, **24**, 418
34. Pankina, Z. A. and Shchukina, M. N. (1970). *Khim. Geteroltsiklich Soedin.*, 245
35. Muhlstadt, M., Krausmann, H. and Fischer, G. (1970). *J. Prakt. Chem.*, **312**, 254
36. Stanovnik, B. and Tisler, M. (1970). *Monatsh. Chem.*, **101**, 303
37. Duncan, R. L. (1971). *Ger. Offen.*, 2 051 961
38. Hans, O. (1970). *U.S. Pat.*, 3 531 482
40. Bowie, R. A. (1970). *Chem. Commun.*, 565
41. Reimlinger, H. (1971). *Chem. Ber.*, **104**, 2801
43. Lister, J. M., Manners, D. S. and Timmis, G. M. (1970). *J. Chem. Soc. C*, 1313
44. O'Brien, D. E., Weinstock, L. T. and Cheng, C. C. (1970). *J. Heterocycl. Chem.*, **7**, 99
45. March, L. C. and Joullie, M. M. (1970). *J. Heterocycl. Chem.*, **7**, 395

46. Vinot, N. (1971). *Bull. Soc. Chim. Fr.*, 2708
47. Winterfeld, K. and Wildersohn, M. (1970). *Arch. Pharm.*, **30**, 44
48. Weinstock, J. (1970). *U.S. Pat.*, 3 505 329
49. Baranova, N. V., Sheinkman, A. K. and Kost, A. N. (1970). *Khim. Geterotsiklich Soedin.*, 1148
50. Clark, J. and Ramsden, T. (1971). *J. Chem. Soc. C*, 1942
51. Yurugi, S. and Kikuchi, S. (1971). *Ger. Offen.*, 2 046 577
52. Matsuara, I. (1970). *Japan pat.*, 7 033 185
53. Bastide, J. and Lamatre, J. (1971). *Bull. Soc. Chim. Fr.*, 1336
54. Tennant, G. and Sutherland, D. R. (1971). *J. Chem. Soc.*, 4, 706
55. Cresswell, R. M., Mentha, J. W. and Seaman, R. (1970). *Ger. Offen.*, 1 904 894
56. Capuano, L., Ebner, W. and Schrepfer, J. (1970). *Chem. Ber.*, **103**, 82
57. Temple, C., Kussner, C. L. and Montgomery, J. A. (1971). *J. Org. Chem.*, **36**, 2974
58. Temple, C., Kussner, C. L. and Montgomery, J. A. (1971). *J. Org. Chem.*, **36**, 3502
59. Shvedov, V. I., Altukhova, L. B. and Grinev, A. N. (1970). *Khim. Geterotsikl Soedin.*, 1048
60. Glover, E. E. and Adamson, J. (1971). *J. Chem. Soc. C*, 2748
61. Schram, K. H. and Townsend, L. B. (1971). *Tetrahedron. Lett.*, 4757
62. George, T., Mehta, D. V. and Dabholkar, D. A. (1971). *J. Org. Chem.*, **36**, 2192
63. Townsend, L. B., Panzica, R. P., Rousseau, R. J., Reddick, S. M. and Robins, R. K. (1970). *J. Org. Chem.*, 631
64. Taylor, E. C. and Evans, B. E. (1971). *J. Chem. Soc. C*, 189
65. Youssefyeh, R. D. and Kalums, A. (1970). *Chem. Commun.*, 1371
66. Parish, W. W. and Broadbent, H. S. (1971). *J. Heterocycl. Chem.*, **8**, 527
67. Townsend, L. B., Long, R. A. and Gerster, J. F. (1970). *J. Heterocycl. Chem.*, **7**, 863
68. Stevens, M. F. G. and Mackenzie, S. M. (1970). *J. Chem. Soc. C*, 2298
69. Stevens, M. F. G. and Stevens, H. N. E. (1970). *J. Chem. Soc. C*, 2289
70. Spirio, V. and Plescia, S. (1971). *Ann. Chim. (Rome)*, **61**, 206
71. Lewis, A. and Shepherd, R. G. (1971). *J. Heterocycl. Chem.* **8**, 47
72. Smith, P. A. S. and Ahmad, Y. (1971). *J. Org. Chem.*, **36**, 2972
73. Albert, A. and Ohta, K. (1970). *J. Chem. Soc. C*, 1540
74. Lewis, A. and Shepherd, R. G. (1971). *J. Heterocycl. Chem.*, **8**, 41
75. Brown, D. J. and Sugimoto, T. (1971). *Aust. J. Chem.*, **24**, 633
76. Cheeseman, G. W. H. and Rafiq, M. (1971). *J. Chem. Soc. C*, 2732
78. Brown, D. J. and Sugimoto. T. (1971). *J. Chem. Soc. C*, 2616
79. Jacobsen, N. W. and Dickinson, R. G. (1970). *Chem. Commun.*, 1719
80. Perveer, F. Y. and Afonina, I. I. (1971). *Zh. Org. Khim.*, **7**, 420
81. Adembri, G., DeSio, F., Nesi, R. and Scotton, M. (1970). *J. Chem. Soc. C*, 1536
82. Yoneda, F., Kanahori, M., Ogiwara, K. and Nishigaki, S. (1970). *J. Heterocycl. Chem.*, **7**, 1443
83. Dahn, H. and Fumeaux, J. P. (1970). *Bull. Soc. Vaudoise Sci. Natur.*, **70**, 313
84. Elguero, J., Jacquier, R., Haq, A. and Tarrago, G. (1970). *Bull. Soc. Chim. Fr.*, 3136
85. Winkley, M. W. (1970). *J. Chem. Soc. C*, 1869
86. Robinson, B. (1970). *Chem. Rev.*, **69**, 227
87. Parrick, J. and Kelly, A. H. (1970). *J. Chem. Soc. C*, 305
88. Sugiura, S., Inoue, S. and Goto, T. (1970). *Yakugaku Zasshi*, **90**, 423
89. Almirante, L., Mugraini, A., De Toma, N. and Murmann, W. (1971). *Boll. Chim. Farm.*, **110**, 322
90. Shaw, G. and Smallwood, B. M. (1970). *J. Chem. Soc. C*, 2206
91. Aryvsina, V. M. and Ordzhonikidze, M. N. (1970). *Khim. Geterotsikl. Soedin*, 525
92. Heinisch, L. (1970). *Z. Chem.*, **10**, 188
93. Pentimalli, L. and Milani, G. (1970). *Gazz. Chim. Ital.*, **100**, 1106
94. Portnoy, S. (1970). *J. Heterocycl. Chem.*, **7**, 703
95. Kreutzberger, A. (1970). *Pharmazie*, **25**, 460
96. Yanai, M., Kurashi, T., Kinoshita, T. and Nishimura, M. (1970). *J. Heterocycl. Chem.*, **7**, 465
97. Reimlinger, H., Lingier, W. R. F., Vaudewalle, J. J. M. and Merényi, R. (1971). *Chem. Ber.*, **104**, 3947
98. Postovskii, I. and Golomolzin, B. V. (1970). *Khim. Geterotsikl. Soedin*, 100
99. Yoshina, S. and Maeba, I. (1970). *Chem. and Pharm. Bull.*, **18**, 842
100. Robba, M., Dore, G. and Bonhomme, M. (1970). *Compt. Rend. Acad. Sci. Ser. C*, **271**, 1090

101. Potts, K. T. and Surapaneni, C. R. (1970). *J. Heterocycl. Chem.* **7**, 1019
102. Sauer, W. and Henseke, G. (1970). *Z. Chem.* **10**, 381
103. Doleschall, G. and Lempert, K. (1970). *Acta Chim. Acad. Sci. Hung.*, **64**, 369
104. Kutsuma, T., Sato, K., Chida, T. and Tsukuno, Y. (1970). *Yakugaku Zasshi*, **90**, 251
105. Brown, D. M. and Sorolla, A. G. (1971). *J. Chem. Soc. C*, 126
106. Bergmann, F. G., Kleiner, M. A. and Rashi, M. (1970). *Brit. Pat.*, 1 201 997
107. Cherkasov, V. M. and Kurilenko, L. K. (1970). *Khim. Geterotsikl. Soedin.*, 1432
108. Dunn, G. L. and Berthold, R. V. (1970). *U.S. Patent.*, 3 536 711
109. Schaeffer, H. J., Odin, E. and Bittner, S. (1971). *J. Pharm. Sci.*, **60**, 1184
110. Rogers, G. T. and Ulbricht, T. L. V. (1971). *J. Chem. Soc. C*, 2364
111. Lespagnol, A., Debaert, M. and Minardvaillant, N. (1970). *Chim. Thier*, **5**, 321
112. Vogt, B. R. (1971). *U.S. Pat.*, 3 590 045
113. Kempter, F. E., Rokos, H. and Pfleiderer, W. (1970). *Chem. Ber.*, **103**, 885
114. Pfleiderer, W. and Kempter, F. E. (1970). *Chem. Ber.*, **103**, 908
115. Scheiffele, E. (1970). *Ger. Pat.*, 1 950 075
116. Anderson, W. K. and Friedman, A. E. (1971). *Can. J. Chem.*, **49**, 668
117. Kobe, J., Stanovnik, B. and Tisler, M. (1970). *Monatsh. fur. Chemie.*, **101**, 724
118. Harper, J. F. and Wibberley, D. G. (1971). *J. Chem. Soc. C*, 2985
119. Brekiesz-Lewandowska, B. and Talik, Z. (1970). *Rocz. Chem.*, **44**, 69
120. Tulchinskaya, L. S., Klebanova, V. D., Polyakova, N. A., Dvoryantseva, G. G. and Berezovskii, V. M. (1970). *Zh. Obshch. Khim.*, **40**, 868
121. Kotva, R., Semonsky, M., Vachek, J. and Jelinek, V. (1970). *Coll. Czech. Chem. Commun.*, **35**, 1610
122. Reimlinger, H. (1970). *Chem. Ber.*, **103**, 1900
123. Reimlinger, H. and Peiren, M. A. (1971). *Chem. Ber.*, **104**, 2237
124. Carboni, S., Dasettimo, A., Ferrarini, P. L. and Ciantelli, P. L. (1970). *J. Heterocycl. Chem.*, **7**, 1037
125. Reimlinger, H., Vandewalle, J. J. M., King, G. S. D., Lingier, W. R. F. and Meréngi, R. (1970). *Chem. Ber.*, **103**, 1918
126. Gladstone, W. A. F., Aylward, J. B. and Norman, R. O. C. (1969). *J. Chem. Soc. C*, 2587
127. Kobe, J., Stanovnik, B. and Tisler, M. (1970). *Tetrahedron*, **26**, 3357
128. Scott, F. L. and O'Mahony, T. A. F. (1970). *Tetrahedron Lett.*, 1841
129. Reimlinger, H., Lingier, W. R. F., Vandewalle, J. J. M. and Merényi, R. (1971). *Chem. Ber.*, **104**, 3965
130. Kreutzberger, A. and Schuecker, R. (1971). *Tetrahedron*, **27**, 3247
131. Zellstoffabrik, W. (1971). *Brit. Pat.*, 1 230 289
132. Middleton, W. J. and Metzger, D. (1970). *J. Org. Chem.*, **35**, 3985
133. Miyaki, M. and Shimizu, B. (1970). *Chem. Pharm. Bull. Japan*, **18**, 1446
134. Dvoryantseva, G. G., U'yanova, T. N., Syrova, G. P., Sheinker, Y. N., Aryuzina, V. M., Sycheva, T. P. and Shchukina, M. N. (1970). *Teov. Eksp. Khim.*, **6**, 23
135. Bergmann, F., Levene, L., Neiman, Z. and Brown, D. J. (1970). *Biochim. Biophys. Acta*, **222**, 191
136. Konrad, G. and Pfleiderer, W. (1970). *Chem. Ber.*, **103**, 722
137. Skoog, F., Burrows, W. J., Armstrong, D. J., Kaminek, M., Bock, R. M., Hecht, S. M., Danman, L. G., Leonard, N. J. and Occolowitz, J. (1970). *Biochemistry*, **9**, 1867
138. Ikehara, M., Uesugi, S. and Yasumoto, M. (1970). *J. Amer. Chem. Soc.*, **92**, 4735
139. Reimlinger, H., Jacquier, R. and Daunis, J. (1971). *Chem. Ber.* **104**, 2702
140. Stamm, H. (1970). *Justus Liebigs Ann. Chem.*, **731**, 174
141. Sugiura, S., Kakoie, H., Inoae, S. and Goto, T. (1970). *Yakugaku. Zasshi*, **90**, 436
142. Sarma, R. H. and Kaplan, N. O. (1970). *Biochemistry*, **9**, 539
143. Sarma, R. H., Moore, M. and Kaplan, N. O. (1970). *Biochemistry*, **9**, 549
144. Kan, L. S. and Li, N. C. (1970). *J. Amer. Chem. Soc.* **92**, 281
145. Kan, L. S. and Li, N. C. (1970). *J. Amer. Chem. Soc.*, **92**, 4823
146. Iwamura, H., Leonard, N. J. and Eisinger, J. (1970). *Proc. Nat. Acad. Sci. U.S.*, **65**, 1025
147. Mcandless, J. M. and Stewart, R. (1970). *Can. J. Chem.*, **48**, 263
148. Beach, R. L. and Plaut, G. W. E. (1970). *Biochemistry*, **9**, 760
149. Clark, J. (1971). *Org. Mass. Spectrom.*, **5**, 447, 913, 1419
150. Haug, P. and Urushibara, T. (1970). *Org. Mass. Spectrom.*, **3**, 1365
151. Shaw, S. J. Desiderio, D. M., Tsuboyama, K. and McCloskey, J. A. (1970). *J. Amer. Chem. Soc.*, **92**, 2510

152. Dolhun, J. J. and Wiebers, J. L. (1970). *Org. Mass. Spectrom.*, **3**, 669
153. Heiss, J., Klaus-Peter, Z. and Voelter, W. (1970). *Org. Mass. Spectrom.*, **3**, 181
154. Potts, K. T., Brugel, E. and Singl, V. P. (1971). *Org. Mass. Spectrom.*, **5**, 1
155. Kennard, O., Isaacs, N. W., Coppola, J. C., Kirby, A. J., Warren, S., Motherwell, W. D. S., Watson, D. G., Wampler, D. L., Chenery, D. H., Larson, A. C., Kerr, K. A. and Rivadisanseverino, L. (1970). *Nature (London)*, **225**, 333
156. Simundza, G., Sakore, T. D. and Sobell, H. M. (1970). *J. Molec. Biol.*, **48**, 263
157. Craven, B. M. and Gartland, G. L. (1970). *J. Pharm. Sci.*, **59**, 1666
158. Nakand, N. I. and Igarashi, S. J. (1970). *Biochemistry*, **9**, 577
159. Pochun, F. and Michelson, A. M. (1970). *Compt. Rend. Acad. Sci. Ser. D,* **270**, 1829
160. Muller, F., Eriksson, L. E. G. and Ehrenberg, A. (1970). *Eur. J. Biochem.*, **12**, 93
161. Paudler, W. W., Pokorny, D. J. and Cornrich, S. J. (1970). *J. Heterocycl. Chem.*, **7**, 291
162. Gines-Sorolla, A., Gryte, C., Cox, M. L. and Parham, J. C. (1971). *J. Org. Chem.*, **36**, 1228
163. Parnham, J. C., Birdsall, N. J. M., Lee, T-C and Delia, T. J. (1971). *J. Org. Chem.*, **36**, 2635
164. Fujii, T., Wu, C. and Itaya, T. (1971). *Chem. Pharm. Bull.*, **19**, 1368
165. Purham, J. C., Winn, T. G. and Brown, G. B. (1971). *J. Org. Chem.*, **36**, 2639
166. Robins, M. J., Khwaja, T. A. and Robins, R. K. (1970). *J. Org. Chem.*, **35**, 636
167. Gracheva, I. N., Veinberg, A. Ya. and Samokhvalov, G. I. (1971). *Zh. Obshch. Khim.*, **41**, 1376
167a. Howard, J., Gurwara, S., Vince, R. and Bittmer, S. (1971). *J. Med. Chem.*, **14**, 367
168. Tolman, R. L., Tolman, G. L., Robbins, R. K. and Townsend, L. B. (1970). *J. Heterocycl. Chem.*, **7**, 799
169. Bariana, D. S. (1971). *J. Med. Chem.*, **14**, 543
170. Roos, K. B. and Salemink, C. A. (1971). *Rec. Trav. Chim.*, **90**, 1181
171. Pentimalli, L. and Passalacqua, V. (1970). *Gazz. Chim. Ital.*, **100**, 110
172. Thompson, B. B., Larocca, J. P. and Gibson, C. A. (1971). *J. Pharm. Sci.*, **60**, 74
173. Irwin, W. J., Wibberley, D. G. and Cooper, G. (1971). *J. Chem. Soc. C*, 3870
174. Kazymov, A. V., Shchelkina, L. P. and Kabirova, N. G. (1971). *Khim. Geterotsikl. Soedin.*, **7**, 279
175. Zupan, M., Stanovnik, B. and Tisler, M. (1971). *J. Heterocycl. Chem.*, **8**, 1
176. Yakhontov, L. N. and Azimov, V. A. (1970). *Khim. Geterotsiklich. Soedin.*, **6**, 32
177. Terent'ev, A. P., Gracheva, R. A., Kotov, A. L. and Solor'eva, L. D. (1970). *Regal. Rostra. Rast. Khim. Sredstvami.*, **5**, and Giner-Sorolla, A. and Burcherol, J. H. (1970). *J. Med. Chem.*, **14**, 816
178. Giner-Sorolla, A. (1971). *J. Heterocycl. Chem.*, **8**, 651 and Kawashima, H. and Kamashiro, I. (1970). *Japan. Pat.*, 7 028 783
179. Lister, H. and Kiburio, J. (1971). *J. Chem. Soc. C*, 1587
180. de Roos, K. B. and Salemink, C. A. (1971). *Rec. Trav. Chim.*, **90**, 1166
181. Paudler, W. W., Porkorny, D. J. and Good, J. J. (1971). *J. Heterocycl. Chem.*, **8**, 37
182. Schmidt, C. L. and Townsend, L. B. (1970). *J. Heterocycl. Chem.*, **7**, 715
183. Arutyunyan, E. A., Gunar, V. I. and Zav'yalov, S. I. (1970). *Izv. Akad. Nauk. SSSR, Ser. Khim.*, 953
184. Hamada, Y., Takeuchi, I. and Hirota, M. (1971). *Chem. Pharm. Bull. Japan*, **19**, 1751
185. Nozaki, H. Yamamoto, Y. and Noyori, R. (1966). *Tetrahedron Lett.*, 1123 and Russell, G. A. and Weiner, S. A. (1966). *J. Org. Chem.*, **31**, 248
185a. Kovacs, G., Meszaros, Z., Bodnar, J. and Kadas, I. (1971). *Hung. Teljes.*, 1161
186. Brown, E. V. and Plasz, A. C. (1970). *J. Heterocycl. Chem.*, **7**, 593
187. Kim, D. H. and Santilli, A. A. (1971). *Tetrahedron Lett.*, 2441
188. Potts, K. T. and Sorm, M. (1971). *J. Org. Chem.*, **36**, 8
189. Sasaki, T., Kanematsu, K. and Vchide, M. (1971). *Bull. Soc. Chem. Japan.*, **44**, 858
190. Taylor, E. C., Thompson, M. J., Perlmann, K., Mengel, R. and Pfleiderer, W. (1971). *J. Org. Chem.*, **36**, 4012
191. Viscontini, M., Fraterschroeder, M., Cogoligreuter, M. and Argentini, M. (1970). *Helv. Chim. Acta*, **53**, 1434
192. Pfleiderer, W. (1971). *Justus Liebigs Ann. Chem.*, 747
193. Neiman, Z. (1970). *J. Chem. Soc. C*, 91
194. Pochon, F., Pascal, Y., Pitha, P. and Michelson, A. M. (1970). *Biochim. Biophys. Acta*, **213**, 273

195. Gelling, I. R. and Wibberley, D. G. (1971). *J. Chem. Soc. C*, 780
196. Clark, J. and Yates, F. S. (1971). *J. Chem. Soc. C*, 371
197. Clark, J. and Yates, F. S. (1971). *J. Chem. Soc. C*, 2278
198. Brown, D. J. and Sugimoto, T. (1970). *J. Chem. Soc. C*, **19**, 2661
199. Ferarini, P. L. (1971). *Ann. Chim. (Rome)*, **61**, 318
200. Stanovnik, B., Tisler, M. and Stefanov, B. (1971). *J. Org. Chem.*, **36**, 3812
201. Stanovnik, B., Tisler, M., Ceglar, M. and Bah, V. (1970). *J. Org. Chem.*, **35**, 1138
202. Gurret, P., Jacquier, R. and Maury, G. (1971). *J. Heterocycl. Chem.*, **8**, 643
203. Hoogzand, C. (1971). *Rec. Trav. Chim.*, **90**, 1225
204. Reimlinger, H., Lingier, W. R. F. and Vandewalle, J. J. M. (1971). *Chem. Ber.*, **104**, 3976
205. Davies, L. S. and Jones, G. (1970). *Tetrahedron Lett.* 1049
206. Golomolzin, B. V. and Postovskii, I. Y. (1970). *Khim. Geterotsiklich. Soedin.*,281
207. Landquist, J. K. (1971). *J. Chem. Soc. C*, 2735
208. Clark, J. and Smith, C. (1971). *J. Chem. Soc. C*, 1948
209. Eistert, B. and Endres, E. (1970). *Justus Liebigs. Ann. Chem.*, **734**, 56
210. Stanovnik, B. and Tisler, M. (1971). *J. Heterocycl. Chem.*, **8**, 785
211. Leonard, N. J., Mcdonald, J. J. and Reichmann, M. E. (1970). *Proc. Nat. Acad. Sci. U.S.*, **67**, 93
212. Elad, D., Rosenthal, I., Salomon, J. and Sperling, J. (1971). *Chem. Commun.*, 49
213. Jones, G. and Davies, L. S. (1970). *Tetrahedron Lett.*, 3475
214. Hubert, A. J. and Reimlinger, H. (1970). *Chem. Ber.*, **103**, 2828
215. McCaustland, D. J. and Cheng, C. C. (1970). *J. Heterocycl. Chem.*, **7**, 467
216. Paudler, W. W. (1970). *U.S. Clearinghouse Fed. Sci. Tech. Inform.*, **70**, 45
217. Temple, C., Rose, J. D., Elliott, R. D. and Montgomery, J. A. (1970). *J. Med. Chem.*, **13**, 853
218. Temple, C., Laseter, A. G., Rose, J. D. and Montgomery, J. A. (1970). *J. Heterocycl. Chem.*, **7**, 1195
219. Uglesic, A. and Seiwerth, R. (1970). *Ger. Offen.*, 1 940 511 and Lesher, G. Y. (1970). *U.S. Pat.*, 3 516 994, and Naito, T., Oshima, Y., Domori, R., Nagasaki, S., Tanaka, Y. and Yoshimura, R. (1970). *Japan. Pat.*, 7 030 337
220. *Annual Reports in Medicinal Chemistry, 1970*, (1971) 64, Ed. C. K. Cain, Academic Press, New York & London
221. Murakami, M., Kanahara, S., Inukai, N., Ozasa, T., Ishida, S. and Imai, K. (1970). *Ger. Offen.*, 1 957 957, and Murakami, M., Kawahara, S., Ishida, S. and Shibanuma, T. (1970). *Japan Pat.*, 7 019 308, and Roch, J., Mueller, E., Narr, B. and Machleidt, H. (1970). *S. African Pat.*, 6 905 505
222. Montgomery, J. A. and Hewson, K. (1970). *J. Med. Chem.*, **13**, 427, and Schaeffer, H. J., Johnson, R. N., Odin, E. and Hansch, C. (1970). *J. Med. Chem.*, **13**, 452
223. Baker, B. R. and Hans-Ulrich, S. (1971). *J. Med. Chem.*, **14**, 802
224. Almirante, L., Mugnaini, A., DeToma, N., Gamba, A., Murmann, W. and Hidalgo, J. (1970). *J. Med. Chem.*, **13**, 1048
225. Wiedemann, F., Thiel, M., Stach, K., Roesch, E. and Hardebeck, K. (1971). *Ger. Offen.*, 1 962 057
226. Czuba, L. J. (1970). *Ann. Reports Med. Chem.*, 64
227. Gupta, C. M., Bhaduri, A. P., Khanna, N. M. and Mukherjee, S. K. (1971). *Indian J Chem.*, **9**, 201
e.g. Weinstock, J., Wilson, J. W., Wiebelhaus, V. D., Maas, A. R., Brennan, F. T. and
228. Sosnowski, G. (1968). *J. Med. Chem.*, **11**, 573, and Zins, G. R. (1970). *Ann. Reports Med. Chem.*, 89

Note: Reference numbers 21, 29, 42 and 77 are not cited.

7
Oxygen Heterocyclic Molecules

A. PELTER
University College of Swansea

7.1 INTRODUCTION

The reader will find that the following chapter differs somewhat in layout and emphasis from its fellows in this volume. The reasons for this are as follows. (a) The chapter attempts to cover the recent advances in knowledge of the chemical and physical properties of *all* oxygen heterocyclic compounds at any oxidation level. (b) Many oxygen heterocyclic compounds are natural products and frequently are of interest for reasons other than purely chemical ones (i.e. physiological, phytochemical, etc.). In view of this, biosynthetic considerations are of great interest and due weight must be given to *in vitro* experiments purporting to serve as models for reactions postulated to occur *in vivo*. (c) Perhaps owing to the emphasis on complex natural products very little fundamental work on the chemical properties of the ring system as such (e.g. radical reactions of pyrones) has been carried out.

It must be pointed out that there has been no attempt to list all the numerous flavanoids and coumarin derivatives characterised in the period under review. Only those compounds of special interest on chemical or other grounds will be mentioned.

7.2 FIVE-MEMBERED RING SYSTEMS CONTAINING ONE OXYGEN HETEROCYCLIC ATOM

7.2.1 Furans

The difficulties of preparing simple 3-substituted furans (1), a grouping possessed by many natural products has been commented on and the following general route investigated[1].

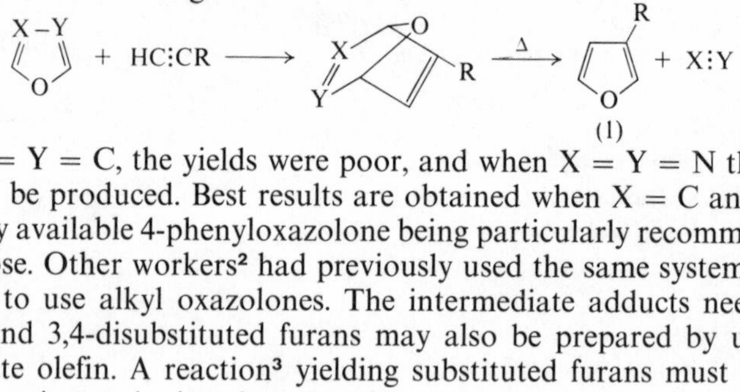

(1)

When $X = Y = C$, the yields were poor, and when $X = Y = N$ the adduct could not be produced. Best results are obtained when $X = C$ and $Y = N$, the readily available 4-phenyloxazolone being particularly recommended for the purpose. Other workers[2] had previously used the same system, but had preferred to use alkyl oxazolones. The intermediate adducts need not be isolated and 3,4-disubstituted furans may also be prepared by use of the appropriate olefin. A reaction[3] yielding substituted furans must involve a most interesting acyl migration, namely

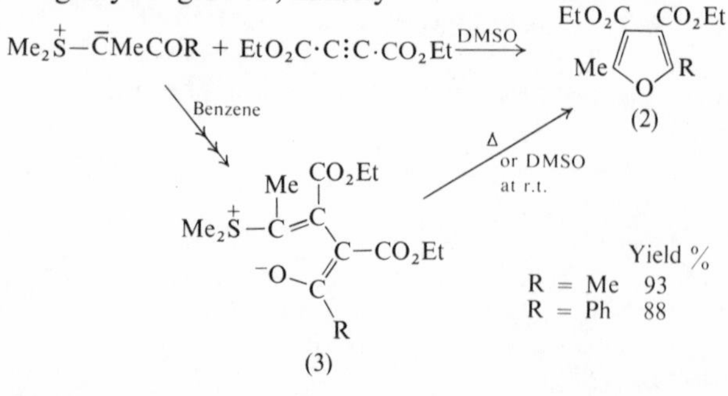

	Yield %
R = Me	93
R = Ph	88

(Ref. 3)

The ylid (3) was isolated when the reaction was carried out in benzene and on further heating or dissolution in dimethyl sulphoxide (DMSO) was converted into the furan (2). A not dissimilar reaction giving high yields of furans involves the reaction of an allenic salt with ketones[4]. The reaction as shown

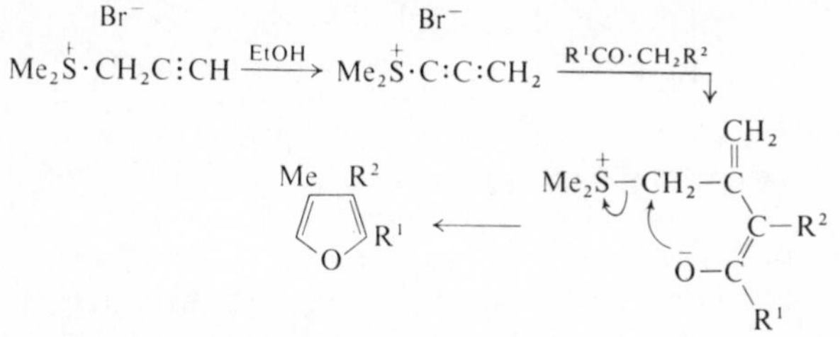

$$Me_2\overset{+}{S}\cdot CH_2C\vdots CH \xrightarrow{EtOH} Me_2\overset{+}{S}\cdot C\vdots C\vdots CH_2 \xrightarrow{R^1CO\cdot CH_2R^2}$$

is very versatile, and a further variation involves the use of sodio-derivatives of β-keto-esters. Furan may be derived from the pyrolysis of pyrone[5]. Coumarin similarly gives benzofuran in high yield, the reactions underlining the marked similarities between pyrolytic and electron impact processes. Isobenzofuran itself has been prepared and characterised[6]. A remarkably simple process can yield substituted iso-indoles and isobenzothiophenes as well as isobenzofurans[7].

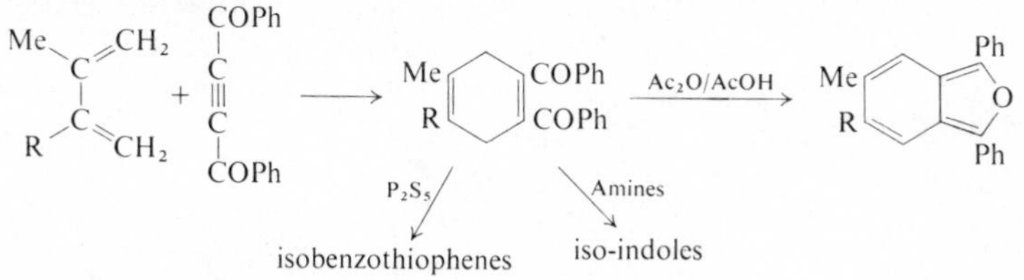

(Ref. 7)

1,3-Diphenylquinoxalino[2,3-c]furan has been prepared and is recommended as a stable but reactive diene[8]. The oxidation of a cyclobutadiene–metal complex is a novel route to a similar system[9].

Hiroaka[10] has been testing his general predictions[11] for the mechanism of the photolysis of furans. The light-induced rearrangement of 2-cyanofuran to the 3-isomer was shown to involve a cyclopropene aldehyde which was isolated as the methanol adduct. Warming induced ring expansion and loss of methanol completed the isomerisation. The mechanism differs completely from that involved in the superficially similar rearrangement (also light-induced) of N-methyl-2-cyanopyrrole to the 3-cyano derivative. This rearrangement proceeds through a cyclobutaneaziridine intermediate. The same scheme has been tested on 2,5-dimethylfuran. Once more an acyclo-propene is the first rearrangement product, and is trapped as an imine adduct which undergoes further rearrangement to a mixture of pyrroles[12]. The generality of the scheme is underlined in that 2,5-dimethylthiophen gives the same mixture of pyrroles on similar treatment[13]. Furthermore, the findings are in line with earlier results[14]. The acetylation and formylation of 3-methylfuran have been systematically investigated[15]. Acetylation gives a mixture of 2- and 5-acetyl-3-methylfuran in a roughly 2:1 proportion. The formylations studied were more specific, the major product (c.90%)

being 2-formyl-3-methylfuran. The Brown–Stock extended selectivity treatment for electrophilic substitution has been applied to furan[16, 17], acylation being studied in detail. The β-position is the least active in both thiophen and furan.

	σ_α^+	σ_β^+
Thiophen	-0.79	-0.52
Furan	-0.93	-0.44

Furans may act directly as sources of aliphatic systems, α-Lithiofuran reacts with trialkylboranes to give allylic alcohols on oxidation[18]. Furan undergoes a Diels–Alder reaction with cyclopropene which is in sharp contrast to the reaction of this olefin with any other diene[19]. The product is a 1:1 mixture of endo- and exo-adducts, ascribed to the lowering of 'flagpole' interactions in the transition state on replacing a methylene group by an oxygen atom. This would suggest that steric rather than electronic factors predominate in this reaction. The endo-isomer is labile and yields a mixture of cycloheptadienones. The reaction of furfuraldehyde with aromatic amines has been conclusively shown to give 2,4-diarylaminocyclopentenones, these structures replacing all those previously in the literature[20]. The reaction of 2-chloromethylfuran with cyanide gives a mixture of 2-cyano-5-methylfuran (4) and 2-cyanomethylfuran (5)[21]. The product (4) arises from the intermediate (6) which was readily observed (by n.m.r.) in the crude mixture. The neutralised mixture of (5) and (6) was quite stable, even up to 80 °C! It is said that pure (6) is more stable than pure (5). On addition of aqueous potassium cyanide the substance (6) rapidly yields (4).

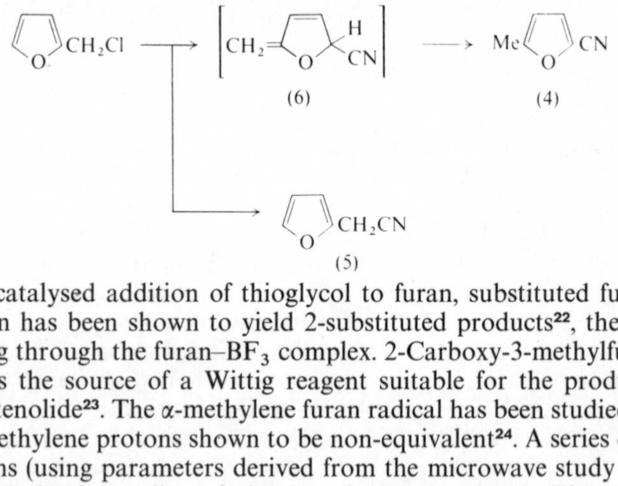

The BF$_3$-catalysed addition of thioglycol to furan, substituted furans and benzofuran has been shown to yield 2-substituted products[22], the reaction proceeding through the furan–BF$_3$ complex. 2-Carboxy-3-methylfuran may be used as the source of a Wittig reagent suitable for the production of ylidenebutenolide[23]. The α-methylene furan radical has been studied by e.s.r. and the methylene protons shown to be non-equivalent[24]. A series of INDO calculations (using parameters derived from the microwave study of furan) emphasised the inequality of the two hydrogen atoms. The rotational barrier about the 2,6-bond (1.40 Å) was 105.3 kJ mol^{-1}. The question of the rotational isomerism of 2-acylfurans has undergone a multidirectional attack (though, unfortunately, an attack without a common nomenclature). Infra-red studies of the system (7) $\rightleftarrows$ (8), have indicated that the parent compound exists as a mixture of anti- and syn-forms in the proportions 1:0.34 in carbon tetrachloride[25]. In chloroform, however, the proportions change to 1:1.14. In the first solvent 3-methylfurfuraldehyde exists only as

(7), whilst in chloroform $a:s$ is $1:0.91$. The results for other furans are summarised, the 2-ketones existing as (7) in all solvents[25]. However n.m.r. studies

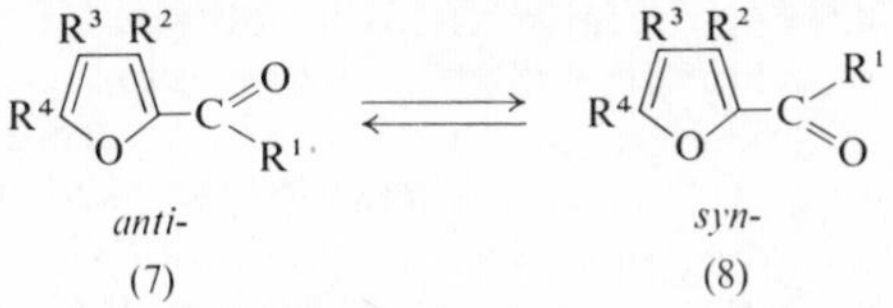

anti-   syn-

(7)   (8)

of furfuraldehyde indicate a 90 % predominance of the syn-form (8) in ether, acetone and (surprisingly) octadeuteriotoluene[26]. It is stated that there may be far more of the anti-form in 2-acetyl- and 2-carbomethoxy-furan. The nuclear Overhauser effect (NOE) has been utilised on the grounds that irradiation of the methyl group of the syn-form (8) of 2-acetylfurans should enhance the 3-H signal. This would not occur in the anti-isomer[27]. At $-120\ ^\circ C$ the spectrum could be clearly analysed and the conclusion was drawn that the syn-form predominates in the parent compound as well as in various derivatives (namely, 4,5-dibromo-, 4-bromo- and 4,5-dideuterio-2-acetylfuran). Attempts have been made[28] to calculate the thermodynamic parameters for the rotational isomerism of 2-acetylfuran based on the shape of the n.m.r. signal due to H-3 between -115 and $+30\ ^\circ C$ and assuming the same entropy differences as for furfuraldehyde. The enthalpy of activation for internal rotation in 2-acetylfuran was calculated to be less than that for furfuraldehyde by c. 20 %. A theoretical approach to the problem for furfuraldehyde using the total $\pi + \sigma$ charges predicts a predominance of the syn-form, though the n.m.r. spectrum in dimethyl ether at room temperature indicates that the anti form is preferred[29]. The discrepancy has been assigned to solvent effects, which certainly require clarification.

The structure of kahweofuran, a constituent of coffee aroma, has been elucidated and it and an isomer have been synthesised[30].

7.2.2 Dihydrofurans

These compounds may serve as the source of both aliphatic and aromatic systems. 2,3-Dihydrofurans with 2-unsaturated substituents are isomerised to benzenoid compounds on heating with palladium on alumina[31]. The reaction is particularly useful if the 2-substituent is a furan ring, as benzofurans

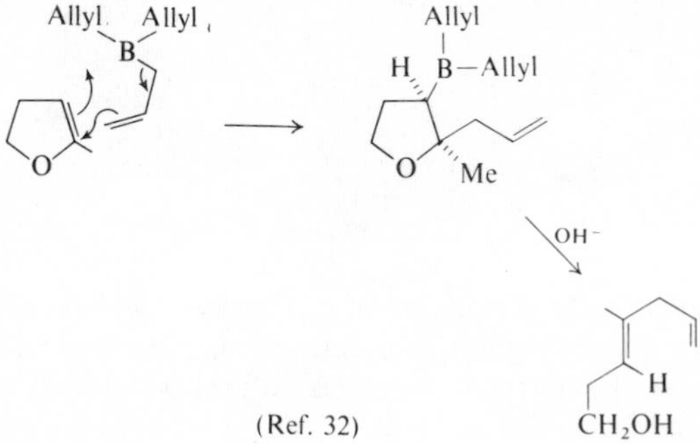

(Ref. 32)

result in good yields. However, if a phenyl ring is on the 2-position, isomerisation does not occur; instead dismutation to furans and tetrahydrofurans occurs. 2-Methyl-4,5-dihydrofuran yields unsaturated alcohols with triallyl borane[32]. The stereochemistry of the product would suggest that the elimination is not a simple concerted one. In this case, dihydrofuran is acting as an enol ether in part of a very general scheme for the production of 1,4-dienes. Reaction of the same dihydrofuran with diphenylacetylene yields a 4–5 fused ring system direct[33]. A thorough n.m.r. analysis of the product was carried out. A new type of photo-oxidation involving a phenolic hydroxy group and a neighbouring olefin gives rise to dihydrofuran systems[34]. It will be most interesting to see whether this reaction is general and whether enolic hydroxy groups can participate.

7.2.3 Tetrahydrofurans

A painstaking and very useful study of the oxidation of alcohols to yield tetrahydrofurans has been carried out[35]. It transpires that silver oxide, frequently used for this reaction, is in fact a very poor reagent and silver acetate is considerably superior. Silver trifluoroacetate, however, yields mainly ketones. Benzyne reacts with tetrahydrofuran[36]. One further interesting addition to the methods available for the stereo-analysis of complex lignans consists in a comparison of the n.m.r. spectra in hexadeuterio-DMSO and in the same solvent containing sodium dimethylsulphoxide. Benzylic protons next to an aromatic ring with a *p*-phenolic group move by *c.* 8 Hz. In this way, it was possible to establish the stereochemistry of pluviatol (9)[37]. The reaction of (10) with disiamylborane to yield (11) and (12) in the proportions shown

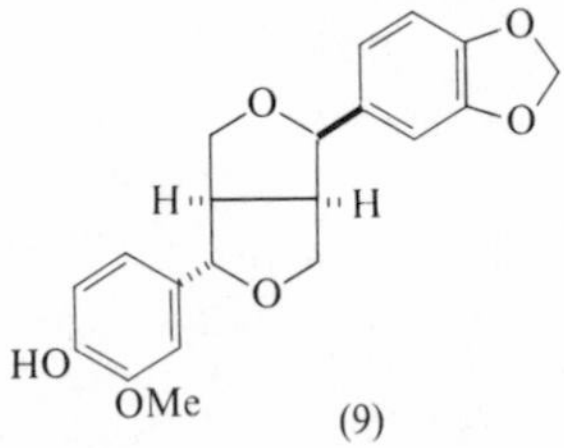

(9)

(Ref. 37)

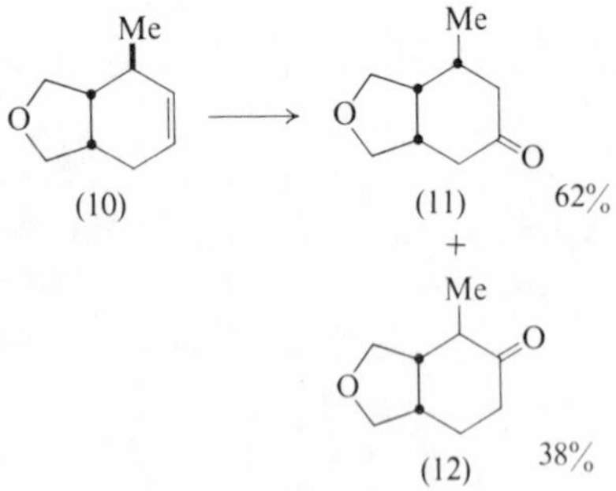

(10)   (11)   62%

+

(12)   38%

(Ref. 38)

is in great contrast to the corresponding reduction of 3-methylcyclohexene. Torsional effects cannot explain this and participation by the oxygen lone pairs in the electrophilic attack is suggested[38]. However, initial complexing of the borane might also be effective.

7.3 SIX-MEMBERED RING SYSTEMS CONTAINING ONE OXYGEN HETEROCYCLIC ATOM

7.3.1 Tetrahydropyrans, dihydropyrans and pyrans

The rosoxides have been synthesised by an intramolecular cyclisation of a primary alcohol group on to a diene system[39]. A further novel method for the production of tetrahydropyrans consists of the addition of an aldehyde to a 1,3-diene under the influence of palladium acetonylacetate and triphenyl-phosphine[42]. The yields are good and by varying proportions of the catalysts, open-chain or cyclised materials may be produced, the former not being the precursors of the latter. (Formaldehyde, however, yields only the pyran.) The rate of reaction varies dramatically with solvent, ethanol being recommended. Oxa-adamantane (13) has been synthesised, the key step being

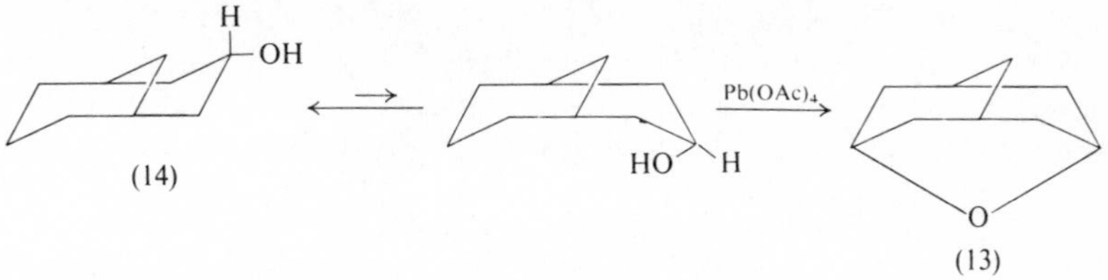

(Ref. 41)

production of the heterocyclic ring by lead tetra-acetate oxidation of the *endo*-alcohol (14). The *exo*-alcohol does not react with this oxidising agent[41]. Tetrahydropyranyl ethers may be used in ways analogous to the uses of enamines, but possess the advantage that they can be produced by Birch reduction with the double bonds in defined positions, because of the known specificities of the reduction and isomerisation processes[42].

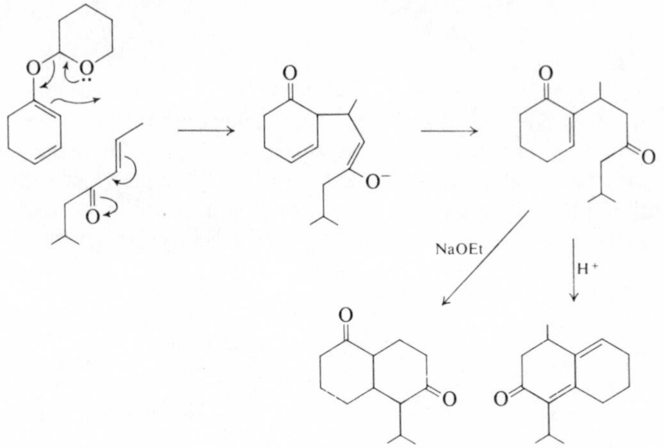

(Ref. 42)

Furthermore, an analogous reaction occurs with simple tetrahydropyranyl enol ethers and ketones[42]. 4-Methoxytetrahydropyran-4-yl ethers have been used as alternatives to the tetrahydropyranyl (THP) protecting group[43]. The advantage is that one derivative only is formed with optically active alcohols. The hydrolysis rate is about one-third that of the corresponding THP ether. 4-Chloro-3-alkyltetrahydropyrans (readily available from terminal olefins, formaldehyde and hydrogen chloride[44]) can be dehydrohalogenated to 3-alkyldihydropyrans. Alternatively, they can be used in Friedel–Crafts reactions to yield 3-alkyl-4-aryltetrahydropyrans[45]. An interesting paper on the conformation of dihydropyrans indicates that they normally exist in the 'sofa' conformation rather than the half-chair. Many useful references are included[46]. The hydroboration of Δ^2-dihydropyran proceeds exclusively at the 3-position[47]. Addition of excess diborane leads to a cleavage reaction, the excess behaving as a Lewis acid. This was proved by the use of trideuterioborane followed by boron trifluoride. Δ^2-Dihydrofuran was also hydroborated at the 3-position only. On photolysis Δ^2-dihydropyran gives formylcyclobutane and products probably derived from this[48]. It would seem that there is no general scheme into which furan and other enol ethers can fit. The photochemical oxidation of dihydropyrans does not proceed through an allylic peroxide; if the allylic position is blocked, the same type of oxidation occurs[40]. In the absence of light a quite different oxidation occurs[50]. It is suggested that photo-oxidation proceeds through

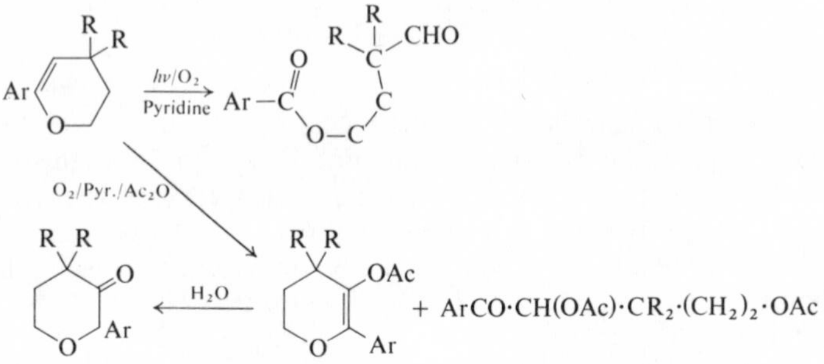

(Ref. 50)

an oxetan intermediate and a general scheme for olefins has been proposed[51]. Those olefins (enamines, enol ethers, styrenes, etc.) in which the carbonium ion (15) may be stabilised would yield ketone fragments. However, Kearns[52] has calculated that olefins of low ionisation potential can add oxygen directly to give dioxetans. It is also noteworthy that Δ^2-dihydropyran itself yields a

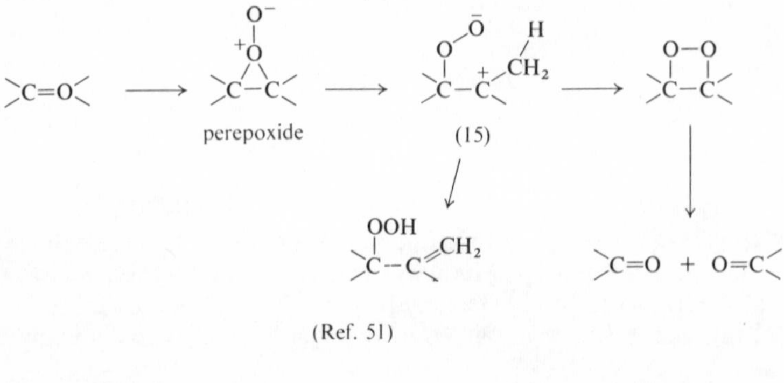

(Ref. 51)

peroxide[53] and presumably the 2-aryl-group is required further to stabilise (15). A most interesting use of dihydropyrans is of potential industrial importance as a synthesis of exaltolides (19)[54] The dihydropyran (16) (readily pro-

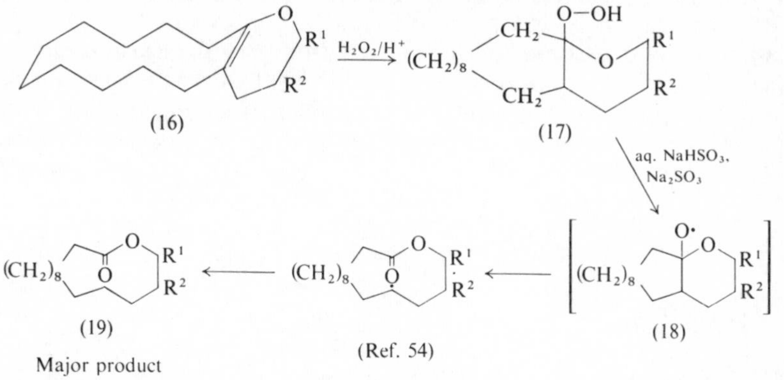

(16)

(17)

aq. NaHSO₃, Na₂SO₃

(19)

Major product

(18)

(Ref. 54)

duced from the medium ring ketone) yields the hydroperoxide (17) in almost quantitative yield. Reduction gives the radical (18) which undergoes cleavage and hydrogen abstraction to give the saturated ketone (19) as the major product. Photochemical oxidation of benzene gives 2-formyl-4H-pyran and quantitative studies of the reaction have been carried out[55]. It has been shown that preparations of 2H-pyrans involving dienones as intermediates always give the cyclised material if a potential 4-substituent is lacking, but if it is present the dienone itself results[56].

7.3.2   2-Pyrones, 4-pyrones and pyrylium salts

A general method for preparing 2-pyrones consists of treating the carboxymethylene Wittig reagent with 1,3-diketones[19]. Two very similar general methods consist of treating diphenylcyclopropenone with the stabilised acylpyridinium ylid (20)[20] or the sulphur ylid (21)[21]. The yields can be excellent and the reaction may proceed along either of the paths shown. In the course of characterisation of a new class of naturally occurring 2-pyrones lacking an oxygen atom at C-4 an interesting synthesis was accomplished. The readily available triacetic lactone (6-methyl-4-hydroxypyrone) was

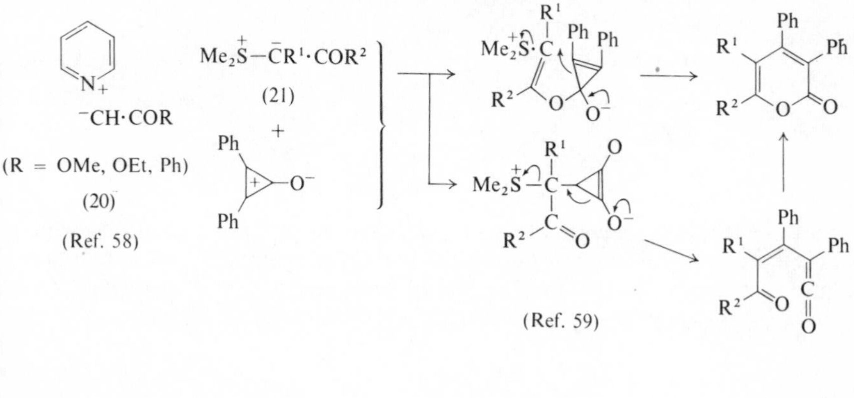

(21)

(R = OMe, OEt, Ph)

(20)

(Ref. 58)

(Ref. 59)

reduced to 6-methylpyrone by a sequence similar to that used in the strychnine synthesis[60]. The product was then condensed with appropriate aldehydes to yield the natural products. The photolysis in methanol of 4-methoxy-6-methylpyran-2-one gives 4-methyl-6,6-dimethoxydihydropyrone in high yield[61], the rearrangement proceeding through a cyclobutene acid. The isolation of arenol (22) and homoarenol (23) from *Helichrysum arenium*[62] and of helipyrone (24)[63] and obtusifolin (25)[64, 65] from *Gnaphalium obtusifolium* indicates the probable polyacetate origin of the pyrone ring and also that its coupling to a phenol via a methylene group may be widespread in these plants.

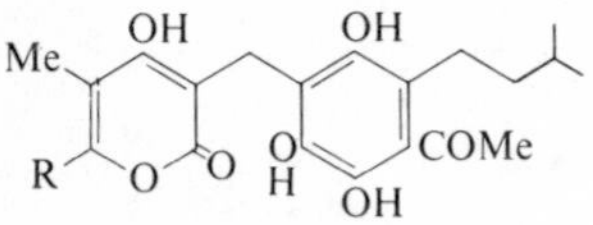

(22) R = Me
(23) R = Et

(Ref. 62)

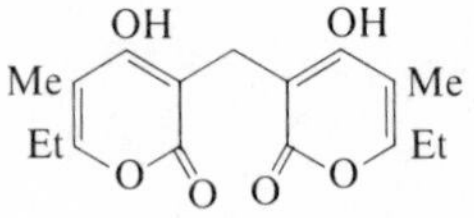

(24)

(Ref. 63)

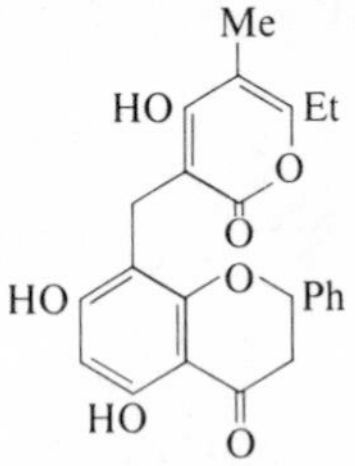

(25)

(Refs. 64, 65)

The synthesis of helipyrone has been reported[66].

2-Pyrones may be photolytically converted into 4-pyrones the reaction being reported to proceed through a cyclopentenone epoxide[67]. Presumably this in turn must be formed by an oxygen migration from a cyclobutane derivative. 2,6-Diphenylpyran-4-one undergoes an interesting reductive dimerisation under the influence of light and anthraquinone dimer[68]. It has been noted that 2,6-dimethylpyran-4-one coordinates europium shift reagents at the carbonyl oxygen atom[69], an effect shown also by flavones even when containing many alkoxy groups[70]. Funicone (26) is of interest as is biosynthesis may be via the complex pyran (27)[71], very similar to other compounds produced from *P. funiculosium* Thom. An important observation is

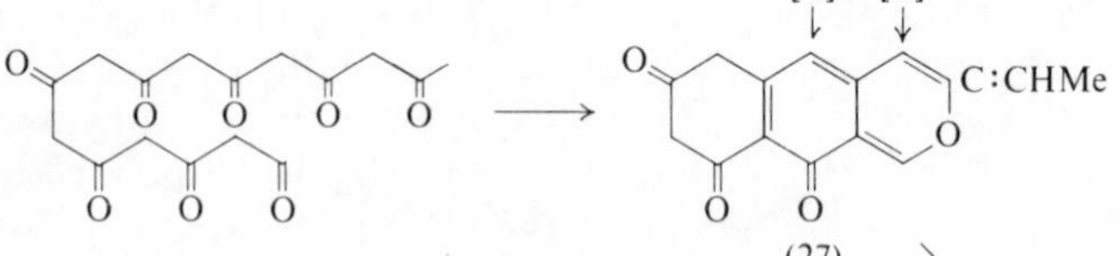

(27)

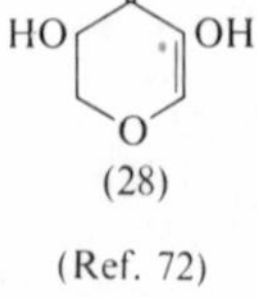

(28)

(Ref. 72)

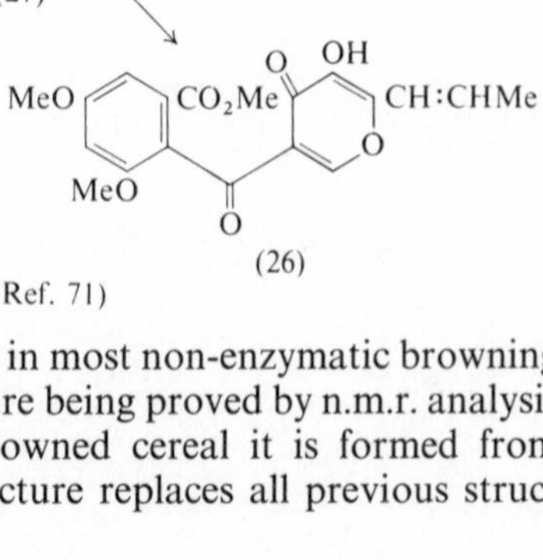

(26)

(Ref. 71)

that the Maillard reaction product, present in most non-enzymatic browning reactions, is represented by (28), the structure being proved by n.m.r. analysis and synthesis[34]. In the production of browned cereal it is formed from 1-deoxy-1-(L-prolino)-D-fructose. This structure replaces all previous structures for reaction products from such mixtures as D-glucose, methylamine and acetic acid[72]. The use of pyrone derivatives as masked sources of polyketide chains continues to attract attention and to throw light on the chemistry of the pyrones themselves. Studies of the di-anion of triacetic lactone have been made in efforts to attack selectively the 6-methyl group[73]. The di-anion gives a complex mixture on alkylation, but reacts cleanly at the 6-methylene position with ketones. The di-lithio salt is also benzoylated and carboxylated at this position. The chemistry of the benzoyl derivative (29) has been studied; a summary is given below.

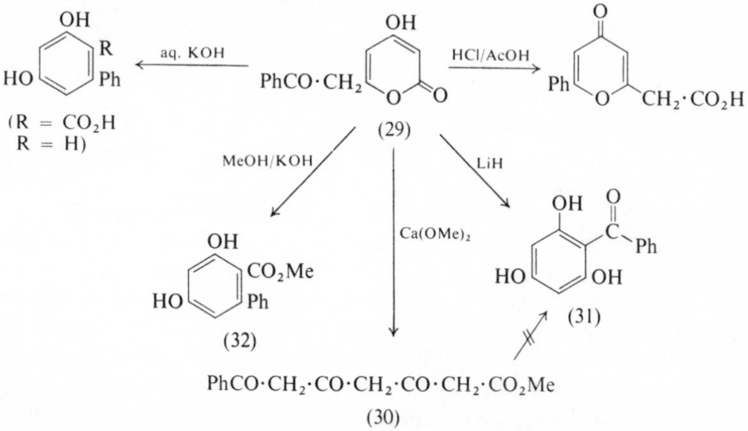

PhCO·CH₂·CO·CH₂·CO·CH₂·CO₂Me

(30)

$$PhCO \cdot CH_2 \cdot CO \cdot CH_2 \cdot CO \cdot CH_2 \cdot CO_2 Me$$

Magnesium methoxide gives a mixture of (31) and (32) but the polyketide (30) is not an intermediate, and a mechanism involving a ketene-anion derived by base attack on (29) has been evolved[74]. The attempts to produce larger masked polyketides continue[75–77] and a new approach has been to use the compound (33) which yields the xanthone (34) with base. Reduction of the ketone group followed by rearrangement gives rise to 4-methoxycoumarin and hydroxynaphthalene systems. 3,6-Diphenyl-4,5-isobenzopyrone is readily

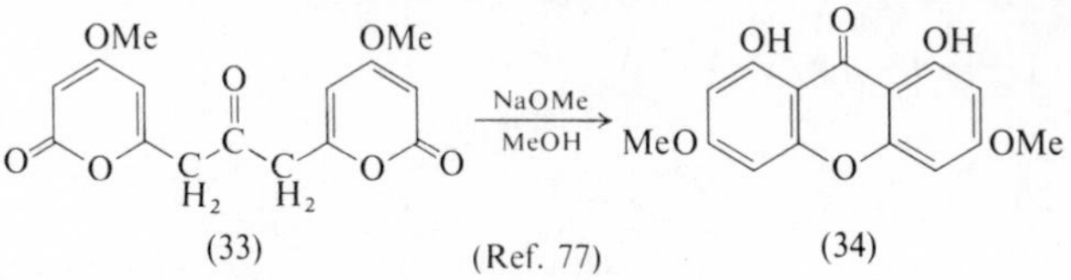

(33) (Ref. 77) (34)

produced[78], but the parent compound is unstable and may be trapped as the N-phenylmaleimide adduct[79]. The 6-methyl- and 6-phenyl derivatives were isolated as iron carbonyl complexes[79]. Some results on the dehydrogenation of steroid dihydropyrones have been reported[80].

2,6-Dimethylpyrylium salts are readily prepared by hydride abstraction (triphenylmethyl hexachloroantimonate being a particularly recommended reagent) from heptan-2,6-dione[81]. The half-lives for acid-catalysed deuterium exchange for 2,6-dimethylpyrylium salts have been thoroughly analysed and a very general scheme proposed. Three groups[82–84] have studied the reactions of hydroxylamine and hydrazine with pyrylium salts. The first two groups have concentrated on 2,4,6-trisubstituted compounds. Direct reaction of 2,4,6-triphenylpyrylium perchlorate with aqueous hydrazine gives an intermediate keto-hydrazide (isolated) which cyclises to a diazocyclo-heptatriene. However in the presence of hydroxide anions the pseudo-base reacts via the diketone followed by a dihydropyrazino derivative, which is cleaved by acid to 2,4-diphenylpyrazine and acetophenone. This behaviour is unique to hydrazine and does not occur with phenylhydrazine or hydroxy-lamine, which gives dihydroisoxazole derivatives. The counter-ion can affect the reaction. 2,4-Diphenyl-6-methylpyrylium iodide with methylhydrazine gives 60% of the compound (35) and only 10% of (36). The borofluoride in

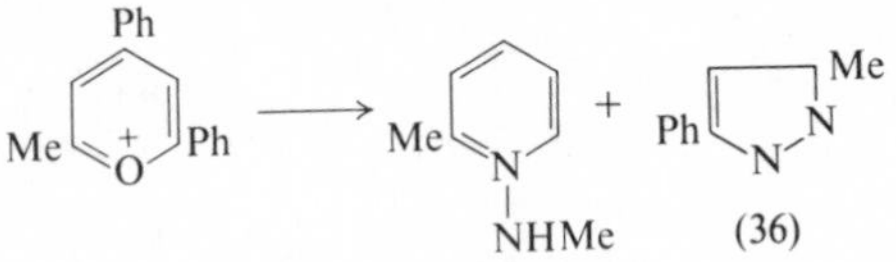

the same conditions gives none of the pyridine derivative and 41% of (36). In acid media hydroxylamine reacts with 2,4,6-triphenylpyrylium salts in a manner highly dependent on the pH. More highly substituted salts, however, give only pyridine N-oxides, also given by 2,4,6-trialkylpyrylium salts at any pH value. A scheme to explain these results has been proposed. The photolytic oxidation of 2,4,6-triphenylpyrylium 3-oxide (37) has been investigated[85] and an ozonide-like intermediate is formed from which the triketone (38) is produced. Cyclisation to (39) is followed by benzoyl migration to give the product (40).

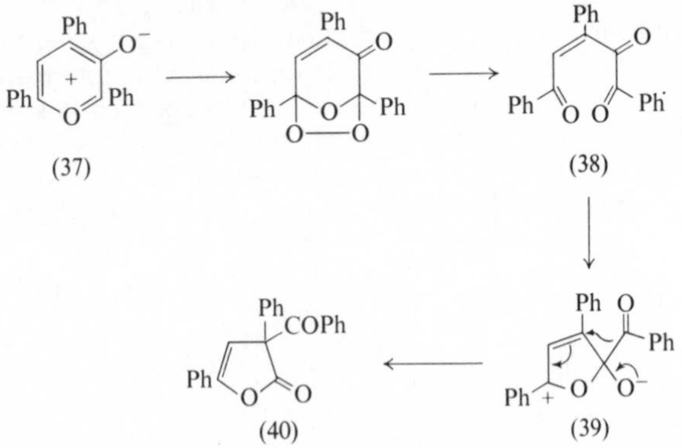

The result differs from the previous investigations of 2,6-diaryl-4,5-benzo-pyrylium 3-oxide which undergoes straightforward oxidative cleavage to a keto-anhydride[86].

7.3.3 Chromenes, chromanones and chromons

Δ^3-Chromene has been converted into the dibromide and the hydrolysis studied. The conformation of the trans isomer of the dibromide and of several related compounds has been established as mainly (41) and of the cis as (42)[87].

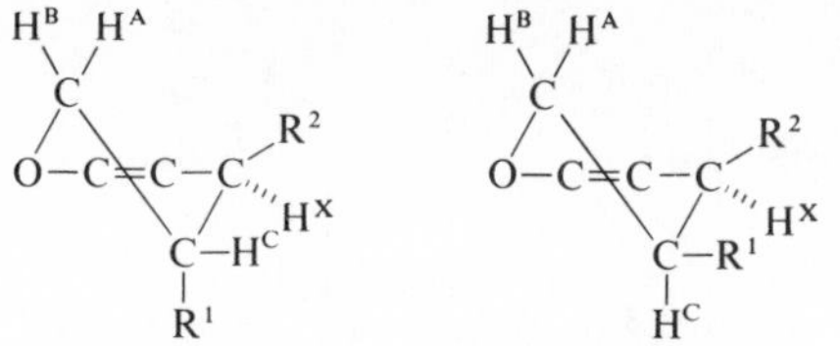

(41)   (42)

(Ref. 49)

Addition of 2-methoxybutadiene to a phenol readily yields a 2-methoxy-3-methylchromanone. Birch reduction followed by base condensation leads to a very efficient synthesis of bicyclo-systems. Acid condensation gives benzenoid products[88]. A very pleasing chromanone synthesis involves an enamine condensation with a salicaldehyde[89]. 2-Methyl-7-methoxychromanones have been used as starting materials for the synthesis of 6-oxasteroids[90]. A very unusual result was obtained from attempts to carry out a Wolff–Kishner reduction on chromanone[91]. The cyclopropane (43) was obtained presumably by attack of the 4-anion on the 2-position as in (44).

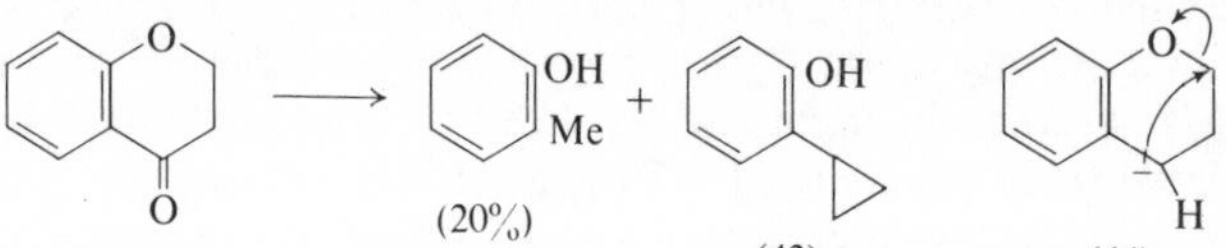

(20%)

(43)   (44)

(40%)

(Ref. 91)

Selenochromanone gives (45) which may well be derived from (46). A new chromone synthesis of good efficiency consists of heating o-hydroxyacetophenone with dimethoxydimethylaminomethane[92].

(45)   (46)

(Ref. 91)

The latter one-carbon unit might well be useful for isoflavone production from o-hydroxydesoxybenzoins. Oxalyl chloride reacts with chromone to yield the 4,4-dichloro derivative[93]. Diborane reduction of chromone gives 3-hydroxychroman in good yield. This is also a minor product in the diborane reduction of coumarin[94]. From *Dianella revoluta* and *Styandra grandis* a series of 2-alkyl-5,7-dihydroxy-6-methylchromones have been isolated. The alkyl groups are saturated C_{27}, C_{29} and C_{31} chains, thus removing the limitation that all naturally occurring chromones possess a 2-methyl or a 2-hydroxymethyl group[95]. It supports the idea of the polyacetate origin of these compounds as against that of the incorporation of a mevalonate unit postulated by Dean. The first simple chromone to be isolated from a lichen or fungal source has been characterised[96], as has the first C-glycosylchromone[97]. Bakarol (47) from the leaves of *Cassia siamea*, proves to be a disguised chromone[98] most probably derived as shown.

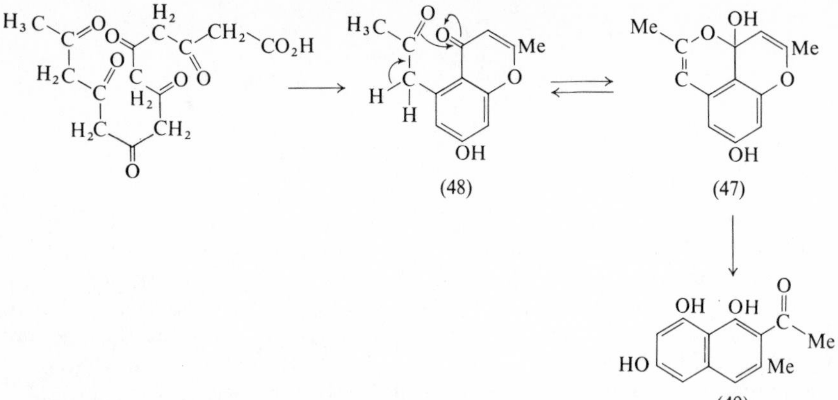

(48)

(47)

(49)

This pathway is made more probable by the synthesis of bakarol from the chromone (48) and by the isolation of (48) from the flowers of the same plant[99]. On hydrolysis bakarol, which is not an artefact, yields (48). The latter shows antibiotic activity against gram-negative organisms. Bakarol may act as a source of a polyketide chain and has been converted[100] into 6-hydroxy-musizin, isolated from *Aphis nerii*[101]. The latter also produces a fascinating series of stable *p*-quinone methides, some of which are derived from (49), e.g. (50) which may be produced from (49) by reaction with diacetyl. A series

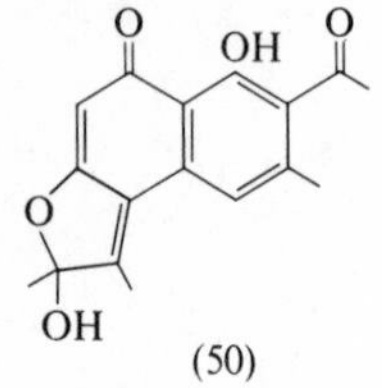

(Ref. 101)

(51)

(50)

(Refs. 101, 102)

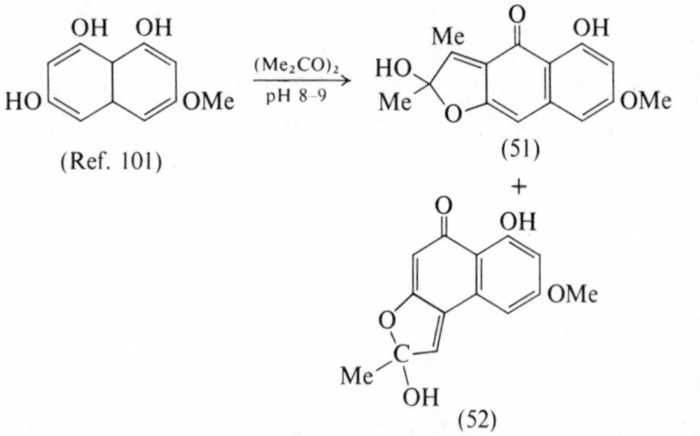

+

(52)

of reactions of various phenols with diacetyl gave stable *o*- as well as *p*-quinone methides, e.g. (51), (52). A series of naphthopyrones has been isolated from the crinoid *Comanthus parvicirrus timorensis*[102].

7.4 FLAVANOIDS

7.4.1 Flavans and flavenes

Quinone methides produced by heating *o*-hydroxymethylene phenols may be trapped by olefins to give disubstituted chromans. Styrene reacts with the parent compound to give flavan. Alternatively, the quinone methide may be generated by heating an *o*-methylphenol with a metal oxide. The reaction would appear to be general[103]. 6-Methyl-7-hydroxyflavans are extremely sensitive to air oxidation. They are readily oxidised in both rings by high potential quinones, this reaction providing a route to dracorhodin, for example[104]. The occurrence of optically active naturally occurring flavans has been reported[104, 105]. The synthesis of optically active 4-aminoflavan has been accomplished[106]. Hydrogenation of 3-methoxyflav-2-enes gives the cis-3-methoxyflavan, the same stereochemistry being obtained with the acetate[101]. The acid hydrolysis of trans,trans-3-acetamido-4-hydroxyflavan has been shown to proceed through a cyclic intermediate with inversion at C-4[108]. The mechanism of the conversion of flavan-3,4-diols into dihydro-flavonols and flavanones in acid media, has been investigated and shown to involve formation of a carbonium ion at C-3, adjacent to the carbonyl group at C-4[109]. This opens the way to many speculations as to the nature of dimerisation processes. The alkylation of catechin with bromoacetic ester has been examined and the relative activities of the phenolic oxygen atoms with regard to alkylation have been delineated[110]. 2-Hydroxyflav-3-ene is produced by irradiation of 2'-hydroxychalcone[111]. The biosynthetic analogy drawn from this experiment seems to be based on a misapprehension of the feeding experiments. Benzyne, generated in a number of ways, reacts with cinnamaldehyde to give flav-3-ene directly[112]. A most unusual and general flavene synthesis involves the reaction of aryloxymagnesium bromides with cinnamaldehyde[113]. This activity of aryloxymagnesium halides may have important implications for aromatic synthesis.

7.4.2 Flavanones, flavone and flavonols

The reaction of flavanone with hydrazoic acid gives the alkyl-migrated lactam in high yield[114]. The conformation and configuration of 3-arylidene flava-nones has been investigated. In both the cis- and trans-compounds the 2-aryl group is axial[115]. ($\pm$)-Peltogynol tetramethyl ether has been synthesised[116]. A most interesting reductive ring contraction has been shown to occur[117].

$$\text{Na(MeOCH}_2\text{CH}_2\text{O)AlH}_2$$

(Ref. 117)

The flav-3-ene is not an intermediate in the reaction. The c.d. and o.r.d. curves have been correlated with the absolute configurations of flavanones, 3-hydroxy-flavanones and their glycosides[118]. With the exception of ($+$)-sakuranetin (2R) and ($-$)-dihydrofisetin (2S, 3S), all the flavanones are 2S, and all the three hydroxyflavanones are 2R, 3R. Flavanones with unusual appendages continue to be discovered. Obtusifolin has previously been mentioned (see p. 212) and two other flavanoligans have been obtained from Silybum marianum. The first, silydianin, has structure (53)[117] and the second, silychristin, has been assigned structure (54)[120].

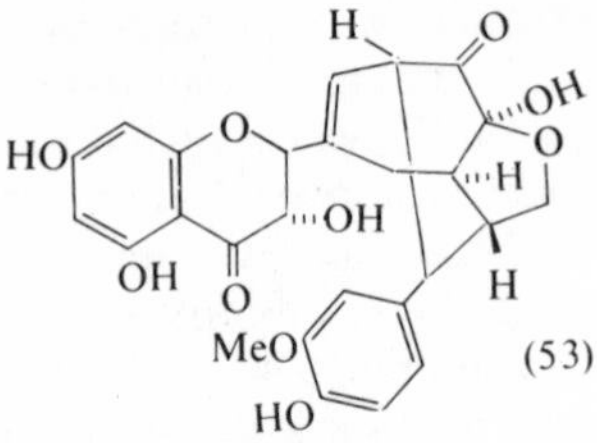

(53)
(Ref. 119)

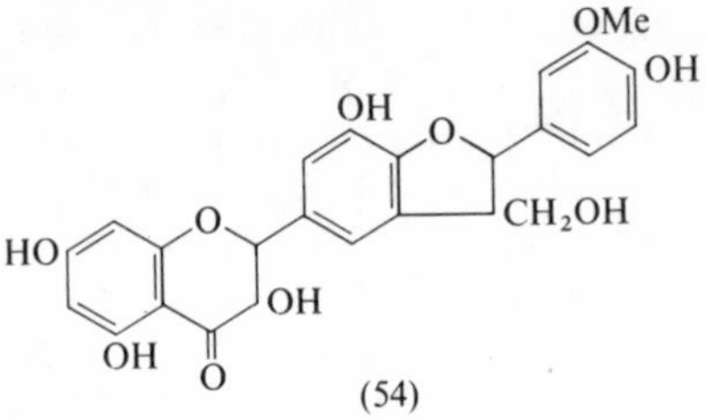

(54)
(Ref. 120)

Like silybin[121], these compounds can be derived from the oxidative coupling of taxifolin and coniferyl alcohol.

A general scheme for the biosynthesis of flavanoids treats the chalcone precursor as a modified C-6—C-3-lignin precursor. Oxidation of the *p*-hydroxy group may give rise to all the major groups of flavanoids of an oxidation level the same or higher than the flavanone. Chemical experiments to support one of the schemes, the production of flavones and aurones were presented[122]. The configuration of aurones has been investigated, with the result that in this case the configuration cannot be deduced from the β-proton, but instead it is the protons in the 2′, 6′ positions that are useful[123]. Normally, it is the *trans*-isomer that is produced. With sulphuryl chloride, flavone gives 2,3,3-trichloroflavanone which on reduction gives 3-chloroflavone, produced directly with thionyl chloride[124]. Under the influence of light, diphenyl-acetylene adds directly to the double bond of flavone to give a dihydro-cyclobutene derivative. A similar reaction occurs with isoflavone[125]. By a photo-Fries migration (55) gives (56), which may be converted with base into 3-hydroxyflavanones or by further irradiation into flavone[126]

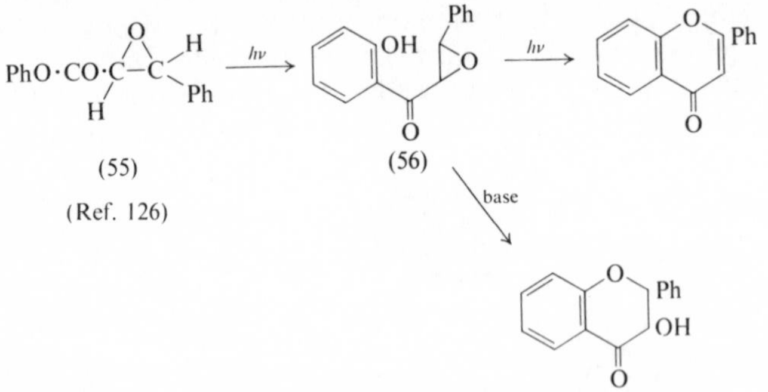

(55)
(Ref. 126)

(56)

The photo-Fries migration in many phenolic cinnamates has been studied[127]. A novel synthesis of a tetrahydroflavone system is as shown[128]. A most beautiful extraction followed by characterisation shows that the flavone present in *Clerodendron trichotomum*. Thumb is actually acacetin-7-β-glucurono-β-(1,2)-D-glucoronide[129]. A new chromonoflavone has been isolated[130]. The unsuitability of ethanol as a co-solvent for the spectroscopic identification of functional groups in hydroxyflavones using aluminium chloride has been pointed out[131]. In this solvent the spectra are sensitive to traces of water and methanol is recommended. Malvidin-3,5-diglycoside is converted into the anhydro-base at pH 7. Heating in acid then not only caused reconversion into the anthocyanidin but also cleavage to a 3-hydroxy-coumarin[132]. Hydrogen peroxide oxidation of the diglycoside is a relatively straightforward cleavage between C-2 and C-3, to yield the diester, malvone. CNDO/2 calculations on flavylium compounds come to the surprising

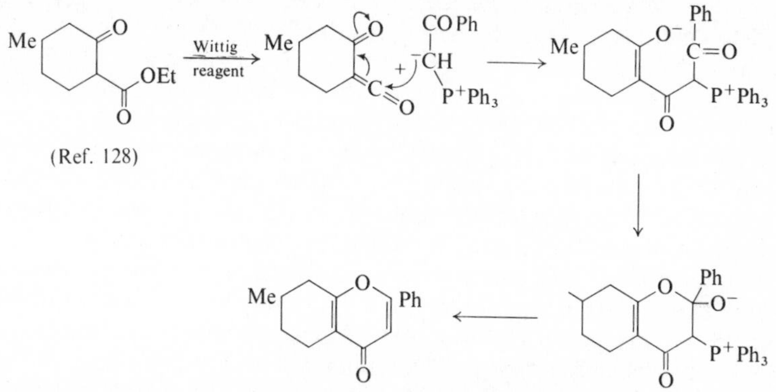

(Ref. 128)

conclusion that the oxonium oxygen is appreciably negative. The same is true of 3-hydroxyflavylium salts[134]. The irradiation of 2'-methoxyflavonol (57) gives the diketone (58) in quantitative yield, presumably through the epoxide (59)[135].

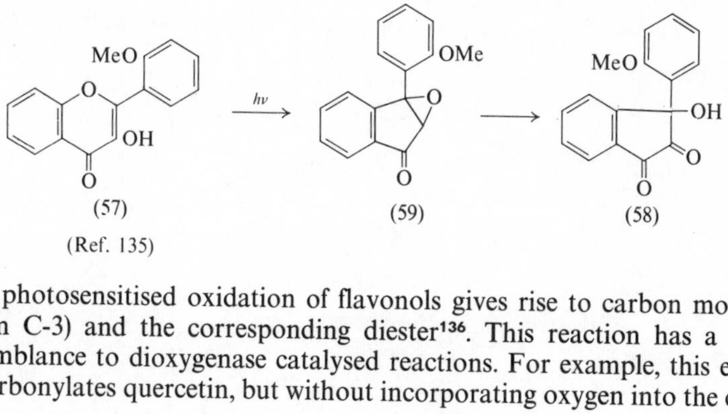

(57)   (59)   (58)

(Ref. 135)

The photosensitised oxidation of flavonols gives rise to carbon monoxide (from C-3) and the corresponding diester[136]. This reaction has a formal resemblance to dioxygenase catalysed reactions. For example, this enzyme decarbonylates quercetin, but without incorporating oxygen into the carbon

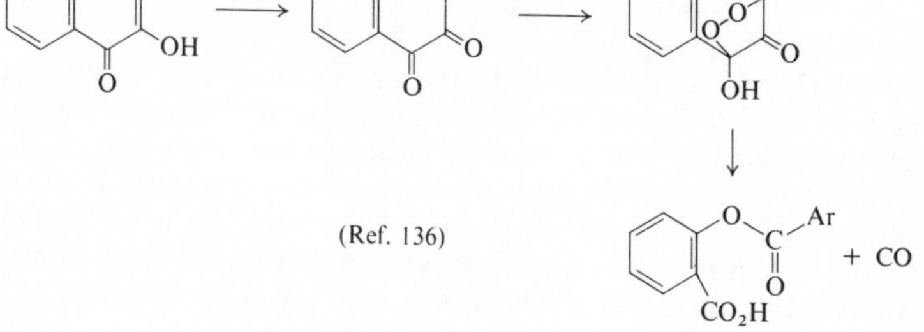

(Ref. 136)

+ CO

monoxide. A 3-hydroxy group is necessary and tracer experiments indicate the following pathway. Manganese dioxide oxidation gives dimers.

7.4.3 Polyflavanoids

One of the outstanding achievements in this field has been the almost complete characterisation of a tetraflavanoid[137]. The difficulties associated with this project are shown by the variety of techniques used. The methoxy and acetyl groups were counted by the use of carefully graduated solvent induced shifts in the 220 MHz n.m.r. spectrum, a point being reached at which all were visible. This also allowed the four overlapping aliphatic systems to be analysed by decoupling experiments at different ratios of C_6D_6–$CDCl_3$. To ascertain the sequence of flavanoid units involved the examination of line broadening at 300 MHz. The overall structure has no less than fourteen chiral centres. The structure given to a trimeric flavanoid, in which one linkage is said to be 4-0-4^1 in no way rests on such sure foundations[138]. A very useful fractionation of the anthocyanidins from *Calluna vulgaris* L. has been accomplished[139]. A biflavanoid carboxylic acid, probably 4,6″-linked has been characterised[140]. The extra carbon atom in the form of a carboxyl group may be the remnant of another flavanoid unit or it may be used as a further linkage point. Another unusual biflavanoid is a 3,8-coumaronylflavanone from *Phyllogeiton zeyheri* Sond[141]. There is a high incidence of rotational isomerism in a 4,6″- or 4,8″-linked natural bi- and triflavanoid tannins, which can further add to the complexity of the n.m.r. spectra[142]. This has been independently observed on model substances and

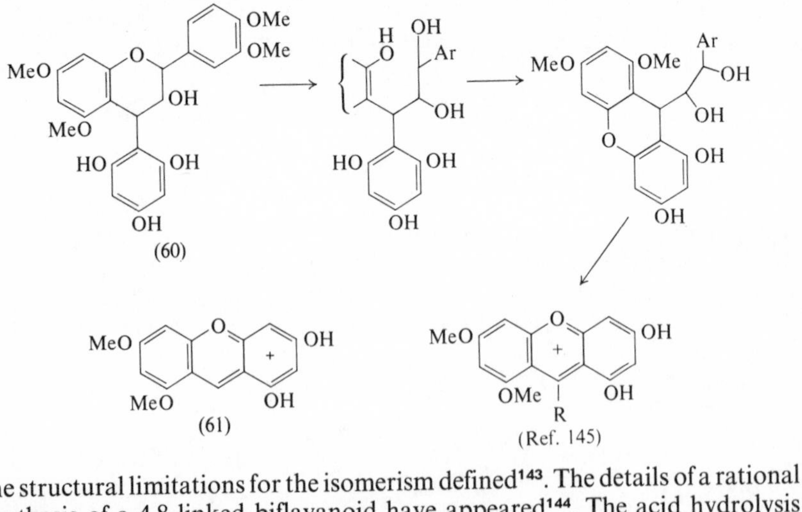

the structural limitations for the isomerism defined[143]. The details of a rational synthesis of a 4,8-linked biflavanoid have appeared[144]. The acid hydrolysis of non-hydrolysable tannins always yields yellow pigments as well as anthocyanidins. These pigments have the same u.v. spectra as hydroxyxanthylium salts. When the reaction was carried out with (60), a compound with exactly the spectrum of (61) was observed and this may arise as shown. The pathway involves a rather unusual loss of a phenolic hydroxy group[145].

The yellow compound which develops during grape juice storage has been isolated and characterised as the glucose derivative of a 1,3,6,8-tetrahydroxy-xanthylium salt[146]. The stereochemical results of the degradation of poly-flavanoid bark fractions by thioglycolic acid have been established[147]. 3,8-Linked biflavanoids from Guttifereae continue to attract attention. The full report of the establishment of structure of the G.B. linked series has appeared[148]. Biflavanoid glycosides have also been fully characterised[149, 150]. Volkensiflavone (talbotoflavone) has been fully characterised[151-154]. De-hydrogenation of the fully methylated derivatives of morelloflavone and volklensiflavone gave the first fully characterised 3,8″-biflavonyls[153]. It is of interest that in the n.m.r. spectrum of the compound derived from morello-flavone there was an upfield signal from a methoxyl group that did not move on addition of deuteriobenzene to the deuterochloroform solution. It would appear that there was internal solvation and that this may be a limitation to the use of the method of solvent-induced methoxyl shifts[155]. The oxidative coupling of apigenin using potassium ferrocyanide has been carried out[156]. The result was that none of the naturally occurring dimer series was obtained, but some important new series of dimers were produced. The view was expressed, mistakenly in the opinion of the reviewer, that therefore processes involving flavonoid radicals were not occurring in natural dimerisations. However, in biological conditions some phenolic groups may well be present as glycosides and the whole process modified. Support is given to this viewpoint by oxidation experiments[157] on 4′,7-dimethoxyapigenin which give the 6,6″-dimer. A rational synthesis of cupressuflavone hexamethyl ether confirms the 8,8″-linkage[158]. Plants of the *Cupressaceae* have been screened for biflavonyls. Interestingly apigenin co-occurs with amentoflavone, cupressuflavone and hinokiflavone in *C. torulosa*[159]. The Cycadaceae are also undergoing examination for biflavonyls[160]. *Cycas revoluta* gives in addition to the normal biflavonyls the first representatives of a new class, 2,3-dihydroamentoflavone and 2,3-dihydrohinokiflavone[161]. This class of compound is also found in *Metasequoia glyptostroboides*[162]. It would be interesting to see whether the dihydro compounds are produced directly by a mixed oxidation or represent the reduction of a flavone ring to a flavanone, a process not previously considered in the various possible biosyntheses of flavanones. The Auracaraceae continue as rich sources of agathisflavones as well as cupressu- and amentoflavone derivatives, the three classes frequently occurring together in the leaves of the same plant[163, 164]. Cephalotaxus species provide mainly amentoflavone derivatives as well as an apigenin glycoside[165].

(Ref. 171)   (62)

Optically active amentoflavone and podocarpusflavone A have been isolated from *Garcinia livingstonii*, a plant from the Guttifereae, botanically far removed from the normal sources of biflavonyls[153]. From *Juniperis chinesis* the isolation of (−)-amentoflavone and hinokiflavone have been reported[166]. The constitution and configuration of the theaflavin pigments of black tea have been completely elucidated[167, 168]. A new black tea pigment has been isolated and characterised[169], as have the theaflavic and epitheaflavic acids[170]. Reduction of a methoxyaurone has yielded a new class of flavonoid dimer (62)[171].

7.4.4   Isoflavanoids

2-Carbomethoxy-7-methoxychromone reacts with an aromatic diazonium salt in the presence of cupric chloride to give a 2-carbethoxyisoflavone[172]. The total synthesis of isomilletone by the reaction of an appropriate enamine with isotubaic chloride has been achieved[173]. A novel synthesis of isoflavones is as shown[174].

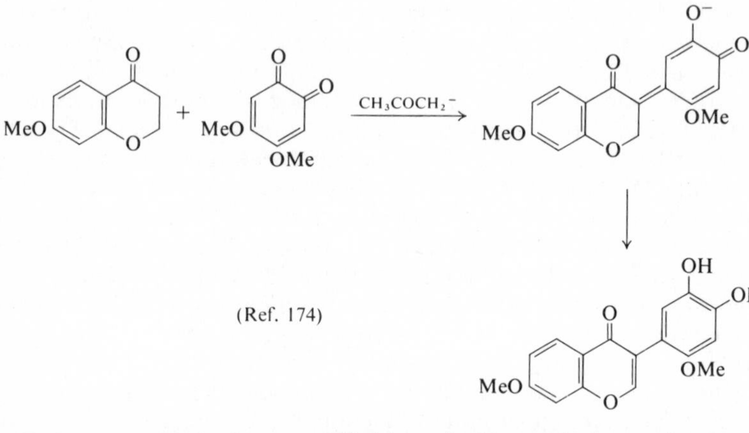

(Ref. 174)

The B-ring may be further modified by suitable oxidation–reduction procedures. The Wanzlick approach to the preparation of coumestrol has been adapted to the preparation of pterocarpan analogues[175]. The photo-oxidation of dehydrorotenoids proceeds mainly at the methylene group next to the oxygen to give a pyronochromone, but to a small extent carbon monoxide is expelled to give a compound of the lisetin type which is hydrolysed to a 3-aryl-4-hydroxycoumarin[176]. The complex and interesting reactions associated with bismethylene transfers to 2′-hydroxy isoflavones have been studied in depth[177]. The extra methylene bridge of a rotenoid has been clearly shown to arise from the methyl group of a 2′-methoxyisoflavone in *A. Fructicosa* seeds. The biosynthesis of isoflavones seems to proceed in accord with the ideas of Pelter, but the opening of the spirocyclopropane–cyclohexadienone intermediate may be carried out with a methylating agent rather than by protonation to yield a methoxyisoflavone directly. This would seem to be the case for rotenoid biosynthesis[178]. The details of the further oxygenation of ring B remain obscure, however, and it is most surprising that the 4′-

methoxyisoflavone is a substrate for this rather than a 4'-hydroxy isoflavone. Perhaps a 4',2'-intermediate may be possible. It has been noted that daidzein is an excellent precursor of coumestrol, but *not* via 4,4',7-trihydroxy-3-phenylcoumarin[180]. A 2'-hydroxy isoflavone may be produced instead. Sterile *Phaseolus aureus* plants can degrade isoflavones, but seemingly only those with a free 4'-hydroxy group[181]. On the other hand, *Cicer arietinium* has a rhizosphere bacterium associated with it which can degrade flavones and isoflavones, but only those with common hydroxylation patterns. The products are unknown[182]. The isolation of repensol from diseased white clover has been reported, the paper giving a good review of the coumestrol literature[183]. A coumestan isopentenylated in both rings A and D has been isolated from *Desmodium gangeticium*[184]. From subterranean clover and red clover, isoflavone-7-O-glycosides and glycosides malonated at C-6 have been characterised[185]. The bulbs of *Eucomis autumnalis* contain both homoiso-flavones and dihydrohomoisoflavones[186].

7.5 COUMARINS

7.5.1 Coumarins and dihydrocoumarins

The electrochemical reduction of coumarin has been studied[187]. In the presence of optically inactive amines a mixture of the ($\pm$) and *meso*-forms of the 4-linked dimer is obtained, as well as dihydrocoumarin. If optically active bases are used, asymmetric induction occurs. This may be as high as 66.9% with yohimbine down to 2.6% with nicotine. 4-Methylcoumarin gives dimers and 4-methyldihydrocoumarin. With sparteine the 4-methyl-dihydrocoumarin produced has undergone 17% of optical induction to give the ($+$)-form. Most surprisingly, the 5-methoxy derivative undergoes

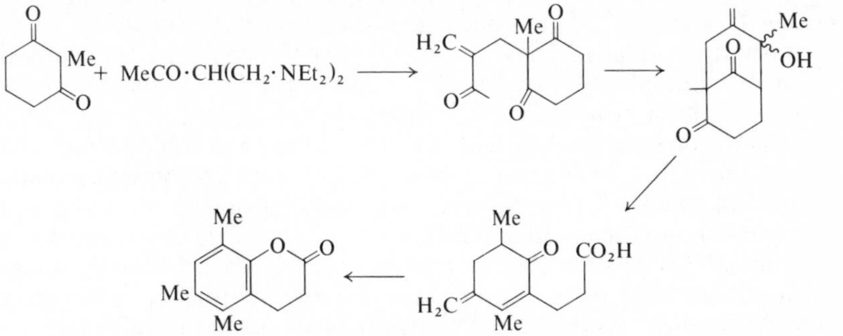

(Ref. 192)

optical induction in the opposite sense using the same base. As a corollary to this work, ($+$)-4-methyldihydrocoumarin was shown to be the *R* form[188]. *o*-Hydroxyacetophenones react with carbon disulphide to give 2-thio-4-hydroxycoumarins[189]. The pK_A of the 6-methyl derivative is 4.30 as compared with 5.03 for the corresponding hydroxycoumarin. Although diazomethane alkylates the sulphur atom to give a chromone derivative, a study of the u.v. data suggests that the compounds exist predominantly in the coumarin form. Furanocoumarins have been studied and it would seem that, in general, activity to both electrophilic substitution and to hydrogenation resides in the furan ring[190]. An instructive synthesis of kotanin and des-methylkotanin, dimeric coumarins, has been accomplished[191]. The dimerisation step was carried out by the action of oxygen on an aryl-lithium in the presence of a cuprous salt. A most unusual but efficient synthesis of dihydro-coumarins involves the following steps[192]. No intermediates need be isolated, and other substituted cyclohexan-1,3-diones may be used. The sequence of steps involved in the biosynthesis of scopolin has been defined[193]. The structure of halfordinin seems to indicate that the hindered C-5 unit is introduced by direct alkylation, rather than rearrangement[194]. The establishment of the structure of nieshoutol provides excellent examples of the use of the NOE and of brosylates to give specific deshielding effects[195].

7.6 INTRODUCTION OF SIDE CHAINS INTO PHENOLS

Very many oxygen heterocyclic compounds arise by the introduction of say isoprenoid or cinnamyl units into phenols and many experiments have been designed to emulate the possible natural processes. In the last two years the accent has been on terpenoid units rather than on C-cinnamylation, though sometimes the experiments are based on the known cinnamylation techniques[196–198].

Cinnamyl pyrophosphate has been prepared and shown to react with resorcinol at pH 7.2 to give a mixture of the benzylstyrene and neoflavonoid. Although the alcohol gives the same mixture it is not an intermediate in the reaction[199]. In a similar fashion, 3-methyl-but-2-en-1-ol or 3-methyl-but-1-en-3-ol (readily available) react with phenols[200] to give chromans. The reaction was later shown to be controllable to give either chromans or iso-pentenyl phenols[201]. The isopentenyl group may be modified to give chromenes eventually. Thus an allylic peroxide could be formed which could be reduced and cyclised to the chromene[202]. An analogy for this process has been provided[202, 203]. It might even be possible to ring close the hydroperoxide directly. Alternatively, the double bond could be epoxidised and converted into the chroman-3-ol and thence into 2,2-dimethyl chromenes, or it could give the five-membered ring alcohol as in the rotenoids or, for example, columsianetin. An excellent study of the control of this reaction has been made[204]. Oxidation at the methyl group has been used[205]. Another approach has been to oxidise the benzyl styrene, chroman, or isopentenyl phenol. The direct oxidation with dichlorodicyanoquinone (DDQ) has been studied and been shown to work excellently to give either 2,2-dimethyl-chromenes or flav-3-enes from the appropriate substrate[206]. Alternatively, the chroman may be oxidised with N-bromosuccinimide (NBS) and the chromene produced by elimination[207]. The chromene may be synthesised directly by use of a C_5-unit at the appropriate oxidation level. Crombie's group have introduced 3-methyl-1,1-dimethoxybutan-3-ol for dimethyl-chromenylation[208], but 3-methylbut-2-enol itself had been used with success[209]. A very important discovery[208] is that hydrogen-bonded phenols are inert to the reagent and compounds such as lonchocarpin are readily produced. The method has already been used to advantage to synthesise various mammea extractives and also to show that the structure of ponnalide requires revision[210]. This work in particular has really arisen from the beautiful work on the citral condensation with phenols. There is the same specificity with regard to hydrogen-bonded phenols and the flemingins[211] become readily available. Even more striking is the direct synthesis of rubranine[212] (63) from pinocembrin[213].

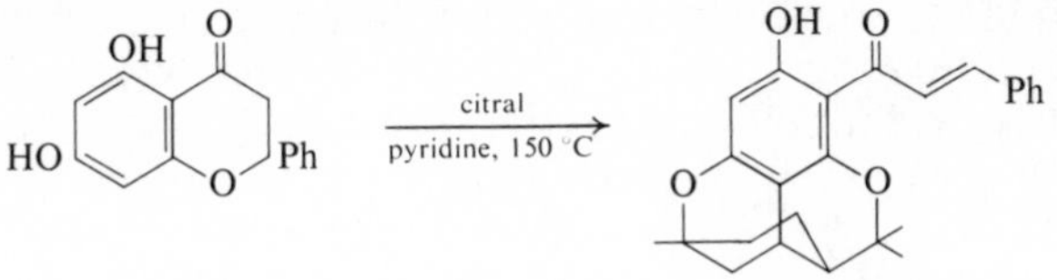

(63)

(Ref. 213)

A somewhat less direct synthesis has also been accomplished[214]. This work on cannabis components[215] may be illustrated by the reaction shown[216].

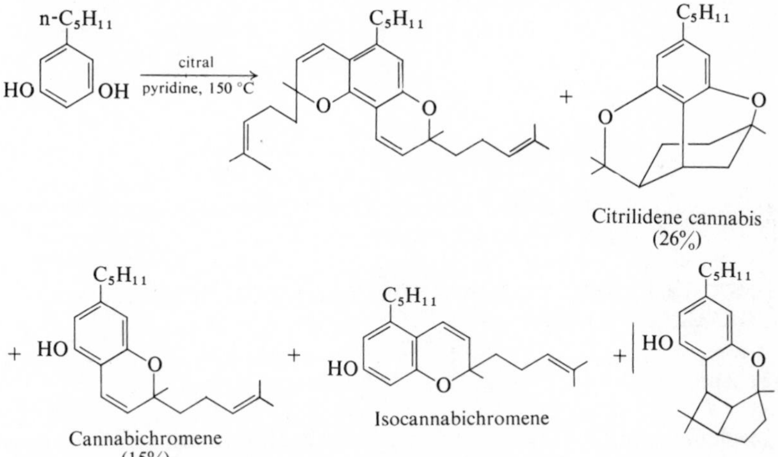

Citrilidene cannabis
(26%)

Cannabichromene
(15%)

Isocannabichromene

(Ref. 216)

Cannabicyclol

A quite separate but equally fruitful approach is to add the C_5-unit to an oxygen atom and then rearrange to give the C-alkylated product. A most successful method is to alkylate the phenol with 3-chloro-3-methylbut-1-yne and to heat the ether with dimethylaniline[217]. 2,2-Dimethylchromenes are produced in high yield[217, 218], it being noted that dimethylformamide (DMF) is an excellent solvent/reagent. Seselin and braylin are readily prepared and it was shown that potassium carbonate may isomerise the initially formed angular flavonochromene to a linear derivative[217]. These same ethers may be used in a different way. Reduction followed by rearrangement gives mainly o-isopentenyl phenols[219, 220]. Criteria for the successful C-isopentenylation of phenols have been established[221]. Treatment of sodium or potassium phenolates in absolutely dry xylene with the bromide is a recommended procedure. The O-ethers result in high yield if DMF and other polar solvents are used. These ethers may be used for Claisen rearrangements to yield the heavily hindered 2,2-dimethylprop-1-ene side chain, frequently found in natural products[222]. It is interesting that 7-O-isopentenylcoumarin gives some product in which the rearranged side chain is at C-3 of the heterocyclic ring. Such compounds occur naturally[223]. The characterisation of the siccanochromenes[224-226] and the tocopherol derivatives recently isolated from marine algae (Tuonia atomaria)[227] shows how readily steroid-like molecules containing heterocyclic oxygen atoms are formed in nature. A very ingenious synthesis of the skeleton of gambogic acid derivatives and the morrelins is based on the idea that these compounds are derived biogenetically from xanthones[228]. An example is given below.

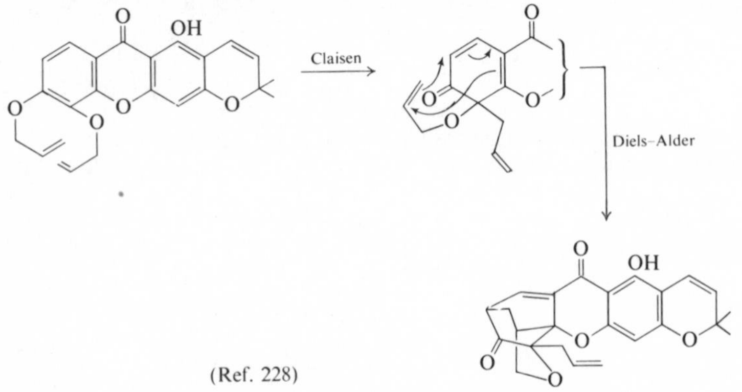

(Ref. 228)

The reaction of phenols with isobutyraldehyde gives coumarans[229].

7.7 HETEROCYCLIC COMPOUNDS WITH TWO HETEROCYCLIC OXYGEN ATOMS

7.7.1 Dioxolanes

The mechanism of 1,3-dioxolane formation by the boron trifluoride catalysed reaction of epoxides with carbonyl compounds has been investigated[230] by the use of ^{18}O-acetone and a previous suggestion[231] confirmed. 2-Substituted dioxolanes are readily converted into esters by photolysis[232] and this could well be used as a highly specific method of oxidation. With trichlorofluoromethane as solvent the product in all cases is a β-chloro-ester but rather

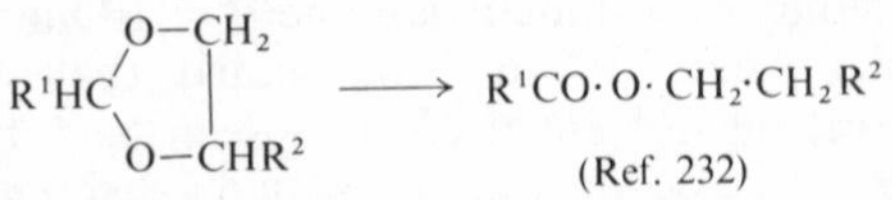

$$R^1HC \overset{O-CH_2}{\underset{O-CHR^2}{\diagdown}} \longrightarrow R^1CO \cdot O \cdot CH_2 \cdot CH_2R^2$$

(Ref. 232)

surprisingly the mechanism is not a simple chlorine abstraction. Stereochemical studies show that there is 100% inversion at the carbon atom bearing the halogen. The reaction of dimethoxydimethylaminomethane with diols gives 2-dimethylaminodioxolanes which with acetic anhydride give olefins with a high degree of regio- and stereo-specificity[233]. A somewhat similar reaction is the sterospecific elimination of 2-acyloxydioxolanes on heating on *o*-xylene[234]. The yields can be excellent. When toluene-*p*-sulphonic acid is added however a diene results[234], presumably by acyl transfer followed by elimination[235]. A somewhat similar reaction is the thermal breakdown of

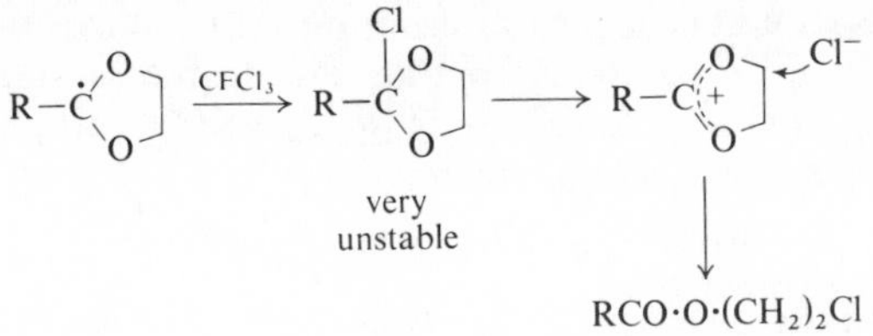
CFCl3
very
unstable
Cl⁻
RCO·O·(CH2)2Cl

(Ref. 232)

2-ethoxydioxolenes to give mixtures of ketenes and acetylenes. With acid present only acetylenes are obtained[235]. When dioxolanes are treated with triphenylmethyl borofluoride followed by water, then ketones result. This provides a breakdown of ketals in the absence of acid, the mechanism of which may be somewhat as shown[236].

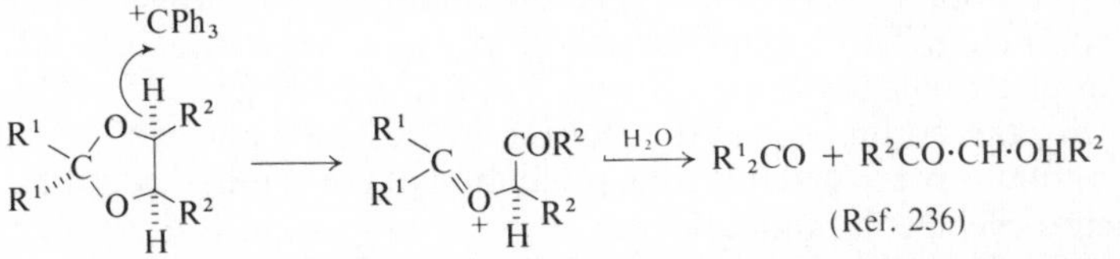
⁺CPh3
H2O
R¹2CO + R²CO·CH·OHR²

(Ref. 236)

1-Oxo-2-methylcylobutan-2-ol exists as a spiro-dioxolan rather than a 1,4-dioxan. This was proved by examination of the pseudo-contact shifts on the ^{13}C spectrum[237]. An extremely careful study of structural assignments to *cis*- and *trans*-4-substituted-2-methyldioxolans has been made[238] which underlines the dangers of assumptions based on purely physical measurements without chemical background. Thus incorrect structures were assigned to *cis*- and *trans*-2,4-dimethyl-1,3-dioxolan[239] because the erroneous assumption was made that methyl groups shield protons in a 1,3-*cis*-arrangement whereas deshielding actually occurs[240]. Moreover, the coupling constants of protons in five-membered rings are of little use in defining ring geometry[241]. A firm base line was provided by the following sequence.

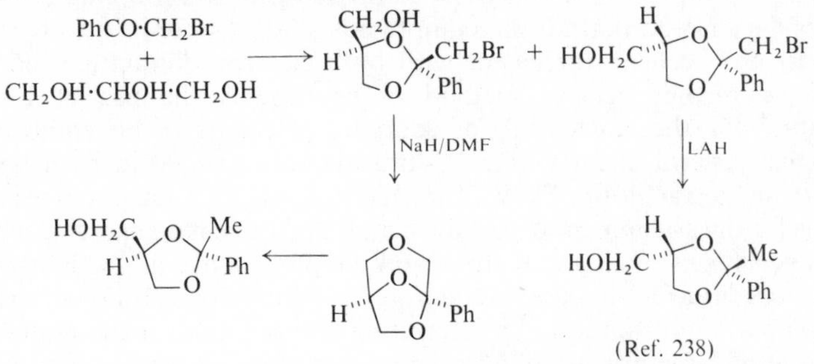
PhCO·CH2Br
+
CH2OH·CHOH·CH2OH
NaH/DMF
LAH

(Ref. 238)

The alcohols of known configuration were then converted into the methyl compounds, carbomethoxy compounds, acetates, etc. It was clear that both 2- and 4-methyl substituents deshield *β-cis* protons and also each other.

This effect is particularly marked for a carbomethoxy group in benzene. Away from this established series, various criteria of stereochemistry can be used. When the plane of the 2-aryl group is parallel to the C-4 — C-5 bond, then a cis-substituent experiences a marked shielding effect. Except in benzylidene compounds this should be a general situation. Also a 4-carbomethoxy group produces a first-order spectrum so that the degree of stagger about the C-4 — C-5 bond can be deduced. A large $J_{4,5'}:J_{4,5}$ ratio indicates a cis-relationship of the 2- and 4-substituents.

7.7.2 1,3-Dioxans

The number of papers on the conformation of these compounds shows little sign of decreasing. Earlier equilibration studies, with the object of determining the conformational free energies of alkyl groups in 1,3-dioxans have been summarised[242, 243]. A general attack on the problem of ^2J $(X \cdot CH_2 \cdot Y)$ for 1,3-diheterocyclic, monocyclic and acyclic systems has been made [244]. The authors attempted to obtain a general equation by iterative fitting of experimental data. Three main contributions were taken into account: (i) the Pauling electronegativities of X and Y, (ii) the bond distances C—X and C—Y, (iii) the mutual orientation of the free orbitals and σ(C—H) bonds. A nonogram is presented allowing prediction of the geminal coupling values in diheterocyclic compounds where X and Y are variously S, O, Se, C. This would supplant the previous treatment which fails for five-membered rings[245] and for X, Y = O, S; N, S; S, S; Se, Se,· etc. The method adopted is related to the more recent method for treating ^2J in Me_nX compounds in which the bond distance is adopted as a parameter for the possibility of overlap between the p-lobes and adjacent C—H bonds[246]. With the ring compounds in particular the angle must also be taken into account and hence an angle function is included. The nonogram can give indications as to the conformation of the ring. It is not yet particularly accurate for acyclic compounds. The crystal structure of 2-p-chlorophenyl-1,3-dioxan is known and this served as a base line for a 220 MHz analysis of 2-alkyl- and 2-aryl-1,3-dioxans[247]. This was rather useful as chemical shift differences for the 4-, 5- and 6-protons are large as compared with the coupling constants. The results showed that all the compounds studied were in the chair form and were in excellent agreement with both electron diffraction and x-ray data[248]. A further general method of promise is the use of $Eu(fod)_3$ reagents[249] for the elucidation of coupling constants of benzodioxans[250]. The n.m.r. spectra of forty-four 1,3-dioxans are recorded in both benzene and carbon tetrachloride[251, 252]. The $\Delta\delta(CCl_4 - C_6H_6)$ may sometimes be suitable for the assignment of certain signals and there is a general discussion on the variations of chemical shifts and coupling constants with structure. Owing to a large 1,3-diaxial interaction 2,2,4-trimethyl-1,3-dioxan exists in a skew-boat conformation. An interesting computation of the enthalpy of formation of 2-methoxytetrahydropyran is in close agreement with experimental data, suggesting the validity of the assumptions made[253]. Whether similar computations will explain some of the observations of conformation of 1,3-dioxans remains to be seen. ΔG° values for the configurational equilibria between *cis*- and *trans*-5-substituted-2-isopropyl are solvent dependent, the substituent sometimes preferring the axial positions (substituent *not* alkyl). For the 5-t-butyl derivative the solvent hardly affects the equilibrium[254, 255]. The n.m.r. and i.r. spectra of some 2,5,5-trisubstituted 1,3-dioxans have been studied[256]. In the case of (64) a hydrogen bond exists, the molecule adopting the conformation shown. This would not seem to be so for (65)[257, 258].

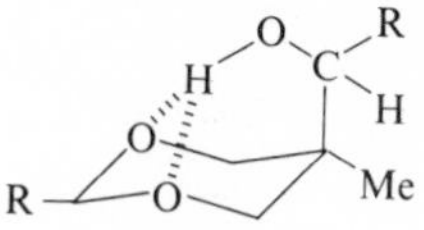

(64)   (65)

(Ref. 256)   (Ref. 257)

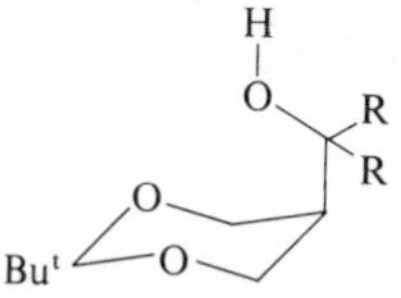

The conformations of the possible stereoisomers of 2,6-dimethyl-4-t-butyl-1,3-dioxan have been studied[259], the compounds being made by unexceptionable routes[260, 261]. A *syn*-axial methyl causes deshielding and helps assignment. The 2,4-*trans*, 2,6-*cis* isomer is not the standard chair but a mixture of two twist forms. 2-Methoxy-1,3-dioxan exists with the methoxy group axial[262], a result readily understood from the 2-methoxytetrahydropyran case previously quoted[253]. Equilibration studies on 5-alkoxy-1,3-dioxans together with dipole measurements have led to results that allow qualitative prediction of the conformational equilibria[263]. Temperature dependence studies have helped to decide 1,3-dioxan conformation in a number of cases[264] as has the n.m.r. chemical shift method[265]. Thermodynamic parameters have been computed for 5-amino- and 5-acetyl-5-methyl-1,3-dioxan[266]. The amino group takes up an axial configuration but the acetoxy group is equatorial.

7.7.3 1,4-Dioxans

The photo-addition of ketenes to 1,2-diketones has been studied[267]. Phenanthraquinone gives either 2-oxo-1,4-dioxan derivatives or spirobutyrolactones. The photo-additions of phenanthraquinone to olefins is not stereospecific for the *cis*- or *trans*-olefins and seems to proceed via a triplet excited state[268]. Although γ-chloro-alcohols give substituted oxetans when heated with base, 1-chloro-2-(β-hydroxyethoxy)-3-hydroxypropane gives only 2-hydroxymethyl-1,4-dioxan under these conditions[269]. 2,3-Dichloro-1,4 dioxan reacts with 1,3-propandiol to give a symmetrical 2,2′-linked bi-1,3-dioxan (the rings of which are in the chair form).

7.8 MISCELLANEOUS

The uses and preparations of macrocyclic polyether sulphides have been investigated[270]. The article also is useful as a review of similar compounds. The reaction of quinones with ketenes serves to introduce very hindered groups via a spirobutyrolactone[271]. The reaction proceeds well with diphenyl-

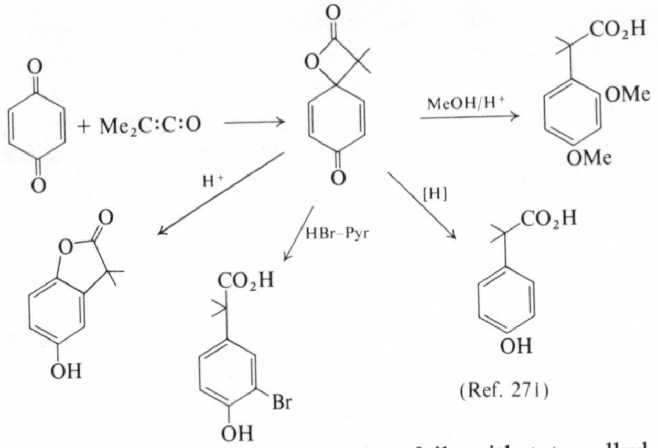

(Ref. 271)

ketene and many substituted quinones, but fails with tetra-alkyl quinones. Studies on *o*-quinone methides have produced many extremely interesting results[272-275], but perhaps pride of place goes to the ingeniously conceived synthesis of the lignan carpanone (66)[276]. This is a one-step reaction as shown.

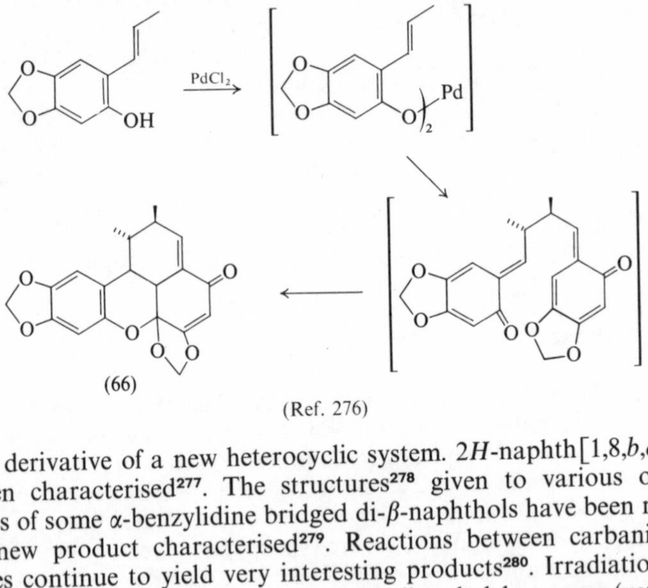

(66)

(Ref. 276)

An oxo derivative of a new heterocyclic system. 2*H*-naphth[1,8,*b,c*]oxepin has been characterised[277]. The structures[278] given to various oxidation products of some α-benzylidine bridged di-β-naphthols have been reviewed, and a new product characterised[279]. Reactions between carbanions and quinones continue to yield very interesting products[280]. Irradiation of iso-pentenyl-quinones gives rise not only to 2,3-dimethylchromenes (yet another method for the production of these compounds) but also to a seven-membered,

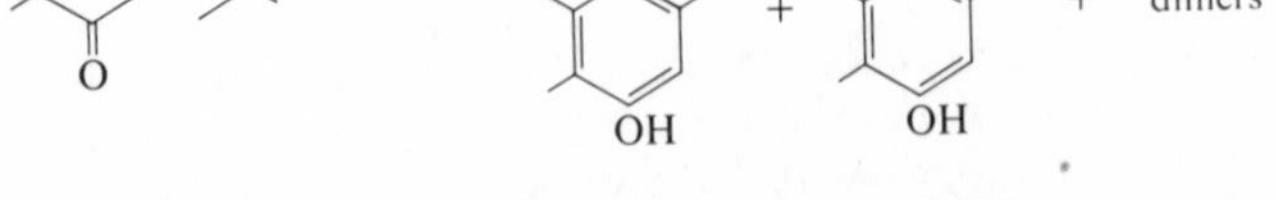

(Ref. 281)

ring system[281]. The reactions of a large number of oxygen heterocyclic compounds with diborane have been studied[282]. The rearrangement of some heterocyclic dienones has been shown to proceed by an aralkyl migration, not an aryloxy migration[283].

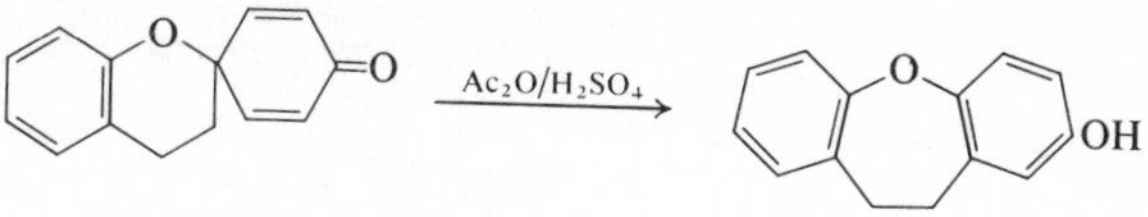

(Ref. 283)

Irradiation of (67) in the presence of acid gives (68), whereas heating gives (69), which is not an intermediate in the production of (68)[284].

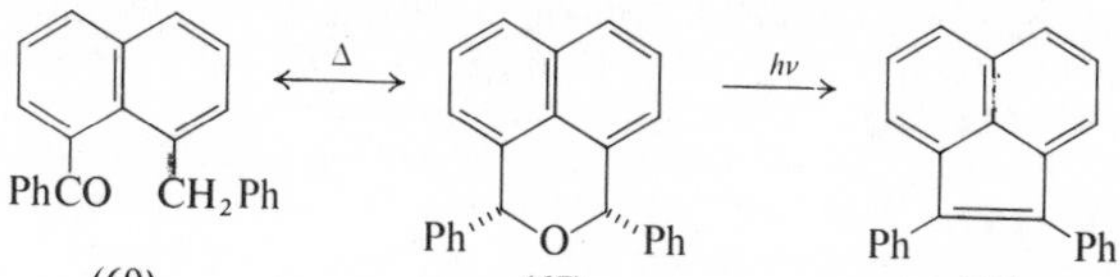

(69)   (67)   (68)

(Ref. 284)

The 10π-heteronin, in which the heterocyclic atom is oxygen, is stated to be non-aromatic. This conclusion is based on thermal stability[285]. The same compound undergoes a reaction that could be the first concerted $2_\pi + 8_\pi$ (though it could also be $2_\pi + 2_\pi + 2_\pi$)[286]. The photochemistry of 1-oxa-2,3,6,7-cycloheptadiene is in line with that of other vinyl ethers[287]. Robustol is an

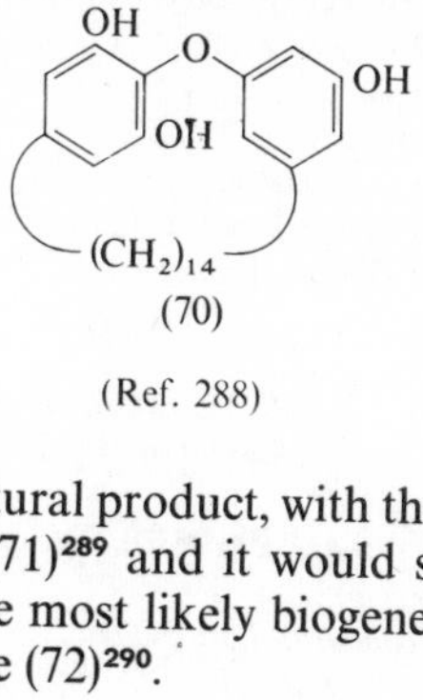

(70)

(Ref. 288)

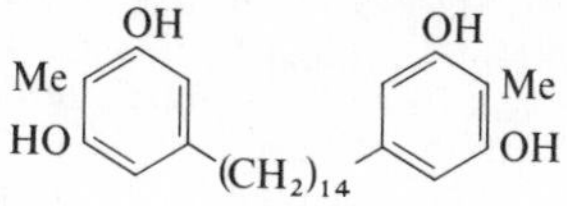

(71)

(Ref. 289)

unusual natural product, with the structure (70)[288]. This is highly reminiscent of striatol (71)[289] and it would seem that phenolic coupling of such a precursor is the most likely biogenesis of (70). Arugosin has been characterised as a mixture (72)[290].

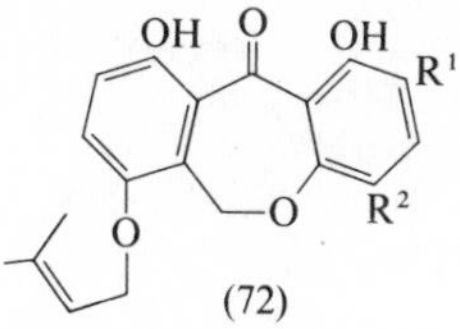

(72)

R^1=CH$_2$·CH:CMe$_2$, R^2=H   2 parts
R^1=H, R^2=CH$_2$·CH:CMe$_2$ 1 part

(Ref. 290)

Acknowledgement

The Author wishes to acknowledge the courtesy extended by the following

Publishers and Societies for permission to reproduce chemical formulae from their journals:

Pergamon Press Ltd.   *Tetrahedron Letters,Tetrahedron,*
                                   Phytochemistry
The Chemical Society   *J. Chem. Soc. C. Chem. Commun.*
The American Chemical Society   *J. Am. Chem. Soc. J. Org. Chem.*
C.S.I.R.O.   *Austral. J. Chem.*
Verlag Helvetica Chemie Acta   *Helv. Chim Acta*
Verlag Chemie G.m.b.H.   *Chem. Ber.*

References

1. Olsen, S. R. and Turner, S. (1971). *J. Chem. Soc.,* 1632
2. Grigg, R. and Jackson, J. L. (1970). *J. Chem. Soc. C,* 552
3. Higo, M. and Mukaiyama, T. (1970). *Tetrahedron Letters,* 2565
4. Batty, J. W., Howes, P. D. and Stirling, C. J. M. (1971). *Chem. Commun.,* 534
5. Brent, D. A., Hribar, J. D. and DeJongh, D: C. (1970). *J. Org. Chem.,* **35,** 135
6. Wege, D. (1971). *Tetrahedron Letters,* 2337
7. White, J. D., Mann, M. E., Kirschenbaum, H. D. and Mitra, A. (1971). *J. Org. Chem.,* **36,** 1048
8. Haddadin, M. J., Yavrouian, A. and Issidorides, C. H. (1970). *Tetrahedron Letters,* 1409
9. Muller, E. and Langer, E. (1970). *Tetrahedron Letters,* 735
10. Hiroaka, H. (1971). *Chem. Commun.,* 1610
11. Hiroaka, H. (1970). *J. Phys. Chem.,* **74,** 574
12. Couture, A. and Lablanche-Combier, A. (1971). *Chem. Commun.,* 891
13. Couture, A. and Lablanche-Combier, A. (1969). *Chem. Commun.,* 524
14. van Tamelen, E. E. and Whitesides, T. H. (1968). *J. Amer. Chem. Soc.,* **90,** 3894
15. Kutney, J. P., Hanssen, A. W. and Nair, G. V. (1971). *Tetrahedron,* **27,** 3323
16. Clementi, S., Linda, P. and Marino, G. (1970). *Tetrahedron Letters,* 1389
17. Ciranni, G. and Clementi, S. (1971). *Tetrahedron Letters,* 3833
18. Suzuki, A., Miyawa, N. and Itoh, M. (1971). *Tetrahedron,* **27,** 2775
19. LaRochelle, R. W. and Trost, B. M. (1970). *Chem. Commun.,* 1353
20. Lewis, K. G. and Mulquiney, C. E. (1970). *Austral. J. Chem.,* **23,** 2315
21. Divalt, S., Chun, M. C. and Joullie, M. J. (1970). *Tetrahedron Letters,* 777
22. Rindone, B. and Scholastico, C. (1971). *J. Chem. Soc. C,* 3339
23. Corrie, J. E. T. (1971). *Tetrahedron Letters,* 4873
24. Kispert, L. D., Quijano, R. C. and Pittmann, C. V., Jr. (1971). *J. Org. Chem.,* **36,** 3837
25. Chadwick, D. J., Chambers, J., Meakins, G. D. and Snowden, R. L. (1971). *Chem. Commun.,* 624
26. Martin, M. L., Rose, J. C., Martin, G. J. and Founari, P. (1970). *Tetrahedron Letters,* 3407
27. Dahlquist, K. I. and Hornfeldt, A. B. (1971). *Tetrahedron Letters,* 3837
28. Arlinger, L. (1970). *Acta Chem. Scand.,* **24,** 662
29. Juchnovski, I. and Kaneti, (1971). *Tetrahedron,* **27,** 4269
30. Buchi, G., Degen, P., Gautschi, F. and Wilhalm, B. (1971). *J. Org. Chem.,* **36,** 199
31. Dana, G., Scribe, P. and Girault, J. P. (1970). *Tetrahedron Letters,* 4137
32. Mikhailov, B. M. and Bubnov, Y. N. (1971). *Tetrahedron Letters,* 2127
33. Servé, M. P. and Rosenberg, H. M. (1970). *J. Org. Chem.,* **35,** 1237
34. Shani, A. and Mechoulam, R. (1970). *Chem. Commun.,* 272
35. Roscher, N. M. (1971). *Chem. Commun.,* 471
36. Wolthuis, E., Bourma, B., Modderman, J. and Sytsma, L. (1970). *Tetrahedron Letters,* 407
37. Corrie, J. E. T., Green, G. H., Ritchie, E. and Taylor, W. C. (1970). *Austral. J. Chem.,* **23,** 133
38. Mundy, B. P., DeBarnardis, A. R. and Olzenberger, R. O. (1971). *J. Org. Chem.,* 3830
39. Echinasi, E. H. (1970). *J. Org. Chem.,* **35,** 1097
40. Manyik, R. M., Walker, W. E., Atkins, K. E. and Hammack, E. S. (1970). *Tetrahedron Letters,* 3813

41. Fisch, M., Smallcombe, S., Gramain, J. C., McKervey, M. A. and Anderson, J. G. (1970). *J. Org. Chem.*, **35**, 1886
42. Birch, A. J., Dickman, J. and MacDonald, P. L. (1970). *Chem. Commun.*, 52
43. Reese, C. B., Saffhill, R. and Sulston, J. E. (1970). *Tetrahedron*, **26**, 1023
44. Stapp, P. R. (1969). *J. Org. Chem.*, **34**, 479
45. Stapp, P. R. and Drake, C. A. (1971). *J. Org. Chem.*, **36**, 522
46. Martin, J. C. (1970). *Bull. Soc. Chim. Fr.*, 277
47. Zweifel, G. and Plamondon, J. (1970). *J. Org. Chem.*, **35**, 898
48. Srinivasan, R. (1970). *J. Org. Chem.*, **35**, 786
49. Atkinson, R. S. (1970). *Chem. Commun.*, 177
50. Atkinson, R. S. (1971). *Chem. Commun.*, 585
51. Atkinson, R. S. (1971). *J. Chem. Soc. C*, 784
52. Kearns, D. R. (1969). *J. Amer. Chem. Soc.*, **91**, 6554
53. Schenck, G. O. (1952). *Angew. Chem.*, **64**, 12
54. Becker, J. and Ohloff, G. (1971). *Helv. Chim. Acta*, **54**, 2889
55. Luria, M. and Stein, G. (1970). *Chem. Commun.*, 1651
56. Duperrier, A. and Dreux, J. (1970). *Tetrahedron Letters*, 3127
57. Sorensen, A. K. and Klitgard, N. A. (1970). *Acta Chem. Scand.*, **24**, 343
58. Sasaki, T., Kanematsu, K. and Kakchi, A. (1971). *J. Org. Chem.*, **36**, 2451
59. Hayasi, Y. and Nozakaki, H. (1971). *Tetrahedron*, **27**, 3085
60. Bittencourt, A. M., Gottlieb, O. R., Mors, W. B., Magalhães, M. T., Mageswaran, S., Ollis, W. D. and Sutherland, I. O. (1971). *Tetrahedron*, **27**, 1043
61. Bedford, C. T., Forrester, J. M. and Money, T. (1970). *Canad. J. Chem.*, **48**, 2645
62. Vrkoc, J., Dolejs, L., Sedmera, P., Vasikova, S. and Sorm. F. (1970). *Tetrahedron Letters*, 3369
63. Opitz, L. and Hänsel, R. (1970). *Tetrahedron Letters*, 5105
64. Narayanan, P., Zechmeister, K., Röhrl, M. and Hoppe, W. (1970). *Tetrahedron Letters*, 3945
65. Hänsel, R., Ohlendorf, D. and Pelter, A. (1970). *Zeit. Natur.*, **25b**, 989
66. Kapper, T. and Schmidt, H. (1970). *Tetrahedron Letters*, 5105
67. Ishibe, N., Odani, M. and Sunami, N. (1971). *Chem. Commun.*, 1034
68. Ishibe, N., Odani, M. and Teramura, K. (1970). *Chem. Commun.*, 371
69. Hart, H. and Love, G. M. (1971). *Tetrahedron Letters*, 625
70. Pelter, A. and Ohlendorf, D., Unpublished observations
71. Merlini, L., Nasini, G. and Selva, A. (1970). *Tetrahedron*, **26**, 2746
72. Mills, F. D., Weisleder, D. and Hodge, J. E. (1970). *Tetrahedron Letters*, 1243
73. Wachter, M. P. and Harris, T. M. (1970). *Tetrahedron*, **26**, 1685
74. Harris, T. M. and Wachter, M. P. (1970). *Tetrahedron*, **26**, 5255
75. Scott, A. I., Guildford, H., Ryan, J. J. and Skingle, D. (1971). *Tetrahedron*, **27**, 3025
76. Scott, A. I., Guildford, H. and Skingle, D. (1971). *Tetrahedron*, **27**, 3039
77. Scott, A. I., Pike, D. G., Ryan, J. J. and Guildford, H. (1971). *Tetrahedron*, **27**, 3051
78. Holland, J. M. and Jones, D. W. (1970). *J. Chem. Soc. C*, 531
79. Holland, J. M. and Jones, D. W. (1970). *J. Chem. Soc, C*, 536
80. Sarel, S., Shalon, Y. and Yanuka, Y. (1970). *Chem. Commun.*, 81
81. Farcasiu, D., Vasilescu, A. and Balaban, A. T. (1971). *Tetrahedron*, **27**, 681
82. Balaban, A. T. (1970). *Tetrahedron*, **26**, 739
83. Sniekus, V. and Kan, G. (1970). *Chem. Commun.*, 1208
84. Pedersen, C. L., Harrit, N. and Buchardt, O. (1970). *Acta Chem. Scand.*, **24**, 3435
85. Wasserman, H. H. and Pavia, D. L. (1970). *Chem. Commun.*, 1459
86. Guld, G. and Price, C. C. (1961). *J. Amer. Chem. Soc.*, **83**, 1770; 1962, **84**, 2090
87. Cotterill, W. D., Cottam, J. and Livingstone, R. (1970). *J. Chem. Soc. C*, 1006
88. Dolby, L. J. and Adler, E. (1971). *Tetrahedron Letters*, 3803
89. Onaka, T. (1971). *Tetrahedron Letters*, 4395
90. Dann, O., Hagedorn, K. W. and Hofman, H. (1971). *Ber.*, **104**, 3313
91. Bellinger, N., Cagniant, D. and Cagniant, P. (1971). *Tetrahedron Letters*, 49
92. Föhlisch, B. (1971). *Ber.*, **104**, 348
93. Föhlisch, B. (1971). *Ber.*, **104**, 350
94. Clark Still, W. and Goldsmith, D. J. (1970). *J. Org. Chem.*, **35**, 2282
95. Cooke, R. G. and Down, J. G. (1970). *Tetrahedron Letters*, 1039

96. Arshad, M., Devlin, J. P. and Ollis, W. D. (1971). *J. Chem. Soc. C,* 1324
97. Haynes, L. J., Holdsworth, D. K. and Russell, R. (1970). *J. Chem. Soc. C,* 2581
98. Bycroft, B. W., Hassaniali-Walzi, A., Johnson, A. W. and King, T. J. (1970). *J. Chem. Soc. C,* 1686
99. Arora, S., Deymann, H., Tiwari, R. D. and Winterfeldt, E. (1971). *Tetrahedron,* **27,** 981
100. Brown, K. S. and Weiss, U. (1971). *Tetrahedron Letters,* 3501
101. Brown, K. S. and Baker, P. M. (1971). *Tetrahedron Letters,* 3505
102. Smith, I. R. and Sutherland, M. D. (1971). *Austral. J. Chem.,* **24,** 1487
103. Bolon, P. A. (1970). *J. Org. Chem.,* **35,** 3666
104. Cardillo, G., Merlini, L., Nasini, G. and Salvadori, P. (1971). *J. Chem. Soc. C,* 3967 (cf. Cardillo, G., Cricchio, R. and Merlini, L. (1971). *Tetrahedron,* **27,** 1875
105. Cooke, R. G. and Down, J. G. (1971). *Austral. J. Chem.,* **24,** 1257 (cf. (1970). *Tetrahedron Letters,* 583, 1037, 1039)
106. Bognar, D., Clark-Lewis, J. W., Liptakne-Tokes, A. and Rokosi, M. (1970). *Austral. J. Chem.,* **23,** 2015
107. Clark-Lewis, J. W. and Jemison, R. W. (1970). *Austral. J. Chem.,* **23,** 315
108. Grouiller, A. and Pacheco, H. (1971). *Tetrahedron Letters,* 4881
109. duPreez, I. C., Fourie, T. G. and Roux, D. G. (1971). *Chem. Commun.,* 333
110. Sears, K. D. and Engen, R. J. (1971). *Chem. Commun.,* 612
111. Dewar, D. and Sutherland, R. G. (1970). *Chem. Commun.,* 272
112. Heany, H. and McCarthy, C. T. (1970). *Chem. Commun.,* 123
113. Casiraghi, G., Casnati, G. and Salerno, G. (1971). *J. Chem. Soc. C,* 2546
114. Misiti, D. and Rimaton, V. (1970). *Tetrahedron Letters,* 947
115. Keane, D. D., Marathe, K. G., O'Sullivan, W. I., Philbin, E. M., Simons, R. M. and Teagner, P. C. (1970). *J. Org. Chem.,* **35,** 2286
116. Clark-Lewis, J. W. and Mahandru, M. M. (1971). *Austral. J. Chem.,* **24,** 549; (cf. (1970). *Chem. Commun.,* 1287)
117. Clark-Lewis, J. W. and Mahandru, M. M. (1971). *Austral. J. Chem.,* **24,** 563
118. Gaffield, W. (1970). *Tetrahedron,* **26,** 4093
119. Abraham, D. J., Takagi, S., Rosenstein, R. D., Shiono, R., Wagner, H., Hörhammer, L., Seligmann, O. and Farnsworth, N. R. (1970). *Tetrahedron Letters,* 2675
120. Wagner, H., Seligmann, O., Hörhammer, L., Seitz, M. and Sonnenbichler, J. (1971). *Tetrahedron Letters,* 1895
121. Pelter, A. and Hänsel, R. (1968). *Tetrahedron Letters,* 2911
122. Pelter, A., Bradshaw, J. and Warren, R. F. (1971). *Phytochem.,* **10,** 835
123. Brady, B. A., Healy, M. M., Kennedy, J. A., O'Sullivan, W. I. and Philbin, E. M. (1970). *Chem. Commun.,* 1435
124. Merchant, J. R. and Regl, D. V. (1970). *Chem. Commun.,* 380
125. Schonberg, A. and Khandelwal, G. D. (1970). *Ber.,* **103,** 2780
126. Ramakrishnan, V. T. and Kagan, J. (1970). *J. Org. Chem.,* **35,** 2898
127. Ramakrishnan, V. T. and Kagan, J. (1970). *J. Org. Chem.,* **35,** 2901
128. Crabbe, P., Diaz, E., Haro, J., Perez, G., Salgado, D. and Santos, E. (1970). *Tetrahedron Letters,* 5069
129. Okigawa, M., Hatanaka, H., Kawano, N., Matsunaga, I. and Tamura, Z. (1970). *Tetrahedron Letters,* 2935
130. Kaneta, M. and Sugiyama, N. (1971). *J. Chem. Soc. C,* 1982. (cf. (1969). *Bull. Chem. Soc. Japan,* **42,** 2084
131. Porter, L. J. and Markham, K. R. (1970). *Phytochem.,* **9,** 1363
132. Hrazdina, G. (1971). *Phytochem.,* **10,** 1125
133. Hrazdina, G. (1970). *Phytochem.,* **9,** 1647
134. Martensson, O. and Warren, C. H. (1970). *Acta Chem. Scand.,* **24,** 2745
135. Matsuura, T., Takemoto, T. and Najashima, R. (1971). *Tetrahedron Letters,* 1539
136. Matsuura, T., Matsushima, H. and Nakashima, R. (1970). *Tetrahedron,* **26,** 435
137. Ferreira, D., Hundt, H. K. L. and Roux, D. G. (1971). *Chem. Commun.,* 1257
138. Seshadri, T. R. and Trikha, R. K. (1971). *Ind. J. Chem.,* **9,** 302
139. Brachet, J. and Paris, R. R. (1970). *Phytochem.,* **9,** 438
140. duPreez, I. C., Rowan, A. C. and Roux, D. G. (1970). *Chem. Commun.,* 492
141. Volsteedt, R. F. and Roux, D. G. (1971). *Tetrahedron Letters,* 1647
142. duPreez, I. C., Rowan, A. C., Roux, D. G. and Feeney, J. (1971). *Chem. Commun,* 315
143. Weinges, K., Marx, H. D. and Goritz, K. (1970). *Ber.,* **103,** 2336

144. Weinges, K., Perner, J. and Marx, H. D. (1970). *Ber.*, **103**, 2344
145. Jurd, L. and Somers, T. C. (1970). *Phytochem.*, **9**, 420
146. Hrazdina, G. and Borgell, A. J. (1971). *Phytochem.*, **10**, 2211
147. Sears, K. D. and Casebier, R. L. (1970). *Phytochem.*, **9**, 1590
148. Jackson, B., Locksley, H. D., Scheinmann, F. and Wolstenholme, W. A. (1971). *J. Chem. Soc. C*, 3791
149. Konoshima, M. and Ikeshiro, Y. (1970). *Tetrahedron Letters*, 1717
150. Konoshima, M., Ikeshiro, Y. and Miyahara, S. (1970). *Tetrahedron Letters*, 4203
151. Herbin, G. A., Jackson, B., Locksley, H. D., Scheinmann, F. and Wolstenholme, W. A. (1970). *Phytochem.*, **9**, 221
152. Joshi, B. S., Kamat, V. N. and Vishwanathan, N. (1970). *Phytochem.*, **9**, 881
153. Pelter, A., Warren, R. F., Chexal, K. K., Handa, B. K. and Rahman, W. (1971). *Tetrahedron*, 1625
154. Locksley, H. D. and Murray, I. G. (1971). *J. Chem. Soc. C*, 1332
155. Pelter, A., Warren, R., Usmani, J. N., Rizvi, R. H., Ilyas, M. and Rahman, A. (1969). *Experientia*, **25**, 351
156. Molyneux, R. J., Waiss, A. C. and Haddon, W. H. (1970). *Tetrahedron*, **26**, 1409
157. Natarajan, S., Murti, V. V. S. and Seshadri, T. R. (1971). *Ind. J. Chem.*, **9**, 383
158. Ahmad, S. and Razaq, S. (1971). *Tetrahedron Letters*, 4631
159. Natarajan, S., Murti, V. V. S. and Seshadri, T. R. (1970). *Phytochem.*, **9**, 575
160. Handa, B. K., Chexal, K. K., Rahman, W., Okigawa, M. and Kawano, N. (1971). *Phytochem.*, **10**, 436
161. Geiger, H. and de Groot-Pfleiderer. (1971). *Phytochem.*, **10**, 1936
162. Beckmann, S., Geiger, H. and de-Groot-Pfleiderer. (1971). *Phytochem.*, **10**, 2465
163. Machima, T., Okigawa, M., Kawano, N., Khan, N. U., Ilyas, M. and Rahman, W. (1970). *Tetrahedron Letters*, 2937
164. Khan, N. U., Ansari, W. H., Usmani, J. N., Ilyas, M. and Rahman, W. (1971). *Phytochem.*, **10**, 2130
165. Khan, N. U., Ilyas, M., Rahman, W., Okigawa, M. and Kawano, N. (1971). *Phytochem.*, **10**, 2541
166. Pelter, A., Warren, R. F., Hameed, N., Ilyas, M. and Rahman, W. (1971). *J. Ind. Chem. Chem. Soc.*, **48**, 204
167. Coxon, D. T., Holmes, A., Ollis, W. D. and Vora, V. C. (1970). *Tetrahedron Letters*, 5237
168. Bryce, T., Collier, P. D., Fowlis, I., Thomas, P. E., Frost, D. and Wilkins, C. K. (1970). *Tetrahedron Letters*, 2789
169. Coxon, D. T., Holmes, A. and Ollis, W. D. (1970). *Tetrahedron Letters*, 5241
170. Coxon, D. T., Holmes, A. and Ollis, W. D. (1970). *Tetrahedron Letters*, 5247
171. Clark-Lewis, J. W. and Jemison, R. W. (1970). *Austral. J. Chem.*, **23**, 89
172. Bhatia, V. P. and Mathu, K. B. L. (1971). *Tetrahedron Letters*, 2371
173. Miyana, M. (1970). *J. Org. Chem.*, **35**, 247
174. Weber-Schilling, C. A. and Wanzlick, H. W. (1971). *Ber.*, **104**, 1518
175. Ferreira, D., Brink, C. V. D. M. and Roux, D. G. (1971). *Phytochem.*, **10**, 1141
176. Chubachi, M. and Hamada, M. (1971). *Tetrahedron Letters*, 3537
177. Crombie, L., Davis, J. S. and Whiting, D. A. (1971). *J. Chem. Soc. C*, 304. (cf. Caplin, G. H., Ollis, W. D. and Sutherland, I. O. (1968). *J. Chem. Soc. C*, 2302)
178. Crombie, L., Dewick, P. M. and Whiting, D. A. (1970). *Chem. Commun.*, 1469
179. Crombie, L., Dewick, P. M. and Whiting, D. A. (1971). *Chem. Commun.*, 1183
180. Dewick, P. M., Barz, W. and Grisebach, H. (1970). *Phytochem.*, **9**, 775
181. Barz, W., Adamek, Ch. and Berlin, J. (1970). *Phytochem.*, **9**, 1735
182. Barz, W. (1970). *Phytochem.*, **9**, 1744
183. Wang, E. and Latch, G. C. M. (1971). *Phytochem.*, **10**, 486
184. Purushothaman, H. K., Kishore, V. M., Narayanaswami, V. and Connolly, J. D. (1971). *J. Chem. Soc.*, 2420
185. Beck, A. B. and Knox, J. R. (1971). *Austral. J. Chem.*, **24**, 1509
186. Sidwell, W. T. and Tamm, Ch. (1970). *Tetrahedron Letters*, 475
187. Gourley, R. N., Grimshaw, J. and Millar, P. G. (1970). *J. Chem. Soc. C*, 2318
188. Grimshaw, J. (1970). *J. Chem. Soc. C*, 2324
189. Dean, F. M., Frankham, D. D., Hatam, N. and Hill, B. W. (1971). *J. Chem. Soc. C*, 218
190. Kaufman, K. D., Worden, L. R., Lude, E. T., Strong, M. K. and Reitz, N. C. (1970). *J. Org. Chem.*, **35**, 157

191. Buchi, G., Klaubert, D. H., Shank, R. C., Weinreb, S. M. and Wogan, G. N. (1971). *J. Org. Chem.*, **36**, 1143
192. Raman, P. V., Sankarappa, S. K. and Swaninathan, S. (1970). *Tetrahedron Letters*, 2353
193. Fritig, B., Hirth, L. and Ourisson, G. (1970). *Phytochem.*, **9**, 1963
194. Macleod, J. K. (1970). *Tetrahedron Letters*, 3611 (cf. (1970). *Tetrahedron Letters*, 1319)
195. Murray, R. D. H. and Ballantyne, M. M. (1970). *Tetrahedron*, **26**, 4473
196. Mageswaran, S., Ollis, W. D., Roberts, R. J. and Sutherland, I. O. (1969). *Tetrahedron Letters*, 2897
197. Jurd, L. (1968). *Experientia*, **24**, 858
198. Jurd, L. (1969). *Tetrahedron*, **25**, 1407
199. Larkin, J., Nonhebel, D. C. and Wood, H. C. S. (1970). *Chem. Commun.*, 455
200. Molyneux, R. J. and Jurd, L. (1970). *Tetrahedron*, **26**, 4743
201. Jurd, L., Steven, K. and Manners, G. (1971). *Tetrahedron Letters*, 2275
202. Fourrey, J. L., Rondest, J. and Polonsky, J. (1970). *Tetrahedron*, **26**, 3839
203. Bowman, R. M., Grundon, M. F. and James, K. J. (1970). *Chem. Commun.*, 667
204. Murray, R. D. H., Sutcliffe, M. and McCabe, P. H. (1971). *Tetrahedron*, **27**, 4901
205. Bohlmann, F. and Franke, H. (1971). *Ber.*, **104**, 3229
206. Cardillo, G., Cricchio, R. and Merlini, L. (1971). *Tetrahedron*, **27**, 1875
207. Grundon, M. F. G. and James, K. J. (1970). *Chem. Commun.*, 1427
208. Bandaranayake, W. M., Crombie, L. and Whiting, D. A. (1971). *J. Chem. Soc. C*, 811
209. Lewis, J. R. and Reary, J. B. (1970). *J. Chem. Soc. C*, 1662
210. Games, D. E. and Haskins, N. J. (1971). *Chem. Commun.*, 1005
211. Cardillo, G., Merlini, L. and Mondelli, R. (1968). *Tetrahedron*, **24**, 497
212. Combes, G., Vassort, Ph. and Winternitz, F. (1970). *Tetrahedron*, **26**, 598
213. Bandaranayake, W. M., Crombie, L. and Whiting, D. A. (1971). *J. Chem. Soc. C*, 804
214. Kane, V. V. and Grayek, T. L. (1971). *Tetrahedron Letters*, 3991
215. Crombie, L. and Ponsford, R. (1971). *J. Chem. Soc. C*, 788
216. Crombie, L. and Ponsford, R. (1971). *J. Chem. Soc. C*, 796
217. Hlubucek, J., Ritchie, E. and Taylor, W. C. (1971). *Austral. J. Chem.*, **24**, 2347 (cf. (1969). *Tetrahedron Letters*, 1369)
218. Mukerjee, S. K., Sarkar, S. C. and Seshadri, T. R. (1970). *Ind. J. Chem.*, **8**, 861
219. Murray, R. D. H., Ballantyne, M. M. and Mathai, K. P. (1971). *Tetrahedron*, **27**, 1245 (cf. Hlubucek, J., Ritchie, E. and Taylor, W. C. (1969). *Chem. Ind. London*, 1780)
220. Murray, R. D. H., Hogg, T. G., Ballantyne, M. M. and McCabe, P. H. (1971). *Tetrahedron Letters*, 3317
221. Hlubucek, J., Ritchie, E. and Taylor, W. C. (1971). *Austral. J. Chem.*, **24**, 2355
222. Ballantyne, M. M., McCabe, P. H. and Murray, R. D. H. (1971). *Tetrahedron*, **27**, 871
223. Russch, J., Szendrei, K., Minker, E. and Novak, I. (1970). *Tetrahedron Letters*, 4305
224. Hirai, K., Suzuki, K. T. and Nozoe, S. (1971). *Tetrahedron*, **27**, 6057. (cf. (1969). *Tetrahedron Letters*, 2457; (1971). *Chem. Commun.*, 527, 528)
225. Nozoe, S. and Suzuki, K. T. (1971). *Tetrahedron*, **27**, 6063
226. Nozoe, S. and Hirai, K. (1971). *Tetrahedron*, **27**, 6073
227. Gonzalez, A. G., Darias, J. and Martin, J. (1971). *Tetrahedron Letters*, 2729
228. Quillinan, A. J. and Scheinmann, F. (1971). *Chem. Commun.*, 966
229. Martin, J. C., Franke, N. W. and Singerman, G. M. (1970). *J. Org. Chem.*, **35**, 2904
230. Blackett, B. N., Coxon, J. M., Hartshorn, M. P., Lewis, A. J., Little, G. R. and Wright, G. J. (1970). *Tetrahedron*, **26**, 1311
231. Yandouskii, Y. N. and Temnikova, T. I. (1968). *J. Org. Chem. USSR*, **4**, 1695
232. Hartgrink, J. W., v.d. Laan, L. C. J., Engberts, J. B. F. N. and deBoer, T. J. (1971). *Tetrahdderon*, **27**, 4323
233. Eastwood, F. W., Harrington, K. L., Josan, J. S. and Pura, J. L. (1970). *Tetrahedron Letters*, 5223
234. v.d. Veek, A. P. M. and v. Putten, F. H. (1970). *Tetrahedron Letters*, 3951. (cf. Scheeren, J. W., v.d. Veek, A. P. M. and Stevens, W. (1969). *Rec. Trav. Chim.*, **88**, 95)
235. Moss, G. I., Grank, G. and Eastwood, F. W. (1970). *Chem. Commun.*, 206
236. Barton, D. H. R., Magnus, D. D., Smith, G. and Zurr, D. (1971). *Chem. Commun.*, 861
237. Duggan, J. C., Urry, W. H. and Schaefer, J. (1971). *Tetrahedron Letters*, 4197
238. Inch, T. O. and Williams, N. (1970). *J. Chem. Soc. C*, 263
239. Anteunis, M. and Alderweireld, F. A. (1964). *Bull. Soc. Chim. Belge*, **73**, 889
240. Altona, C. and v.d. Veek, A. P. M. (1968). *Tetrahedron*, **24**, 4377

241. Inch, T. D. (1969). *Ann. Rev. N. M. R. Spectroscopy*, **2,** 71
242. Eliel, E. L. (1970). *Acc. Chem. Research*, **3,** 1
243. Eliel, E. L. (1970). *Bull. Soc. Chim. Fr.*, 517
244. Anteunis, M., Swaelens, G. and Gelan, J. (1971). *Tetrahedron*, **27,** 1917
245. Cookson, R. and Crabb, T. (1968). *Tetrahedron*, **26,** 2385; 2397
246. Lacey, M., Macdonald, C., Pross, A., Shannon, J. and Sternhell, S. (1970). *Austral. J. Chem.*, **23,** 1421
247. Eliel, E. L. and Buys, H. R. (1970). *Tetrahedron Letters*, 2779
248. Romers, C., Altona, C., Buys, H. R. and Havinga, E. (1969). *Topics in Stereochemistry*, **4**, 39. (New York: Interscience)
249. Rondeau, R. E. and Sievers, R. E. (1971). *J. Amer. Chem. Soc.*, **93,** 1522
250. Caputo, J. and Martin, A. R. (1971). *Tetrahedron Letters*, 4547
251. Pihlaja, K. and Äyras, P. (1970). *Acta Chem. Scand.*, **24,** 531
252. Pihlaja, K. and Äyras, P. (1970). *Acta Chem. Scand.*, **24,** 204
253. Pihlaja, K. and Tuomi, M. (1970). *Acta Chem. Scand.*, **24,** 366
254. Eliel, E. L. and Kaloustian, M. K. (1970). *Chem. Commun.*, 290
255. Eliel, E. L. and Rileanu, D. I. C. (1970). *Chem. Commun.*, 291
256. Crabb, T. A. and Newton, R. F. (1970). *Tetrahedron*, **26,** 693
257. Eliel, E. L. and Banks, H. D. (1970). *J. Amer. Chem. Soc.*, **92,** 4730
258. Dratler, R. and Laszlo, P. (1970). *Tetrahedron Letters*, 2670
259. Tavernier, D. and Anteunis, M. (1971). *Tetrahedron*, **27,** 1677
260. Eliel, E. L. and Nader, F. W. (1970). *J. Amer. Chem. Soc.*, **92,** 584
261. Eliel, E. L. and Giza, C. (1968), *J. Org. Chem.*, **33,** 3754
262. Nader, F. W. and Eliel, E. L. (1970). *J. Amer. Chem. Soc.*, **92,** 3050
263. Hutchinson, B. J., Jones, R. A. Y., Katritzky, A. R. Record, K. A. F. and Brignell, P. J. (1970). *J. Chem. Soc. B*, 1224
264. Jones, V. I. P. and Ladd, J. A. (1971). *J. Chem. Soc. B*, 567
265. Pihlaja, K. and Tenhosaari, A. (1970). *Suomen Kemistilehti*, **43,** 175
266. Coene, E. and Anteunis, M. (1970). *Tetrahedron Letters*, 595
267. Burpitt, R. D., Brannock, K. C., Nations, R. G. and Martin, J. C. (1971). *J. Org. Chem.*, **36,** 2222
268. Chow, Y. L. (1970). *Canad. J. Chem.*, **48,** 3045
269. Wojtowicz, J. A., Polak, R. J. and Zaslowski, J. A. (1971). *J. Org. Chem.*, **36,** 2232
270. Pedersen, C. J. (1971). *J. Org. Chem.*, **36,** 254
271. Chitwood, J. L., Glen Gott, P., Krutak, J. J. and Martin, J. C. (1971). *J. Org. Chem.*, **36,** 2216
272. Adler, C., Brasen, S. and Miyaki, S. M. (1971). *Acta Chem. Scand.*, **25,** 2055
273. McIntosh, C. L. and Chapman, O. L. (1971). *Chem. Commun.*, 771
274. McIntosh, C. L. and Chapman, O. L. (1971). *Chem. Commun.*, 383
275. Kirby, G. W., Tan, S. L. and Uff, B. C., (1970). *Chem. Commun.*, 1138
276. Chapman, O. L., Engel, M. R., Springer, J. P. and Clardy, J. C. (1971). *J. Amer. Chem. Soc.*, **93**, 6696
277. Buu-Hoi, N. P., Saint Raf, G. and Perche, J. C. (1970). *J. Chem. Soc. C*, 1327
278. Dean, F. M. and Locksley, H. D. (1963). *J. Chem. Soc.*, 393
279. Bennett, D. J., Dean, F. M. and Price, A. W. (1970). *J. Chem. Soc. C*, 1557
280. Dean, F. M., Hindley, K. B. and Houghton, L. E. (1971). *J. Chem. Soc. C*, 1171
281. Creed, D., Werbin, H. and Strom, T. E. (1971). *J. Amer. Chem. Soc.*, 502
282. Kirkiachanon, B. S. and Ranlais, D. (1970). *Bull. Soc. Chim. Fr.*, 1139
283. Choudhury, A. M., Schofield, K. and Ward, R. S. (1970). *J. Chem. Soc. C*, 2543
284. Schulz, A. G. and Schlesinger, R. H. (1970). *Chem. Commun.*, 1044
285. Anastassiou, A. G., Eachus, S. W., Cellura, R. P. and Gebrian, J. H. (1970). *Chem. Commun.*, 1133
286. Anastassiou, A. G. and Cellura, R. P. (1970). *Chem. Commun.*, 484
287. Cockcroft, R. D., Waali, E. E. and Rhoads, S. J. (1970). *Tetrahedron Letters*, 3539
288. Cannon, J. R., Chow, P. W., Metcalf, B. W., Power, A. J. and Fuller, M. W. (1970). *Tetrahedron Letters*, 352
289. Ramussen, M., Ridley, D. D., Ritchie, E. and Taylor, W. C. (1968). *Austral. J. Chem.*, **21,** 2989
290. Ballantine, J. A., Francis, D. J., Hassall, C. H. and Wright, J. L. C. (1970). *J. Chem. Soc. C*, 1175

8
Aromatic Sulphur Heterocycles

W. CARRUTHERS
University of Exeter

8.7 PROPERTIES OF FUNCTIONAL GROUPS

Alkoxy compounds, amines, bromothiophenes, thienylcarbinols,
formylthiophenes, hydroxythiophenes and benzo[b]thiophenes,
lithium derivatives and Grignard reagents, methyl derivatives,
nitro compounds, allyl sulphides

8.1 RING SYNTHESIS

During the period under review there has been continued interest in the synthesis of heteroaromatic sulphur compounds derived from thiophene and other sulphur-containing aromatic systems.

8.1.1 Thiophene derivatives

The equilibrium yields of thiophene formed in the reaction of C_4 hydrocarbons with hydrogen sulphide and sulphur dioxide have been studied[1].

A versatile new route to 2-acyl- or 2-aroyl-thiophenes from monothio-β-diketones and α-bromocarbonyl compounds has been described[2]. In the presence of triethylamine, condensation leads to a hydroxydihydrothiophene derivative which is readily dehydrated in the presence of acid to give the thiophene in almost quantitative yield. This is more convenient than the earlier related route from 2-thiocarbonylenamines and α-bromocarbonyl compounds[3]. Smutny[3] has observed the formation of 2-acylthiophenes from ethylene-1,1-dithiolates and α-halogenocarbonyl compounds via base catalysed reaction between the active methylene group in the intermediate (1) and one of the substituents X and Y.

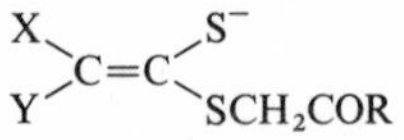

$$R = CH_3, C_6H_5$$

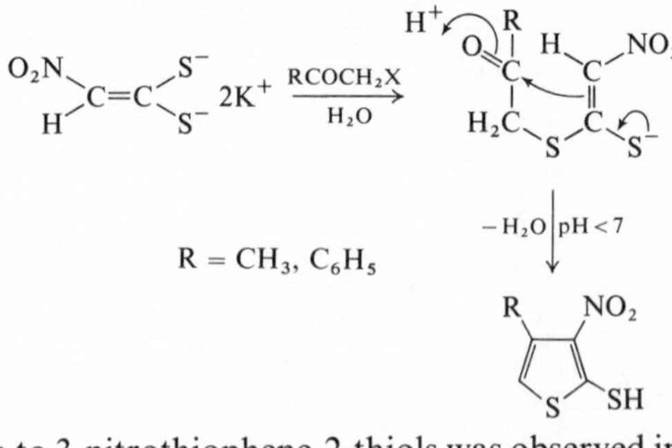

An alternative cyclisation to 3-nitrothiophene-2-thiols was observed in the reaction of dipotassium 2-nitroethylene-1,1-dithiolate with α-halogenocarbonyl compounds[4].

The preparation of aminothiophenes is often attended by difficulties, but certain derivatives of 2,4-diaminothiophene are conveniently obtained from readily available α-chlorothioacetanilides in refluxing methanol[5]. A novel route to 3-hydroxythiophenes by reaction of allenic nitriles with mercaptoacetates has been described[6]. Thus, reaction of 4-methylpenta-2,3-dienonitrile with a suspension of the sodium salt of methyl mercaptoacetate in ether at room temperature precipitates the sodium salt of 4-hydroxy-2-isopropylthiophene-2-carbonitrile in almost quantitative yield. With methyl o-mercaptobenzoate good yields of the corresponding thiochroman-4-ones are obtained. 2-Cyano- and 2-phenyl-3-hydroxythiophenes have been obtained by reaction of enethiols with chloroacetonitrile or benzyl chloride and Dieckmann cyclisation of the initial product[7].

The synthesis of a number of heterophanes, including the thiophene derivative (2) from appropriate 1,4-dicarbonyl compounds has been reported[8].

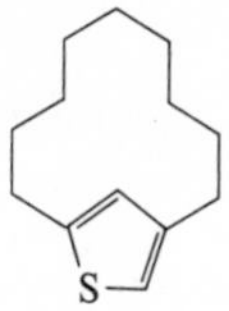

(2) 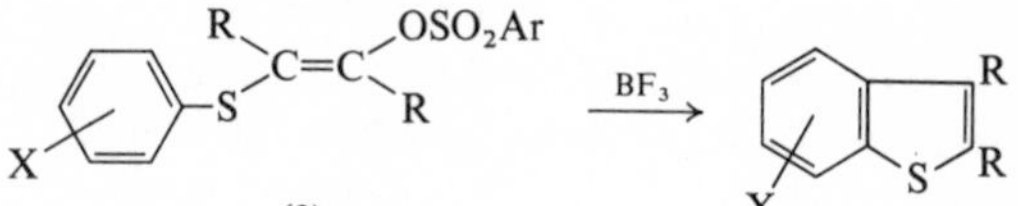   (3)

8.1.2   Benzo[b] and benzo[c]thiophenes

A comprehensive review of methods of synthesis of the benzo[b]thiophene ring system up to 1968 has been published[9]. Most of these procedures involved arylthioacetaldehyde acetals, arylthioacetones, aryl phenacyl sulphides and S-arylthioglycolic acids, and further examples have been recorded recently[10]. An improved method for cyclisation of arylthioacetaldehyde acetals uses the new reagent BF_3–$(CF_3CO)_2O$ [11]. Cyclisation of p-tolylthioacetone with polyphosphoric acid leads, by rearrangement, to a mixture of 2,5- and 3,5-dimethylbenzo[b]thiophene[12].

The mechanism of the oxidative cyclisation of β-aryl-α-mercaptoacrylic acids to benzo[b]thiophene-2-carboxylic acids with iodine or chlorine, which is preparatively important, has been studied in detail[13]. It is suggested that, in aprotic solvents, a sulphenyl halide rather than the disulphide postulated earlier, is formed as a reaction intermediate. It then cyclises to the thiophene derivative probably by a polar mechanism, although a free radical mechanism may operate under some conditions.

Italian workers have described a novel route to 2,3-diarylbenzo[b]thiophenes from arylthiovinyl sulphonic esters (structure (3), R^1, R^2 = aryl, X = H, Me, OMe, Cl, Br)[14]. Benzo[b]thiophene derivatives were not formed under these conditions when R = H or alkyl. When the substituent X is para to the vinylthio group, cyclisation affords 6-substituted benzothiophenes by rearrangement, and not the expected 5-substituted compounds. In contrast meta substituted esters give the expected 4- and 6-substituted benzothiophenes. Experiments with the para-deutero compound, however, gave the 5- and 6-deuterated benzothiophene in almost equal amounts. It is suggested that the cyclisation occurs by two distinct pathways one or other of which is favoured depending on the substitution pattern in the arylthio ring.

The conversion of aryl allyl sulphides into derivatives of 2,3-dihydrobenzo[b]thiophene and thiachroman by heating in amine solvents has been known for some years, and probably involves a 'thio-Claisen' rearrangement of the starting materials to o-allylthiophenols which cyclise under the reaction conditions. A new route to the benzo[b]thiophene nucleus by 'thio-Claisen' rearrangement of prop-2-ynyl phenyl sulphides is reported[15]. Thus, phenyl prop-2-ynyl sulphide in quinoline at 200 °C affords mainly 2-methylbenzo-[b]thiophene along with some phenyl allenyl sulphide, and but-2-ynyl phenyl sulphide gives 2-ethyl- and 2,3-dimethylbenzo[b]thiophene. It is apparent from the complexity of the products that the 'thio-Claisen' rearrangement is preceded by a thiopropynylic rearrangement establishing the equilibrium $Ph \cdot S \cdot CH_2 \cdot C = CR \rightleftharpoons Ph \cdot S \cdot CR = C = CH_2$, both components of which subsequently undergo 'thio-Claisen' rearrangement.

A new synthesis of 3-alkyl- or aryl-benzo[b]thiophenes, based on the reaction of o-mercaptoketones, or their xanthates, with dimethylsulphonium methylide has been reported[16]. Other derivatives are formed by reaction of some β-styryl derivatives with thionyl chloride or sulphur monochloride in the presence of pyridine. Cinnamic acid, for example, affords 3-chloro-benzo[b]thiophene-2-carboxylic acid[17].

Non-symmetrical benzo[c]thiophenes have not been easy to obtain. A reaction leading to such derivatives by a novel rearrangement has been described[18]; sublimation of the enamine (4) gave the benzo[c]thiophene in 19% yield.

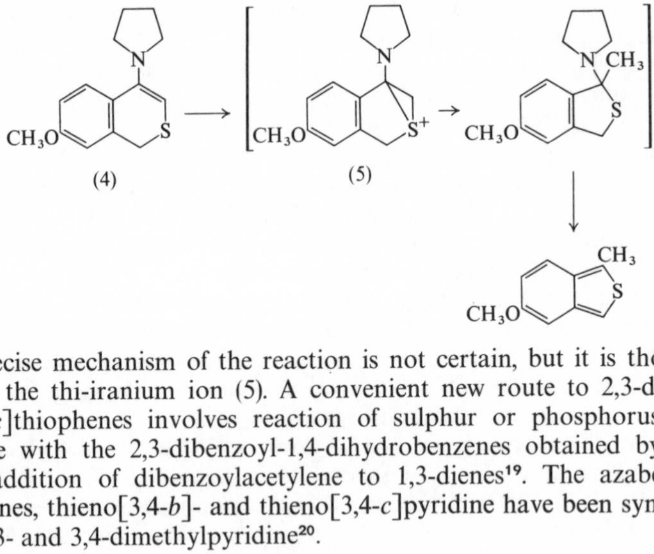

The precise mechanism of the reaction is not certain, but it is thought to involve the thi-iranium ion (5). A convenient new route to 2,3-diphenyl-benzo[c]thiophenes involves reaction of sulphur or phosphorus pentasulphide with the 2,3-dibenzoyl-1,4-dihydrobenzenes obtained by Diels–Alder addition of dibenzoylacetylene to 1,3-dienes[19]. The azabenzo[c]-thiophenes, thieno[3,4-b]- and thieno[3,4-c]pyridine have been synthesised from 2,3- and 3,4-dimethylpyridine[20].

8.1.3 Polycyclic thiophene derivatives

A number of new syntheses of condensed polycyclic thiophene derivatives have been reported, but there has been little systematic work.

The synthesis of condensed thiophenes by direct insertion of sulphur into polycyclic aromatic hydrocarbons is being studied[21]. Treatment of phenanthrene with hydrogen sulphide over an alumina catalyst at 630 °C gave phenanthro[4,5-bcd]thiophene in 39% yield. Reaction of o-, m- and

p-xylenes with sulphur in the presence of aluminium chloride gave tetra-methylthianthrenes[22], and cadalene and eudalene formed condensed thiophene derivatives on reaction with sulphur at 260 °C[23]. Further studies of the rearrangements during Elbs pyrolysis of 3-aroylbenzo[b]thiophenes to derivatives of benzo[b]naphtho[2,1-d]thiophene have been reported[24]. [1]Benzothieno[2,3-b][1]benzothiophene was obtained by a novel route by reaction of 1,1-diphenylethylene with four equivalents of sulphur at 240 °C[25]. In an interesting new reaction, polycyclic thiophene derivatives have been obtained by reaction of sulphur with the rhodium complex of bis-acetylenic compounds[26].

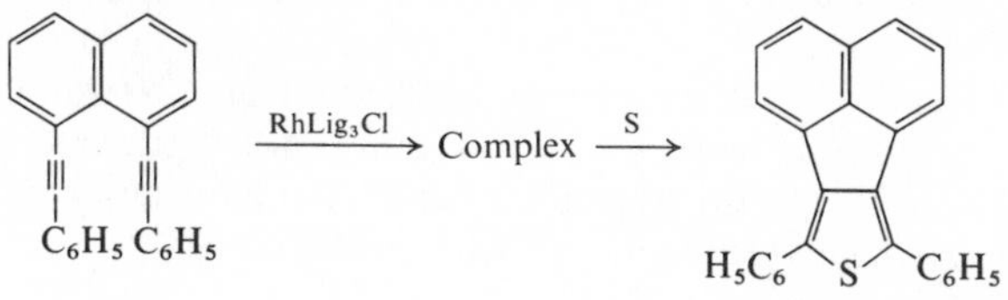

$$\xrightarrow{\text{RhLig}_3\text{Cl}} \text{Complex} \xrightarrow{\text{S}}$$

C$_6$H$_5$ C$_6$H$_5$

H$_5$C$_6$ S C$_6$H$_5$

53%

Condensed thiophene derivatives have been obtained by photocyclisation of thiophene analogues of stilbene[27]. In more recent work it is shown further that photocyclisation of iodophenyl- and iodothienyl-thienyl ethylenes provides a good route to naphthothiophenes and benzodithiophenes[28]. A full account has been given of the synthesis and resolution of some S-heterohelicenes[29]. S-Analogues of penta-, hexa- and hepta-helicene were obtained by photocyclisation of appropriate precursors.

Benzo[1,2-c; 3,4-c']dithiophene (6) has been prepared for comparison with phenanthrene[30] and as part of a study of the chemical properties of thiophene analogues of fluorene the six cyclopentadithiophenes were prepared by reduction of the ketones obtained by Ullmann cyclisation of the appropriate di-iododithienyl ketones[31].

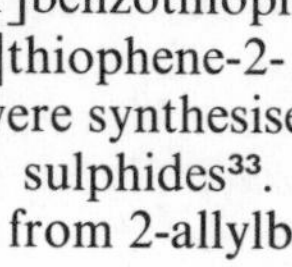

(6) 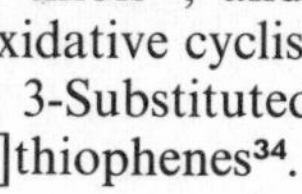 (7) 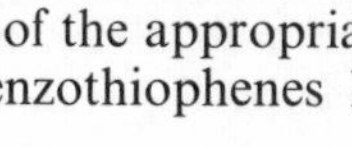 (8)

A number of derivatives of thieno[2,3-b][1]benzothiophene and thieno-[3,2-b][1]benzothiophene have been prepared from the appropriate benzo[b]thiophene-2- and -3-thiols[32], and the six isomeric dithienothio-phenes were synthesised by oxidative cyclisation of the appropriate dilithio dithienyl sulphides[33]. 1- and 3-Substituted dibenzothiophenes have been obtained from 2-allylbenzo[b]thiophenes[34].

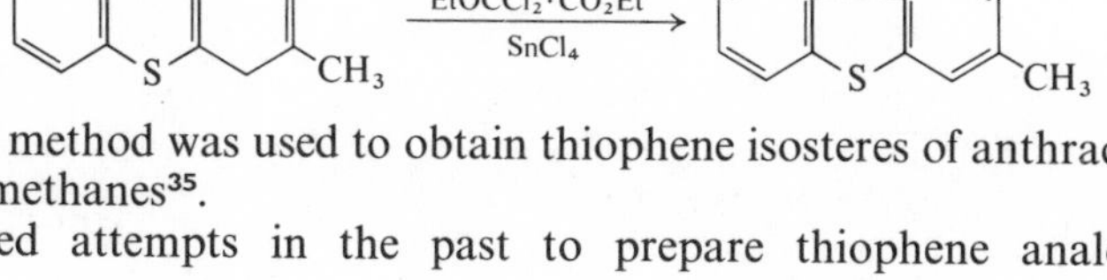

CO$_2$C$_2$H$_5$

$$\xrightarrow[\text{SnCl}_4]{\text{EtOCCl}_2 \cdot \text{CO}_2\text{Et}}$$

CH$_3$

CH$_3$

A similar method was used to obtain thiophene isosteres of anthracene from dithienylmethanes[35].

Repeated attempts in the past to prepare thiophene analogues of benzocyclobutene or biphenylene have been unsuccessful, but recently the biphenylene analogue (7) has been obtained by a Wittig reaction between 1,2-benzocyclobutadiene-quinone and the phosphorus ylid derived from bis(chloromethyl) sulphide[36].

Naphtho[2,3-c]thiophene (8) and the 4-aza derivative thieno[3,4-b]-quinoline could not be obtained by pyrolysis of the corresponding 1,3-dihydrosulphoxides, although their transient existence was shown by the formation of adducts with N-phenylmaleimide[37].

8.1.4 Thiopyrylium salts and thiabenzene derivatives

Thiopyrylium salts are conveniently obtained from 2H-thiopyran-2-ones and -4-ones by reaction with lithium aluminium hydride and treatment of the resulting carbinols with perchloric acid[38]. An alternative route involves oxidation of 2H-thiopyran-2-thiones with peracetic acid[38].

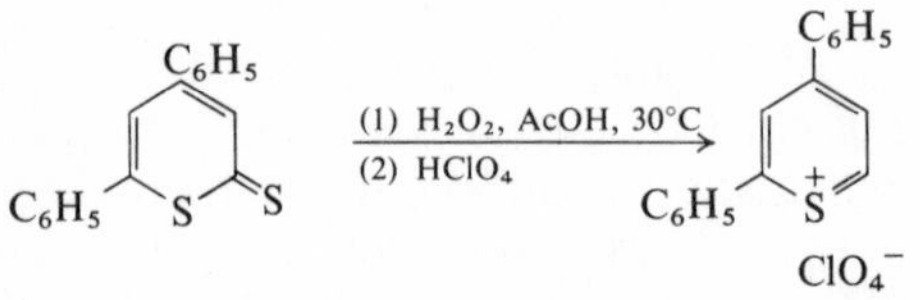

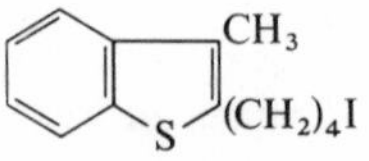

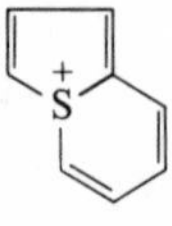

A polycyclic thiopyrylium salt has been obtained by treatment of the corresponding 2H-derivative with perchloric acid; a thi-iranium ion is suggested as an intermediate[39]. Derivatives of indolo[2,3-b]thiopyrylium chloride were obtained by reaction of indoline-2-thione with 1,3-diketones in alcoholic hydrogen chloride. Treatment of the salts with base liberated the thiopyrano-[2,3-b]indole, a new heterocyclic pseudoazulene system[40].

Reaction of the iodide (9) with AgPF₆ gave the corresponding benzo-thiophenium salt, a hydro analogue of the potentially aromatic thiopyrano-[1,2-a]thiophene ion (10)[41]. The $\gamma\gamma'$-dithiopyrylium cation, which is iso-π-electronic with the recently synthesised ditropylium cation, has been obtained from thiopyrylium iodide[42].

Interest in thiabenzene and related compounds continues unabated (compare pp. 247 and 252). These compounds are usually obtained by reaction of thiopyrylium perchlorates with phenyl-lithium. Thus, 1,2-diphenyl-2-thia-naphthalene was obtained from 2-thianaphthalenium perchlorate as shown below, utilising the fact that while phenyl-lithium generally attacks thio-pyrylium salts at sulphur, Grignard reagents react at carbon[44].

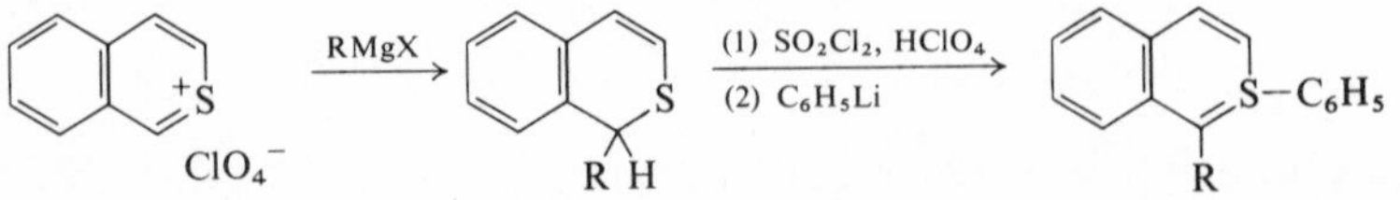

$$R = C_6H_5, \text{ } t\text{-}C_4H_9$$

However, phenylethynyl-lithium reacted with 2,4,6-triphenylthiopyrylium perchlorate to give products formed by coupling at carbon[45], possibly by rearrangement of the desired 1-phenylethynyl-2,4,6-triphenylthiabenzene.

1-Phenyl-2-thia-1,2-dihydronaphthalenes are also obtained by photo-cycloaddition of thiobenzophenone to acetylenic compounds[43].

Several thiabenzene 1-oxides have been synthesised, in all cases by routes involving dimethyloxosulphonium methylide. In an extension of this procedure 1-methyl-3,5-disubstituted thiabenzene oxides have been obtained by reaction of dimethyloxosulphonium methylide with 1,3-disubstituted propyn-2-ones[46].

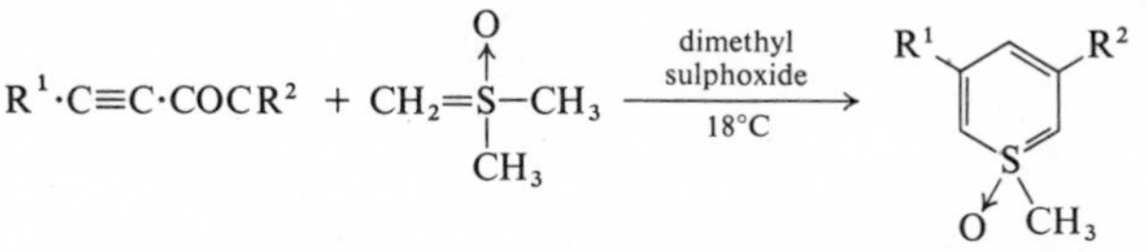

$$R^1{\cdot}C{\equiv}C{\cdot}COCR^2 + CH_2{=}\overset{\overset{O}{\uparrow}}{\underset{CH_3}{S}}{-}CH_3 \xrightarrow[18°C]{\substack{\text{dimethyl}\\\text{sulphoxide}}}$$

The reaction proceeds by Michael addition of the methylide to the acetylenic ketone to form an allylide which is readily converted into the thiabenzene oxide.

1-Methyl-3,5-diphenylthiabenzene oxide has been converted into 1-methyl-3,5-diphenylthiabenzene[47].

8.1.5   6a-Thiathiophthenes and dithiolium salts

A useful new synthesis of symmetrically substituted 6a-thiathiophthenes proceeds from 4H-thiapyran-4-thiones by ring opening with sodium sulphide in an aprotic solvent and oxidation of the resulting anion by potassium ferricyanide[48].

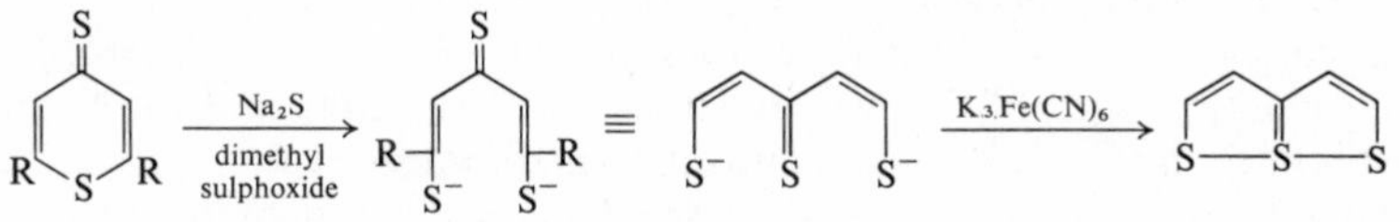

Another route to this structure involves heating 3-methyl-1,2-dithiolium salts with methyl dithiobenzoate[49].

Variously-substituted 3-acylmethylenethio-1,2-dithiolium salts are obtained by reaction of α-bromoketones with 1,2-dithiol-3-thiones[50]. A method for the preparation of 1,2-dithiolium salts and their conversion into thion-thiophthenes which is claimed to be suitable for large scale preparations, has been described[51].

8.1.6   Thia-annulenes

The period under review has seen the publication of a number of reports relating to the synthesis of sulphur analogues of the annulenes.

Theoretical work[52] predicts that the 8π-electron thiepin ring system should be anti-aromatic. This ring system is known from earlier work on annelated derivatives to undergo very ready extrusion of sulphur with formation of a benzene ring. Recently, however, the stable derivative (11) has been obtained[53].

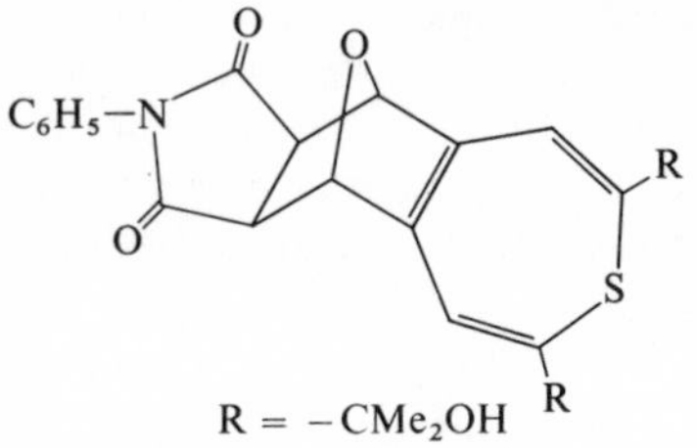

$$R = -CMe_2OH$$

(11)

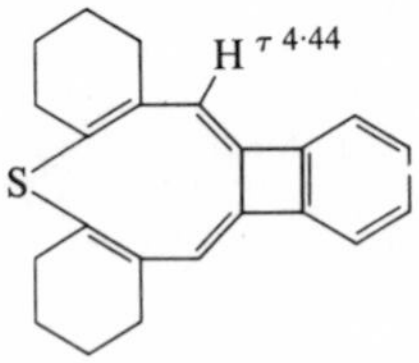

(12) 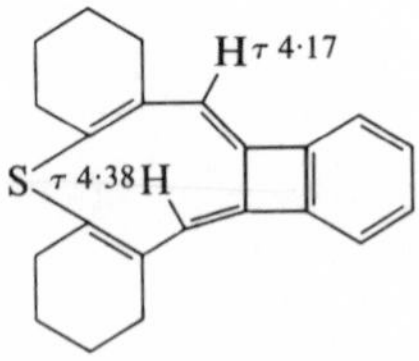   (13)

A new convenient synthesis of 4,5-dihydrothiepin-1,1-dioxides proceeds from reaction of sulphur dioxide with diazoalkenes[54]. Thus vinyldiazomethane and sulphur dioxide gave 4,5-dihydrothiepin dioxide in 29% yield, presumably through formation of an episulphone followed by Cope rearrangement. The dihydro compound was readily dehydrogenated to thiepin dioxide. Other examples were studied and the reaction appears to be general. Since episulphones can also be prepared from sulphenes and diazoalkanes the method is of wide scope.

The 10π-electron thionin, which is potentially aromatic, has not yet been prepared, but the two derivatives (12) and (13) have been obtained by a Wittig reaction with bis(2-formyl cyclohexenyl) sulphide and the triphenyl phosphorane derived from α,α'-dibromobenzocyclobutene[55]. Neither compound gave any evidence in the n.m.r. spectrum of a diamagnetic ring current indicative of an aromatic thionin ring, and the molecules are probably non-planar. Derivatives of thia[11]- and thia[13]-annulene have been prepared by similar methods. These compounds also showed no evidence of a ring current in the macrocyclic ring[56].

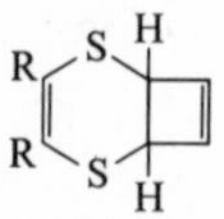

(14) 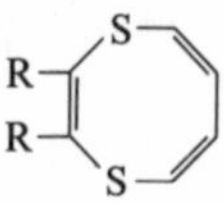   (15) 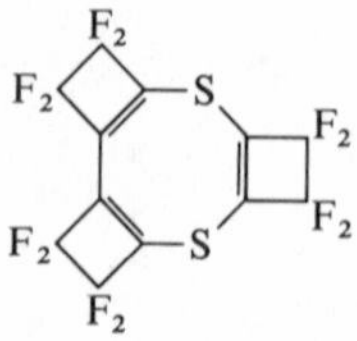   (16) 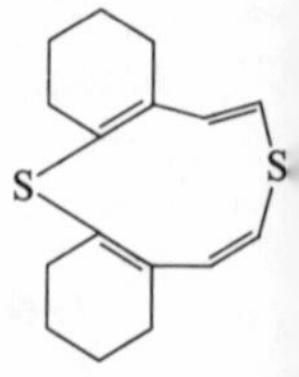   (17)

Attempts to prepare derivatives of the 10π-electron system 1,4-dithiocin (as 15) by pyrolysis or photolysis of bicyclic isomers (14) led mainly to extrusion of sulphur. The benz-derivative (R,R = benzene ring) was obtained but gave no evidence for the existence of a diamagnetic ring current in the hetero-ring[57]. On the other hand, ^{19}F n.m.r. chemical shift data for the compound (16), obtained from reaction of 1-lithio-2-chloro-3,3,4,4-tetra-fluorocyclobutene with sulphur mono- or di-chloride, are interpreted in terms of a ring current in the hetero ring[58].

The 12π-electron bisheteroannulene (17) has also been obtained but, in contrast to the carbocyclic[12]-annulene, gives no indication in the n.m.r. spectrum of the existence of a paramagnetic ring current[59].

Attempts have also been made to prepare thia[17]-annulene itself. Precursors were obtained by Glaser cyclisation of appropriate $\alpha\omega$-terminal acetylenes, but could not be converted into the fully conjugated hetero-annulene[60].

Synthesis of porphin analogues containing furan and thiophene rings in place of pyrrole rings has been reported[61]. A review of macrocycles containing the thiophene ring system has been published[62] (see also Section 8.6, page 255).

8.2 STRUCTURE AND THEORETICAL STUDIES

The question of the participation of sulphur 3d orbitals in the ground state of thiophene has been of interest for many years. Recent molecular orbital calculations by two groups of workers indicate that 3d orbitals play little part in bonding in the molecule[63]. Extensive calculations covering heats of atomisation, resonance energies, ionisation potentials and bond lengths for a number of thiophene compounds have been published[64]; the calculated values are in good agreement with experiment. Calculated resonance energies of the isomeric benzothiophenes differ considerably and reflect the difference in stability of the two systems. For the bithienyls little interaction between the rings is predicted, the calculated resonance energies being roughly twice that for thiophene. The results for thiepin, benzo[d]thiepin and thieno[cd]thiepin suggest that these compounds are not likely to be aromatic. SCF MO calculations for the isomeric thienopyrroles suggest that the -[2,3-b]- and -[3,2-b]- compounds are more stable than the -[2,3-c]- and -[3,4-c]- isomers[65].

Interest continues in the type of bonding in 'tetravalent sulphur' hetero-cycles such as the thiabenzenes and related thiacyclic systems. These are generally regarded as neutral covalent structures and it has been suggested that, in most cases, bonding to sulphur is of the $3p\pi$–$2p\pi$ type, except in cases where there is strong steric congestion round the sulphur atom, when $3d\pi$–$2p\pi$ bonding is preferred[66]. However, recent n.m.r. and deuterium exchange studies with 1-methyl-3,5-diphenylthiabenzene and the corresponding 1-oxide indicate the absence of a ring current in these compounds and hence of through-conjugation at sulphur, and it is suggested that these compounds have an ylid type structure involving weak $2p\pi$–$3d\pi$ orbital overlap at sulphur[67].

The ^{1}H n.m.r. spectrum of thi-inyl S,S-dioxide anions (18) are consistent with an extensive π-electron delocalisation within the heterocyclic ring, in agreement with earlier observations of the remarkable acidity of the neutral precursors[68].

Results of a study of the u.v. and n.m.r. spectra and pK_{R+} values of a series of 2-p-substituted-phenyl-4-phenyl-1,3-dithiolium perchlorates showed that a phenyl group at C-2 largely stabilises the 1,3-dithiolium cation by conjugation. Substituent effects indicate that the positive charge in the 1,3-dithiolium ring is largely localised in the S—C—S group and that the most important limiting structures are (19) and (20) [69].

(18)   (19)   (20)   (21)

A review of structure and bonding in 1,2-dithiolium cations has been published[70]. x-Ray structure studies on various derivatives show that in these compounds the 1,2-dithiolium ring is planar and is stabilised through delocalised π-bonding. The effect of alkyl groups on the u.v. spectrum and polarographic properties of 1,2- and 1,3-dithiolium ions was studied and compared with the result of SCF—LCI calculations. The non-uniform charge distribution of dithiolium cations was confirmed by n.m.r. data[71].

There has been much discussion concerning the structure of 6a-thia-thiophthenes (21) (review[72]). Experimental evidence favours a symmetrical structure in solution, with equal S—S distances for symmetrical substituted compounds. However, x-ray diffraction studies suggest that in certain derivatives in the solid phase the S—S distances may be unequal[73]. x-Ray crystal structure studies with 2,5-dimethylthiathiophthene have been interpreted in terms of a symmetrical structure with equal S—S distances[74], but more recent work on the electronic polarisation spectrum has been taken as showing an unsymmetrical structure in the electronic ground state[75]. Theoretical studies of the parent compound and some simple derivatives, using all valence electron CNDO/2SCF MO calculations show significant σ polarisation with a shift of σ electrons from the central S to the terminal S atoms. The d-orbital population on S_{6a} is significantly higher than on S_1 and S_6. Localisation energies for prototypical electrophilic and nucleophilic substitution of the parent molecule indicated, in agreement with experiment, that C-3 is the preferred position for electrophilic attack and C-2 for nucleophilic attack. Previous work has shown that there is partial covalent bonding between all the S atoms in (22); in the symmetrical compound (22, Me_3C for C_6H_5) the S—S distances closely resemble those in (22) [76].

The n.m.r. spectra of symmetrically substituted 6a-selenathiophenes shows that in solution these compounds also possess real or time-averaged C_{2v} symmetry[77]. x-Ray crystallographic studies show that in the parent compound the two S—Se bonds have the same length[78].

x-Ray crystallographic study of the aromatic 2,2'-bi-1,3-dithiole showed that the bridging double bond was longer than the ring double bonds, and the overall molecule is not quite planar. Comparison with the results of theoretical studies suggests the participation of sulphur d-orbitals in the bonding[79].

Calculated delocalisation energies for the azulenodihydrothiophene–heptafulvenothiophene system predict a greater π-electron stability in the former, in contrast to the case with the isosteric benzene analogue. This prediction is borne out by the relative stabilities of alkyl derivatives; but substitution of a strongly conjugating group at C-2 reverses the stability order[80].

Comparison of the energy barriers to conformational inversion of the seven-membered ring in 2,3,6,7-dibenzothiepin and related compounds shows that the change in delocalisation energies associated with the seven-membered ring system is due to increased Ar—C=C—Ar and Ar—X—Ar conjugation and is not a special property of the cyclic conjugated system[81].

8.3 PHYSICAL PROPERTIES

8.3.1 Molecular dimensions; x-ray crystallographic studies

2-Acetyl-5-bromothiophene is shown by x-ray crystallography to have the s-*trans* conformation of the carbonyl group[82], in agreement with conclusions drawn from the n.m.r. spectra of 2-acylthiophenes[83] and of 2-acetyl- and 2-formyl-5-fluorothiophene[84]. The enthalpy and entropy changes for the internal rotation of thiophene-2-aldehydes have been determined[85].

Three-dimensional x-ray analysis of 2,5-diphenyl-1,4-dithi-in 1-oxide shows that the hetero ring is arranged in a boat conformation with the sulphinyl oxygen in the axial position[86].

The absolute configuration of the heterohelicene (23) has been determined by x-ray diffraction[87]. Like other heterohelicenes and the helicenes themselves it exists in the form of a shallow spiral. Optically active specimens of several typical sulphur heterohelicenes have been obtained[88].

Other polycondensed thiophene derivatives studied by x-ray crystallography include dibenzothiophene[89] and [1]benzothieno[2,3-*b*]benzothiophene[25].

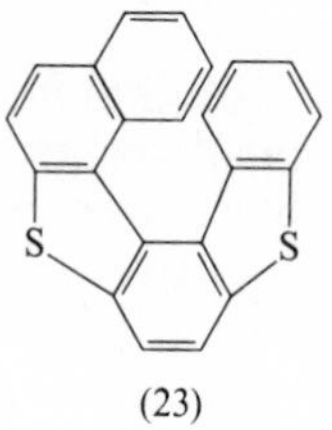

(23) 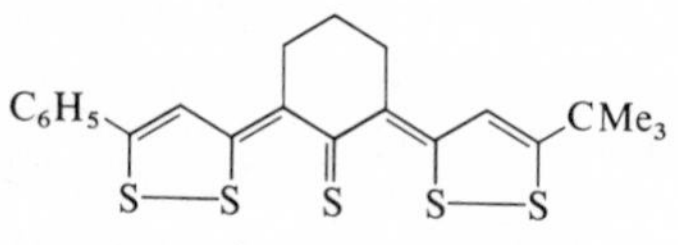   (22) 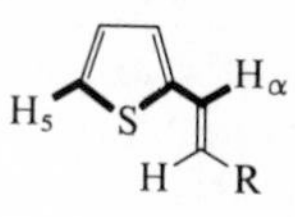   (24)

8.3.2 Ultraviolet, infrared and electron spin resonance spectra

In the ultraviolet absorption spectra of polyenyl substituted thiophenes the contribution of a heterocyclic ring is roughly equal to that of a butadiene unit[90]. Cyclopropyl substituents have a pronounced effect on the u.v. spectra of thiophenes, approximating to that of a phenyl substituent[91]. The u.v. spectra of 1-alkylthiophenium salts show a marked resemblance to those of the corresponding 1-oxides and 1,1-dioxides[92], and that of phenanthro [4,5-bcd]thiophene is closely similar to that of its hydrocarbon isostere, pyrene[21]. The u.v. spectra of four dithienothiophenes were in excellent agreement with those calculated by SCF MO methods[33]. CNDO/2 studies on the electronic spectra of the isomeric thienothiophenes have been reported[93], as have electronic spectra of a number of S-heterohelicenes[87].

Examination of the infrared spectrum of 2-deuteroformylthiophene and 2-[18O]formylthiophene gave new information on the assignment of bands in the 1600–1700 cm^{-1} region[94].

A review of recent progress in the study of thiophene derivatives by electron spin resonance spectroscopy has been published[95]. The e.s.r. spectra of benzo-[b]thiophene-2,3-semidione[96], of dibenzo[b,f]thiepin sulphoxide and thi-oxanthone sulphoxide radical anion[97] and of the radical-anions obtained from a number of polycyclic thiophene compounds with Na–K alloy[98] have been studied.

8.3.3 Nuclear magnetic resonance spectra

The effect of solvent and concentration on the n.m.r. spectra of thiophene and a number of condensed polycyclic thiophene derivatives has been examined[99]. Evidence was obtained for a specific association of chloroform with the S heterocycles and for the interaction of acetone with the S atoms. The chemical shifts of the methylene protons can be used to distinguish between 2- and 3-methylene groups in alkyl thiophenes[100]. In the ^{19}F n.m.r. spectra of the polyfluorothiophenes the α-fluorines resonate to high field of the β-fluorines; large 2–5 F–F couplings are observed[101]. Calculated and observed chemical shifts for a number of derivatives of 2,3-dibromobenzo-[b]thiophene[102], nitro-3-acetylbenzo[b]thiophenes[103] and of the four iso-meric thianaphthenothiapyrylium perchlorates[104] have been reported.

The ^{1}H n.m.r. spectra of a series of S-alkylthiophenium salts are consistent with a pyramidal arrangement of the S—C bonds, the non-coplanarity causing anisotropic effects[92].

N.M.R. coupling constants for 2-acetyl- and 2-formyl-5-fluorothiophene as well as for 2,5-diformylthiophene indicate that the principal rotamer at room temperature and below is the one in which the carbonyl group eclipses the sulphur atom[105]. In 3-substituted 2-formylthiophenes changes in the aldehyde conformation with the nature of the ortho substituent were ob-served[106]. Similarly, a series of 2-thienylethylenes (24, R = NO$_2$, CO$_2$H, COCH$_3$) are considered to have the transoid conformation shown on the basis of their n.m.r. spectra, which show relatively large long-range coupling constants[107] ($J_{5,\alpha} = 0.7$–0.8 Hz). Long-range H–F coupling constants in 2-thienylcarbonyl fluoride decrease in the order $J_{F,H(5)} > J_{F,H(4)} > J_{F,H(3)}$[108].

Cross-ring coupling over five and six bonds is observed in the ^{1}H n.m.r. spectra of a number of condensed polycyclic thiophene derivatives[109]. High resolution ^{1}H n.m.r. spectra of some polynuclear thiophene derivatives have been reported[110]. In the n.m.r. spectra of heterohelicenes, protons at-tached to terminal rings in compounds which are helical occur at unusually high field because they are in the shielding zone of the underlying aromatic ring[111].

8.3.4 Mass spectrometry

Recent work using [2-^{13}C]-thiophene establishes that the thiophene ring undergoes carbon scrambling under electron impact[112, 113]. The mass spectra of a number of S-alkylthiophenium salts were similar to those of the parent thiophene, due to ready loss of the S-alkyl group. In some cases expansion of the thiophene ring was observed[92]. Similarly, fragmentation of 1-phenyl-thiabenzene, 9-phenyl-9-thia-anthracene and 9,10-diphenyl-9-thia-anthracene occurs principally at the S-phenyl group, producing the corresponding thiopyrylium ions which show up as major peaks in the spectra[114].

In the mass spectrometer the S-heterohelicenes undergo a rearrangement, with loss of ethylene, similar to that shown in the conversion of hexahelicene into coronene[111].

8.3.5 Charge-transfer complexes

The charge-transfer interactions of thiophene and other five-membered heterocycles with tetracyanoethylene, chloranil and maleic anhydride were studied by electronic spectroscopy[115]. In the iodine complexes of γ-pyrone and 1-thia-γ-pyrone the binding site is the carbonyl oxygen[116]. Charge transfer excited states are thought to be responsible for the fluorescence of 5-acyl-2,2'-bithienyls in hydrogen-bonding solvents[117].

8.4 SUBSTITUTION REACTIONS

8.4.1 Electrophilic substitution

8.4.1.1 Thiophene derivatives

A study of the bromination of thiophene in aqueous acetic acid and of the protodetritiation of 2- and 3- tritiothiophene suggests that the mechanism of these reactions, and probably of other electrophilic substitutions of thiophene, are the same as those of benzene derivatives. The sulphur atom appears to have little effect on the mechanism except for making the molecule more reactive[118]. Similarly, bromination and chlorination of a range of thiophene derivatives showed the same characteristics as those of the corresponding benzene derivatives[119]. Kinetic studies show that in nitration also the reactions of thiophenes are not different from those of benzene, but along with some other hetero-aromatic compounds thiophene has a greatly enhanced susceptibility to nitrosation[120]. Furan, thiophene and pyrrole show considerable differences in sensitivity to substituent effects in trifluoro-acetylation[121], and thiophene was less reactive than selenophene in a range of electrophilic substitution reactions[122].

Reaction of 3-methylthiophene with N-bromosuccinimide gives 2-bromo-3-methylthiophene; bromine affords the 5-bromo compound[123]. Similarly, bromination of 2-benzoylthiophene leads to the 4- or 5-bromo derivative, depending on the conditions[124]. In the presence of Friedel–Crafts catalysts, or iron powder, thiophene reacts with sulphuryl chloride to give a mixture of chlorinated 2,2'-dithienyls; reaction without a catalyst gave chlorothiophenes[125]. Fluorination of thiophene and tetrahydrothiophene with potassium tetrafluorocobaltate affords mainly polyfluoro-thiolanes and -3-thiolenes[126].

It has long been known that thiophene derivatives are polymerised by acids. The product obtained by action of polyphosphoric acid on thiophene has now been shown to consist principally of a mixture of *cis*- and *trans*-2,4-di-2-thienyltetrahydrothiophene and 4,7-di-2-thienyl-1,2,3,4-tetrahydrobenzo[*b*]thiophene[127].

2-Thienylcopper preparations are obtained by reaction of 2-thienyllithium with cuprous iodide or bromide. They react smoothly with iodo- or bromo-arenes in pyridine or quinoline, providing a valuable route to 2-arylthiophenes[128]. With 1,3,5-trinitrobenzene a Meisenheimer complex was formed and on oxidation with benzoquinone was transformed into 2-(2,4,6-trinitrophenyl)thiophene[129].

Examples of amine- and ether-directed lithiation of certain thiophene derivatives have been reported[130]. For 5-methyl-N,N-dimethyl-2-thenylamine, specific lithiation at position 3, directed by the dimethylamino group, was easily effected. N,N-Dimethyl-3-thenylamine and methyl-3-thenyl ether, were exclusively lithiated at C-2.

8.4.1.2 Thiabenzene derivatives

The C-2, C-4 and C-6 positions of thiabenzene 1-oxides are carbanionic in character. In d-trifluoroacetic acid rapid exchange of ring protons at these positions took place, probably through equilibration with the corresponding 2H- and 4H-thi-inium 1-oxides[46].

8.4.1.3 Benzo[b]thiophene derivatives

A recent review of the chemistry of benzo[*b*]thiophene[9] has emphasised the need for a re-examination of much of the earlier work on its substitution reactions. In contradiction to an earlier report it is found that nitration of 3-acetyl- and 3-formyl-benzo[*b*]thiophene under various conditions leads to substitution in the benzene nucleus with no detectable reaction in the thiophene ring; attempted bromination gave no reaction even at 60 °C [103]. Bromination of 2,3-dibromobenzo[*b*]thiophene gave the 2,3,6-tribromo compound. Nitration was less specific and gave a mixture containing mainly the 2,3-dibromo-6-nitro- and 2,3,6-tribromo-compounds[102]. Other studies have been concerned with the bromination of 2-phenylbenzo[*b*]thiophene[131], the bromination and Friedel–Crafts acylation of 6-methoxy-3-methylbenzo[*b*]thiophene[132], the nitration of benzo[*b*]thiophene-2-carboxylic acid and its 3-methyl derivative[133] and the nitration and bromination of 2-bromo-3-methylbenzo[b]thiophene[134].

In a recent study it is shown that 2- and 3-[$^3H_1$]methylbenzo[b]thiophene undergo detritiation in anhydrous trifluoroacetic acid, presumably through reversible attachment of a proton to the ring carbon atom *ortho* to the [$^3H_1$]methyl group followed by loss of a proton from the latter to give an olefinic species[135].

8.4.1.4 Polycyclic thiophene derivatives

Studies have also been made of various electrophilic substitution reactions in [1]benzothieno[2,3-b]benzothiophene[25], benzo[1,2-b: 5,4-b′]dithiophene[35], thieno[2,3-b] [1]benzothiophene and thieno[3,2-b] [1]benzothiophene[136], thieno-[2,3-b]-, -[3,2-b]-, -[2,3-c]- and -[3,2-c]- pyridine[137] and thieno[3,2-f]-quinoline[138]. Nitration, sulphonation, and Friedel–Crafts acetylation of thieno-[2,3-c]- and -[3,2-c]-pyridine took place mainly at C-3 as in benzo[b]thiophene, and nitration of thieno[2,3-b]- and -[3,2-b]-pyridine likewise afforded in each case the 3-nitro compound. These results are consistent with values for the reactivity indices calculated by simple Hückel theory[139].

In agreement with theoretical predictions Vilsmeier formylation of thieno[3,4-b]thiophene gave a mixture of the 4- and 6-aldehydes; the same products were obtained by metallation with n-butyl-lithium and reaction with dimethylformamide[140]. Direct formylation of thieno-[2,3-b] and -[3,2-b]-thiophene gave the 2-aldehyde[141], and further studies on Vilsmeier formylation confirm that 6a-thiathiophthene undergoes electrophilic substitution at C-3 [142].

Metallation of 4H-indeno[1,2-b]thiophene with butyl-lithium and reaction with carbon dioxide gave the 4-carboxylic acid exclusively, but similar treatment of 8H-indeno[1,2-c]thiophene led to a mixture of the 1-, 3-, and 4-carboxylic acids[143]. The relative rates of metallation of some tetramethyl-cyclopentadithiophenes have been studied[144].

8.4.2 Nucleophilic substitution

A study of the kinetics of the reaction of a number of bromonitrothiophenes with sodium selenophenoxide and of 5-substituted 2-nitro-3-bromothiophens with sodium thiophenoxide gave the following order of reactivities: 3-bromo-4-nitro ≪ 2-bromo-3-nitro < 2-bromo-5-nitro < 3-bromo-2-nitro, and confirmed the validity of the Hammett relationship for reactions occurring directly on the thiophene nucleus[145]. The relative activity of 2-nitro-3-X- and 3-nitro-2-X-thiophenes was variable and depended on the nature of X [146]. Reaction of 2,3-dinitrothiophene with several anions gave a mixture of products formed by replacement of either nitro group[147]. Nucleophilic displacement of fluorine from tetrafluorothiophene with methoxide gives 2,3,4-trifluoro-5-methoxythiophene[101]. The specific rate and equilibrium constant for formation of the Meisenheimer complex from 2-methoxy-3,5-dinitrothiophene and methoxide ion have been determined[148].

Contrary to an earlier report reaction of (25) with hydroxide ion leads not to a ring-opened product but to the hydroxythiopyrone (26) through nucleophilic attack at C-2 [149].

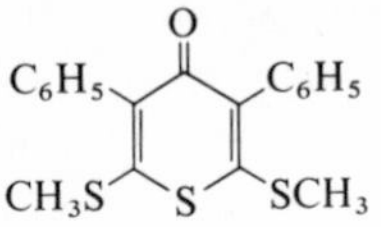

(25) 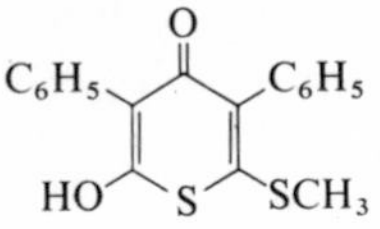   (26) 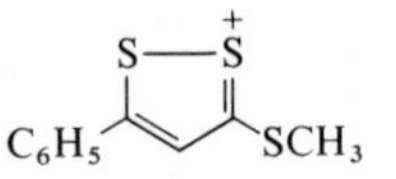   (27) 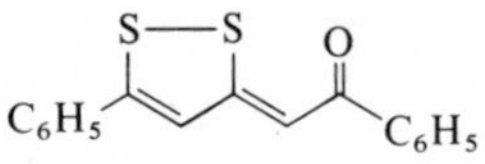   (28)

Dithiolylidene ketones (28) have been obtained previously by reactions involving nucleophilic attack at C-3 of 1,2-dithiolium salts with suitable carbanions. A more convenient route is exemplified in the conversion of (27) into (28) with sodium benzoylacetate in boiling ethanol[150]. Another route to compounds of type (28) proceeds from 1,2-dithiol-3-thiones[50], or 4H-thiopyran-4-thiones[48].

8.4.3   Free radical substitution

Phenylation of thiophene in solution with a number of different reagents afforded in most cases a mixture of 2- and 3-phenylthiophenes in the ratio 9:1 [151]. A different ratio was obtained by gas-phase phenylation. Thiophene was five times more reactive than benzene in arylations at 600 °C [152].

8.4.4   Reactions at sulphur

S-Alkyl-thiophenium, -benzo[b]thiophenium and -dibenzothiophenium salts have been obtained by reaction of the thiophene with alkyl halides and silver tetrafluoroborate. The salts are readily solvolysed; they are powerful alkylating agents[92].

8.5   RING-MODIFYING PROCESSES

8.5.1   Additions

An unusual 1,4-addition of maleic anhydride to the thiophene nucleus of a polycyclic compound has been observed[153].

8.5.2   Ring-opening reactions

A number of novel reactions leading to opening of the thiophene ring have been reported. Anodic oxidation of thiophene derivatives in acidified methanol leads to 1,4-dicarbonyl compounds[154]. Photosensitised oxidation of 2,5-dimethylthiophene gave a mixture of the cis-sulphene (29) and trans-hex-3-en-2,5-dione, possibly by way of the thio-ozonide (30) or the sulphox-onium ion (31) [155].

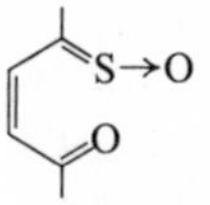

MeO$_2$C

H$_2$N

S

CH=CH·CH=C·CO$_2$Me

CN

(29) 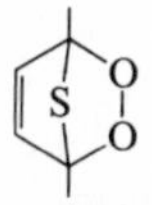   (30) 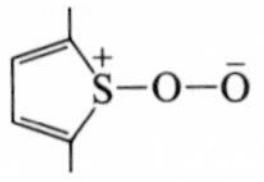   (31) 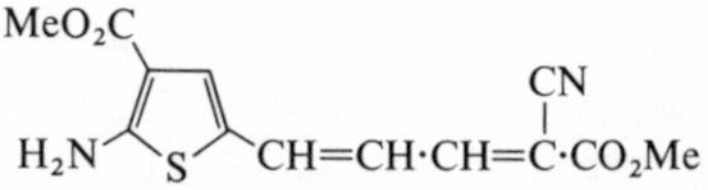   (32)

3-Thienyl-lithium derivatives decompose in refluxing ether to give ring-opened products[156]. 3-Benzo[b]thienyl-lithium compounds behave similarly at temperatures not much in excess of $-70\,°C$[157]. The coloured product (32) obtained by reaction of methyl α-cyano-β-(2-thienyl)acrylate with morpho-line or piperidine at room temperature is thought to arise by attack of the base on the thiophene ring followed or accompanied by opening of the ring and recyclisation in a different way[158].

1-Methyl-3,5-diphenylthiabenzene 1-oxide on reduction with trichloro-silane gave 3,5-diphenyl-[2H]- and -[4H]-thiapyran[47]. With hydrogen and platinum or zinc and acetic acid ring-opened products were obtained[46].

Reaction of 4-aryl-3-methylthio-1,2-dithiolium salts with secondary ali-phatic amines leads to ring opening, through nucleophilic attack of nitrogen at C-5, to give methyl 3-dialkylamino-2-arylpropendithioates[159]. Other reactions involving 1,2-dithiolium salts which lead to modification of the nucleus have been recorded[50, 142, 150, 160].

8.6 PHOTOCHEMISTRY

The photochemistry of sulphur hetero-aromatic systems has not been so extensively studied as that of some other systems, and a recent review[161] of the photochemistry of heterocyclics contains comparatively few references to work with sulphur compounds.

Much of the work done has been concerned with the photo-rearrangement of 2-aryl- and 2-alkyl-thiophenes to the 3-isomers. Systematic studies by Wynberg and his colleagues have established the main features of the reaction[162]. Best yields are obtained with aryl derivatives. Simple alkyl-thiophenes may also rearrange, although usually in somewhat lower yield[163]. Nothing conclusive is yet known about the mechanism of the rearrangement. The reaction is irreversible and is thought to involve the first excited singlet state of the thiophene. Labelling experiments have established that the migrating group remains attached to the same carbon atom of the nucleus in the product and that only C-2 and C-3 of the thiophene ring are inter-changed. A reaction pathway involving either a cyclopropene thioaldehyde or a bridged structure in which sulphur has an expanded valence shell seem to accommodate most of the experimental facts[162, 163]. The cyclo-propenylthioaldehyde intermediate has also been invoked to explain the formation of pyrrole derivatives when thiophenes are irradiated in presence of primary amines. 2,5-Dimethylthiophene and n-propylamine, for example, give N-propyl-2,5-dimethylpyrrole and a trace of N-propyl-2,4-dimethylpyrrole[164-167]. Irradiation of benzo[b]thiophene in propylamine or piperidine gave, not an indole derivative, but the 3-alkylamino-2,3-dihydrobenzo[b]-thiophene[168].

Photosensitised addition of dimethylmaleic anhydride to thiophene to form a cyclobutane derivative has been known for some time. Sensitised irradiation in cis-1,2-dichloroethylene likewise gives a dichlorocyclobutane derivative; comparable products are formed in high yield from benzo[b]-thiophene and several of its derivatives[169]. Photochemical dimerisation of heterocyclic nuclei to form cyclobutane derivatives is well documented[161]. Irradiation of benzo[b]thiophene dioxide in benzene solution has been shown to give two dimers, not one as previously supposed, identified as the anti-head-to-head and anti-head-to-tail compounds[170]. Similarly, irradiation of 2,6-diphenyl-4H-thiopyran-4-one in non-polar solvents gave an anti-head-to-tail dimer; but in polar solvents in presence of oxygen photo-oxidised products were formed[171]. In contrast, irradiation of 2,6-dimethyl-4H-thiopyran-4-one in benzene or methanol under nitrogen gave only a cage dimer. A different kind of reaction, involving extrusion of sulphur was observed in the irradiation of 2,6-bis(alkylthio)-3,5-diphenyl-4H-thiopyran-4-ones[172].

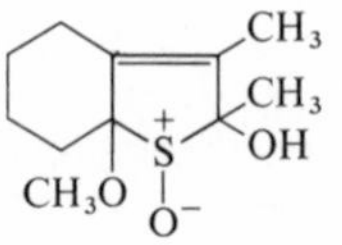

(33) 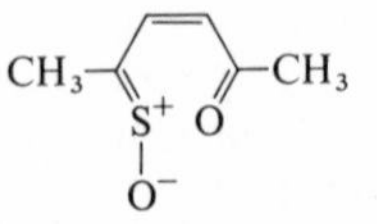　　　　　　　(34) 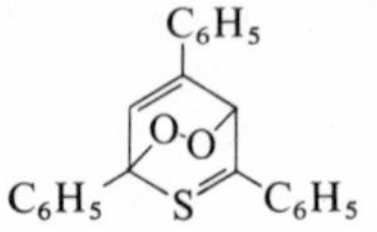　　　　　　　(35)

Rupture of the thiophene ring by photo-oxidation has been reported by two groups of workers. Alkyl-substituted thiophenes, in contrast to thiophene itself, react with singlet oxygen forming sulphines as products or intermediates[155]. Thus, 2,3-dimethyl-4,5,6,7-tetrahydrobenzo[b]thiophene on irradiation in methanol in presence of oxygen and methylene blue gave the product (33); in the reaction with 2,5-dimethylthiophene, methanol was not incorporated and the sulphine (34) could be isolated. A thio-ozonide of the type postulated as an intermediate in these reactions has actually been isolated in the sensitised photo-oxidation of a polycyclic thiopyran[173].

Irradiation of 2,4,6-triphenylthiopyrylium perchlorate in methanol in the presence of oxygen led to destruction of the ring and formation of benzaldehyde. Reaction is believed to proceed by attack of ground state oxygen on the ion, and the endoperoxide (35) seems a possible intermediate[174].

4H-Thiopyran-4-thiones are converted into thiopyrones by photosensitised oxidation, probably by way of an intermediate formed by addition of singlet oxygen to the thioketone[175].

8.7　PROPERTIES OF FUNCTIONAL GROUPS

In line with earlier experiments with other alkoxy derivatives, treatment of 2-t-butoxy-3-thienyl carbinols with toluene-p-sulphonic acid caused dehydration and dealkylation, producing a mixture of 3-alkenyl-3-thiolene-2-ones and 3-alkylidene-4-thiolene-2-ones[177]. Reaction of α-methoxy-3-benzo-

[b]thienylacetic acid with hydrogen bromide gave the expected α-bromo compound, but under the same conditions the 2-isomer formed 3-bromo-2-benzo[b]thienylacetic acid[178].

5-Aminobenzo[b]thiophene has been used to prepare several polycyclic nitrogen-containing systems[179]. Reaction of 4,5-diaminobenzo[b]thiophene with a number of 1,2-diketones and ortho-quinones gave derived polycyclic azines[180]. A series of 2,4-bis(arylamino)thiophenes was found to undergo rapid exchange of the C-5 proton in acid media. The i.r. and n.m.r. spectra suggest that they should be regarded as enamines rather than as aromatic amines[5].

α-Bromothiophenes, on treatment with sodamide in liquid ammonia, rearrange to β-bromothiophenes, providing a convenient and useful route to these otherwise difficult accessible derivatives. Thus, 2-bromothiophene gives 3-bromothiophene in 73% yield, and 2-bromo-3-methylthiophene is converted into 3-bromo-4-methylthiophene in 67% yield[181]. Reaction is believed to take place by transbromination via intermediate carbanions (compare[182]).

Reaction of 2-thienylcarbinol with 1-ethoxy-2-methylbutadiene in the presence of mercuric acetate gave the aldehyde 2-methyl-5-thienylpent-2-enal, presumably by a Claisen–Cope rearrangement of the intermediate thenyl allyl ether[183].

Conversion of 2- and 3-formylthiophene into the 2- and 3-cyclopropyl derivatives by way of the trimethyl-(3-thienyl-n-propyl)ammonium compounds has been described[91]. 2,3-Diformylthiophene reacts with ketones RCH_2COCH_2R in an alkaline medium to form thiophene analogues of benzotropone[184]. 2,2′-Diformyl-3,3′-bithienyl derivatives readily undergo an intramolecular Cannizzaro reaction with methanolic potassium hydroxide[185]. Reactions of this type are uncommon in the biphenyl series.

It is well known that 3-hydroxythiophene is very unstable and exists as a mixture of 4-thiolene-3-one and 3-hydroxythiophene. It is found spectroscopically, however, that 2-cyano-3-hydroxy- and 3-hydroxy-2-phenyl-thiophene exist as intramolecularly hydrogen-bonded thiophenes similar to the 2-alkoxycarbonyl- or 2-acyl-3-hydroxythiophenes[186]. The tautomeric structures of some potential dihydroxythiophenes have been determined[187]. 2,3-Dihydroxythiophene and a number of 4- and 5-substituted derivatives are shown by u.v., i.r. and n.m.r. spectroscopy to exist as the corresponding 3-hydroxy-3-thiolene-2-ones. 2,4- and 3,4-Dihydroxythiophene were similarly shown to have the corresponding hydroxy-thiolene-2-(and -3-)-one forms respectively, but the 2,5-diethoxycarbonyl derivative of the latter is a true dihydroxythiophene. 2,5-Dihydroxythiophenes all existed as thiosuccinic anhydrides. 2-Hydroxybenzo[b]thiophene is known to exist exclusively in the keto form both in the solid phase and in solution. It can be converted into the 2-methoxy compound in 90% yield by reaction with sodium hydride and dimethyl sulphate in hexamethylphosphoramide. With dimethyl sulphate in methanol ring opening occurs[188]. Reaction of a substituted 2-hydroxythiapyrone with diazomethane gave the isomeric enol ethers[149].

An improved route to benzo[b]thiophene-2(3H)- and -3(2H)-one from 2- and 3-benzo[b]thienyl-lithium, by hydrogen peroxide oxidation of the derived cyclotriboroxanes, has been described[188].

A new method for the synthesis of symmetrical dithienyl sulphides was found in the reaction of thienyl-lithium with bis(phenylsulphonyl) sulphide[189]. Benzo[b]thiophene-2-thiols are also readily obtained by reaction of the appropriate 2-lithio derivative with sulphur, or by reduction of the corresponding sulphonyl chloride with lithium aluminium hydride[32]. By reaction with diorganochlorosilanes, thienyl Grignard reagents are converted into diorgano-(2-thienyl)-silanes[190].

The methyl group in 3-methyl-2-nitrothiophene is unexpectedly active, and gives a 73% yield of 2-nitro-3-vinylthiophene in reaction with formaldehyde[191]. The methylene group of cyclopentadithiophenes is susceptible to oxidation with oxygen in presence of $KOBu^t$–Bu^tOH–Me_2SO; ring cleavage occurs with formation of the bithienylcarboxylic acid[192].

2-Aryl-3-nitro- and 2-(o-nitrophenyl)-benzo[b]thiophenes give the corresponding benzo[b]thieno[3,2-b]indole on reaction with triethyl phosphite presumably by a nitrene insertion reaction[193].

In the solid state and in solution mercapto-aldimines of the thiophene series exist mainly in the hydrogen-bonded thione form[194].

Allyl 2- and 3-thienyl sulphides undergo thio-Claisen rearrangement on heating in quinoline solution. With allyl 3-thienyl sulphide the primary rearrangement product, 2-allylthiophene-3-thiol, could be isolated, but in other cases further reaction ensued to give cyclic sulphides and allyl thienyl disulphides[176].

References

1. Afanas'eva, Yu. A., Ryashentseva, M. A., Minachev, Kh. M. and Levitskii, I. I. (1970). *Bull. Acad. Sci. U.S.S.R. Div. Chem. Science,* 1892
2. Takaku, M., Hayasi, Y. and Nozaki, H. (1970). *Bull. Chem. Soc. Jap.,* **43,** 1917
3. Smutny, E. J. (1969). *J. Amer. Chem. Soc.,* **91,** 208
4. Henriksen, L. and Autrup, H. (1970). *Acta Chem. Scand.,* **24,** 2629
5. Chupp, J. P. (1970). *J. Heterocycl. Chem.,* **7,** 285
6. Kay, I. T. and Punja, N. (1970). *J. Chem. Soc. C,* 2409
7. Hedegaard, B., Mortensen, J, Z. and Lawesson, S-O. (1971). *Tetrahedron,* **27,** 3853
8. Bradamante, S., Fusco, R., Marchesini, A. and Pagani, G. (1970). *Tetrahedron Lett.,* 11; Fujita, S., Kawaguti, T. and Nozaki, H. (1971). *Tetrahedron Lett.,* 1119
9. Iddon, B. and Scrowston, R. M. (1970). *Advances in Heterocyclic Chemistry,* Vol. 11, 177, (A. R. Katritzky and A. J. Boulton editors) (London: Academic Press)
10. Dickinson, R. P. and Iddon, B. (1970). *J. Chem. Soc. C,* 2592; Campaigne, E., Dinner, A. and Neiss, E. S. (1970). *J. Heterocycl. Chem.,* **7,** 695
11. Bevis, M. J., Forbes, E. J., Naik, N. N. and Uff, B. C. (1971). *Tetrahedron,* **27,** 1253
12. Dickinson, R. P. and Iddon, B. (1971). *J. Chem. Soc. C,* 2504; Kost, A. N., Budylin, V. A., Matveeva, E. D. and Sterligo, D. O. (1970). *J. Org. Chem. USSR,* **6,** 1516
13. Chakrabarti, P. M. and Chapman, N. B. (1970). *J. Chem. Soc. C,* 914
14. Capozzi, G., Melloni, G. and Modena, G. (1970). *J. Org. Chem.,* **35,** 1217
15. Kwart, H. and George, T. J. (1970). *Chem. Commun.,* 433
16. Bravo, P. Guadiano, G. and Zubiani, M. G. (1970). *J. Heterocycl. Chem.,* **7,** 967
17. Nakagawa, S., Okumura, J., Sakai, F., Hoshi, H. and Naito, T. (1970). *Tetrahedron Lett.,* 3719
18. Deckers, F. H. M., Speckamp, W. N. and Huisman, H. O. (1970). *Chem. Commun.,* 1521
19. White, J. D., Mann, M. E., Kirshenbaum, H. D. and Mitra, A. (1971). *J. Org. Chem.,* **36,** 1048
20. Klemm, L. H., Johnson, W. O. and White, D. V. (1970). *J. Heterocycl. Chem.,* **7,** 463
21. Klemm, L. H., McCoy, D. R. and Olson, D. R. (1970). *J. Heterocycl. Chem.,* **7,** 1347

22. Buu-Hoi, N. P., Servoin-Sidoine, J. and Saint-Ruf, G. (1971). *Bull. Soc. Chim. France,* 2060
23. Kurokawa, S. (1970). *Bull. Chem. Soc. Jap.,* **43,** 1454
24. Marie, Cl., Buu-Hoi, N. P. and Jacquignon, P. (1971). *J. Chem. Soc. C,* 431
25. Dayagi, S., Goldberg, I. and Shmueli, U. (1970). *Tetrahedron,* **26,** 411
26. Müller, E., Thomas, R., Sauerbier, M., Langer, E. and Streichfuss, D. (1971). *Tetrahedron Lett.,* 521
27. Blackburn, E. V. and Timmons, C. J. (1969). *Quart. Rev. Chem. Soc.,* **23,** 482
28. de Luca, G., Martelli, G., Spagnolo, P. and Tiecco, M. (1970). *J. Chem. Soc. C,* 2504
29. Groen, M. B., Schadenberg, H. and Wynberg, H. (1971). *J. Org. Chem.,* **36,** 2797; Lehman, P. G. and Wynberg, H. (1971). *Rec. Trav. Chim.,* **90,** 1113; Wynberg, H. (1971). *Accounts Chem. Res.,* **4,** 65
30. MacDowell, D. W. H. and Maxwell, M. H. (1970). *J. Org. Chem.,* **35,** 799
31. Jordens, P., Rawson, G. and Wynberg, H. (1970). *J. Chem. Soc. C,* 273; Wiersema, A. and Gronowitz, S. (1970). *Acta Chem. Scand.,* **24,** 2593
32. Chapman, N. B., Hughes, C. G. and Scrowston, R. M. (1970). *J. Chem. Soc. C,* 2431
33. de Jong, F. and Janssen, M. J. (1971). *J. Org. Chem.,* 1645, 1998
34. Ashby, J., Ayad, M. and Meth-Cohn, O. (1971). *Chem. Commun.,* 1251
35. Ahmed, M., Ashby, J. and Meth-Cohn, O. (1970). *Chem. Commun.,* 1094
36. Garratt, P. J. and Vollhardt, K. P. C. (1970). *Chem. Commun.,* 109
37. MacDowell, D. W. H., Jeffries, A. T. and Meyers, M. B. (1971). *J. Org. Chem.,* **36,** 1416
38. McKinnon, D. M. (1970). *Can. J. Chem.,* **48,** 3388; Young, T. E. and Hamel, C. R. (1970). *J. Org. Chem.,* **35,** 816, 821
39. de Waard, E. R., Vloon, W. J. and Huisman, H. O. (1970). *Chem. Commun.,* 841
40. Bourdais, J. (1970). *Tetrahedron Lett.,* 2895
41. Cotruvo, J. A. and Degani, I. (1971). *Chem. Commun.,* 436
42. Yoshida, Z-I., Yoneda, S., Sugimoto, T. and Kikukawa, O. (1971). *Tetrahedron Lett.,* 3999
43. Ohno, A., Koisumi, T. and Ohnishi, Y. (1971). *Bull. Chem. Soc. Jap.,* **44,** 2511
44. Price, C. C. and Follweiler, D. H. (1969). *J. Org. Chem.,* **34,** 3202
45. Price, C. C., Follweiler, J., Pirelahi, H. and Siskin, M. (1971). *J. Org. Chem.,* **36,** 791
46. Hortmann, A. G. and Harris, R. L. (1971). *J. Amer. Chem. Soc.,* **93,** 2471
47. Hortmann, A. G. and Harris, R. L. (1970). *J. Amer. Chem. Soc.,* **92,** 1803
48. Dingwall, J. G., Reid, D. H. and Symon, J. D. (1970). *J. Chem. Soc. C,* 2412
49. Brown, E. I. G., Leaver, D. and McKinnon, D. M. (1970). *J. Chem. Soc. C,* 1202
50. Caillaud, G. and Mollier, Y. (1970). *Bull. Soc. Chim. France,* 2018; (1971), 331
51. Klinsberg, E. (1971). *Synthesis,* 243
52. Dewar, M. J. S. and Trinajstic, N. (1970). *J. Amer. Chem. Soc.,* **92,** 1453
53. Hoffman, J. M. and Schlessinger, R. H. (1970). *J. Amer. Chem. Soc.,* **92,** 5263
54. Mock, W. L. (1970). *Chem. Commun.,* 1254; Paquette, L. A. and Maiorana, S. (1971). *Chem. Commun.,* 313
55. Garratt, P. J., Holmes, A. B., Sondheimer, F. and Vollhardt, K. P. C. (1970). *J. Amer. Chem. Soc.,* **92,** 4492
56. Holmes, A. B. and Sondheimer, F. (1970). *J. Amer. Chem. Soc.,* **92,** 5284; (1971). *Chem. Commun.,* 1434
57. Coffen, D. L., Poon, Y. C. and Lee, M. L. (1971). *J. Amer. Chem. Soc.,* **93,** 4627
58. Riley, M. O. and Park, J. D. (1971). *Tetrahedron Lett.,* 2871
59. Garratt, P. J., Holmes, A. B., Sondheimer, F. and Vollhardt, K. P. C. (1971). *Chem. Commun.,* 947
60. Carruthers, W. and Pellatt, M. G. (1971). *J. Chem. Soc. C,* 1485
61. Broadhurst, M. J., Grigg, R. and Johnson, A. W. (1971). *J. Chem. Soc. C,* 3681
62. Meth-Cohn, O. (1970). *Quart. Reports on Sulphur Chem.,* **5,** 129
63. Clark, D. T. and Armstrong, D. R. (1970). *Chem. Commun.,* 319; Gelius, U., Roos, B. and Siegbahn, P. (1970). *Chem. Phys. Lett.,* **4,** 471
64. Dewar, M. J. S. and Trinajstic, N. (1970). *J. Amer. Chem. Soc.,* **92,** 1453
65. Klasinc, L. and Trinajstic, N. (1971). *Tetrahedron,* **27,** 4045
66. Price, C. C., Hori, M., Parasaran, T. and Polk, M. (1963). *J. Amer. Chem. Soc.,* **85,** 2278
67. Hortmann, A. G. and Harris, R. L. (1970). *J. Amer. Chem. Soc.,* **92,** 1803; (1971). **93,** 2471
68. Bradamante, S., Mangia, A. and Pagani, G. (1970). *Tetrahedron Lett.,* 3381
69. Hirai, K. (1971). *Tetrahedron,* **27,** 4003
70. Hordvik, A. (1970). *Quart. Reports on Sulphur Chem.,* **5,** 21

71. Fabian, K., Hartmann, H., Fabian, J. and Mayer, R. (1971). *Tetrahedron*, 27, 4705
72. Klingsberg, E. (1969). *Quart. Rev. Chem. Soc.*, 23, 537
73. Hansen, L. K. and Hordvik, A. (1970). *Acta Chem. Scand.*, 24, 2246; Hordvik, A. and Salthre, L. J. (1970). ibid., 2261
74. Leung, F. and Nyburg, S. C. (1969). *Chem. Commun.*, 137
75. Gleiter, R., Schmidt, D. and Behringer, H. (1971). *Chem. Commun.*, 525
76. Kristensen, R. and Sletter, J. (1971). *Acta Chem. Scand.*, 25, 2366
77. Reid, D. H. (1971). *J. Chem. Soc. C*, 3187
78. Hordvik, A. and Jolshamn, K. (1971). *Acta Chem. Scand.*, 25, 1895
79. Cooper, W. F., Kenny, N. C., Edmonds, J. W., Nagel, A., Wudl, F. and Coppens, P. (1971). *Chem. Commun.*, 889
80. Ammon, H. L., Replogle, L. L., Watts, P. H., Katsumoto, K. and Stewart, J. M. (1971). *J. Amer. Chem. Soc.*, 93, 2196
81. Nógrádi, M., Ollis, W. D. and Sutherland, I. O. (1970). *Chem. Commun.*, 158
82. Steurman, H. J. and Schenk, H. (1970). *Rec. Trav. Chim.*, 89, 392
83. Kaper, L. and de Boer, Th. J. (1970). *Rec. Trav. Chim.*, 89, 825
84. Schuetz, R. D. and Nilles, G. P. (1971). *J. Org. Chem.*, 36, 2188
85. Chadwick, D. J. and Meakins, G. D. (1970). *Chem. Commun.*, 637
86. Bandoli, G., Panattoni, C., Clemente, D. A., Tondello, E., Dondoni, A. and Mangini, A. (1971). *J. Chem. Soc. B*, 1407
87. Groen, M. B., Stulen, G., Visser, G. J. and Wynberg, H. (1970). *J. Amer. Chem. Soc.*, 92, 7218; Groen, M. B. and Wynberg, H. (1971). *J. Amer. Chem. Soc.*, 93, 2968
88. Wynberg, H. (1971). *Accounts Chem. Research*, 4, 65
89. Schaffrin, R. M. and Trotter, J. (1970). *J. Chem. Soc. A*, 1561
90. van Reijendam, J. W., Heeres, G. J. and Janssen, M. J. (1970). *Tetrahedron*, 26, 1291; van Reijendam, J. W. and Janssen, M. J. (1970). ibid., 1303
91. Kellog, R. M. and Buter, J. (1971). *J. Org. Chem.*, 36, 2236
92. Acheson, R. M. and Harrison, D. R. (1970). *J. Chem. Soc. C*, 1764
93. Tajiri, A., Asano, T. and Nakajima, T. (1971). *Tetrahedron Lett.*, 1785
94. Andrieu, C., Pinel, R. and Mollier, Y. (1971). *Bull. Soc. Chim. France*, 1314
95. Lunazzi, L., Mangini, A., Pedulli, G. F. and Tiecco, M. (1971). *Gazzetta*, 101, 10
96. Russell, G. A., Myers, C. L., Bruni, P., Neugebauer, F. A. and Blankespoor, R. (1970). *J. Amer. Chem. Soc.*, 92, 2762
97. Trifunac, A. and Kaiser, E. T. (1970). *J. Phys. Chem.*, 74, 2236
98. Lunazzi, L., Placucci, G., Tiecco, M. and Martelli, G. (1971). *J. Chem. Soc. B*, 1820
99. Ewing, D. F. and Scrowston, R. M. (1971). *Org. Magnetic Resonance*, 3, 405
100. Sone, T. and Takahashi, K. (1971). *Org. Magnetic Resonance*, 3, 527
101. Burdon, J., Campbell, J. G., Parsons, I. W. and Tatlow, J. C. (1971). *J. Chem. Soc. C*, 352
102. Cooper, J., Ewing, D. F., Scrowston, R. M. and Westwood, R. (1970). *J. Chem. Soc. C*, 1949
103. Brophy, G. C., Sternhell, S., Brown, N. M. D., Brown, I., Armstrong, K. J. and Martin-Smith, M. (1970). *J. Chem. Soc. C*, 933
104. Young, T. E. and Hamel, C. R. (1970). *J. Org. Chem.*, 35, 821
105. Schuetz, R. D. and Nilles, G. P. (1971). *J. Org. Chem.*, 36, 2188; Huckerby, T. N. (1971). *Tetrahedron Lett.*, 3497
106. Roquis, B. and Fournie-Zaluski, M. C. (1971). *Org. Magnetic Resonance*, 3, 305
107. Huckerby, T. N. (1971). *Tetrahedron Lett.*, 353
108. Schaumburg, K. (1971). *Can. J. Chem.*, 49, 1146
109. Bugge, A., Gestblom, B. and Hartmann, O. (1970). *Acta Chem. Scand.*, 24, 105, 1953; Wynberg, H. and Feijen, J. (1970). *Rev. Trav. Chim.*, 89, 77; Dressler, M. L. and Joullie, M. M. (1970). *J. Heterocycl. Chem.*, 7, 1257
110. Bartle, K. D., Jones, D. W. and Matthews, R. S. (1971). *Tetrahedron*, 27, 5177; Bartle, K. D., Jones, D. W., Matthews, R. S., Birch, A. and Crombie, D. A. (1971). *J. Chem. Soc. B*, 2092
111. Groen, M. B., Schadenberg, H. and Wynberg, H. (1971). *J. Org. Chem.*, 36, 2797
112. de Jong, F., Sinnige, H. J. M. and Janssen, M. J. (1970). *Rec. Trav. Chim.*, 89, 225
113. van Thuyl, J., Klebe, K. J. and van Houte, J. J. (1970). *Organic Mass Spectrometry*, 3, 1539
114. Price, C. C., Follweiler, J., Pirelahi, H. and Siskin, M. (1971). *J. Org. Chem.*, 36, 791
115. Yoshida, Z. and Kobayashi, T. (1970). *Tetrahedron*, 26, 267

116. Kulevsky, N. and Liu, G. J. (1970). *J. Phys. Chem.*, **74,** 751
117. Atkinson, R. E. and Hardy, F. E. (1971). *J. Chem. Soc. B*, 357
118. Butler, A. R. and Hendry, J. B. (1970). *J. Chem. Soc. B*, 170, 852
119. Butler, A. R. and Hendry, J. B. (1970). *J. Chem. Soc. B*, 848
120. Butler, A. R. and Hendry, J. B. (1971). *J. Chem. Soc. B*, 102
121. Clementi, S. and Marino, G. (1970). *Chem. Commun.*, 1642
122. Linda, P. and Marino, G. (1970). *J. Chem. Soc. B*, 43
123. Wiersema, A. and Gronowitz, S. (1970). *Acta Chem. Scand.*, **24,** 2593
124. Vol'kenshtein, Yu. B., Karmanova, I. B. and Gol'dfarb, Ya. L. (1970). *Bull. Acad. Sci. U.S.S.R. Div. Chem. Science*, 2592
125. Sone, T., Sakai, K. and Kuroda, K. (1970). *Bull. Chem. Soc. Japan*, **43,** 1411
126. Burdon, J., Parsons, I. W. and Tatlow, J. C. (1971). *J. Chem. Soc. C*, 346; Burdon, J., Campbell, J. G., Parsons, I. W. and Tatlow, J. C. (1971). *J. Chem. Soc. C*, 352
127. Curtis, R. F., Jones, D. M. and Thomas, W. A. (1971). *J. Chem. Soc. C*, 234
128. Nilsson, M. and Ullenius, C. (1970). *Acta Chem. Scand.*, 2379
129. Nilsson, M., Ullenius, C. and Wennerström, O. (1971). *Tetrahedron Lett.*, 2713
130. Slocum, D. W. and Gierer, P. L. (1971). *Chem. Commun.*, 305
131. Dickinson, R. P. and Iddon, B. (1970). *J. Chem. Soc. C*, 2592
132. Campaigne, E., Dinner, A. and Neiss, E. S. (1970). *J. Heterocycl. Chem.*, **7,** 695
133. Cooper, J. and Scrowston, R. M. (1971). *J. Chem. Soc. C*, 3405
134. Cooper, J. and Scrowston, R. M. (1971). *J. Chem. Soc. C*, 3052
135. Eaborn, C. and Wright, G. J. (1971). *J. Chem. Soc. B*, 2262
136. Chapman, N. B., Hughes, C. G. and Scrowston, R. M. (1971). *J. Chem. Soc. C*, 463
137. Dressler, M. L. and Joullié, M. M. (1970). *J. Heterocycl. Chem.*, **7,** 1257; Klemm, L. H., Zell, R., Barnish, I. T., Klemm, R. A., Klopfenstein, C. E. and McCoy, D. R. (1970). *J. Heterocycl. Chem.*, 373
138. Chapman, N. B., Clarke, K. and Sharma, K. S. (1970). *J. Chem. Soc. C*, 2334
139. Robba, M., Lecompte, J-M. and Cugon de Sévricourt, M. (1971). *Tetrahedron*, **27,** 487
140. Wynberg, H. and Feijen, J. (1970). *Rev. Trav. Chim.*, **89,** 77
141. Bugge, A. (1971). *Acta Chem. Scand.*, **25,** 27
142. Duguay, G., Reid, D. H., Wade, K. O. and Webster, R. G. (1971). *J. Chem. Soc. C*, 2829
143. MacDowell, D. W. H. and Jeffries, A. T. (1970). *J. Org. Chem.*, **35,** 871; (1971). *J. Org. Chem.*, **36,** 1053
144. Wiersema, A. K. and Gronowitz, S. (1971). *Acta Chem. Scand.*, **25,** 1195
145. Guanti, G., Dell'Erba, C. and Spinelli, D. (1970). *Gazzetta*, **100,** 184; *J. Heterocycl. Chem.*, **7,** 1333
146. Guanti, G., Dell'Erba, C. and Macero, P. (1971). *J. Heterocycl. Chem.*, **8,** 537
147. Dell'Erba, C. and Guanti, G. (1970). *Gazzetta*, **100,** 223
148. Doddi, G., Illuminati, G. and Stegel, F. (1971). *J. Org. Chem.*, **36,** 1918
149. Teague, H. J. and Tucker, W. P. (1970). *J. Org. Chem.*, **35,** 1968
150. Brown, E. I. G., Leaver, D. and McKinnon, D. M. (1970). *J. Chem. Soc. C*, 1202
151. Camaggi, C. M., Leardini, R., Tiecco, M. and Tundo, A. (1970). *J. Chem. Soc. B*, 1683
152. Fields, E. K. and Meyerson, S. (1970). *J. Org. Chem.*, **35,** 67
153. Wiersema, A. K. and Gronowitz, S. (1970). *Acta Chem. Scand.*, **24,** 2653
154. Šrogl, J., Jander, M. and Valentová, U. (1970). *Coll. Czech. Chem. Commun.*, **35,** 148; Jander, M., Šrogl, J., Janoušová, A., Kubelka, V. and Holik, M. (1970). *Col. Czech. Chem. Commun.*, 2635
155. Skold, C. N. and Schlessinger, R. H. (1970). *Tetrahedron Lett.*, 791; Wasserman, H. H. and Strehlow, W. (1970). *Tetrahedron Lett.*, 795
156. Gronowitz, S. and Frejd, T. (1970). *Acta Chem. Scand.*, **24,** 2656; Jakobsen, H. J. (1970); *Acta Chem. Scand.*, 2663
157. Dickinson, R. P. and Iddon, B. (1970). *Tetrahedron Lett.*, 975; (1970). *J. Chem. Soc. C*, 2592; (1971). *J. Chem. Soc. C*, 182, 3447
158. Yasuda, H. and Midorikawa, H. (1971). *J. Org. Chem.*, **36,** 2196
159. Le Costumer, G. and Mollier, Y. (1971). *Bull. Chem. Soc. France*, 2958; see also Duguay, G. and Quimou, H. (1970). ibid., 1918
160. Caillaud, G. and Mollier, Y. (1971). *Bull. Soc. Chim. France*, 2326
161. Reid, S. T. (1970). *Advances in Heterocyclic Chemistry*, Vol. 11, 1 (A. R. Katritzky and A. J. Boulton, editors). (London: Academic Press)
162. Wynberg, H. (1971). *Accounts Chem. Research*, **4,** 65

163. Kellogg, R. M., Dik, J. K., van Driel, H. and Wynberg, H. (1970). *J. Org. Chem.*, **35**, 2737
164. Couture, A. and Lablache-Combier, A. (1969). *Chem. Commun.*, 524
165. Couture, A. and Lablache-Combier, A. (1971). *Tetrahedron*, **27**, 1059
166. Lablache-Combier, A. and Remy, M-A. (1971). *Bull. Soc. Chim. France*, 679
167. Couture, A. and Lablache-Combier, A. (1971). *Chem. Commun.*, 891
168. Grandclaudon, P. and Lablache-Combier, A. (1971). *Chem. Commun.*, 892
169. Neckers, D. C., Dopper, J. H. and Wynberg, H. (1970). *J. Org. Chem.*, **35**, 1582
170. Harpp, D. N. and Heitner, C. (1970). *J. Org. Chem.*, **35**, 3256
171. Sugiyama, N., Sato, Y., Kataoka, H., Koshima, C. and Yamada, K. (1969). *Bull. Chem. Soc. Japan*, **42**, 3005
172. Ishibe, N. and Odani, M. (1971). *Chem. Commun.*, 702
173. Hoffmann, J. M. and Schlessinger, R. H. (1970). *Tetrahedron Lett.*, 797
174. Yoshida, Z., Sugimoto, T. and Yoneda, S. (1971). *Tetrahedron Lett.*, 4259
175. Ishibe, N., Odani, M. and Sunami, M. (1971). *Chem. Commun.*, 118
176. Mortensen, J. Z., Hedgegaard, B. and Lawesson, S-O. (1971). *Tetrahedron*, **27**, 3831
177. Pedersen, E. B. and Lawesson, S-O. (1970). *Tetrahedron*, **26**, 2959
178. Gronowitz, S., Rehnö, J. and Sandström, J. (1970). *Acta Chem. Scand.*, **24**, 304
179. Buu-Hoï, N. P., Dufour, M., Jacquignon, P. and Martani, A. (1971). *J. Chem. Soc. C*, 2428
180. Chapman, N. B., Clarke, K. and Sharma, K. S. (1971). *J. Chem. Soc. C*, 919
181. Reinicke, M. G., Adickes, H. W. and Pyun, C. (1971). *J. Org. Chem.*, **36**, 2690
182. Reinicke, M. G. and Adickes, H. W. (1968). *J. Amer. Chem. Soc.*, **90**, 511
183. Thomas, A. F. and Ozainne, M. (1970). *J. Chem. Soc. C*, 220
184. Guillard, R. and Fournari, P. (1971). *Bull. Soc. Chim. France*, 1437
185. Wiklund, E. and Håkansson, R. (1970). *Acta Chem. Scand.*, **24**, 341
186. Hedegaard, B., Mortensen, J. Z. and Lawesson, S-O. (1971). *Tetrahedron*, **27**, 3853
187. Mortensen, J. Z., Hedegaard, B. and Lawesson, S-O. (1971). *Tetrahedron*, **27**, 3839
188. Dickinson, R. P. and Iddon, B. (1970). *J. Chem. Soc. C*, 1926
189. de Jong, F. and Janssen, M. J. (1971). *J. Org. Chem.*, **36**, 1645
190. Kartasheva, L. I., Nametkin, N. S. and Chernysheva, T. I. (1970). *J. Gen. Chem. USSR*, **40**, 1253
191. Srinivasan, K., Balasubramanian, K. K. and Swaminathan, S. (1971). *Chem. Ind.*, 398
192. Rawson, G. and Wynberg, H. (1971). *Rec. Trav. Chim.*, **90**, 39
193. Chippendale, K. E., Iddon, B. and Suschitzky, H. (1971). *Chem. Commun.*, 203
194. Bogdanov, V. S., Kalik, M. A., Yakoveev, I. P. and Gol'dfarb, Ya. L. (1970). *J. Gen. Chem. USSR*, **40**, 2085

9
Heterocyclic Compounds Containing both Nitrogen and Oxygen, and Nitrogen and Sulphur

R. M. SCROWSTON
University of Hull

9.1   INTRODUCTION

The present article aims to highlight the significant advances which have been made in the chemistry of nitrogen–oxygen and nitrogen–sulphur heterocyclic compounds during the years 1970 and 1971. Because of the vast number of structures which fall within this category, it has been necessary to restrict the coverage to the well-known classes of compounds. Only 5- and 6-membered aromatic systems are included; particular emphasis is placed on the former. Hydro-derivatives are reviewed in Chapter 11. Heterocyclic compounds and their benzo-homologues are usually discussed under a single heading. Owing to their unique properties, mesoionic compounds are considered collectively (Section 9.6).

Several general trends in heterocyclic research are reflected in the review. More and more photochemical transformations are being investigated, and the results are being rationalised on a sound mechanistic basis. A general mechanism for the photorearrangement of 5-membered aromatic hetero-cyclic compounds has been announced[1] and previous work on this subject has been reviewed[2]. Increasing use is being made of the Woodward–Hoffmann orbital symmetry rules in discussing the synthesis and reactions of hetero-cyclic compounds. The importance of electrocyclic and sigmatropic reactions in 5-membered heterocyclic chemistry has been reviewed[3]. Another review[4] surveys the 1,3-dipolar cyclo-addition reactions of nitrile oxides; further important cyclo-addition reactions will be discussed in the sequel.

Within the framework of the present chapter the most notable advances have been made in isothiazole chemistry, probably because new synthetic procedures have made such compounds more readily accessible.

9.2   5-MEMBERED RINGS CONTAINING ONE NITROGEN AND ONE OXYGEN ATOM

9.2.1   Oxazoles

9.2.1.1   Ring synthesis

Modifications to some well-known[5] routes to oxazoles and benzoxazoles have been announced. For example, hot polyphosphoric acid is superior to other reagents for the cyclisation of the α-acetamido-ketones, $AcCHR^1\cdot NHAc$, to the oxazoles (1a)[6]. 2,5-Diphenyloxazole is obtained (80%) by cyclodehydration of N-phenacylbenzamide, $PhCO\cdot CH_2\cdot NH\cdot COPh$, which is conveniently prepared by Friedel–Crafts acylation of benzene with benzamidoacetyl chloride[7]. 2-Alkylbenzoxazoles (2) may be prepared (50–75%) under mild conditions[8] from o-aminophenol and the appropriate carboxylic acid (RCO_2H) by using polyphosphoric acid ethyl ester as the condensing agent. The formation of (2) by reaction of the o-azidophenyl esters (3a) with triethyl phosphite[9], is of mechanistic interest because the cyclisation proceeds via a phosphorimidate intermediate (3b), whereas the

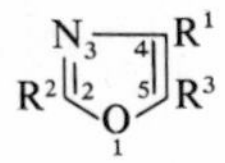

(1)

(a) $R^2 = R^3 = Me$
(b) $R^3 = H$

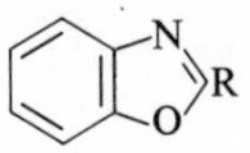

(2)

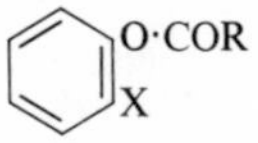

(3)

(a) $X = N_3$
(b) $X = N:P(OEt)_3$

analogous cyclisation of o-nitrophenyl esters[10] is believed to involved a nitrene precursor of the type $Ar\overset{-}{N}$—O—$\overset{+}{P}(OEt)_3$.

A new synthesis of oxazoles (22–70% yields) involves the single-step cyclocondensation of α-azido-esters or -ketones with acyl halides in the presence of triphenylphosphine[11]:

$$R'CH(N_3)\cdot COR^3 + R^2CO\cdot X \xrightarrow{Ph_3P} (1) + HX + N_2 + Ph_3PO \qquad (9.1)$$

$$(X = Cl \text{ or } Br)$$

Oxazoles lacking a 2-substituent (1b), which previously[5] were rather inaccessible, can be obtained readily by α-metalation of an alkyl isocyanide with either potassium t-butoxide or n-butyl-lithium, and treatment of the resulting anion with an acyl chloride[12]. The intermediate 2-acylisocyanides cyclise spontaneously via the enol form, to give the oxazoles (1b) in high yields (50–80%) (9.2). 5-Amino-oxazole-4-carbonitrile, a useful intermediate

$$R^1\bar{C}H\cdot NC \xrightarrow[\text{(b)}\ H^+]{\text{(a)}\ R^3CO\cdot Cl} R^1CH(NC)\cdot COR^3 \longrightarrow (1b) \qquad (9.2)$$

for the preparation of some bicyclic systems containing the oxazole nucleus, is obtained conveniently by reduction of phenylazomalononitrile with zinc and formic acid[13].

Two new syntheses involve aziridine intermediates. In the first, oxidation of N-aminophthalimide or variously substituted N-aminosuccinimides (of general formula X—NH$_2$) with lead tetra-acetate in the presence of gem-disubstituted olefins, ArCH:CYZ (where Y and Z are electron-withdrawing groups), gives the aziridines (4)[14]. These cyclise, either spontaneously or on heating, to give a wide range of oxazoles; for example, (4a) affords 5-amino-2-phenyloxazole-4-carbonitrile in unspecified yield when heated in ethanol. In the second[15], 2-aroylaziridines (4b) react with 1-nitroso-2-naphthol. The reaction is of theoretical significance because it proceeds via an initial 1,3-dipolar addition in both orientations to the N=O bond. The resulting oxadiazolidines undergo spontaneous 1,3-cleavage to nitrones, which then cyclise to give good yields of the 2-aryl- and 2-aroyl-naphtho[1,2-d]oxazoles (5a) and (5b), respectively.

Heating 4-acyl-2-oxazolin-5-ones (6) causes cycloelimination of carbon dioxide and formation of trisubstituted oxazoles (1) in high yields (70–95%)[16]. The isomeric 2-acyl-3-oxazolin-5-ones (7) undergo a similar reaction, but

(4)

(a) Ar^1 = Ph, X = phthalimido,
    Y = CN, Z = CO·NH$_2$
(b) X = cyclohexyl, Y = H
    Z = COAr^2

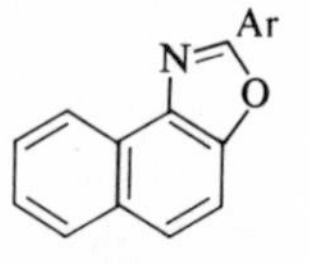

(5)

(a) Ar = aryl
(b) Ar = aroyl

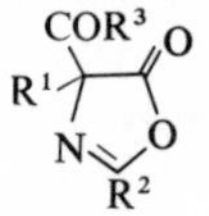

(6)

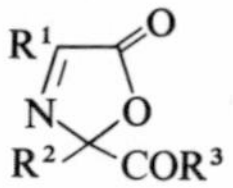

(7)

the 2- and 4-substituents (R^2 and R^1, respectively) of the final product (1) are formally interchanged[16]. This interesting reaction has been explained in terms of a C-acylnitrile ylide intermediate, e.g. $R'\cdot C\colon\overset{+}{N}—\bar{C}R^2\cdot COR^3$ from (7).

9.2.1.2 Spectroscopic and theoretical studies

The mass spectra of benzoxazole[17], and its 2-phenyl-[18], 2-amino-, and 2-alkylamino-derivatives[17] are characterised by very stable molecular ion peaks. Further fragmentation invariably proceeds with loss of CO; loss of HCN or RCN (where R is a 2-substituent) from the molecular ion is also a significant process[17, 18], except for 2-amino-derivatives[17].

Chemical shifts and coupling constants have been obtained from the n.m.r. spectra of a series of 2-methylbenzoxazoles, and the effects on these of ring substituents have been correlated with the corresponding effects in benzene derivatives[19]. A careful study of the u.v. spectra of benzoxazole and benzothiazole in the vapour phase has revealed no evidence for a π^*-n transition[20].

An SCF–MO study of the tautomerism of 2-phenylamino-oxazole supports the experimental finding that the amino-form predominates in ethanolic solution[21].

9.2.1.3 Ring-modifying processes: addition

5-Ethoxyoxazole and its 4-methyl derivative undergo Diels–Alder addition with $MeO_2C\cdot C\dot{:}C\cdot CO_2Me$ or diphenylcyclopropenone[22], but the expected adducts eliminate HCN or MeCN spontaneously to give dimethyl 2-ethoxy-furan-3,4-dicarboxylate and 2-ethoxy-3,5-diphenyl-4-pyrone, respectively. Other phenyl- or alkoxy-substituted oxazoles give related 3-substituted- or 3,4-disubstituted-furans by reaction with acetylenic aldehydes[23], ketones[23, 24], acids[24], or esters[22, 25]. As well as providing a useful synthesis of otherwise rather inaccessible 3-(oxo-substituted)furans, these reactions provide the first example of a retrodiene reaction involving elision of HCN or a nitrile.

2-Oxazolin-5-ones (8) undergo Michael addition to activated multiple bonds, the site of attack being determined by the nature of the acceptor molecule[26]. For example, benzoylacetylene adds exclusively to the 4-position, whereas acrylonitrile reacts solely in the 2-position to afford 3-oxazolin-5-ones, e.g. (9) from (8a). The adduct (9) can be hydrolysed to give a readily separable mixture of $Me_2CH\cdot CO\cdot CO_2H$, $R^1CO\cdot CH_2\cdot CH_2\cdot CN$, and $R^1CO\cdot CH_2\cdot CH_2\cdot CO_2H$. Because (8a) is obtained in high yield from the acid, R^1CO_2H, by reaction of the corresponding acyl chloride with valine and cyclisation of the product with acetic anhydride, this sequence provides a useful means for the three-carbon homologation of a carboxylic acid[27]. The adduct (10), obtained by treatment of the oxazoline (8b) with N-cinnamyl-

(8)   (9)   (10)

(a) $R^2 = Me_2CH$
(b) $R^1 = Ph, R^2 = Me$

ideneaniline, is probably formed by Diels–Alder addition to the isomeric open-chain ketene, $PhCO\cdot NH\cdot CMe\dot{:}C\dot{:}O$[28], but the possibility of Michael addition to (8b), followed by rearrangement, cannot be excluded.

9.2.2 Isoxazoles

9.2.2.1 Ring synthesis

The well-known[4] 1,3-cyclo-addition reaction between acetylenes, $Ar^2C\vdots CH$, and nitrile oxides, $Ar^1C\vdots N \rightarrow O$, continues to be used widely for the preparation of 3,5-disubstituted isoxazoles (11). The question as to whether the reaction proceeds by a concerted mechanism of the type outlined[29] for other 1,3-dipolar cyclo-additions, or through an intermediate acetylenic oxime, $Ar^1C(\vdots NOH)\cdot C\vdots CAr^2$, is still being debated. Battaglia et al.[30, 31] support the former mechanism, and have produced kinetic evidence to show that, although an oxime is definitely formed in the reaction, it is not the precursor of (11). Other workers[32, 33] believe that both mechanisms operate, and that which mechanism predominates may depend on the nature of the substituents, Ar^1 and Ar^2 [33].

Reaction of nitrile oxides $ArC\vdots N \rightarrow O$, with suitably substituted olefins gives rise to 3,5-disubstituted 2-isoxazolines (12), which are easily converted into the corresponding isoxazoles (13). Vinyl or isopropenyl acetates give the acetoxy-derivatives (12a), which are heated without isolation to give (13a) in good yield[34]; 1,1-dichloroethylene gives the 5-chloro-compounds (13b) directly[35]; β-acylenamines, $R^1CO\cdot CH\vdots CR\cdot NMe_2$, give the 4-acyl compounds (13c) in yields of 20–65%, provided that conditions are care-

(11) (12) (13)

(a) X = OAc, R = H or Me, Y = H
(b) X = R = Cl, Y = H
(c) X = NMe$_2$, Y = COR^1

fully defined to reduce the formation of by-products, which arise by reaction of $ArC\vdots N \rightarrow O$ with the liberated dimethylamine[36].

Benzoyl cyanide N-oxide, $PhCO\cdot C\vdots N \rightarrow O$, is less reactive than benzonitrile oxide in 1,3-cyclo-addition reactions. Nevertheless, it reacts quantitatively with phenylacetylene to give 3-benzoyl-5-phenylisoxazole, and with styrene to give the corresponding 2-isoxazoline[37]. The latter can then be aromatised with N-bromosuccinimide. With other acetylenic dipolarophiles the yields are often lower because of the thermal instability of the products.

In an unusual variant of the nitrile oxide synthesis, 3-aryl-5-phenylisoxazoles are formed when the acetylenic component is replaced by the ylide, $PhCO\cdot\bar{C}H\cdot\overset{+}{P}Ph_3$ [38]. It seems to the writer that this reaction may involve the elimination of Ph_3PO from an initially formed oxazaphosphole to give a 2-aryl-3-benzoylazirine (cf. Ref. 4), which can then rearrange to the isomeric isoxazole.

A new general, unequivocal, synthesis of 3,5-diarylisoxazoles involves treatment of the oxime of a ketone possessing an α-hydrogen atom, e.g. $Ar^1CMe\vdots NOH$, with two equivalents of n-butyl-lithium at 0 °C. The resulting

1,4-dianion, $Ar^1C(\bar{C}H_2){:}N\bar{O}$, is arylated with an aromatic ester, Ar^2CO_2Me, and cyclised with acid, to give the isoxazole (11)[39]. In a related procedure[40], the dianion is carbonated, and the resulting carboxy-oxime intermediate is then acid-cyclised to give the corresponding 3-aryl-2-isoxazolin-5-one (14) (30–60%).

3-(R-substituted)-5-ethoxy-4-ethoxycarbonylisoxazoles are formed in

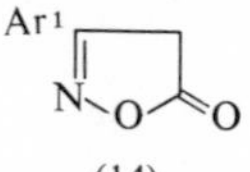

(14) 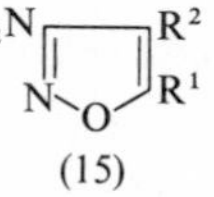   (15)

high yields by thermolysis of the azides, $(EtO_2C)_2C{:}CR{\cdot}N_3$[41]. The latter are obtained from the corresponding chloro-compounds, which are readily prepared by treatment of the diethyl acylmalonates, $(EtO_2C)_2CH{\cdot}COR$, with PCl_5.

3-Amino-isoxazoles (15) are conveniently prepared (60–80%) by treatment of the 2,3-dibromocarbonitriles, $R^1CHBr{\cdot}CBr(CN)R^2$, with N-hydroxyurea in methanolic NaOH[42].

9.2.2.2 Spectroscopic studies

The problem of identifying unambiguously each of the two possible 3,5-diarylisoxazoles (11) which may arise from the much used[43] reaction between hydroxylamine and an unsymmetrical 1,3-diketone has been investigated by mass spectrometry[39, 44]. It is found that the spectra of the two isomers are significantly different, and that the 5-substituent is readily identified because of its expulsion as an abundant $[Ar^2CO]^+$ ion. Confirmatory evidence is available from the n.m.r. spectra[45], in which the chemical shift of 4-H is virtually unaffected by the presence of a *meta-* or *para*-substituted phenyl group in the 3-position, but is affected appreciably by a similarly substituted 5-aryl group.

Skeletal rearrangement ions are observed in the mass spectra of 5-alkoxy-isoxazoles and of related compounds having a heteroatom at C-5[46, 47]; these have been discussed in terms of the isomeric azirines. The need to recognise that the mass spectra of such isoxazoles may appear more complex because of thermal isomerisation to azirines prior to electron bombardment has been stressed[46].

3-Hydroxy-[48] and 4-hydroxy-isoxazoles[49] have been shown spectroscopically to exist in solution mainly as the hydroxy-tautomers, irrespective of substituents or of solvents.

9.2.2.3 Ring-modifying processes: ring-opening

(a) *Isoxazoles* — Nishiwaki and his co-workers have investigated thoroughly the thermal rearrangements of the 5-amino-, 5-alkoxy-, and 5-alkylthio-isoxazoles (16) to their valence isomers, the 2H-azirines (17). The alkoxy- (16a) and alkylthio-compounds (16b) rearrange without complication at $c.$ 200 °C to give (17a) and (17b), respectively[46]. Remarkably high yields (50–70%) of (17a) are obtained. The azirine ((17a); Ar = Ph, X = OEt) rearranges photochemically, but not thermally, to 5-ethoxy-2-phenyloxazole[46]. The amines (16c) give the corresponding carboxamides (17c) when heated in dilute 2-ethoxyethanol solution; when heated in air,

(16)   (17) 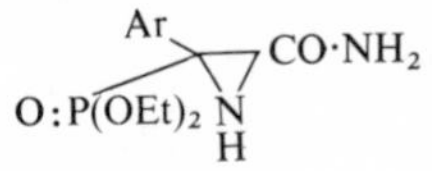   (18)

(a) R = H, X = O-Alkyl
(b) R = H, X = S-Alkyl
(c) R = H, X = NH$_2$
(d) R = Me, X = NH$_2$, Ar = Ph

(16c) and (17c) each give the same diarylpyrazine-2,5-dicarboxamide[50]. Isoxazole (16d) was shown earlier[51] to isomerise at 180 °C to 4-methyl-5-phenyl-4-imidazolin-2-one. It is now[52] shown that the rearrangement of (16d) to the imidazolin-2-one competes with intramolecular rearrangement to the corresponding azirine (17d). The azirine (17d) cannot be converted thermally into the imidazolin-2-one, and therefore is not a precursor of the latter[52]. When the amine (16d) is heated in dilute solution, the azirine (17d) only is obtained, but in low yield (28%)[52]. 5-Amino-3-arylisoxazoles (16c) react with boiling triethyl phosphite to give diethyl 3-carbamoyl-2-phenyl-aziridin-2-ylphosphonates (18) as the only identifiable products (<20% yield); these probably arise via the azirines (17c)[53].

The well-known[43] hydrogenolysis of isoxazoles (11; R^1 and R^2 instead of Ar^1 and Ar^2) to give the vinylogous amides, $R^2CO \cdot CH:CR^1 \cdot NH_2$, has been exploited as the basis of several elegant new synthetic procedures. In an example of a new general synthesis of 3-acylpyridines[54], 4-chloromethyl-3,5-

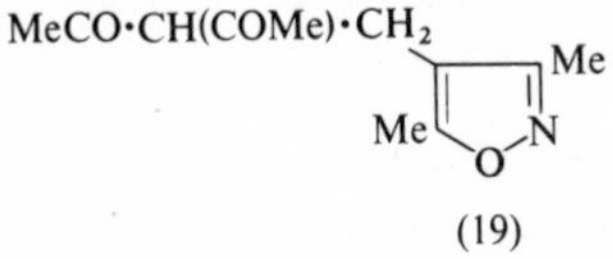

(19) 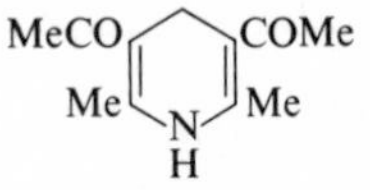   (20) 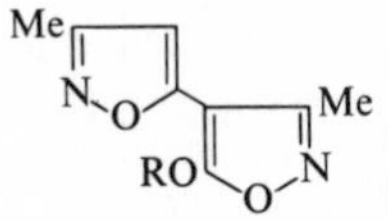   (21)

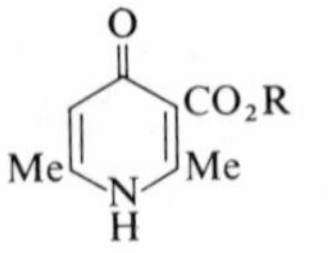

(22) 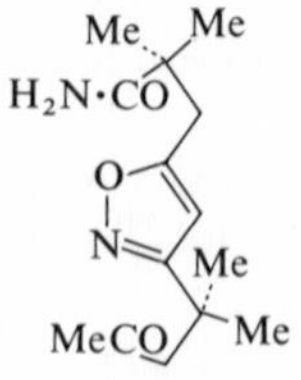   (23) 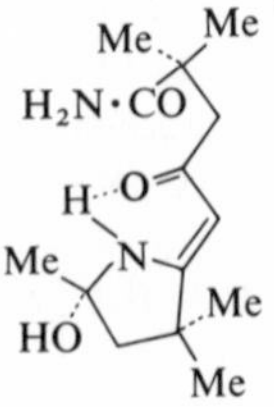   (24)

dimethylisoxazole is first treated with acetylacetone. The open-chain intermediate obtained by hydrogenation of the resulting 4-(3-oxoalkyl)isoxazole (19) cyclises spontaneously to afford a high yield (74%) of the 1,4-dihydropyridine (20), which is easily converted into the corresponding pyridine by oxidation. In a related procedure[55], the 4,5′-bisisoxazole (21) is hydrogenated in the presence of a palladium catalyst; under these conditions only the alkoxy-substituted ring is cleaved. Continued hydrogenation in the presence of nickel promotes cleavage of the remaining ring, and the product is cyclised by heating to give the 4-pyridone (22) in high yield. Two groups of workers[56-58] have made elegant use of the isoxazole cleavage reaction for preparing semi-corrins, as potential intermediates in the synthesis of vitamin B_{12}. In a typical example[58] of the approach, the isoxazole (23) is hydrogenated, and the nature of the 3-substituent is such that the ring-opened product cyclises spontaneously to give the pyrrole-based monocyclic compound (24). Cyclo-elimination of water from (24) then affords the desired semicorrin.

(b) *Anthranils* (2,1-*benzisoxazoles*)* − Anthranils provide some interesting examples of the general rearrangement[59]:

$$(9.3)$$

3-Methyl-7-nitroanthranil (25a) is converted into 4-acetylbenzofurazan (26a) by trimethyl phosphite[59]. If the 7-nitro-group is first deoxygenated to the corresponding nitroso-compound, this is a clear cut example of the general reaction (9.3) (i.e. $D:E = N:O$). However, it has also been observed[60] that the 7-nitroanthranil (25b) rearranges on heating to the benzofurazan N-oxide (26b). It might seem, therefore, that the first reaction, (25a) → (26a), could proceed via deoxygenation of the isomeric benzofurazan N-oxide (26a; $R^3 = \rightarrow O$). The thermal rearrangement[61] of 7-acetyl-3-methylanthranil is of theoretical signifiance because it provides the first example of a *reversible* rearrangement of the type shown in equation (9.3) and of a sigmatropic shift of order [1,9]. The rearrangement product in this case is, of course, identical with the starting material, but the course of the reaction can be followed by observing the coalescence on heating of the two methyl signals in the n.m.r. spectrum, and the reproduction of the original spectrum on cooling.

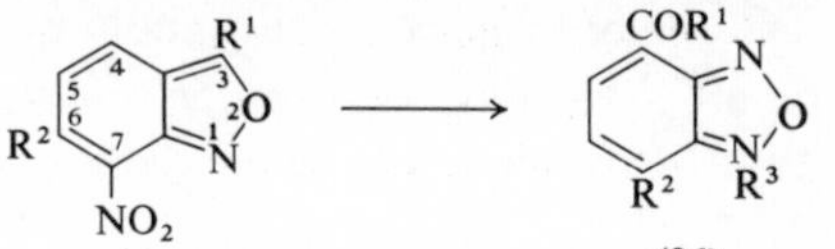

(a) R^1 = Me, R^2 = H
(b) R^1 = H, R^2 = Cl

(a) R^1 = Me, R^2 = H, no R^3
(b) R^1 = H, R^2 = Cl, R^3 = → O

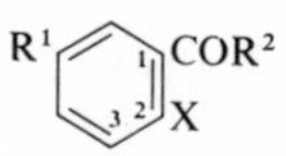

(a) X = NH_2
(b) X = OH

9.2.2.4 Photochemistry

(a) *Isoxazoles* − Photo-irradiation of 5-amino-3-phenylisoxazole in ether gives the $2H$-azirine (17c; Ar = Ph) as the primary product[50]. This undergoes

*According to IUPAC nomenclature the two benzo-derivatives of isoxazole are called benz[d]-isoxazole and benz[c]isoxazole, but *Chemical Abstracts* and the *Ring Index* prefer to use the names 1,2-benzisoxazole and 2,1-benzisoxazole, respectively. The *Chemical Abstracts* system is used in the present discussion for the former isomer; to avoid confusion, the latter isomer is referred to here as anthranil (as is the practice of the Chemical Society).

further rearrangement to the two ring-chain tautomers, $PhCO \cdot NH \cdot CH_2 \cdot CN$ and 5-amino-2-phenyloxazole, the equilibrium being in favour of the former[50].

(b) 1,2-*Benzisoxazoles and anthranils* — Photolysis of 3-(R^2-substituted) anthranils (R^2 = H, alkyl, or aryl) in strongly acidic solutions gives the carbonyl compounds (27a) as the major products[62, 63]. The nature of the substituent, R^1, in (27a) depends on the reaction medium: in 66% aqueous sulphuric acid, R^1 = OH[62, 63]; in methanolic sulphuric acid, R^1 = OMe[62]; in concentrated hydrochloric acid, R^1 = Cl[63]. In some cases[62], the isomeric compounds (27a; R^1 in 3-position) are minor products. Photolysis of 3-substituted 1,2-benzisoxazoles in concentrated sulphuric acid gives the 2,5-dihydroxy-compounds (27b; R^1 = OH), together with small amounts of the 2,3-dihydroxy-isomers[62].

9.3 5-MEMBERED RINGS CONTAINING ONE NITROGEN AND ONE SULPHUR ATOM

9.3.1 Thiazoles

9.3.1.1 Ring synthesis

Surprisingly few new synthetic procedures have been developed during the period 1970–1971; those which have been announced relate almost exclusively to 2-aminothiazoles, which can often be obtained more conveniently by the well-known[64] reaction of an α-halogeno carbonyl compound with thiourea. In a useful modification of this method[65, 66], $\alpha\beta$-unsaturated ketones, $ArCH:CH \cdot COMe$, react with thiourea or N-phenylthiourea in the presence of bromine or iodine to give high yields (70–90%) of the thiazoles ((28); R = H or Ph, respectively), in which the 4-substituent has been formed by saturation of the double bond of the starting ketone, supposedly by the HBr (or HI) which is generated in the reaction. A new synthesis[67] of 2-amino-thiazoles is based on the reaction of β-acylvinylphosphonium salts, $Ph_3\overset{+}{P} \cdot CH:CH \cdot COR \; X^-$, with thiourea. The positively charged phosphorus atom, rather than the carbonyl group, determines the site of addition, so that

$ArCH_2 \cdot CH_2 \underset{S}{\overset{N}{\bigsqcup}} NHR$ $Ph_3P^+ \cdot CH_2 \underset{S}{\overset{R}{\bigsqcup}} NH_2$ $\overset{+}{N}R_2 \; HSO_4^-$

(28)   (29) $S \diagdown O / Ph$

(30)

the reaction proceeds via the intermediates, $Ph_3\overset{+}{P} \cdot \overset{-}{C}H(COR) \cdot SC(:\overset{+}{N}H_2) \cdot NH_2$ X^-. Proton transfer from nitrogen to carbon gives the corresponding imine, which cyclises spontaneously to give the (5-thiazolylmethyl)phosphonium salts (29) (75–80%). Alkaline hydrolysis gives the corresponding 5-methyl compounds. Thiocyanatamine ($NH_2 \cdot SCN$) reacts with a ketone, $R^1CO \cdot CH_2R^2$, to give 2-amino-4-(R^1-substituted)-5-(R^2-substituted)thiazoles in yields of up to 88%[68]. Intermediate enamines, $R^1C(NH_2):CR^2(SCN)$, have been isolated in some cases. The readily available[69] N-(5-phenyl-1,3-oxathiol-2-ylidene) tertiary iminium salts (30) undergo an easy ring cleavage reaction when treated with aqueous ammonia[70]. The products, $PhCO \cdot CH_2 \cdot S \cdot C(:NH) \cdot NR_2$, undergo a spontaneous cyclodehydration reaction to give 4-phenyl-2-(tertiary amino)thiazoles (83–91 %).

The reaction of thiourea with a one molar excess of p-benzoquinone (and other quinones) in the presence of hydrochloric acid provides a useful route to 2-amino-6-hydroxybenzothiazole ($>70\%$ yield)[71]. S-(2,5-Dihydroxyphenyl)thiouronium chloride is probably an intermediate in the reaction because it can be obtained by treatment of the quinone with an excess of thiourea, and it is converted into the observed product by the action of more quinone.

9.3.1.2 Spectroscopic studies

The mass spectra of [2-^{2}H]-, [2-^{13}C]-, and [4-^{2}H]-thiazole confirm unambiguously the previous observation[72] that the elimination of HCN from the molecular ion of thiazole takes place by cleavage of the 1,2- and 3,4-bonds[73]. The fragmentation of 2-arylthiazoles centres around similar cleavage reactions, but in this case another characteristic fragment ion, $ArC\overset{\cdot}{:}S^+$, arises from fission of the 1,5- and 2,3-bonds[74].

A careful analysis of the i.r. spectra of benzothiazole and a wide range of its derivatives has enabled the ν(CH) stretching and the γ(CH) out-of-plane deformation vibrations to be assigned[75].

9.3.1.3 Substitution reactions

(a) *Electrophilic* — Nitration of either the hydrochloride or the BF_3 complex of 2-methylthiazole with nitronium tetrafluoroborate or BF_3—N_2O_4 affords a mixture of the 4- (11.5%) and 5-nitro-isomers (48%)[76]. Conventional procedures[77] give only 3–4% of the latter. 4- And 5-alkylthiazoles are nitrated (HNO_3–H_2SO_4) in the 5- and 4-positions, respectively[78]. 2,5-Dimethylthiazole is mercurated in the 5-position with mercury(II) acetate and acetic acid; the product reacts with bromine in carbon tetrachloride to give the corresponding 5-bromo-compound[79].

Surprisingly, oxidation of 2-methylbenzothiazole with permaleic acid gives the N-oxide, rather than the S,S-dioxide[80]. Benzothiazole N-oxide cannot be prepared by direct oxidation of the parent heterocycle, but it can be obtained by decarboxylation of the corresponding 2-carboxylic acid, prepared from the 2-methyl derivative by a series of standard reactions[81].

(b) *Nucleophilic* — The reactions between a range of 4- or 5-substituted 2-chlorothiazoles and the benzenethiolate ion in methanol have been investigated kinetically[82]. 4- Or 5-halogenothiazoles, which were previously stated[64] to be unreactive towards nucleophiles, are now found to react with methoxide ion at rates comparable with those of the 2-halogeno-isomers[83].

2-Chlorobenzothiazole does not give the expected 2-NH·NMe₂ derivative when treated with N,N-dimethylhydrazine[84]. Remarkably, the major product is the 2-NMe·NH₂ compound (77%), formed by *dealkylation* of the dimethylhydrazine. The *deamination* products, 2-dimethylamino- and

2-methylamino-benzothiazole, are also formed in yields of 20% and 1%, respectively; the former is the major product from the reaction of 2-chloro-benzoxazole with N,N-dimethylhydrazine.

(c) *Radical*—Thiazole and several of its 4- and 5-alkyl derivatives have been treated with cyclohexyl radicals[85, 86]. Because of the strong nucleophilic character of these radicals, the electron-depleted 2-position is always sub-substituted preferentially, but some substitution can also occur in a methyl group. 2-Thiazolyl radicals, produced by the photolysis of 2-iodothiazole, react with alkylbenzenes to give a mixture of isomeric alkyl 2-thiazolyl-benzenes[87]. The isomer distribution and the relative substitution rates follow the characteristic pattern for homolytic aromatic arylation. The 2-position in benzothiazole is the most reactive towards phenyl radicals[88] (*ex* dibenzoyl peroxide) and acyl radicals[89] (*ex* aldehydes). The latter reaction provides a useful synthetic route to 2-acyl derivatives; it can also be used to detect the formation of acyl radicals in the oxidation of aldehydes with various reagents. 2-Chloro- and 2-methylbenzothiazole react preferentially in the 4-position with phenyl radicals[88].

9.3.1.4 Ring-modifying processes: ring-opening

A new synthesis of aldehydes has been developed[90], starting from the commercially available 2,4-dimethylthiazole. This is metalated in the 2-methyl group by n-butyl-lithium, and the product is treated with an alkyl halide, RX. The resulting $(2\text{-}CH_2R)\text{-}4$-methylthiazole is N-methylated with trimethyl-oxonium fluoroborate and then reduced with borohydride, to give the $(2\text{-}CH_2R)$-3,4-dimethylthiazolidine. When treated with aqueous mercury(II) chloride, this undergoes ring cleavage to give the aldehyde, $RCH_2 \cdot CHO$, in 50–75% yield. An analogous synthesis of aldehydes[91] starts from dihydro-1,3-oxazines, but the present method has the advantage that strongly acidic conditions are avoided at all stages.

9.3.1.5 Photochemistry

2-Phenylthiazole is isomerised by u.v. radiation to a mixture of 3-phenyliso-thiazole and 4-phenylthiazole[92]; by varying the conditions either of the two isomers can be made to predominate. Under identical conditions, 4-phenylthiazole is unaffected, and the 5-phenyl isomer is isomerised slowly to 4-phenylisothiazole[92]. U.V. irradiation of 2,5-diphenylthiazole gives mainly 3,4-diphenylisothiazole (32%), together with small amounts (c. 8% each) of 4,5-diphenylthiazole and its photocyclisation product, phenanthro-[9,10-d]thiazole[93]. Under similar conditions, 2,4-diphenylthiazole is largely unchanged, but some (3.3%) 3,5-diphenylisothiazole can be isolated[93].

9.3.2 Isothiazoles

9.3.2.1 Ring synthesis

Isothiazoles are now attracting considerable interest, and many elegant methods for their synthesis are being devised. 4-Carbonitriles are obtained by relatively simple procedures. For example, the cyclisations of $(NC)_2C$: $C(SNa)\cdot NH_2$[94] and $(NC)_2C:C(SH)\cdot NH\cdot COR$[95] to 5-amino-3-chloro-isothi-azole-4-carbonitrile and the corresponding 5-N-acyl derivatives, respectively, are readily accomplished by means of chlorine. Arylidenemalononitriles, $ArCH:C(CN)_2$, are converted into 5-aryl-3-chloroisothiazole-4-carbonitriles by treatment with S_2Cl_2, SCl_2, or $SOCl_2$ [96]. The enamine, $PhC(NH_2):CH\cdot CN$, is attacked at the α-carbon atom by the Vilsmeier complex from dimethyl-formamide and $POCl_3$. The product, $PhC(:NH)\cdot C(CN):CH\cdot NMe_2$ (as a salt), is then treated with NaSH to effect replacement of NMe_2 by SH, and the resulting thiol is cyclised with iodine: 3-Phenylisothiazole-4-carbonitrile is obtained in yields of up to 60%[97]. With Vilsmeier complexes from other amides, isothiazoles are often formed in very low yields, probably because the complex acylates the enamine preferentially on the nitrogen atom, rather than on the carbon atom.

The readily available 3,3'-dithiopropionamide, $(-S\cdot CH_2\cdot CH_2\cdot CO\cdot NH_2)_2$, can be cyclised to 3-hydroxyisothiazole (75%) with either chlorine or SO_2Cl_2[98]. 3-Hydroxyisothiazole had been obtained previously[99] by cyclisa-tion of $XS\cdot CH:CH\cdot CO\cdot NH_2$ (X = CN). Such a compound (X = Cl), formed by successive chlorination and dehydrochlorination of the starting amide, is a likely intermediate in the present reaction[98].

3-Amino- and 3-methoxy-isothiazole can be prepared from 3,3'-dithio-propionitrile, $(-S\cdot CH_2\cdot CH_2\cdot CN)_2$ by a method similar to that just described[100]. Treatment of the nitrile with methanolic hydrogen chloride gives the methyl imidate hydrochloride, which reacts with chlorine to give the 3-methoxy-compound (58%). Alternatively, treatment of the methyl imidate salt with ammonia gives the corresponding amidine salt, which yields the 3-amino-compound (26%) on cyclisation with chlorine[100].

The first direct ring synthesis of a 5-hydroxyisothiazole (31, or its tautomer) has been accomplished[101] by the condensation of a β-aminocrotonic ester, $MeC(NH_2):CH\cdot CO_2R$ (R = Me or Et), with perchloromethanethiol in the presence of base.

The nitrile oxide synthesis[4] of isoxazoles has no useful analogy in iso-thiazole chemistry because of the instability of nitrile sulphides. However, $PhCN \rightarrow S$ is a probable product of the thermolysis of the oxathiazolone (32)[102] and of the photolysis of the 1,3,2-oxathiazolium 5-oxide (33)[103], since the addition of $MeO_2C\cdot C:C\cdot CO_2Me$ to each of these reactions gives the expected isothiazole derivative. Another preparation of isothiazoles from (33) will be described in Section 9.6.3.

(31)   (32)   (33)   (34)

3-Substituted 2,1-benzisothiazoles (34; R = Me or Ph) are obtained in high yields by pyrolysis of the appropriate 2-azidoaryl thioketones[104]. The latter are conveniently obtained from the corresponding ketones[104].

9.3.2.2 Ring-modifying processes: ring-opening

Such reactions are often observed when 3-chloro-1,2-benzisothiazole (35) is treated with nucleophiles. Sodium cyanide in aqueous acetone gives *o*-cyanophenyl thiocyanate (62%), bis-(*o*-cyanophenyl) disulphide (37a) (22%), and 2-acetyl-3-aminobenzo[*b*]thiophen (38a) (6%); n-butyl-lithium gives *o*-(n-butylthio)benzonitrile (90%); sodium thiophenoxide gives (37a) (50–55%), (37b) (*c*. 5%), and diphenyl disulphide (46–50%)[105]; hydrazine, zinc and hydrochloric acid, or nitrous acid give (37a)[106]. These reactions have been discussed in terms of nucleophilic attack by X⁻ at either sulphur (equation 9.4) or chlorine (equation 9.5)[105, 106]. The formation of (38a) must involve nucleophilic attack by the carbanion derived from acetone. Certain 1,3-dicarbonyl compounds react with (35) in the presence of base to give

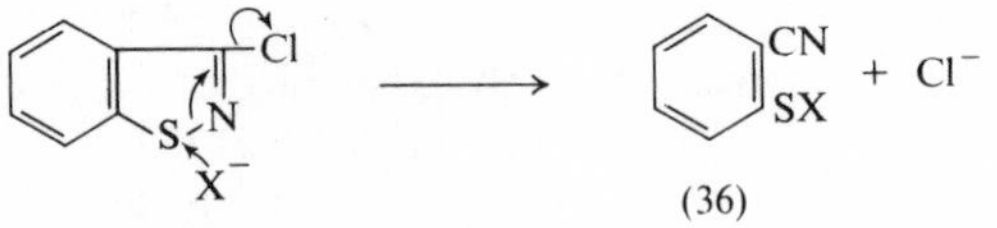

$$(9.4)$$

$$(9.5)$$

(35)

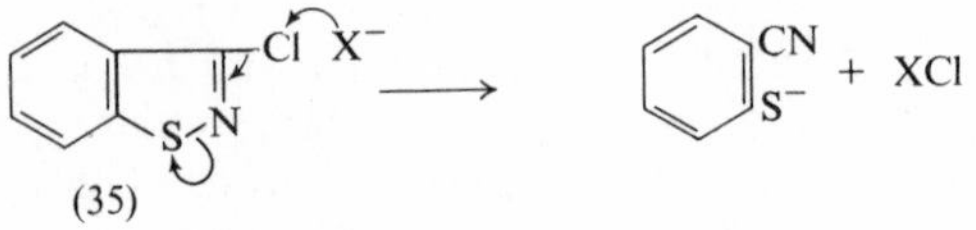

(36)

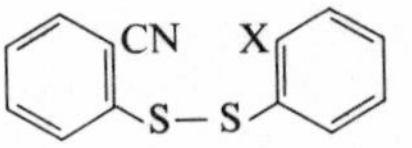

(37)

(38)

(a) X = CN
(b) X = H

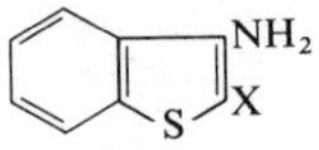

(a) X = COMe
(b) X = CO₂Et

3-aminobenzo[*b*]thiophens in high yields: acetylacetone gives (38a) (71%), ethyl acetoacetate gives (38b) (71%), and diethyl malonate (under certain conditions) gives (38b) (70%)[107]. These reactions probably involve attack of a carbanion at sulphur to give an *S*-substituted *o*-cyanothiophenol. For example, the anion from diethyl malonate would give (β6 X = CH(CO₂Et)₂), which could then lose an ethoxycarbonyl group in a retro-Claisen reaction and cyclise, but not necessarily in this order[107]. Other nucleophiles, e.g. EtŌ[105], ethyl cyanoacetate, and diethyl malonate (under certain conditions)[107] substitute normally in (35) without ring cleavage.

Ring-opening reactions of non-annellated isothiazoles have been studied less widely. 3,5-Dimethylisothiazole is cleaved by n-butyl-lithium by addition across the S—N bond, to give MeCO·CH:CMe·SBuⁿ [108]; similar cleavage of the 3-methyl compound had been observed previously[109]. 4-Methylisothiazole is lithiated mainly in the 5-position without ring opening, but a small amount of BuⁿS·CH:CMe·CN, which is also formed in the reaction, is believed to result from decomposition (cf. Equation 9.5) of the 3-lithio-derivative[108].

9.3.2.3 Photochemistry

The photochemistry of the phenylisothiazoles has been examined. 3-Phenyl-isothiazole gives 4-phenylthiazole[92, 110], the 4-phenyl isomer is unchanged[92], and the 5-phenyl isomer is slowly converted into 3-phenylisothiazole, either alone[110], or admixed with 4-phenylthiazole[92]. 3,5-Diphenylisothiazole is converted into 2,4-diphenylthiazole[110]. The reactions probably proceed via a tricyclic sulphonium cation[110].

9.4 5-MEMBERED RINGS CONTAINING ONE OXYGEN AND TWO NITROGEN ATOMS

9.4.1 Furazans and furoxans

9.4.1.1 Ring synthesis

Recently-announced syntheses are concerned mainly with furoxans (39). Nitrosation (NO_2^-/HCl) of dimethylphenacylsulphonium bromide, $PhCO \cdot CH_2 \cdot \overset{+}{S}Me_2$ Br^-, gives (39a), which undoubtedly is formed by the generation and spontaneous dimerisation of benzoyl cyanide N-oxide[111]. In this respect the reaction is analogous to the dimerisation of substituted benzonitrile oxides[112], for which a one-step concerted mechanism is now postulated[113] as an alternative to the previously[29] suggested carbene mechanism. The disubstituted furoxans (39b) are obtained (62–100%) by treatment of α-diazosulphones, α-diazoketones, or ethyl diazoacetate with N_2O_3 [114].

The pyrolysis of substituted methyl N-(o-nitrophenyl)carbamates, o-$O_2N \cdot C_6H_4 \cdot NH \cdot CO \cdot OMe$, provides a convenient general method for preparing benzofurazans[115]. This reaction resembles the well-known[112] thermolysis of o-nitroaryl azides to benzofuroxans, in the sense that both reactions proceed via related nitrene intermediates.

9.4.1.2 Structure of benzofuroxan

The cyclisation of o-dinitrosobenzene to benzofuroxan, and hence the inter-conversion of the two isomers of benzofuroxan, has been discussed theoretically by Hoffmann et al.[116]. Cyclisation of two coplanar o-nitroso-groups is shown to be an energetically unfavourable process. One nitroso-group is, therefore, rotated 90 degrees out of plane to permit a bonding interaction between an oxygen lone pair and the low-lying nitroso-group π^* level.

9.4.1.3 Nucleophilic substitution reactions of benzofurazans

Two independent studies[117, 118] have shown that 4-halogeno-5-nitrobenzo-furazans and the corresponding 5-halogeno-4-nitro-compounds are unusu-ally reactive towards nucleophiles, probably because of the high 4,5-bond order, which means that these compounds can be regarded as activated halogeno-alkanes rather than as true aryl halides. Conversely, because of the low 5,6-bond order, the 5-halogeno-6-nitrobenzofurazans are practically inert towards nucleophiles.

9.4.1.4 Photochemistry of benzofurazan

Photolysis of benzofurazan in methanol or benzene gives $MeO_2C\cdot NH\cdot(CH\vdots CH)_2\cdot CN$ or the N-acylazepine, cis-$AzCO\cdot(CH\vdots CH)_2\cdot CN$-$cis$ respectively[119]. When photolysed in the presence of triethyl phosphite, substituted benzofurazans give high yields of the correspondingly substituted dinitriles, cis-$NC\cdot CH\vdots CH\cdot CH\vdots CH\cdot CN$-$cis$, which are formed by reduction of the intermediate N-oxides[120].

9.4.2 1, 3, 4-Oxadiazoles

9.4.2.1 Ring synthesis

2-Amino-5-aryl-1,3,4-oxadiazoles may be prepared in yields of $>80\%$ by treatment of aldehydic semicarbazones, $ArCH\vdots N\cdot NH\cdot CO\cdot NH_2$, with lead

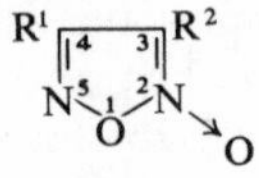

(39)

(a) $R^1 = R^2 = COPh$
(b) $R^1 = R^2 = SO_2Ar$,
    COAr or CO_2Et

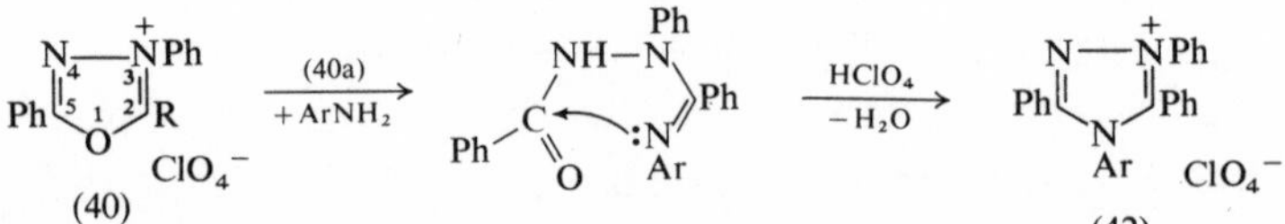

(40)

(a) R = Ph
(b) R = CH_2Ph

(41)

(42)

Mechanism A

tetra-acetate[121]. 3-Aryl-1,3,4-oxadiazolium salts of unambiguous orientation may be obtained by cyclisation of N,N'-diacyl-N-arylhydrazines with acetic anhydride and perchloric acid, e.g. (40a) from $PhCO\cdot NPh\cdot NH\cdot COPh$[122].

9.4.2.2 Ring-modifying processes: ring-opening

The elegant rationalisation by Boyd and his co-workers of the complex ring transformations which can occur when 2,3,4-trisubstituted 1,3,4-oxadiazolium salts* react with nucleophiles is worthy of special mention. Such reactions can be of two mechanistic types[123]. In type A, one atom of the nucleophile is incorporated in the new ring and the cyclisation involves electrophilic attack by an amide carbonyl group. The reaction of (40a) with aromatic amines, $ArNH_2$, to give the s-triazolium salts (42)[124] provides a good illustration of the mechanism. There is ample evidence that in general the nucleophile attacks C-2; (40a) reacts with aniline to give the amidrazone (41; Ar = Ph), which can then cyclise by the method just described to give the observed product (42; Ar = Ph)[124].

(40a) Reacts with aqueous sodium sulphide to give N'-benzoyl-N-phenyl-N-thiobenzoylhydrazine, $PhCO·NH·NPh·CSPh$, which in the presence of acetic anhydride and perchloric acid cyclises (Mechanism A) to give 2,3,5-triphenyl-1,3,4-thiadiazolium perchlorate and some of the original oxadiazolium salt[125].

Hetzheim et al. have described the following transformations of 2-amino-3-phenacyl-1,3,4-oxadiazolium bromides (43):

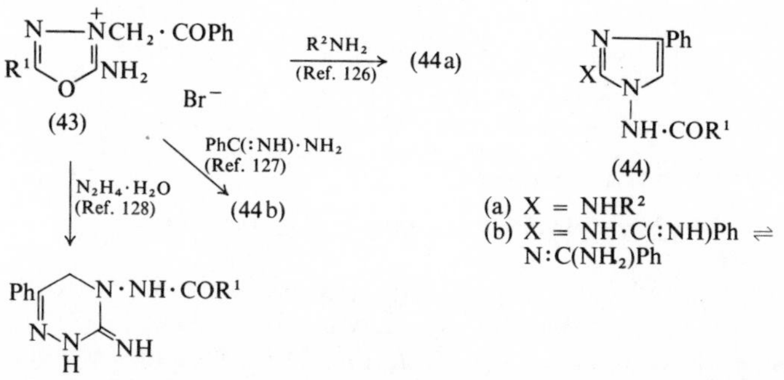

It seems likely that these reactions can be explained in terms of Mechanism A, except that recyclisation must involve a ketonic, and not an amidic, carbonyl group.

In Mechanism B, two atoms of the reagent serve to build the new ring, and the cyclisation proceeds by nucleophilic attack by a nitrogen atom[123]. This is illustrated clearly by the reaction of (40a) with ethyl cyanoacetate in the presence of ethanolic triethylamine to yield the 3-aminopyrazole (47)[129]. The crystalline intermediate, $PhCO·NH·NPh·CPh:C(CN)·CO_2Et$, can lose a proton to give the anion (45), which can cyclise to (46) as shown. The observed product (47) is then formed by protonation of (46) and transfer of the benzoyl group to ethanol. Related pyrazoles are formed when (40a) reacts with 1,3-dicarbonyl compounds or malononitrile[129]. 3-Amino-1,5-diphenyl-s-triazole is formed by the type B mechanism when (40a) is treated with cyanamide in the presence of triethylamine[123].

2-Benzyl(or 2-alkyl)-1,3,4-oxadiazolium perchlorates are deprotonated by triethylamine to give benzylidene (or alkylidene) oxadiazolines, e.g. (48) from (40b), which, in the presence of acetyl chloride, dimerise (cf. Mechanism B) to 3-($\beta\beta$-diacylhydrazino)pyrazoles, e.g. (49) from (48)[130].

---

*For simplicity, the ensuing discussion will be confined to the reactions of the 2,3,5-triphenyl derivative (40a).

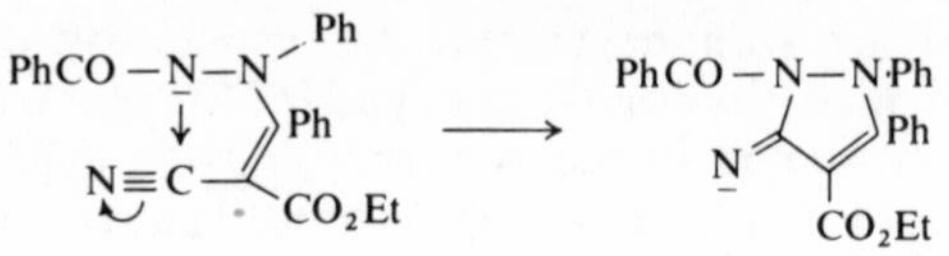

PhCO—N—N
   ↓   Ph
 N≡C
        CO₂Et

(45)

⟶

PhCO—N—N·Ph
        Ph
 N
        CO₂Et

(46)

Mechanism B

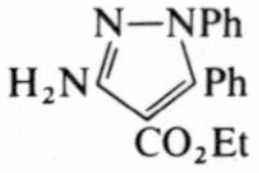

N—NPh
H₂N   Ph
   CO₂Et

(47)

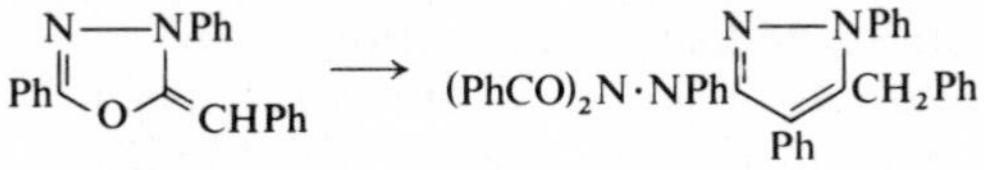

N——NPh
Ph   O   CHPh

(48)

⟶

N——NPh
(PhCO)₂N·NPh   CH₂Ph
        Ph

(49)

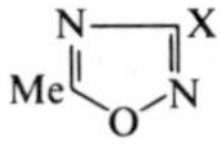

N——X
Me   O   N

(50)

Y=X
ZH   A
N   B
   D

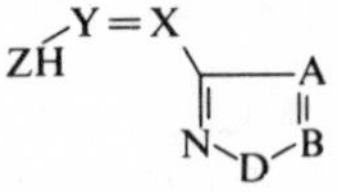

(51)

(a) X = NH·CO·NHPh
(b) X = N:CH·NHAr
(c) X = C(:NOH)Ar

9.4.3   1, 2, 4-Oxadiazoles

9.4.3.1.   Ring-modifying processes: ring-opening

The oxadiazole (50a) is converted by ethanolic sodium hydroxide, but not by heat alone, into 3-acetamido-1-phenyl-s-triazol-5(4H)-one[131]; (50b) is converted by heat alone, or by alkali, into a 3-acetamido-1-aryl-s-triazole[132]; (50c) is converted by boiling 6M-hydrochloric acid into a 3-amino-4-arylfurazan[133], presumably via the 3-acetamido-compound. The authors do not make clear that these are further examples of rearrangements[134] common to 5-membered nitrogen heterocyclic compounds of the type (51).

9.5   5-MEMBERED RINGS CONTAINING ONE SULPHUR AND TWO NITROGEN ATOMS

Discussion will be confined mainly to 1,2,3- and 1,3,4-thiadiazoles. Apart from a detailed study of the mass spectra of a range of substituted 5-amino-1,2,4-thiadiazoles[135] and the definite recognition of 1,2,5-thiadiazole N-oxides[136] (the sulphur analogues of furoxans), little attention has been paid to the remaining two isomers.

9.5.1   1, 2, 3-Thiadiazoles

9.5.1.1   Ring synthesis

1,2,3-Benzothiadiazoles may be prepared from diazotised 7-aminobenzo-thiazoles[137] or 7-amino-1,2-benzisothiazoles[138] by unusual ring transformations. Although such diazo-compounds can decompose by the normal processes, they can also undergo rearrangement reactions in which the diazonium cation reacts with the sulphur atom to form a 1,2,3-thiadiazole ring, with the remainder of the original hetero-ring appearing as a 7-acyl-amino-group[137] (equation 9.6) or a 7-oxo-group[138] (equation 9.7), respectively. The formamido-compounds (52a) are often hydrolysed to the corresponding amines (52b) under the conditions of the reaction[137, 139]. The diazonium salts derived from amines (52b) can undergo a similar rearrangement to give isomeric diazo-compounds, in which the substituents, R^1 and R^2, are interchanged[140].

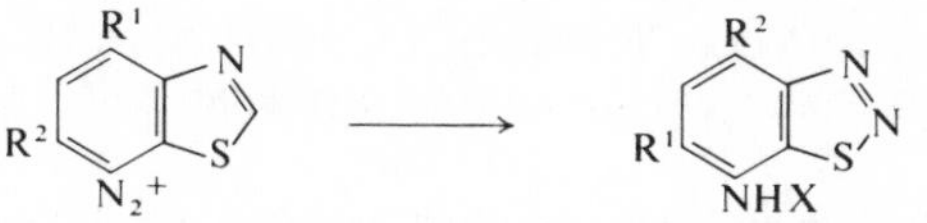

(9.6)

(52)

(a) X = CHO
(b) X = H

(9.7)

CHO(via—CH:NH)

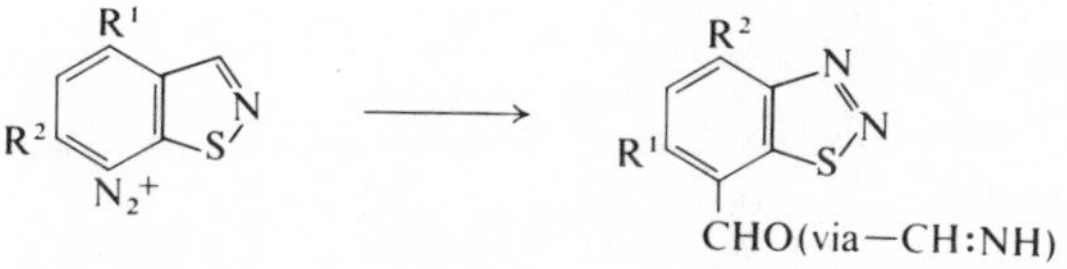

Diazotised 4-amino-3-methylisothiazole reacts with thiourea to give 4-acetyl-1,2,3-thiadiazole[141]. The overall reaction bears a close formal similarity to reaction (9.7). In this case, however, it seems that the geometry of the system must preclude direct attack of the diazo-group on the sulphur atom.

9.5.1.2 Spectroscopy and photochemistry

In the mass spectrometer, the molecular ions of 4-R^1, 5-R^2-disubstituted 1,2,3-thiadiazoles first lose nitrogen[142, 143]. Rearrangement of the resulting radical-ion to the thioketene radical-ion, $[R^1R^2C:C:S]^{\cdot+}$, is an important subsequent process[142].

The photolysis of 1,2,3-thiadiazoles gives a variety of products, which are formed by a radical process, following an initial loss of nitrogen[144].

9.5.2 1, 3, 4-Thiadiazoles

9.5.2.1 Ring synthesis

The cyclisation of thiosemicarbazide derivatives,

$$RNH \cdot CS \cdot NH \cdot NH \cdot C(:Z) \cdot NHY,$$

to give 1,3,4-thiadiazoles and s-triazoles in acid and alkaline solutions, respectively, is well known[145]. Kurzer has now prepared and cyclised a wide range of related compounds, $ArCO \cdot NH \cdot CS \cdot NH \cdot NH \cdot C(:Z) \cdot NHY$ (53). The thiadiazoles (54) obtained from the acid cyclisations are listed in Table 9.1. It can be seen that the ease with which the group, —NHY, is eliminated

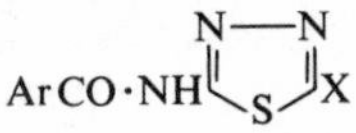

$$ArCO \cdot NH \underset{S}{\overset{N\!-\!N}{\left\langle \quad \right\rangle}} X$$

(54)

from the general adduct (53) during thiadiazole formation decreases in the order $NH_2 > NH \cdot NH_2 > NHPh$.

The thiocarbonohydrazide, $PhNH \cdot NH \cdot CS \cdot NH \cdot NH \cdot C(:NAr) \cdot NHAr$, behaves unusually in that a 2-arylamino-5-phenylazo-1,3,4-thiadiazole is

Table 9.1 1,3,4-Thiadiazoles (54) obtained by acid cyclisation of $ArCO \cdot NH \cdot CS \cdot NH \cdot NH \cdot C(:Z) \cdot NHY$ (53)

Starting material (53)	Product(s) (54)		Molecule eliminated during cyclisation	Ref.
Y = H, Z = NH	X = NH_2		NH_3	146
Y = Ph, Z = NH	X = NHPh	(major)	NH_3	146
	X = NH_2	(minor)	$PhNH_2$	
Y = NH_2, Z = NH	X = $NH \cdot NH_2$	(major)	NH_3	147
	X = NH_2	(minor)	N_2H_4	
Y = $N{:}CMe_2$, Z = NH	X = $NH \cdot N{:}CMe_2$		NH_3	147
*Y = NH_2, Z = NAr^1	X = $NHAr^1$		N_2H_4	147
Y = N:CHPh, Z = O	X = OH		$PhCH{:}N \cdot NH_2$	148
Y = NHPh, Z = O	X = OH		$PhNH \cdot NH_2$	148
*Y = NH_2, Z = O	X = OH		N_2H_4	148
Y = H, Z = O	X = OH		NH_3	148
Y = NH_2, Z = S	X = SH		N_2H_4	149

*Starting material prepared *in situ*: decomposes spontaneously

formed under *alkaline* conditions[150]. The reaction probably involves an initial dehydrogenation of the $PhNH \cdot NH$-group to give a phenylazo-compound, cyclisation of which can then give only a thiadiazole[150].

9.6 MESOIONIC COMPOUNDS

9.6.1 Mesoionic oxazol-5-ones

The preparation and reactions of 3-methyl-2,4-diaryloxazolium-5-oxides ('münchnones'), e.g. (55a), which were described previously in a series of preliminary communications (1964 onwards), have now been presented in full by the Munich School.

N-Benzoyl-N-methylphenylglycine dissolves in acetic anhydride at 55 °C to give (55a)[151]; at 90 °C the ketone, $MeCO \cdot CHPh \cdot NMe \cdot COPh$, is formed by the Dakin–West reaction[152]. Münchnones are very unstable, and are often prepared *in situ* by treatment of the appropriate N-aroyl-amino-acid with acetic anhydride. Pyrolysis of (55a) proceeds via the valence tautomer, $O{:}C{:}CPh{\cdot}NMe{\cdot}COPh$, which dimerises with loss of CO_2 to give the allene, $PhCO{\cdot}NMe{\cdot}CPh{:}C{:}CPh{\cdot}NMe{\cdot}COPh$[151].

The main practical importance of the münchnones lies undoubtedly in their ability to undergo 1,3-dipolar cyclo-addition reactions with acetylenic or olefinic dipolarophiles. The mechanism of such reactions has been investigated kinetically[153]. The postulated intermediates, (56) and (58),

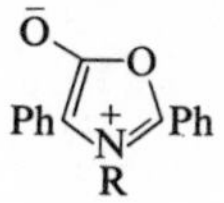

(55)

(a) R = Me
(b) R = H

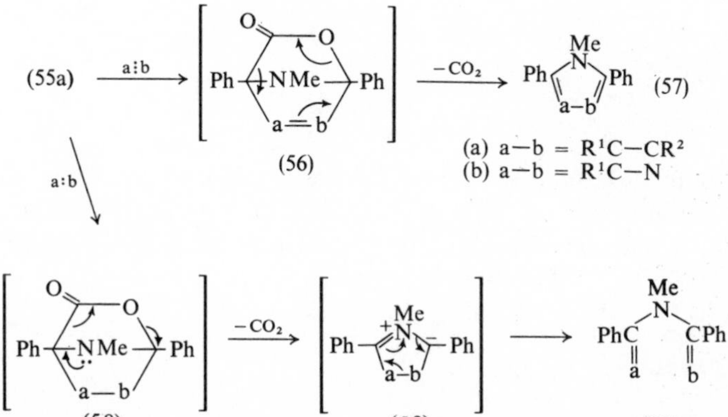

(56)

(a) a—b = R^1C—CR^2
(b) a—b = R^1C—N

(57)

(58)   (59)   (60)

respectively, lose CO_2 spontaneously as shown, to afford a wide variety of products. Thus (55a) reacts with acetylenes, $R^1C{:}CR^2$, to give high yields of N-methylpyrroles (57a)[154] and with nitriles, R^1CN (where R^1 is an electron-withdrawing group), to give imidazoles (57b)[155]. Aldehydes (a:b = $ArCH{:}O$)[156], thiocarbonyl compounds (a:b = $PhCR{:}S$)[157], nitrosobenzene (a:b = $PhN{:}O$)[155], and azo-compounds (a:b = $ArN{:}N{\cdot}CN$)[155] give the expected products (60). Alkenes and $\alpha\beta$-unsaturated esters, nitriles, and ketones (a:b = $R^1CH{:}CR^2R^3$) give the expected cyclic azomethine ylide (59), which either isomerises to a 2-pyrroline or adds a second molecule of the dipolarophile[158]. The product (59), formed by treatment of (55a) with carbon disulphide (a:b = $S{:}C{=}S$) can be seen to be the sulphur analogue of the starting material[157].

Azomethines, $R^1R^2C{:}NR^3$, form 1:1 adducts (61) with (55a), which arise from a $2+2 \rightarrow 4$ cyclo-addition with the valence tautomer, $O{:}C{:}CPh{\cdot}NMe{\cdot}COPh$, and not from a 1,3-dipolar cyclo-addition [159].

The cyclic azomethine ylide (62), formed by deprotonation of the corresponding Reissert salt, is essentially a münchnone imine, and as such can undergo cyclo-addition reactions in the usual way[160]. The reaction with $PhC{:}C{\cdot}CO_2Et$ is of particular significance because it provides the first example of a bridged intermediate of type (56) which is capable of isolation.

The azlactones (63) undergo cyclo-addition reactions[25, 161], in which they behave as the tautomeric oxazolium 5-oxides, e.g. (55b) from (63; $R^1 = R^2 =$ Ph). Products analogous to those from the münchnones are obtained when azlactones react with acetylenes[25], or with $\alpha\beta$-unsaturated esters and nitriles[161],

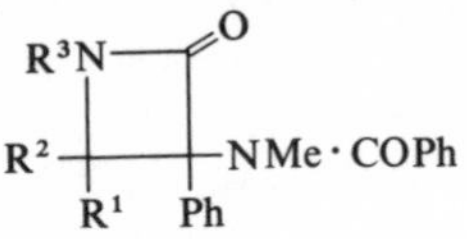

(61) 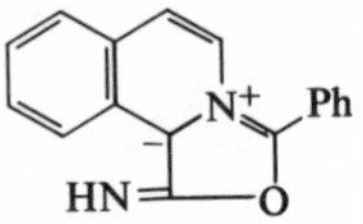   (62) 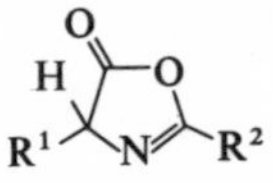   (63)

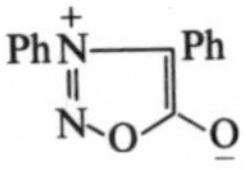

(64) 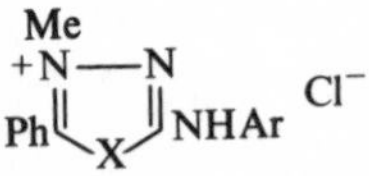   (65)

(a) X = S
(b) X = O

except that in the latter case a 1-pyrroline, rather than a 2-pyrroline, is formed. Cyclo-addition reactions with azlactones suffer from the disadvantage that the products may be sufficiently reactive to undergo further reaction with a second molecule of dipolarophile.

9.6.2 Sydnones

Further 1,3-dipolar cyclo-addition reactions of sydnones have been described[162, 163].

3,4-Diphenylsydnone (64) reacts with $MeO_2C \cdot C \vdots C \cdot CO_2Me$ in the presence of u.v. radiation to give dimethyl 1,3-diphenylpyrazole-4,5-dicarboxylate[164-166], whereas it gives dimethyl 1,5-diphenylpyrazole-3,4-dicarboxylate in the thermal reaction. It is believed[164-166] that the photoreaction proceeds via the nitrileimine, $PhC \vdots \overset{+}{N} - \overset{-}{N}Ph$, which is formed from the isomeric 2,3-diphenyl-diazirine. The photolysis of (64) in the absence of dipolarophiles also proceeds via similar intermediates, to give, *inter alia*, 2,4,5-triphenyl-1,2,3-triazole[164, 165].

9.6.3 Miscellaneous

The mesoionic 1,3,2-oxathiazolium 5-oxide (33) undergoes the expected cyclo-addition reactions: with $MeO_2C \cdot C \vdots C \cdot CO_2Me$, for example, it gives dimethyl 5-phenylisothiazole-3,4-dicarboxylate[167].

The isocyanide dichlorides, $ArN \vdots CCl_2$, react with N-methyl-N-thio-benzoylhydrazine, $PhCS \cdot NMe \cdot NH_2$, to yield the 1,3,4-thiadiazolium chlorides (65a)[168]. The oxygen analogues (65b) are formed similarly from $PhCO \cdot NMe \cdot NH_2$[169]. The cations of (65) may be deprotonated to give the corresponding mesoionic compounds[168, 169].

9.7 6-MEMBERED RINGS CONTAINING NITROGEN AND SULPHUR: PHENOTHIAZINES

Cadogan and his colleagues[170-172] have made elegant studies of the rearrangements which occur when phenothiazines and related compounds are prepared by deoxygenation of aryl 2-nitroaryl sulphides with triethyl phosphite or by pyrolysis of the corresponding 2-azido-derivatives. Otherwise the basic chemistry of phenothiazines has been rather neglected, which is somewhat surprising in view of the prominent place occupied by these compounds in the patent literature.

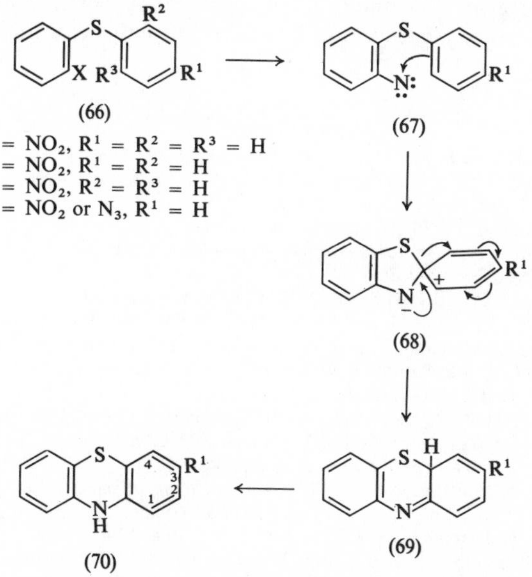

(66)

(a) $X = NO_2$, $R^1 = R^2 = R^3 = H$
(b) $X = NO_2$, $R^1 = R^2 = H$
(c) $X = NO_2$, $R^2 = R^3 = H$
(d) $X = NO_2$ or N_3, $R^1 = H$

(67)

(68)

(70)

(69)

(a) $R^1 = Me$

Cyclisation of phenyl 2-nitrophenyl sulphide (66a) with triethyl phosphite affords phenothiazine (54%)[170]. Similar treatment of the 4-methyl derivative (66c, $R^1 = Me$) gives 3-methylphenothiazine (70a), and not the expected 2-methyl compound. The driving force of the rearrangement lies in the propensity of the initially formed nitrene (67) for reaction via a 5-membered intermediate (68). In an effort to regain aromatic stability, the spirodiene (68) undergoes a 1,2-sigmatropic shift to the hydroaromatic intermediate (69), followed by a prototropic shift to give the observed product (70)[170]. The rearrangement, (66) → (70), seems quite general: it occurs as described for the 4-unsubstituted compound (66a) (shown by isotopic labelling)[170], for compounds (66b) in which one *ortho*-position in the aryl group is blocked (to give a 1-substituted phenothiazine)[171], and for the pyrolysis of the azido-compounds (66; $X = N_3$)[170, 173].

In cases (66d) where both *ortho*-positions in the aryl group are blocked, the spiro-intermediate, cf. (68), is formed as expected in both the nitro- and azido-reactions. Its subsequent transformations may involve deep-seated rearrangements, details of which are discussed in the original work[171]. The case of the dichloro-compounds (66d; $R^2 = R^3 = Cl$) is of interest because the patterns of products from the nitro- and azido-compounds are different, suggesting the participation in the former case of a nitrene precursor, $Ar\bar{N}$—O—$\overset{+}{P}(OEt)_3$, as well as, or instead of, a nitrene[171]. In all other cases the patterns of products from both reactions are similar, suggesting a nitrene as the simplest common intermediate.

2-Nitrenearyl aryl ethers do not normally[174] cyclise to phenoxazines. However, by heating 2-azidophenyl 2,6-dimethoxyphenyl ether, in which both *ortho*-positions in the aryl group are blocked, a mixture of 4-methoxyphenoxazine and 1,2-dimethoxyphenoxazine is obtained[172]. The corresponding sulphide gives 1,2-dimethoxyphenothiazine, but it gives the 1-, rather than the 4-methoxy-compound[171], in accord with theoretical predictions[172].

References

1. van Tamelen, E. E. and Whitesides, T. H. (1971). *J. Amer. Chem. Soc.*, **93**, 6129
2. Lablache-Combier, A. and Remy, M.-A. (1971). *Bull. Soc. Chim. Fr.*, 679
3. Elguero, J. (1971). *Bull. Soc. Chim. Fr.*, 1925
4. Grundmann, C. (1970). *Synthesis*, 344
5. Cornforth, J. W. (1957). *Heterocyclic Compounds*, Vol. 5, Ch. 5 and 6, (R. C. Elderfield, editor). (New York: Wiley)
6. Wiegand, E. E. and Rathburn, D. W. (1970). *Synthesis*, 648
7. Paul, S. D., Juneja, P. S. and Dhane, D. L. (1970). *Ind. J. Chem.*, **8**, 760
8. Kanaoka, Y., Hamada, T. and Yonemitsu, O. (1970). *Chem. Pharm. Bull.*, **18**, 587
9. Leyshon, L. J. and Saunders, D. G. (1971). *Chem. Commun.*, 1608
10. Saunders, D. G. (1969). *Chem. Commun.*, 680
11. Zbiral, E., Bauer, E. and Stroh, J. (1971). *Monatsh. Chem.*, **102**, 168
12. Schöllkopf, U. and Schröder, R. (1971). *Angew. Chem., Int. Ed. Engl.*, **10**, 333
13. Ohtsuka, Y. (1970). *Bull. Chem. Soc. Jap.*, **43**, 187
14. Foucaud, A. and Baudru, M. (1970). *Compt. Rend. Acad. Sci. Ser. C*, **271**, 1613
15. Lown, J. W. and Moser, J. P. (1970). *Chem. Commun.*, 247; *Can. J. Chem.*, **48**, 2227
16. Höfle, G. and Steglich, W. (1971). *Chem. Ber.*, **104**, 1408
17. Ogura, H., Sugimoto, S. and Itoh, T. (1970). *Org. Mass Spectrom.*, **3**, 1341
18. Barker, G. and Ellis, G. P. (1971). *J. Chem. Soc. C*, 1482
19. Katritzky, A. R. and Takeuchi, Y. (1970). *Org. Magn. Resonance*, **2**, 569
20. Gordon, R. D. and Yang, R. F. (1970). *Can. J. Chem.*, **48**, 1722
21. Bodor, N., Schwartz, I. and Trinajstić, N. (1971). *Z. Naturforsch. B*, **26**, 400
22. Grigg, R. and Jackson, J. L. (1970). *J. Chem. Soc. C*, 552
23. Kondrat'eva, G. Ya., Medvedskaya, L. B., Ivanova, Z. N. and Shmelev, L. V. (1971) *Izv. Akad. Nauk SSSR, Ser. Khim.*, 1363
24. Ohlsen, S. R. and Turner, S. (1971). *J. Chem. Soc. C*, 1632
25. Gotthardt, H., Huisgen, R. and Bayer, H. O. (1970). *J. Amer. Chem. Soc.*, **92**, 4340
26. Steglich, W., Gruber, P., Höfle, G. and König, W. (1971). *Angew. Chem., Int. Ed. Engl.*, **10**, 653
27. Steglich, W. and Gruber, P. (1971). *Angew. Chem., Int. Ed. Engl.*, **10**, 655
28. Mohan, S., Kumar, B. and Sandhu, J. S. (1971). *Chem. Ind. (London)*, 671
29. Huisgen, R. (1963). *Angew. Chem., Int. Ed. Engl.*, **2**, 565, 633
30. Battaglia, A. and Dondoni, A. (1970). *Tetrahedron Lett.*, 1221
31. Battaglia, A., Dondoni, A. and Mangini, A. (1971). *J. Chem. Soc. B*, 554
32. Beltrame, P., Sartirana, P. and Vintani, C. (1971). *J. Chem. Soc. B*, 814
33. Morrocchi, S., Ricca, A., Selva, A. and Zanarotti, A. (1970). *Atti. Accad. Naz. Lincei. Cl. Sci. Fis., Mat. Natur., Rend.*, **48**, 231
34. Micetich, R. G. (1970). *Can. J. Chem.*, **48**, 467
35. Micetich, R. G. (1970). *Org. Prep. Proceed.*, **2**, 225
36. Caramella, P. and Cereda, E. (1971). *Synthesis*, 433
37. Sasaki, T., Yoshioka, T. and Suzuki, Y. (1971). *Bull. Chem. Soc. Jap.*, **44**, 185

38. Sasaki, T., Yoshioka, T. and Suzuki, Y. (1970). *Yuki Gosei Kagaku Kyokai Shi*, **28,** 1054
39. Beam, C. F., Dyer, M. C. D., Schwarz, R. A. and Hauser, C. R. (1970). *J. Org. Chem.*, **35,** 1806
40. Griffiths, J. S., Beam, C. F. and Hauser, C. R. (1971). *J. Chem. Soc. C*, 974
41. Friedrich, K. and Thieme, H. K. (1970). *Chem. Ber.*, **103,** 1982
42. Klötzer, W., Bretschneider, H., Fitz, E., Reiner, R. and Bader, G. (1970). *Monatsh. Chem.*, **101,** 1109
43. Barnes, R. A. Reference 5, Ch. 7
44. Nonhebel, D. C. (1970). *Org. Mass Spectrom.*, **3,** 1519
45. Battaglia, A., Dondoni, A. and Taddei, F. (1970). *J. Heterocycl. Chem.*, **7,** 721
46. Nishiwaki, T., Kitamura, T. and Nakano, A. (1970). *Tetrahedron*, **26,** 453
47. Nishiwaki, T. (1971). *Org. Mass Spectrom.*, **5,** 123
48. Jacquier, R., Petrus, C., Petrus, F. and Verducci, J. (1970). *Bull. Soc. Chim. Fr.*, 1978
49. Bianchi, G., Cook, M. J. and Katritzky, A. R. (1971). *Tetrahedron*, **27,** 6133
50. Nishiwaki, T., Nakano, A. and Matsuoka, H. (1970). *J. Chem. Soc. C*, 1825
51. Kanö, H. (1952). *J. Pharm. Soc. Jap.*, **72,** 150
52. Nishiwaki, T., Saito, T., Onomura, S. and Kondo, K. (1971). *J. Chem. Soc. C*, 2644
53. Nishiwaki, T. and Saito, T. (1971). *J. Chem. Soc. C*, 3021
54. Stork, G., Ohashi, M., Kamachi, H. and Kakisawa, H. (1971). *J. Org. Chem.*, **36,** 2784
55. Caramella, P., Metelli, R. and Grünanger, P. (1971). *Tetrahedron*, **27,** 379
56. Traverso, G., Barco, A. and Pollini, G. P. (1971). *Chem. Commun.*, 926
57. Stevens, R. V., Christensen, C. G., Edmonson, W. L., Kaplan, M., Reid, E. B. and Wentland, M. P. (1971). *J. Amer. Chem. Soc.*, **93,** 6629
58. Stevens, R. V., DuPree, L. E., Edmonson, W. L., Magid, L. L. and Wentland, M. P. (1971). *J. Amer. Chem. Soc.*, **93,** 6637
59. Boulton, A. J., Fletcher, I. J. and Katritzky, A. R. (1971). *J. Chem. Soc. C*, 1193
60. Boulton, A. J. and Brown, R. C. (1970). *J. Org. Chem.*, **35,** 1662
61. Parry, K. P. and Rees, C. W. (1971). *Chem. Commun.*, 833
62. Georgarakis, Doppler, Th., Märky, M., Hansen, H.-J. and Schmid, H. (1971). *Helv. Chim. Acta*, **54,** 2916
63. Giovannini, E., Rosales, J. and de Souza, B. (1971). *Helv. Chim. Acta*, **54,** 2111
64. Sprague, J. M. and Land, A. H. Reference 5, Ch. 8
65. Das, B. C. and Mahapatra, G. N. (1970). *J. Indian Chem. Soc.*, **47,** 98
66. Tripathy, H., Dash, B. C. and Mahapatra, G. N. (1970). *Indian J. Chem.*, **8,** 586
67. Zbiral, E. (1970). *Tetrahedron Lett.*, 5107
68. Schmitz, E. and Striegler, H. (1970). *J. Prakt. Chem.*, **312,** 359
69. Hirai, K. (1971). *Tetrahedron Lett.*, 1137
70. Hirai, K. and Ishiba, T. (1971). *Chem. Commun.*, 1318
71. Lau, P. T. S. and Gompf, T. E. (1970). *J. Org. Chem.*, **35,** 4103
72. Clarke, G. M., Grigg, R. and Williams, D. H. (1966). *J. Chem. Soc. B*, 339
73. Bojesen, I. N., Høg, J. H., Nielsen, J. T., Petersen, I. B. and Schaumburg, K. (1971) *Acta. Chem. Scand.*, **25,** 2739
74. Rix, M. J. and Webster, B. R. (1971). *Org. Mass Spectrom.*, **5,** 311
75. Panizzi, J.-C., Davidovics, G., Guglielmetti, R., Mille, G., Metzger, J. and Chouteau, J. (1971). *Can. J. Chem.*, **49,** 956
76. Asato, G. (1970). *U.S. Pat.* 3 497 523
77. Babo, H.v. and Prijs, B. (1950). *Helv. Chim. Acta*, **33,** 306
78. Friedmann, A., Bouin, D. and Metzger, J. (1970). *Bull. Soc. Chim. Fr.*, 3155
79. Gusinskaya, S. L., Telly, V. Yu. and Makagonova, T. P. (1970). *Khim. Geterotsikl. Soedin.*, 345
80. Takahashi, S. and Kanō, H. (1969). *Chem. Pharm. Bull.*, **17,** 1598
81. Takahashi, S., Hashimoto, S. and Kanō, H. (1970). *Chem. Pharm. Bull.*, **18,** 1176
82. Bosco, M., Forlani, L., Liturri, V., Riccio, P. and Todesco, P. E. (1971). *J. Chem. Soc. B*, 1373
83. Bosco, M., Forlani, L., Todesco, P. E. and Troisi, L. (1971). *Chem. Commun.*, 1093
84. Nagarajan, K., Kulkarni, C. L. and Shah, R. K. (1971). *Indian J. Chem.*, **9,** 748
85. Baule, M., Vernin, G., Dou, H. J.-M. and Metzger, J. (1971). *Bull. Soc. Chim. Fr.*, 2083
86. Vernin, G., Dou, H. J.-M. and Metzger, J. (1971). *Compt. Rend. Acad. Sci., Ser. C*, **272** 854
87. Vernin, G., Jauffred, R., Dou, H. J.-M. and Metzger, J. (1970). *J. Chem. Soc. B*, 1678

88. Vernin, G., Dou, H. J.-M., Loridan, G. and Metzger, J. (1970). *Bull. Soc. Chim. Fr.*, 2705
89. Caronna, T., Galli, R., Malatesta, V. and Minisci, F. (1971). *J. Chem. Soc. C*, 1747
90. Altman, L. J. and Richheimer, S. L. (1971). *Tetrahedron Lett.*, 4709
91. Meyers, A. I., Nabeya, A., Adickes, H. W. and Politzer, I. R. (1969). *J. Amer. Chem. Soc.*, **91**, 763
92. Vernin, G., Poite, J.-C., Metzger, J., Aune, J.-P. and Dou, H. J.-M. (1971). *Bull. Soc. Chim. Fr.*, 1103
93. Kojima, M. and Maeda, M. (1970). *Chem. Commun.*, 386
94. Merck, E. (1970). *Fr. Demande* 2 014 527
95. Joos, A. (1970). *Ger. Offen.*, 1 924 830
96. Nakagawa, S., Okumura, J., Sakai, F., Hoshi, H. and Naito, T. (1970). *Tetrahedron Lett.*, 3719
97. Crenshaw, R. R. and Partyka, R. A. (1970). *J. Heterocyclic Chem.*, **7**, 871
98. Lewis, S. N., Miller, G. A., Hausman, M. and Szamborski, E. C. (1971). *J. Heterocyclic Chem.*, **8**, 571
99. Crow, W. D. and Leonard, N. J. (1964). *Tetrahedron Lett.*, 1477
100. Miller, G. A. and Hausman, M. (1971). *J. Heterocyclic Chem.*, **8**, 657
101. Waite, J. A. and Wooldridge, K. R. H. (1972). *Tetrahedron Lett.*, 327
102. Franz, J. E. and Black, L. L. (1970). *Tetrahedron Lett.*, 1381
103. Gotthardt, H. (1971). *Tetrahedron Lett.*, 1277
104. Ashby, J. and Suschitzky, H. (1971). *Tetrahedron Lett.*, 1315
105. Carrington, D. E. L., Clarke, K. and Scrowston, R. M. (1971). *J. Chem. Soc. C*, 3262
106. Hünig, S., Kiesslich, G. and Quast, H. (1971). *Justus Liebigs Ann. Chem.*, **748**, 201
107. Carrington, D. E. L., Clarke, K. and Scrowston, R. M. (1971). *J. Chem. Soc. C*, 3903
108. Micetich, R. G. (1970). *Can. J. Chem.*, **48**, 2006
109. Caton, M. P. L., Jones, D. H., Slack, R. and Wooldridge, K. R. H. (1964). *J. Chem. Soc.*, 446
110. Ohashi, M., Iio, A. and Yonezawa, T. (1970). *Chem. Commun.*, 1148
111. Otsuji, Y., Tsujii, Y., Yoshida, A. and Imoto, E. (1971). *Bull. Chem. Soc. Jap.*, **44**, 223
112. Boyer, J. H. (1961). *Heterocyclic Compounds*, Vol. 7, Ch. 6. (R. C. Elderfield, editor). (New York: Wiley)
113. Barbaro, G., Battaglia, A. and Dondoni, A. (1970). *J. Chem. Soc. B*, 588
114. Engberts, J. B. F. N. and Engbersen, J. F. J. (1971). *Syn. Commun.*, **1**, 121
115. Prokipcak, J. M. and Forte, P. A. (1970). *Can. J. Chem.*, **48**, 3059
116. Hoffmann, R., Gleiter, R. and Mallory, F. B. (1970). *J. Amer. Chem. Soc.*, **92**, 1460
117. Dal Monte, D., Sandri, E., Di Nunno, L., Florio, S. and Todesco, P. E. (1971). *J. Chem. Soc. B*, 2209
118. Boulton, J. J. K. and Kirby, P. (1970). *Chem. Commun.*, 1618
119. Georgarakis, M., Rosenkranz, H. J. and Schmid, H. (1971). *Helv. Chim. Acta*, **54**, 819
120. Mukai, T. and Nitta, M. (1970). *Chem. Commun.*, 1192
121. Lambe, T. M., Butler, R. N. and Scott, F. L. (1971). *Chem. and Ind. (London)*, 996
122. Boyd, G. V. and Dando, S. R. (1970). *J. Chem. Soc. C*, 1397
123. Boyd, G. V. and Dando, S. R. (1971). *J. Chem. Soc. C*, 3873
124. Boyd, G. V. and Summers, A. J. H. (1971). *J. Chem. Soc. C*, 409
125. Boyd, G. V. and Summers, A. J. H. (1971). *J. Chem. Soc. C*, 2311
126. Hetzheim, A., Pusch, H. and Beyer, H. (1970). *Chem. Ber.*, **103**, 3533
127. Hetzheim, A. and Manthey, G. (1970). *Chem. Ber.*, **103**, 2845
128. Hetzheim, A. and Singelmann, J. (1971). *Justus Liebigs Ann. Chem.*, **749**, 125
129. Boyd, G. V. and Dando, S. R. (1971). *J. Chem. Soc. C*, 225
130. Boyd, G. V. and Dando, S. R. (1971). *J. Chem. Soc. C*, 2314
131. Ruccia, M. and Vivona, N. (1970). *Chem. Commun.*, 866
132. Ruccia, M., Vivona, N. and Cusmano, G. (1971). *J. Heterocyclic Chem.*, **8**, 137
133. Lehmann, C., Renk, E. and Gagneux, A.,(1970). *Swiss Pat.*, 498 135
134. Boulton, A. J., Katritzky, A. R. and Hamid, A. M. (1967). *J. Chem. Soc. C*, 2005
135. Miller, A. H. and Pancirov, R. J. (1971). *J. Heterocyclic Chem.*, **8**, 163
136. Pilgram, K. (1970). *J. Org. Chem.*, **35**, 1165
137. Haddock, E., Kirby, P. and Johnson, A. W. (1971). *J. Chem. Soc. C*, 3642
138. Haddock, E., Kirby, P. and Johnson, A. W. (1971). *J. Chem. Soc. C*, 3994
139. Davies, J. H. and Kirby, P. (1967). *J. Chem. Soc. C*, 321
140. Haddock, E., Kirby, P. and Johnson, A. W. (1970). *J. Chem. Soc. C*, 2514

141. Lee, F. T. and Volpp, G. P. (1970). *J. Heterocylic Chem.*, **7**, 415
142. Zeller, K.-P., Meier, H. and Müller, Eu. (1971). *Org. Mass Spectrom.*, **5**, 373
143. Millard, B. J. and Pain, D. L. (1970). *J. Chem. Soc. C*, 2042
144. Krauss, P., Zeller, K.-P., Meier, H. and Müller, E. (1971). *Tetrahedron*, **27**, 5953
145. Sherman, W. A. Reference 112, Ch. 7
146. Kurzer, F. (1970). *J. Chem. Soc. C*, 1805
147. Kurzer, F. (1970). *J. Chem. Soc. C*, 1813
148. Kurzer, F. (1971). *J. Chem. Soc. C*, 2927
149. Kurzer, F. (1971). *J. Chem. Soc. C*, 2932
150. Kurzer, F. and Wilkinson, M. (1970). *J. Chem. Soc. C*, 26
151. Bayer, H. O., Huisgen, R., Knorr, R. and Schaefer, F. C. (1970). *Chem. Ber.*, **103**, 2581
152. Knorr, R. and Huisgen, R. (1970). *Chem. Ber.*, **103**, 2598
153. Knorr, R., Huisgen, R. and Staudinger, G. K. (1970). *Chem. Ber.*, **103**, 2639
154. Huisgen, R., Gotthardt, H., Bayer, H. O. and Schaefer, F. C. (1970). *Chem. Ber.*, **103**, 2611
155. Brunn, E., Funke, E., Gotthardt, H. and Huisgen, R. (1971). *Chem. Ber.*, **104**, 1562
156. Huisgen, R., Funke, E., Gotthardt, H. and Panke, H.-L. (1971). *Chem. Ber.*, **104**, 1532
157. Funke, E., Huisgen, R. and Schaefer, F. C. (1971). *Chem. Ber.*, **104**, 1550
158. Gotthardt, H. and Huisgen, R. (1970). *Chem. Ber.*, **103**, 2625
159. Funke, E. and Huisgen, R. (1971). *Chem. Ber.*, **104**, 3222
160. McEwen, W. E., Mineo, I. C. and Shen, Y. H. (1971). *J. Amer. Chem. Soc.*, **93**, 4479
161. Huisgen, R., Gotthardt, H. and Bayer, H. O. (1970). *Chem. Ber.*, **103**, 2368
162. Sasaki, T. and Kanematsu, K. (1971). *J. Chem. Soc. C*, 2147
163. Potts, K. T., Husain, S. and Husain, S. (1970). *Chem. Commun.*, 1360
164. Angadiyavar, C. S. and George, M. V. (1971). *J. Org. Chem.*, **36**, 1589
165. Märky, M., Hansen, H.-J. and Schmid, H. (1971). *Helv. Chim. Acta*, **54**, 1275
166. Gotthardt, H. and Reiter, F. (1971). *Tetrahedron Lett.*, 2749
167. Gotthardt, H. (1971). *Tetrahedron Lett.*, 1281
168. Ollis, W. D. and Ramsden, C. A. (1971). *Chem. Commun.*, 1222
169. Ollis, W. D. and Ramsden, C. A. (1971). *Chem. Commun.*, 1223
170. Cadogan, J. I. G., Kulik, S., Thomson, C. and Todd, M. J. (1970). *J. Chem. Soc. C*, 2437
171. Cadogan, J. I. G. and Kulik, S. (1971). *J. Chem. Soc. C*, 2621
172. Cadogan, J. I. G. and Lim, P. K. K. (1971). *Chem. Commun.*, 1431
173. Messer, M. and Farge, D. (1968). *Bull. Soc. Chim. Fr.*, 2832
174. Smith, P. A. S., Brown, B. B., Putney, R. K. and Reinisch, R. F. (1953). *J. Amer. Chem. Soc.*, **75**, 6335

10
Saturated and Partially Saturated Systems with Bridge-head Nitrogen Atoms and Saturated Five-and Six-membered Ring Heterocyclic Compounds

TREVOR A. CRABB
Portsmouth Polytechnic

10.1 INTRODUCTION AND SCOPE OF REVIEW ON BRIDGE-HEAD NITROGEN SYSTEMS

Apart from their potential use in medicine, systems possessing nitrogen at a ring fusion are of interest at present because of the conformational instability of the tertiary nitrogen atom at the ring fusion and, in certain systems, of the influence of interactions arising from the presence of more than one heteroatom on the position of conformational equilibrium. Accordingly stereochemical features of these systems will be emphasised in this review whenever results which indicate structures have been published. Only the simpler systems have been covered and the large variety of alkaloids possessing bridge-head nitrogen have been excluded from the review. The classification of the systems follows the method adopted in the comprehensive review[1] on bridge-head nitrogen systems and in the recent review[2] describing the use of infrared (i.r.) and nuclear magnetic resonance (n.m.r.) spectroscopy in stereochemical studies on saturated systems with bridge-head nitrogen.

10.2 FUSED 6/6 RING SYSTEMS WITH NO ADDITIONAL HETEROATOM

10.2.1 Substituted quinolizidines

Contrary to an earlier report[3] Hofmann degradation of both cis- and trans-quinolizidine methiodides gave[4] a 50:50 mixture of N-methylazacyclodecene and N-methylpiperidylbutene. A series of 1-aryl-1-hydroxyquinolizidines has been synthesised[5] by Grignard reactions on 1-ketoquinolizidine. In each case both diastereoisomers show Bohlmann bands indicative of the trans-conformation and one isomer must exist with the aryl group equatorial. A very dilute solution of the other isomer (0.002 M in CCl_4) shows strong intramolecularly hydrogen bonded OH absorption in the i.r. (3475 cm^{-1}) as well as free OH (3609 cm^{-1}) regions and analysis indicates this compound to exist as 60% trans-fused ring conformation with axial phenyl in equilibrium with 40% cis-fused ring conformation with equatorial phenyl. Dehydration of the alcohols gave the $\Delta^{1,2}$-dehydroquinolizidines, which were reduced to the arylquinolizidines, all the isomers of which apparently adopt the trans-fused ring conformation.

Both diastereoisomers of 1-ethylquinolizidine have been obtained[6] by lithium aluminium hydride reduction of the corresponding ethyl-6-quinolizidones prepared via 2-ethyl-2-(β-cyanoethyl)cyclohexane-1,3-dione. Trans-fused ring conformations were assigned to both ethylquinolizidines on the basis of Bohlmann i.r. absorption but the axial/equatorial nature of the ethyl groups was not determined. The condensation of ethyl 5-aryl-3-ketopent-4-enoate (from a cinnamoyl chloride–ethyl acetoacetate synthetic sequence) with Δ^1-piperideine in neutral solution gives in high yield a 4-aryl-1-carbethoxyquinolizidin-2-one[7], which was assigned a trans ring fusion (Bohlmann bands) with the aryl group equatorial. This was converted into two diastereoisomers of 4-aryl-2-hydroxyquinolizidine, the configurations of which were determined from the chemical shifts of the carbinol CH protons (δ 4.12 axial OH; 3.51 equatorial OH). 9a-Hydroxymethylquinolizidine has been prepared[8] by lithium aluminium hydride reduction of the product of the reaction between $\Delta^{1,9a}$-hexahydroquinolizine and trichloroacetic acid in benzene. I.R. hydrogen bonding studies (dilute CCl_4 solution) suggest that the compound exists as an equilibrium mixture of trans-fused and intramolecularly H-bonded cis-fused conformations.

10.2.2 Benzo- and indolo-quinolizidines

A synthesis of hexahydrobenzo[a]quinolizin-3-one from 1-ethoxycarbonyl-methyl-1,2,3,4-tetrahydroisoquinoline has been described[9]. A signal in the n.m.r. spectrum at δ 4.73 with an area equivalent to 0.25 protons was considered to arise from the cis-quinolizidine form and this led to the postulation of an equilibrium mixture of 75% *trans* and 25% *cis* fused conformations. However, bridge-head nitrogen systems of this type interconvert rapidly on the n.m.r. time-scale and the conclusion regarding the position of equilibrium must be based on the n.m.r. spectrum of an impure sample. By a synthetic sequence from the homoveratrylamides of α-alkyl-γ-acetylbutyric acids a series of 3-alkyl-9,10-dimethoxy-11b-methyl-1,2,3,4,6,7-hexahydro-11b*H*-benzo[*a*]quinolizines has been obtained[10]. The angular methyl group is found to be *trans* to the alkyl group when this is methyl or ethyl but *cis* when the alkyl group is larger than ethyl.

A range of phenanthra[*b*]quinolizidines has been synthesised[11] via a standard Pschorr sequence and the 1-hydroxy isomers assigned configurations from H-bonding studies.

Reductive cyclisation of diethyl 1-(3,4-dihydroisoquinol-1-yl)glutarate gave[12] ethyl-4-oxo-1,2,3,6,7,11b-hexahydrobenz[*a*]quinolizine-1-carboxylate. Treatment of the corresponding carboxylic acid with polyphosphoric acid gave the indenoquinolizine (1).

A careful study of conditions required to control the sodium borohydride reduction of 1-(2-indol-3-ylethyl)pyridinium bromides has been made[13]. Reduction of 1-(2-indol-3-ylethyl)pyridinium bromide in sodium hydroxide solution in the presence of methanol and ether followed by acid cyclisation of the resultant dihydropyridine gave 1,4,6,7,12,12b-hexahydroindolo[2,3-*a*]-quinolizine, in 40% yield. The synthetic utility of conversion of the dihydro-pyridines into nitriles by reduction in presence of sodium cyanide is illustrated

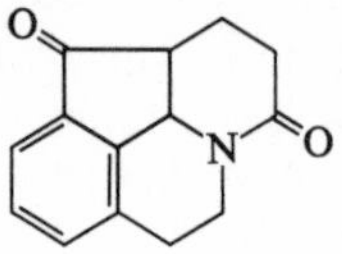

(1) 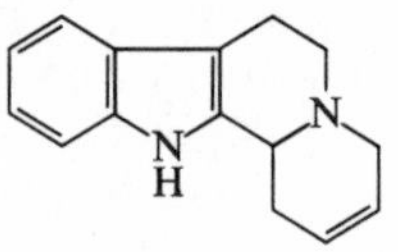   (2) 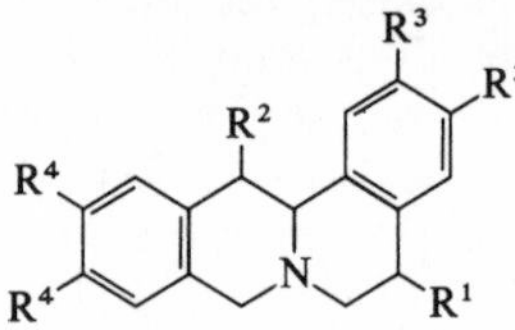   (3)

by an increase in yield (50%) of (2). The two diastereoisomers of 2-benzoyl-octahydroindolo[2,3-*a*]quinolizine have been synthesised[14]. One isomer showed strong Bohlmann bands and no signal below δ 3.5 in the n.m.r. spectrum indicating the *trans*-C–D ring fusion and the other showed weaker Bohlmann absorption and signals at δ 3.96 characteristic of the 12a proton in *cis*-C–D indoloquinolizidine.

Tetrahydro-1*H*,5*H*-benzo[*ij*]quinolizines may be obtained[15] from primary aromatic amines by reaction with formaldehyde and olefins.

10.2.3 Dibenzoquinolizidines

5,6,8,9-Tetrahydro-13b*H*-dibenzo[*ah*]quinolizines have been prepared[16] by cyclisation (6 N HCl) of 1-aryl-3,4-dihydro-2-(hydroxyiminoethyl)-1-iso-quinolinium bromides followed by reduction. The *trans*-fused ring stereo-chemistry was tentatively assigned to one of these compounds but the absence of Bohlmann bands in its i.r. spectrum points to the *cis*-ring fusion. The corresponding 5-hydroxy-tetrahydrodibenzo[*ag*]quinolizines (3) R^1 = OH, R^2 = H) both adopt the *trans*-fused conformation[17], whereas in the case of the 13-methyl isomers ((3) R^1 = H, R^2 = Me) one isomer adopts the *cis*-ring fusion (fast rate of N-methylation; δ Me 1.39) and the other the *trans*-ring fusion (slow rate of N-methylation; δ Me 0.91)[18]. It would appear from these results that a detailed study of the conformational preferences of the dibenzoquinolizidines is required. Work on other dibenzoquinolizidine systems has been published[19]. Benzofurano[*a*]benzo[*e*]quinolizidine has been obtained from the product of reaction between tryptamine and iso-chromanone[20].

10.2.4 Other polycyclic systems containing a quinolizidine moiety

1,2,6,7,7a,8,9,10-Octahydro-3H,5H-benzo[*ij*]quinolizine has been prepared by the annelation of $\Delta^{1,8a}$-octahydroquinoline with 1-chloro-3-iodopropane in the presence of lithium di-isopropylamide[21]. The structures of the de-hydroquinolizidinones obtained from the reaction between 3-oxosteroids and ethyl 2-piperidine acetate have been assigned[22]. An excellent route to the 10-azasteroid system has been described[23] and the isomers obtained show intense Bohlmann bands in the i.r. The stereochemistry of the four 8-aza-estrones have been defined[24] and the inapplicability of the Bhacca–Williams correlation[25] to this system discussed.

10.3   FUSED 6/6 RING SYSTEMS WITH ADDITIONAL HETEROATOMS

10.3.1   Systems containing a 1,3-arrangement of heteroatoms

Recent work has been largely concerned with the conformational analysis of this type of system and the destabilising influence present in those con-formations possessing parallel lone pairs of electrons on the heteroatoms. This latter effect, which has been discussed in detail and described as the generalised anomeric[26] or the Edward–Lemieux effect[27], is also important in the study of the 1,3-heterosystems discussed in Section 10.10.

All four racemic isomers of perhydrobenzo[*e*]pyrido[1,2-*c*][1,3]oxazine and some of the cyclopentano[*e*]- and cycloheptano[*e*]-compounds cor-responding to these have been synthesised[28]. Geminal coupling constant data for the N—CH$_2$—O protons show that whereas two of the isomers exist almost entirely in the conformations shown (4 and 5) the remaining two isomers exist as *c.* 85% *trans*-fused conformations in equilibrium with *cis* A/B conformations. The *trans*-fused ring conformation is also adopted by the ethylene ketal of perhydropyrido[1,2-*c*][1,3]oxazin-4-one[29]. The rates of hydrolysis of some 3-substituted perhydropyrido[1,2-*c*][1,3]oxazines have been studied[30].

Comparison of the n.m.r. parameters of the N—CH$_2$—S protons in some hexahydro-1H,3H-pyrido[1,2-*c*][1,3]thiazines with those of the N—CH$_2$—O protons in the analogous oxa-compounds points to marked deviations from chair geometry for the thiazine ring[31] as a consequence of the 'long' C—S bond. The stereochemistry of a series of substituted octahydro-1H-pyrido-[1,2-c]pyrimidines[32] and a series of methyl substituted perhydrodipyrido-[1,2-c,2',1'-f]pyrimidines[33] have been studied in detail. In the case of the

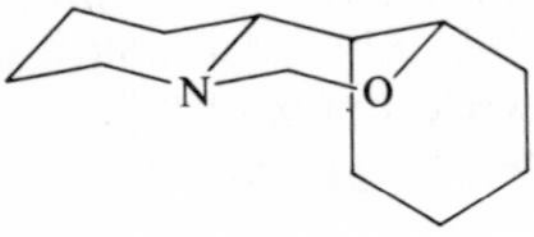

(4) 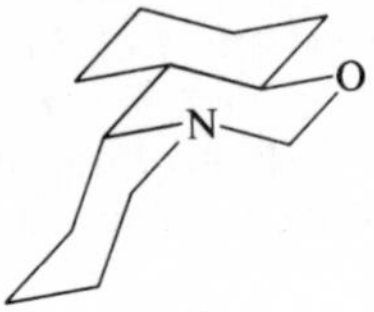   (5) 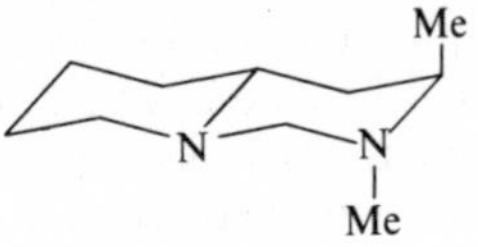   (6)

bicyclic compounds one of the 2,3-dimethyl isomers exists as (6) with N—Me axial and the other with both Me groups equatorial. The *trans-syn-trans* conformation predominates for the various methyl substituted *syn*-perhydro-dipyrido[1,2-c]pyrimidines but the *cis*-1H,12aH-1-methyl- and the *cis*-3H,12aH-3-methyl-*anti*-isomers exist as equilibrium mixtures containing

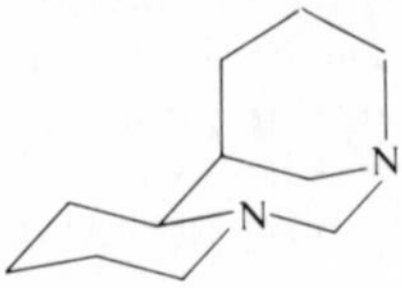

(7) 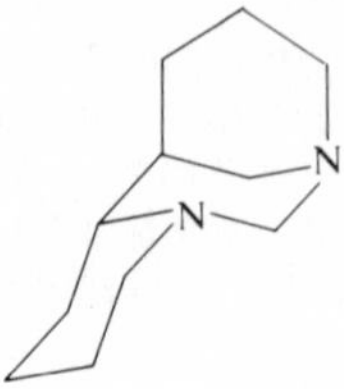   (8)

c. 30% *cis-anti-cis* conformation in equilibrium with the *trans-anti-cis* conformation.

The synthesis of *syn*- and *anti*-perhydro-7,11-methanopyrido[1,2-c][1,3]-diazocine has been achieved[34] from a sequence commencing with the reaction between 3-cyanopyridine and the Grignard reagent from 1-bromo-4-ethoxybutane. The product is converted into 3-(2-piperidyl)pyridine which on reduction and treatment with formaldehyde gives (7) and (8).

A consideration of J_{gem} for the N—CH$_2$—N protons in these compounds and in a series of other tricyclic 1,3-diaza systems produced a correlation between J_{gem} in N—CH$_2$ groups and the dihedral angle between the nitrogen

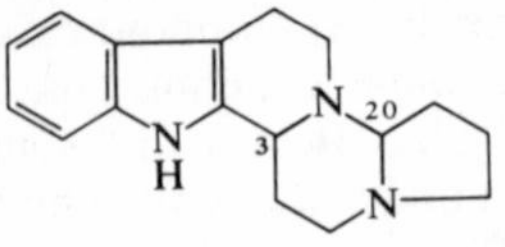

(9) 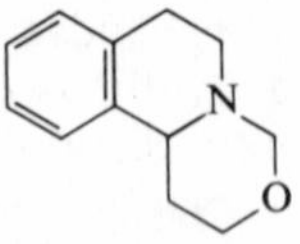   (10) 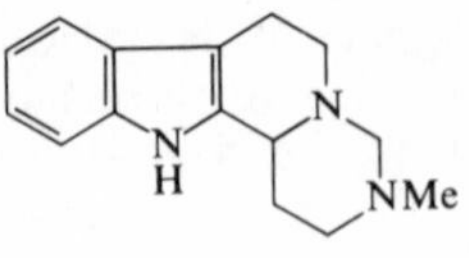   (11)

lone pair and the adjacent CH bond[34] which should prove useful in the conformational analysis of saturated *N*-heterocyclic compounds.

A range of octahydro-4H-pyrido[1,2-a]pyrimidines has been described[35]. The 3-deuterio-, 20-deuterio-, and 3,20-dideuterio-derivatives of elaeocarpidine (9) have been prepared and their i.r. spectra studied. The evidence points to the *trans-anti-trans* stereochemistry[36]. Whereas 4,4a,5,6-tetrahydro-1H,3H-[1,3]oxazino[3,4-*a*]quinoline and 3,4,11,11a-tetrahydro-1H,6H-[1,3]oxazino[3,4-*b*]isoquinoline preferentially adopt the *trans*-ring fusion, 1,6,7,11b-tetrahydro-2H,4H-[1,3]oxazino[4,3-*a*]isoquinoline (10) exists predominantly in the *cis*-ring fused conformation[37]. This difference in conformational preference has been discussed in terms of strain in the various conformations. The indolo compound corresponding to (10) also exists in a *cis*-ring-fusion but the related compound (11) exists at room temperature as an equilibrium mixture containing *c.* 50:50 *cis*- and *trans*-fused ring conformations[38].

10.4 FUSED 6/5 RING SYSTEMS WITH NO ADDITIONAL HETEROATOMS

10.4.1 Substituted indolizidines

The 220 MHz n.m.r. spectrum of 2,2-dideuterioindolizidine[39] shows that $J_{3a-3e} = -9.5$ Hz and $J_{5a-5e} = -11.0$ Hz. This latter coupling is close to that for J_{4a-4e} in quinolizidine (-11.3 Hz) and shows that any change in the hybridisation at nitrogen necessary to reduce strain in *trans*-fused indolizidine is not detectable by means of the geminal coupling constant as was previously postulated[40].

Configurations have been assigned[41] to the epimeric 1-hydroxy-1-phenylindolizidines and 1-hydroxy-1-methylindolizidines from H-bonding studies (i.r. spectroscopy). All four compounds were found to be predominantly *trans*-fused (Bohlmann bands) and the configurations of their

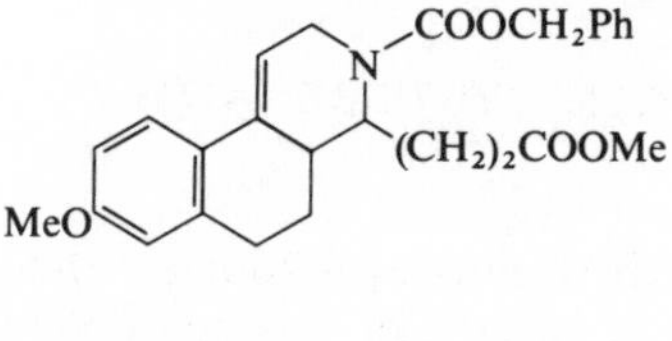

(12) 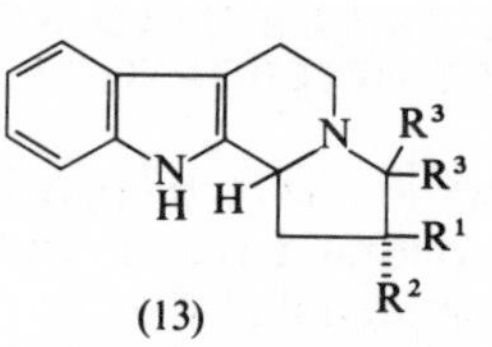 (13) 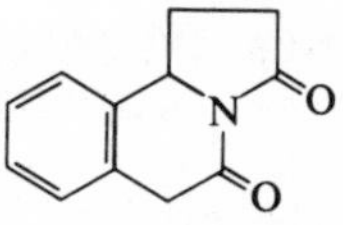 (14) *

methiodides were established from the δ values of the N-methyl signals in the n.m.r. spectrum since in the *cis*-conformation the N-methyl resonance is normally *c.* 0.2 p.p.m. to lower field than in the *trans*-conformation.

Photochemical and peroxide initiated conversion of N-chloroazacyclononan-2-one into indolizidin-3-one and indolizidin-5-one has been described[42].

Treatment of methyl 3-chloro-2-phenylpropionate with ethyl 2-pyrrolidinyl acetate gave a diester which was cyclised to a 7-oxo-6-phenylindolizidine in which the phenyl group was assumed to be *cis* to the angular hydrogen[43].

The synthesis of 1,3-dioxo-indolizidines carrying ester side chains at C(2) has been accomplished[44] by standard methods. The 2-carbethoxy compound on decarbethoxylation and treatment with lithium aluminium hydride yields a 1-hydroxy-2-methylindolizidine.

An interesting route to 13-aza-estrone (unknown B–C stereochemistry) via the diester (12) has been described[45]. The latter compound is prepared by the condensation of a biscarbamate with 2,3-dihydro-6-methoxy-1-vinylnaphthalene in the presence of a Lewis acid. Variation in the nature of the biscarbamate suggests a route to a variety of other systems.

10.4.2 Benzo- and indolo-indolizidines

cis-Fused ring conformations have been assigned to both isomers of the indoloindolizidine (13) (R^1,R^2 = Et,$(CH_2)_3OH$; R^3 = H) on the basis of the absence of Bohlmann bands in their i.r. spectra and the low-field absorption of the angular proton in both compounds. The Me protons absorb at different fields in the n.m.r. spectra of both isomers and on this basis the stereochemistry was assigned[46].

(13; R^3, R^3 = O; R^1 = R^2 = H) has been prepared from tryptamine and 3-pyrrolin-2-one[47], and by an alternative sequence starting from the reaction between tryptamine and dichloromalealdehyde acid[48].

Treatment of 1-ethoxycarbonylethyl-1,2-dihydroisoquinolin-3(4H)-one with polyphosphoric acid gave the pyrrolo[2,1-*a*]isoquinoline-dione (14) by cyclisation on to the amide nitrogen rather than the aromatic ring[49]. 5,6,8,12b-Tetrahydro-8-isoindolo[1,2-*a*]isoquinolone results from the action of polyphosphoric acid on 3-(β-phenylethylamino)phthalide[50].

On the basis of a difference in chemical shift of 1.45 p.p.m. between the methylene protons adjacent to lactam nitrogen in the n.m.r. spectrum of the methyl ether of 13-aza-18-norequilenin a half-chair conformation has been proposed[51] for ring C.

10.5 FUSED 6/5 BRIDGE-HEAD NITROGEN SYSTEMS WITH ADDITIONAL HETEROATOMS

The original claim[52] to have prepared octahydroimidazo[1,5-*a*]pyridine has been substantiated and the i.r. and n.m.r. spectra of this compound published[53]. It is remarkable[52] that no conclusions concerning the stereochemistry of the system were drawn from the spectral data since a *cis*-fused ring conformation was originally claimed[52]. The published i.r. spectrum in fact shows strong Bohlmann bands and the n.m.r. spectrum a large difference in chemical shift between the C(3) methylene protons. This, in conjunction with data on *N*-alkylated derivatives of this system[54], shows that the preferred conformation of octahydroimidazo[1,5-*a*]pyridine in CCl_4 at the n.m.r. probe temperature is that with a *trans*-ring fusion and a pseudoaxial NH bond. N.M.R. evidence has been put forward for ring–chain tautomerism in 3-phenyl- and 3-t-butyl-octahydroimidazo[1,5-*a*]pyridine which is not shown by the corresponding propyl and benzyl compounds.

An excellent route to the imidazo[1,5-*a*]pyridine system from hydantoins has been described[55]. Treatment of the hydantoins with t-butyl hypochlorite followed by triethylamine (or other such amine) gives the dehydrohydantoins which are found to be active Diels–Alder dienophiles. For example 5-phenyl-

3-methyldehydrohydantoin reacts with 2,3-dimethylbutadiene to give (15). Similar compounds are obtained from 5-methoxyhydantoin and 1,3-dienes[56]. The hydantoin (from $PhCO(CH_2)_2CH_2OH$ with ammonium carbonate and potassium cyanide) is cyclised to 1,3-dioxo-8a-phenylperhydroimidazo-[1,5-a]pyridine by toluene-p-sulphonic acid[57] and 1,3-dioxo-2-methyl-perhydroimidazo[1,5-a]pyridine is readily obtained[58] by successive reaction of ethyl pipecolinate with methyl isocyanate and ethanolic potassium hydroxide.

1,2,5,6-Tetrahydro-4H-imidazo[ij]quinoline-2-one has been prepared from 8-aminotetrahydroquinoline dihydrochloride by heating with urea[59]. Two examples of the hexahydroimidazo[1,5-a]quinoline system have been

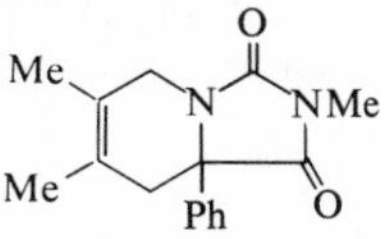

(15) 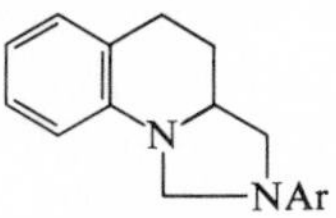   (16)

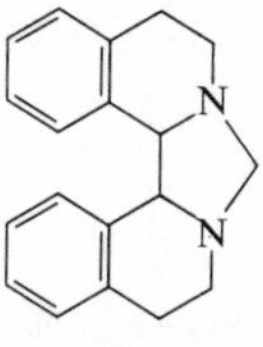

(17)

described[60, 61], and one of these[61] (16) shows a geminal coupling constant for the C(3) methylene protons of -4 Hz. This is close to that observed[62] for the corresponding protons in trans-fused N-phenylperhydroimidazo[1,2-c]-pyridines, suggesting a trans-fused ring conformation for (16). Details concerning some related hexahydroimidazo[2,1-a]-[63] and hexahydro-imidazo[2,1-c]-isoquinolines[64] have been given.

Meso- and d,l-5,6,7,9,10,11,15b,15c-octahydro-8H-di-isoquino[2,1-c:1',2'-e]imidazole (17) have been synthesised[65] from 2,2'-diacetyl-1,1',2,2'-tetra-hydro-1,1'-di-isoquinolines (by $Zn–Ac_2O$ treatment of isoquinoline). The d,l-isomer gave a singlet (δ 4.12) for the N—CH_2—N protons and the meso-isomer an AB quartet (δ 3.78 and 4.51 with $J = -7.7$ Hz). A comparison of the geminal coupling constant in the meso-compound with that in the trans-syn-trans- and trans-syn-cis-perhydropyrido[1,2-c:2',1'-e]imidazoles[66]

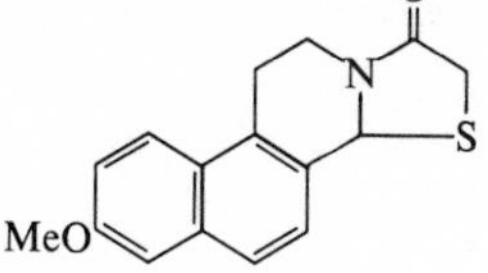

(18) 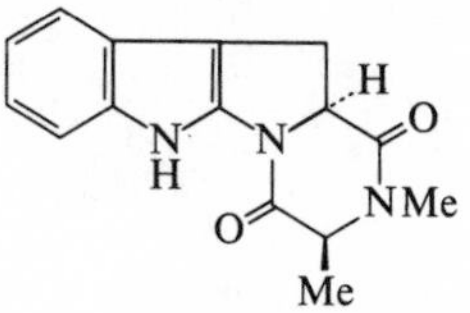   (19) 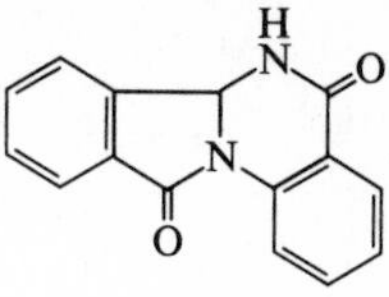   (20)

($J = -3.5$ and -8.0 Hz respectively) suggests that the meso-isomer adopts a trans-syn-cis conformation.

Two systems containing a second heteroatom other than nitrogen have been reported, 3,3a,4,10-tetrahydro-1H-oxazolo[3,4-b]-β-carboline by for-maldehyde condensation with the tetrahydrocarboline carbinol[67], and the steroid (18) by the action of mercaptoacetic acid on 1,2-dihydro-8-methoxy-benzo[f]isoquinoline[68].

Four types of systems possessing an indolizidine nucleus containing an additional nitrogen atom in the six-membered ring have been synthesised; (19) by the action of t-butyl hypochlorite on N-methyl-L-alanyl-L-tryptophan diketopiperazine[69]; an indolo[2,1-b]quinazolone[70] by refluxing 1-(2-acetamido-5-chlorobenzoyl)indole with ethanolic 6 N hydrochloric acid; (20) via anthranilamide[71]; and perhydropyrrolo[2,1-c]pyrazines by cyclisation of α-aminoalkanoyl-L-glutamic acids[72].

3-Aryl-N-(3-hydroxypropyl)hexahydropyridazine is converted by thionyl chloride in chloroform solution into the pyrazolo[1,2-a]pyridazine[73] and a benzo-derivative of the same heterocyclic ring system is prepared from β-bromopropionyl chloride and tetrahydrophthalazin-1-one[74]. The Bohlmann criterion has been shown[75] to be applicable to the 2,3,5,10-tetrahydro-1H-pyrazolo[1,2-b]phthalazine system and variable-temperature n.m.r. spectral data[75] is in accord with the existence of these phthalazines in the *trans*-ring fused conformation, interconverting via synchronous inversion of both nitrogens.

10.6 FUSED 5/5 BRIDGE-HEAD NITROGEN SYSTEMS

Both isomers of 1-methyl-7-ketopyrrolizidine have been isolated and the *trans*-7,7a-H configuration assigned to the most stable isomer[76].

Irradiation of N-chloroacetyl-3,4-dimethoxyphenethylamine with a high-pressure mercury lamp gave among other products 11 % of the pyrrolizinone (21) which is readily converted into 2-keto-8-methoxy-1,2,4,5-tetrahydro-pyrrolo[3,2,1-hi]indole on warming with alumina in toluene[77].

The carbene cyclisation reaction on the tosylhydrazone of the appropriately substituted 2-(N-pyrrolidinyl)acetophenone gives[78] 7-nitro-2,3,9,9a-tetra-hydropyrrolo[1,2-a]indole. The expected imidazo[3,4-a]indole derivative

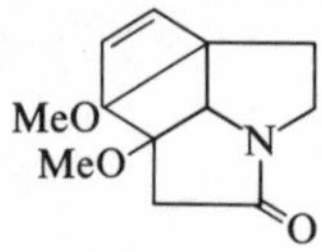

(21) 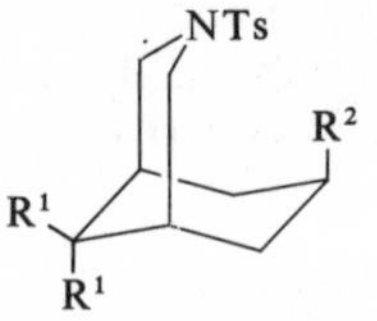   (22) 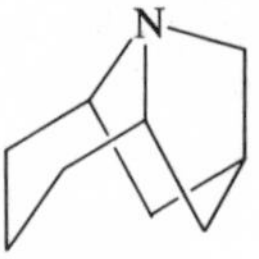   (23)

was obtained from ethyl 2-indoline carboxylate by successive reaction with phenylisocyanate and sodium methoxide[79]. Reduced pyrazolo[1,2]pyrazoles have been obtained by cyclisation of a symmetrically-substituted hydrazine derivative[80] and from the reaction between 4-methylpyrazolidine with the diacid chloride of methylmalonic acid[81].

10.7 BRIDGED RING SYSTEMS

An excellent route has been described to 1-aza-adamantane[82] via the intermediate (22) ($R^1,R^1 = O$. $R^2 = CO_2H$) obtained[83] from N-toluene-p-sulphonylpiperid-4-one and ethyl α-bromomethylacrylate. This intermediate was readily converted into (22) (R^1 = H, R^2 = CH_2OH) which produced 1-aza-adamantane on treatment with HCl–AcOH. Use of other unsaturated bromomethyl esters in the sequence[84] permits the synthesis of substituted derivatives of this system. The azatricyclononane (23) has been obtained[85] as the minor product of the photolytic decomposition of the sodium salt of pseudopelletierine tosylhydrazone.

1-Azabicyclo[3,2,1]octan-6-one has been prepared by cyclisation of N-benzyl-3-hydroxyacetylpiperidine followed by hydrogenolysis of the benzyl group. Quinuclidin-3-one has been similarly obtained[29] and 1-aza-

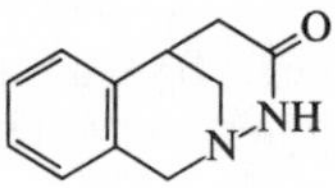

(24) 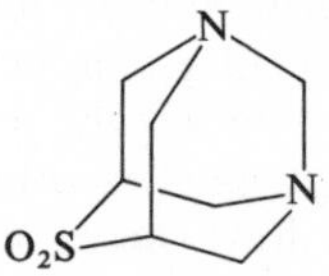   (25)

bicyclo[4,3,1]decan-3-one by a Dieckmann condensation[86]. Treatment of N,N-bisbenzylaminoacetals with 6 N HCl gave dibenzo[cf]-1-azabicyclo-[3,3,1]nonanes[87]. A nitrosation-reduction sequence on methyl 1,2,3,4-tetrahydroisoquinoline 4-acetate has given[88] 2,6-methano-2,3-benzodiazo-cine (24) and a route to (25) from N-benzenesulphonyldiallylamine has been described[89].

10.8 INTRODUCTION AND SCOPE OF REVIEW ON SATURATED FIVE- AND SIX-MEMBERED RING HETEROCYCLES

In this section emphasis has been placed on the simply substituted saturated five- and six-membered monocyclic systems, but the more important polycyclic and bridged compounds incorporating these ring systems have also been included. The 1,3-dioxane system, although closely related to the other 1,3-heterosystems discussed in Section 10.10, is covered in Vol. 4, Chap. 7. In the period under review interesting work has been carried out aimed at a determination of the conformational free energies of substituents located in saturated heterocyclic systems, since these are often very different from the values determined on cyclohexane systems and there is also a variation in this quantity according to the position of the substituent in the ring. The differences between conformational analysis in saturated heterocyclic and in saturated carbocyclic systems have been reviewed[90].

Interest in the anomeric effect in many of these monohetero-systems continues and other work has been concerned with the importance of the generalised anomeric effect in influencing the position of conformational equilibrium in 1,3-heterosystems. These aspects which tend to be given at best only scant attention in general periodical reviews on heterocyclic chemistry have accordingly been given emphasis in the present review.

10.9 SIX-MEMBERED RING SYSTEMS CONTAINING ONE HETEROATOM

10.9.1 Tetrahydropyrans

An examination of protonated tetrahydropyran has been carried out in $FSO_3H-SbF_5-SO_2$ solution by low-temperature n.m.r. spectroscopy[91] and the reactivity of the chlorine atom in 4-chlorotetrahydropyran studied[92].

The most interesting stereochemical feature of the simply substituted tetrahydropyran system is the preference for an axial position of an electron attracting group in the 2-position (anomeric effect). A number of 2-alkoxytetrahydropyrans have been observed for conformational preferences by

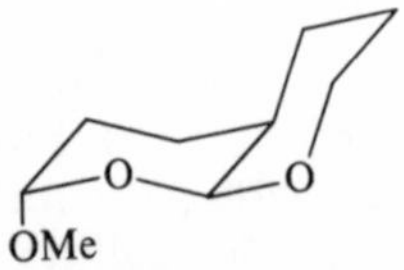

(26) 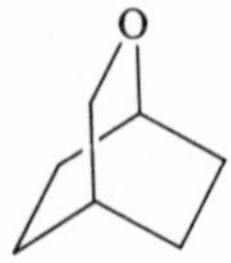   (27) 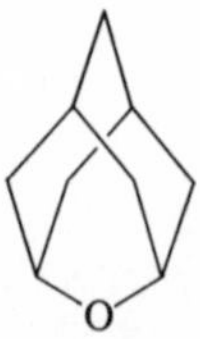   (28)

n.m.r. spectroscopy[93] and by dipole moment measurement[94], and in a related system the axial MeO conformation (26) predominates[95] over the alternative chair–chair conformation. Other 2-substituted tetrahydropyrans have been studied by n.m.r. spectroscopy and it has been shown that the equatorial position is preferred for a 2-acetamido group in contrast to the axial position for azido- and t-butylperoxy-groups[96]. The anomeric alkythio-group has been shown to possess a slight preference for the axial position[97].

cis- And trans-2,5-dimethoxytetrahydropyran have been prepared via a hydroboration sequence on 2-methoxy-3,4-dihydro-2H-pyran and the stereochemistry of these and some related compounds established[98]. N.M.R.

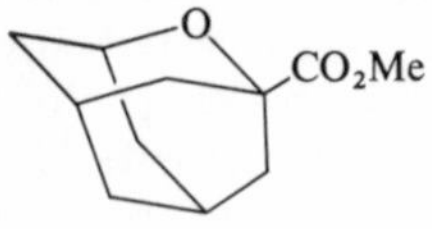

(29) 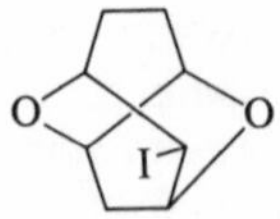   (30) 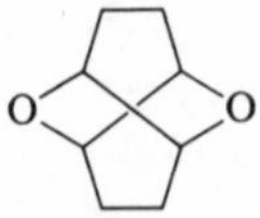   (31)

spectroscopy has also been utilised in the assignment of configuration to isomers of the methyl esters of 2,4-dimethoxytetrahydropyran-6-carboxylic acid[99].

A number of bridged systems incorporating the tetrahydropyran moiety have been synthesised, 2-oxabicyclo[2,2,2]octane, (27)[100] by a 1,4-transannular elimination on cis-4-tosyloxymethylene-1-cyclohexanol; 9-substituted 3-oxabicyclo[3,3,1]nonanes by a Prins reaction on 1-phenylcyclohexene[101]; the oxa-adamantane (28)[102] by irradiation of *endo*-bicyclo[3,3,1]-nonan-3-ol in the presence of mercuric oxide and iodine; *trans*-9-chloro-*cis*-3-oxabicyclo[3,3,1]nonane[103] as the major product resulting from treatment of cyclohexene with paraformaldehyde and hydrogen chloride at $-70\,°C$; the 2-oxa-adamantane (29)[104] by a route involving the conversion of 2-methyladamantan-2-ol into 5-iodo-5-iodomethyl-4-oxahomoadamantane by lead tetra-acetate–iodine in benzene; and the 2,7-dioxatwistane (31) via rearrangement of 10-iodo-2,7-dioxaisotwistane (30)[105].

10.9.2 Thians

A study[106] of the position of conformational equilibrium in 2-alkoxy- and 2-alkythio-thians shows the former system to exist as *c.* 80% axial conformation and the latter as 35–50% axial conformation.

The configurations of thian-1-imines and their *N*-substituted derivatives may be established from the n.m.r. spectral parameters of the α-methylene-group protons. The chemical-shift difference between these protons is greater and the geminal coupling constant numerically smaller for the equatorial imine configuration[107].

The rates of base-catalysed H–D exchange of α-protons in 3,5-diphenyl-4-hydroxythian sulphone in aqueous dimethyl sulphoxide show $k_{eq}/k_{ax} = 1.6$ [108]. 3,5-Diphenyl-3-hydroxythian oxide may be obtained[109] from benzylidine acetophenone and the dimethyl sulphoxide anion in liquid ammonia. Thiacyclobutan-2-one and *t*-butyl-4-thiahexanoate have been obtained by irradiation of thian-4-one[110]. The action of 75% sulphuric acid on 3-(2-mercaptocyclopentyl)allene gives a cyclopentanothiolan in contrast to the action of u.v. radiation which gives a cyclopentanothian[111]. 2-Thia-5-α-cholestane has been prepared via A-seco-2-norcholestan-1,3-diol[112].

10.9.3 Piperidines

Work on piperidine and its derivatives has been largely concerned with the orientation of the lone pair of electrons on nitrogen and the determination of the conformational free energy of *N*-alkyl groups, the orientation of quaternisation, the determination of the conformational free energy of substituents, and other less specifically aimed configurational and conformational determinations.

10.9.3.1 *Orientation of the nitrogen lone pair and conformational free energy of N-alkyl groups*

Previous work on the position of conformational equilibrium in piperidine itself has been summarised and discussed[113] and dipole moment measurements[113] on 4-*p*-chlorophenylpiperidine show that in cyclohexane solution at 25 °C piperidine exists as 67% axial lone pair corresponding to a ΔG^0 of 1.67 kJ mol^{-1}. An explanation of this conformational preference has been offered in terms of a small net attraction between the axial lone pair and a *syn*-axial C—H bond. Similar dipole moment measurements[114] on, for example, 4-*p*-chlorophenyl-1-alkylpiperidines have given ΔG^0 values (in cyclohexane at 25 °C) of 2.72 ± 0.63, 3.98, and 6.03 kJ mol^{-1} for *N*-methyl, *N*-ethyl and *N*-isopropyl groups respectively all in favour of equatorial orientation of the alkyl substituent. The small ΔG^0 for the methyl group (cf. 7.1 kJ mol^{-1} in cyclohexane) has been attributed to ready deformation at the nitrogen atom and the large difference between the values for ethyl and isopropyl (cf. 7.5 and 8.8 kJ mol^{-1} for the same groups in cyclohexane) to the interactions arising between the C(2) and C(6) equatorial bonds and the more bulky axial isopropyl group when it attempts to lean away from the *syn*-axial C—H bonds.

Lambert[115] has measured the difference in chemical shift between the C(6) methylene protons (δ_{ae}) in a number of piperidines and significantly δ_{ae} (in CH$_2$Cl$_2$ solution) increases from 0.48 for piperidine to 0.66 p.p.m. for 3,3-dimethylpiperidine-2,2,5,5-d_4. This suggests that the percentage of equatorial N—H is greater in the latter compound than in piperidine since an axial lone pair preferentially shields an α-axial proton enhancing the value of δ_{ae}.

An empirical correlation between n.m.r. chemical shifts of protons α to nitrogen and conformation in amines has led to the view[116] that piperidine exists as c. 66% equatorial lone pair. This is in disagreement with other estimates but since the effect of remote substituents on chemical shifts is not predictable it seems prudent to favour the results of dipole-moment measurement[113].

A new method[117] of attacking this problem has been through an examination of the paramagnetic shifts of the α-methylene protons in substituted piperidines induced by nickel and cobalt acetylacetonates. Assuming that the contact shifts $\Delta \nu_{eq}$ and $\Delta \nu_{ax}$ in an equatorial lone pair piperidine and $\Delta \nu_{eq}$ in the corresponding axial lone pair piperidine are equal, since these protons are in the same steric relationship to the nitrogen lone pair, the ratio of the observed axial to equatorial proton contact shifts were equated to $P_e + (1 - P_e) [\Delta \nu_{ax}/\Delta \nu_{eq}]_{ax}$ where P_e is the mole fraction of the lone-pair equatorial conformer and the term in square brackets the ratio of the contact shifts for 1,4-dimethylpiperidine in which the lone pair is assumed to be predominantly axial. It is then found that 3- and 4-methylpiperidine exist as 88% lone-pair equatorial, so that this method must be measuring the conformational preference of the nitrogen lone pair complexed with Ni(acac)$_2$. The conformational dependence of the contact shifts for the β- and γ-methylene protons has also been determined[118].

The equilibrium constants for the configurational equilibrium between *cis*- and *trans*-3,5-dimethyl-1-(*p*-substituted phenyl)-4-piperidone (measured by the analysis of the Me signals in the n.m.r. spectra) are very similar to those found for the equilibrium between the corresponding *N*-t-butyl compounds showing that this n.m.r. method is not sufficiently sensitive to detect the small change in hybridisation of the nitrogen lone pair arising from the change in *N*-substituent from alkyl to aryl[119]. The barrier to nitrogen inversion in N-methyl-4-piperidone has been found[120] by variable-temperature n.m.r. studies to be 36.0 ± 1.2 kJ mol^{-1}.

10.9.3.2 Orientation of quaternisation

X-Ray work has shown[121] that the major isomer from the reaction of 1-ethyl-4-phenylpiperidine with methyl iodide is the *cis*-methyl-phenyl compound formed by preferential axial attack. Previous work on quaternisations has been reviewed[122]. Since in the n.m.r. spectra of 1,1-dimethylpiperidinium salts the axial N-methyl signals are broader than the equatorial N-methyl signals as a result of long-range coupling, it was concluded that an axial N-methyl absorbs at higher field than an equatorial N-methyl. Utilising this observation it was possible from line-width studies to show that in simple piperidines methylation occurs by axial approach. Ethoxycarbonylmethylation occurs by equatorial attack in simple piperidines as it does in the tropanes, and equatorial methylation also occurs in piperidine hindered by a β-axial substituent.

In connection with n.m.r. results[122] on the N,N-dialkyl salts it is interesting to note the chemical shifts recorded[123] for a series of protonated N-methyl piperidines. Here it is found that the equatorial N-methyl protons absorb at higher field than an axial N-methyl group except when the molecule bears an α-substituent.

The rate constant for nucleophilic attack by thiophenate anion at the benzyl group in 1-benzylpiperidine benzobromide is approximately doubled by the presence of a 4-t-butyl group[124].

10.9.3.3 Conformational free energies of substituents

From a study[125] of the n.m.r. band widths of the C(4) proton signals, ΔG^0 for the equilibrium (32) $\rightleftharpoons$ (33) has been determined in several solvents. ΔG^0 values in D$_2$O at 31 °C were as follows: X = OH $+0.04$ to -0.96; X = Cl $+1.2$ to $+1.4$; X = OAc $+1.9$; X = OBz $+2.5$ kJ mol^{-1}. A comparison of these ΔG^0 values with those of the corresponding 4,4-dimethylcyclohexanes or monosubstituted cyclohexanes shows a stabilisation of the axial

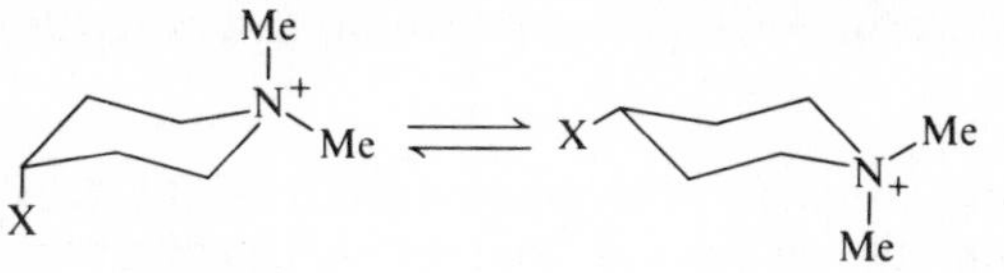

(32)   (33)

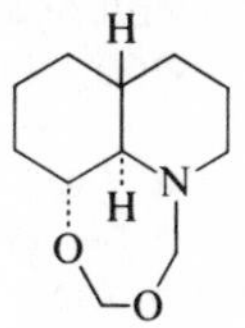

(34)

conformer (32) relative to (33) and this has been attributed to a transannular electrostatic interaction. In a similar way[126] the conformational free energies (kJ mol^{-1} measured in CDCl$_3$ at 31 °C) of the hydroxy (2.3), chloro (1.4), and acetoxy (2.3) groups in the 4-substituted piperidines have been found to be similar to the values in substituted cyclohexanes. These values were different from those in the corresponding piperidinium salts and electrostatic attraction favouring the axial conformer was also invoked to explain this difference.

The conformational free energies of C(2) and C(3) methyl substituents in some N-methyl-4-piperidones have been determined[127] by equilibration in basic solution. An unusually low corrected ΔG^0 (-1.96 kJ mol^{-1}) for the C(2) methyl probably arises from interaction with the N-methyl group. The 2-alkyl group in N-carbodithoic acid sodium salts of 2-alkylpiperidines has been found[128] to adopt the axial position.

10.9.3.4 General chemistry of piperidine

Acyclic diaminoketones have been isolated[129] as by-products in the reaction of N,N-dimethyl quaternary salts of 4-piperidones with primary amines to give N-substituted 4-piperidones. This has led to the proposition of a mechanism for this exchange process through an acyclic olefin. The major products of the exchange reaction on the trans-2,5-dimethylpiperidones retain the stereochemistry, and the product of reaction with t-butylamine is considered to exist in a skew-boat conformation on the basis of n.m.r. evidence (vicinal couplings involving C(2)H and C(6) methylene and J_{gem} of -15 Hz for C(3) methylene).

The stereochemistry of the reduction (NaBH$_4$, Al(OCHMe$_2$)$_3$, Li–NH$_3$) of a series of piperidones has been described[130]. The hydroboration sequence applied to N-methyl-Δ^3-piperideine with ($-$)3-pinanylborane yields 70% of a mixture containing 71% R-3-hydroxy-N-methylpiperidine and 29% of the 4-hydroxy compound[131]. The hydroboration of other N-substituted 1,2,3,6-tetrahydropyridines has been studied[132] and the predominant product is the 3-ol. Chlorination of 1-acetylpiperidine gives 1-acetyl-2,3,3-trichloropiperidine[133]. The chlorination of the N-acetylpipecolines has also been studied[134]. Isomeric 1-methyl-3-hydroxypiperidin-2-ones and -6-ones have been synthesised and characterised[135]. The conversion under acid conditions of 4-(α-hydroxyalkyl)-4-phenylpiperidines into 4-aralkyl-1,2,5,6-tetrahydropyridines has been described[136].

10.9.4 Polycyclic and bridged systems containing the piperidine ring

Oxidation with chromium trioxide of a mixture of N-methyldecahydroquinolin-8-ols gives a mixture of the cis- and trans-N-methyldecahydroquinolin-8-ones. Individual lithium aluminium hydride reduction of these gives two of the isomeric N-methyldecahydroquinolin-8-ols[137]. A third isomer is described[137] and some 4-hydroxy-N-methyldecahydroisoquinolines have been obtained[138]. One isomer of decahydroquinolin-8-ol reacts with formaldehyde to produce the perhydro-1,3,5-dioxazepino[6,7-i,5,6-j]quinoline (34)[139]. Various N-substituted perhydroacridines have been obtained[140–143] and their stereochemistry established. A new synthesis of 3-azabicyclo[3,3,1]-nonan-7-one from 3,5-dinitroanisole via a sodium borohydride reduction–

Mannich condensation route has been described[144]. Spectral data on one of the derived alcohols gave evidence of the chair–chair conformation. N.M.R. correlations also demonstrate the chair–chair conformation for N-tosyl-3-azabicyclo[3,3,1]nonane[145]. Irradiation of the ethylene ketal of 3-(N-chloro-N-methylaminomethyl)cyclohexanone in trifluoroacetic acid gives 2-methyl-2-azabicyclo[2,2,2]octan-6-one and the effect of the ketal group on the direction of the reaction is discussed[146].

Aza-annelation of imines with acrylamide as a route to polycyclic systems possessing the piperidine nucleus has been described[147] and some syntheses of an azatwistane system have appeared[148].

10.10 SIX-MEMBERED RING SYSTEMS CONTAINING TWO HETEROATOMS

10.10.1 1,3-Heterosystems: 1,3-hexahydropyrimidines, tetrahydro-1,3-oxazines, 1,3-oxathians

The 1,3-dioxan system is reviewed in Chapter 3. A molecular orbital calculation has provided some justification for the 'rabbit-ears effect' in hexahydropyrimidine[149]. Dipole-moment measurements on hexahydro-pyrimidines have provided estimates of ΔG^0 for the equilibrium (35) $\rightleftharpoons$ (36). For the compounds I, II, III, and IV, ΔG^0 (corrected) values of -1.4 ± 0.2, 1.9 ± 0.3, 2.3 ± 0.3, $\geqslant7.5\,\text{kJ mol}^{-1}$ were obtained[150]. Using (37) (both lone pairs axial) and (38) (one lone pair axial, one lone pair equatorial) as model

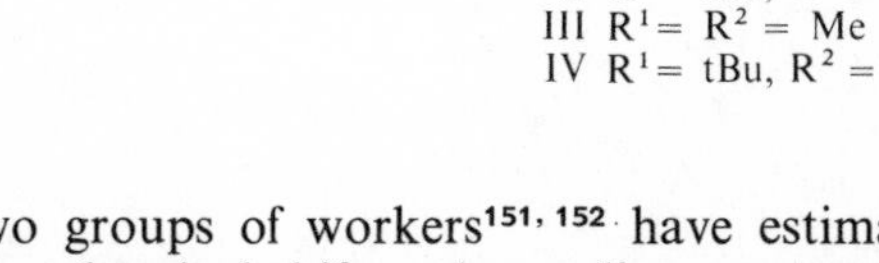

(35)   (36)

I $R^1 = $ tBu, $R^2 = $ H
II $R^1 = $ tBu, $R^2 = $ Me
III $R^1 = R^2 = $ Me
IV $R^1 = $ tBu, $R^2 = $ iPr

compounds two groups of workers[151, 152] have estimated ΔG^0 values for (35) $\rightleftharpoons$ (36) from chemical-shift and coupling-constant values for the C(2) methylene protons. Results from the n.m.r. and dipole measurements for III are in good agreement.

The cis- and trans-(+)- and cis- and trans-(−)-decahydroquinazolines have been synthesised and their absolute configurations established[153]. An examination[154] of the n.m.r. parameters of the C(2) methylene protons has indicated the predominant conformations of some trans-decahydroquinazo-lines, including (39) for the parent compound. The 1,3-dimethyl compound has a similar J_{gem} (-9.4 Hz) to that in 1,3,5-trimethylhexahydropyrimidine and the 3-methyl compound a similar J_{gem} (-11.2 Hz) to that in (38). At

−70 °C in CFCl₃ solution the n.m.r. spectrum of 1-methylhexahydropyrimi-
dine showed coupling between the C(2) protons and the NH proton. The
magnitudes of these (13 and 3 Hz) are in accord with the axial NH conformation[26]. A similar conformation was established for tetrahydro-1,3-
oxazine and the 3-methyl compound[26] at low temperature. Approximate
calculations based on dipole moment measurements on 3-alkyltetrahydro-
1,3-oxazines have been made and for the 3-methyl compound at 25 °C a value
of ΔG^0 of 0.8 kJ mol⁻¹ in favour of the equatorial methyl group obtained[155].

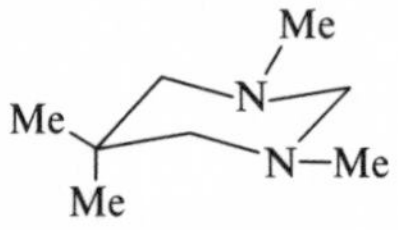

(37) 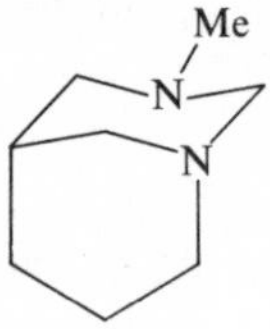   (38) 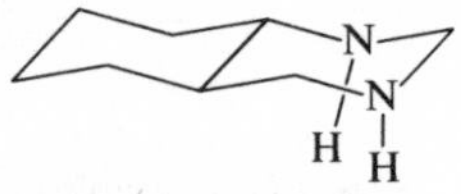   (39)

As an extension of this work[156] the difference in chemical shifts (Δae) between
axial and equatorial C(2) methylene protons and the corresponding geminal
coupling constants in a number of hexahydropyrimidines and tetrahydro-
1,3-oxazines were compared with the percentage of lone-pair axial conformations determined from these dipole-moment studies. For the Δae values a
qualitative correlation existed but there was no correlation between J_{gem}
and conformation in these systems. This apparent breakdown in correlation
must be due to distortion of the heterocyclic system or an effect on J_{gem} from
the N-alkyl groups. The n.m.r. spectrum of 3,5-dimethyl-5-nitrotetrahydro-
1,3-oxazine has been found to be markedly concentration dependent[157].

The temperature variation in the n.m.r. spectra of a series of bishexa-
hydropyrimidylmethanes has been explained in terms of the existence of two
diastereoisomeric sets of conformations which interconvert by ring inversion[158]. A number of substituted 1,3-oxathians have been synthesised[159, 160]
and the n.m.r. spectral features of this system compared with those of the
related 1,3-dioxans and 1,3-dithians[160]. ΔH^0 for the system cis-2-r-4-trans-6-
trimethyl-1,3-oxathian ⇌ trans-2-r-4-6-trimethyl-1,3-oxathian and has been
found[161] to be −4786 ± 92 J mol⁻¹. The utility of 1,3-dithians in aldehyde
synthesis has been explored[162].

10.10.2 1,2- And 1,4-heterosystems:

The axial,axial-, axial,equatorial- and equatorial,equatorial-conformations
of 1,2-dimethylhexahydropyridazine have been found to be approximately
equally populated[163] and the temperature-dependent n.m.r. spectrum of this
compound has been re-interpreted[164]. High yields of 1,2-dialkyl- and 1,2-
diaryl-hexahydropyridazines may be obtained[165] by treatment of the
corresponding 3,6-diones with diborane at 65 °C.

The percentage of axial alkoxy group in 2-alkoxy-1,4-oxathians is less
than in the corresponding 1,4-dioxans. This has been discussed in terms of

interactions between the sulphur orbitals and the alkoxy oxygen orbitals ('hockey-sticks effect')[166].

A conformational study on 1,4-oxathian-2-one by n.m.r. spectroscopy has been described[167]. Reaction between chloroacetone and 3-mercaptopropane-1,2-diol gives 1-methyl-2,8-dioxa-6-thiabicyclo[3,2,1]octane[168]. The sulphoxide oxygen atom has been shown to favour the axial position in 1,2-oxathian 2-oxide providing another example of the importance of the generalised anomeric effect[169]. Metallation experiments on 4-t-butyl-1,2-oxathian 1,1-dioxide with n-butyl-lithium show a large preference of lithium for the 6-equatorial positon[170].

10.11   FIVE-MEMBERED RING SYSTEMS CONTAINING ONE HETEROATOM

10.11.1   Reduced furans and thiophenes

Protonated tetrahydrofurans and tetrahydrothiophenes have been studied by low-temperature n.m.r. spectroscopy[91]. Good yields of insertion products have been obtained[171] from tetrahydrofuran and some methyl-substituted derivatives using phenyl(bromodichloromethyl)mercury as a source of dichlorocarbene. The effect of substituents on the chemical shifts of the C(2) protons in the n.m.r. spectra of a series of substituted tetrahydrofurans has been analysed[172]. $NaBH_4$ reduction of cyclic anhydrides gives good yields of γ-lactones with hydride attack occurring at the carbonyl group adjacent to the most sterically crowded carbon atom[173]. trans-6-Chloro-cis-3-oxabicyclo[4,3,0]nonane has been obtained as the minor product of the reaction described[103] in Section 10.9.1 exo- And endo-9-oxabicyclo[4,2,1]nonan-2-ol have been prepared[174] by Meerwein–Ponndorf–Verley reduction of the corresponding ketone obtained by bromination of 5-hydroxycyclo-octanone and substituted 9-oxabicyclo[4,2,1]nonanes by lead tetra-acetate treatment of some 4- and 5-phenylcyclo-octanols[175]. 3,7-Dioxa[3,3,3]propellane has been obtained[176] from 1,1,2,2-tetrakis(hydroxymethyl)cyclopentane.

Treatment of 2-benzoyltetrahydrothiophene with zinc amalgam and hydrochloric acid gives 2-phenylthian[177].

1,3-Cycloaddition of diazomethane to substituted 2,5-dihydrothiophene 1,1-dioxides gives 3a,4,6,6a-tetrahydro-3H-thieno[3,4-c]pyrazole 5,5-dioxides which are readily converted into substituted 3-thiabicyclo[3,1,0]hexane 3,3-dioxides by irradiation[178].

Lithium aluminium hydride reduction of hexahydro-1H-1,5:2,4-dimethanocyclopenta[c]thiolium bromide (from 2-thia-1,2-dihydro-endo-dicyclopentadiene and HBr) gives 2-methyl-6-thiatricyclo[3,2,1,1^{3,8}]nonane[179]. Derivatives of this system have also been prepared. Photochemically some 2,5-dihydrothiophenes obtained from dimethyl acetylenedicarboxylate and thiocarbonyl-ylides, give rise to unstable vinyl episulphides[180].

The benzene-induced solvent shifts (n.m.r.) of the Me group protons in cis- and trans-2-methyltetrahydrothiophene oxide is greater for the trans isomer[181]. The cis- and trans-isomers of 2,5-dimethyltetrahydrothiophene oxide have been characterised[182]. Irradiation of 1-methyl-4-mercapto-methylcyclohex-1-ene gives a predominance of thiabicyclo[3,2,1]octane[183].

10.11.2 Reduced pyrroles

Free energies of activation for inversion at the nitrogen atom in a series of N-substituted five-membered heterocycles have been determined by variable-temperature n.m.r. studies[184]. In 3,3-dimethylpyrrolidines carrying a nitrogen substituent X, the barriers to inversion are greater for X = Br, Cl, OH, ND_2 than for CD_3, but the magnitude of these substituent effects is not as great as that observed in the corresponding three-membered ring systems. The activation parameter for N-inversion in N-chloropyrrolidine has been found to be 57.8 kJ mol^{-1} [185]. The factors affecting inversion barriers in such systems have been extensively reviewed[186].

Several papers have dealt with the magnitude of the barrier to rotation about the N—CO bond in a variety of systems possessing the N-acetyl-pyrrolidine moiety and the usual method of variable-temperature n.m.r. spectroscopy has been employed. A discussion of the application of this technique to such problems of hindered rotation forms part of a review article[187]. The free-energy barrier to N—CO rotation in a range of N-acyl-pyrrolidines has been estimated as c. 71–88 kJ mol^{-1} [188] and a detailed study of the variable temperature n.m.r. spectra of N-acylprolines made[189].

trans-3-Aryl-2-carboxypyrrolidin-5-one may be obtained[190] with high stereoselectivity from ethyl β-arylacrylates and N-acetylglycine ethyl ester. The cis-isomer is prepared by decarboxylation of the 2,2-dicarboxylic acid[191]. Pyrrolidones have successfully been prepared by catalytic carbonylation of cyclopropylamines[192] and conversion of pyrrolidones into succinimides may readily be achieved by peracetic acid–Mn(III) [193].

10.11.3 Reduced pyrroles incorporated into larger systems

A range of bridged-ring systems incorporating the pyrrolidine nucleus has been described. 7-Azabicyclo[2,2,1]heptane has been obtained by an improved synthesis and some of the properties of this system explored[194].

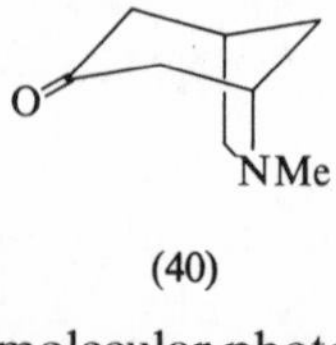

(40) 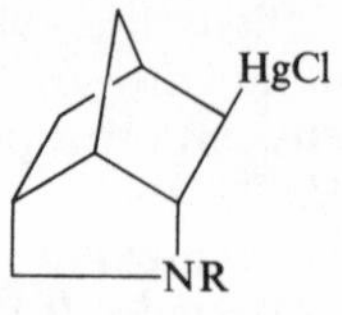   (41) 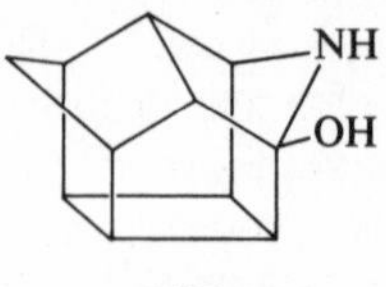   (42)

Intramolecular photo-addition of N-nitroso-4-vinylpiperidine gives 7-sub-stituted-1-azabicyclo[2,2,1]heptane where the substituent is CH=NOH or $CH_2N(OH)NO$ [195], and applications of this type of cyclisation to the production of other five-membered ring azacyclics have been established[196]. N-Methyl-6-azabicyclo[3,2,1]octa-3-one (40) arises[197] from the intramolecular Michael cyclisation of 5-(N-methylaminomethyl)cyclohex-2-enone.

2-Oxa-5-azabicyclo[2,2,1]heptane derivatives have been synthesised from hydroxy-L-proline[198] and (41) by aminomercuration of the appropriately-substituted methylaminonorbornene[199].

Irradiation of N-carbomethoxy-2,3-homo-1H-azepine gave[200] among other products 5-carbomethoxy-5-azatricyclo[4,2,0,0^{2,4}]oct-7-ene. The synthesis of N-sulphonyl-1,2-dimethyl-4-bromo-6-azabicyclo[3,2,1]oct-2-ene has been accomplished[201] by the action of bromine on N-sulphonyl-1,6-dimethyl-7-azabicyclo[4,2,0]oct-3-ene.

The aza-bird-cage compound (42) (and the analogous oxa-system) has been synthesised via the photocycloaddition product obtained from the cyclohexa-1,3-diene-p-benzoquinone Diels–Alder adduct[202].

10.11.4 Hydrocarbazoles

The isomers of some 3-substituted cis-hexahydrocarbazoles have been separated[203] and their configurations assigned from their n.m.r. spectra. The major products of hydrogenation of the 3-substituted tetrahydrocarbazoles are the syn-3-substituted cis-hexahydrocarbazoles. The n.m.r. spectra of the N,N-dimethyl quaternary iodides of these compounds have been examined[204] and coupling constants have been interpreted as suggesting a shallow boat for the syn compounds. Rates of quaternisation and pK_a values for N-methyl-cis-hexahydrocarbazoles substituted at positions 3 and 6 have been obtained[205]. Treatment of 4-(indol-3-yl)butan-2-one with boron trifluoride etherate gives[206] 3,4,4a,9a-tetrahydro-2(1H)-carbazolone. N.M.R. studies have led to the assignment of stereochemistry to some 4a-methyl-1,2,3,4,4a,9a-hexahydrocarbazoles[207].

10.12 FIVE-MEMBERED RING SYSTEMS CONTAINING TWO HETEROATOMS

10.12.1 1,3-Heterosystems: 1,3-oxathiolans, 1,3-dioxolans, imidazolidines

The lack of information regarding the conformational analysis of five-membered heterocycles, due in large part to the conformational mobility of such systems, is now being corrected. Calculated coupling constants between the C(4) and C(5) methylene protons in 2-substituted 1,3-oxathiolans for the envelope conformation with the C(5) atom as the flap are consistent with the observed values and this together with other n.m.r. results suggests that although rapid pseudo-rotation is occurring the preferred conformation is an envelope with either C(5) or O as the flap[208]. A slightly distorted envelope is consistent with n.m.r. data on a series of 1,3-oxathiolans[209]. Acetic anhydride reacts with 2,2-dialkyl-1,3-oxathiolan-5-one S-oxides to give the corresponding 4-acetoxy-2,2-dialkyl-1,3-oxathiolan-5-ones in which the acetoxy group is cis to the position originally occupied by the sulphoxide group[210]. Free energy differences between cis- and trans-2,4-dialkyl-1,3-dioxolans have been determined[211, 212] by equilibration studies and found to favour the *cis*-isomers by $1.3–2.1$ kJ mol^{-1}, whereas the anti-isomers of 2-isopropyl- and 2-t-butyl-*r*-4,*cis*-5-di-t-butyl-1,3-dioxolan are favoured $(0.04$ and 3.85 kJ mol$^{-}1$ respectively)[211].

Unlike 2,4-dimethyl-1,3-dioxolan the *trans*-isomer of 4-methoxycarbonyl-2-methyl-1,3-dioxolan is more stable than the *cis*-isomer[213]. This is perhaps a parallel to the observed reduction in percentage of *cis*-isomer when one of the alkyl groups in a 2,4-dialkyl-1,3-dioxolan is replaced by $—CH_2Br$ which has been discussed in terms of an anomeric effect[212]. Configurations have been assigned to a series of 4-substituted 2-phenyl-1,3-dioxolans in part from a study of their n.m.r. spectra[213].

In 2-alkyl-1,3-dioxolans skew interactions between the methyl groups and the oxygen atoms are minimised[214]. The condensation products of tartaric acid with two moles of formaldehyde have been shown to be bi-dioxolan-4-yl-5,5'-diones[215]. Photochemical treatment of 2-alkyldioxolans and oxathiolans has produced[216] 2-chloromethylcarboxylic esters and thiocarboxylic esters. A detailed n.m.r. study (including the determination of some ^{13}C parameters) has been made on a series of imidazolidines and some information regarding conformation obtained[217].

10.12.2 1,2-Heterosystems: isoxazolidines, 1,2-dithiolans, 1,2-oxathiolans, pyrazolidines

Isoxazolidines have been obtained[218] by addition of nitronic esters (from nitroalkanes and diazomethane) to maleic anhydride and to methyl fumarate. The n.m.r. spectrum of *cis*-1,8-dimethyl-2-oxa-1-azabicyclo[3,3,0]octane in toluene-d_8 at $-50\,^\circ C$ was consistent with the existence of the two nitrogen invertomers[219], and 3-phenyl-4-benzoyl-1,2-dithiolan has been prepared[220] by the action of sodium hydrosulphide on α-bromomethylchalcone. From the reaction between *cis*- and *trans*-2,4-diphenylthietan 1,1-dioxides with t-butoxymagnesium bromide *cis*- and *trans*-3,5-diphenyl-1,2-oxathiolan oxides have been obtained and the conformations of these discussed in relation to their n.m.r. spectra[221]. Hydride reduction of pyrazolinium salts to pyrazolidines has been studied[222, 223] and the synthesis of 1,2-dimethyl-4-hydroxypyrazolidine from *N,N*-dimethylhydrazine and 1,3-dihalogeno-2-hydroxypropane described[224].

References

1. Mosby, W. L. (1961). *Heterocyclic Systems with Bridgehead Nitrogen Atoms*. (New York: Interscience)
2. Crabb, T. A., Newton, R. F. and Jackson, D. (1971). *Chem. Rev.*, **71**, 109
3. Schofield, K. and Wells, R. J. (1963). *Chem. Ind. (London)*, 572
4. Sugimoto, K., Ohme, K., Akiba, M. and Ohki, S. (1970). *Chem. Pharm. Bull.*, **18**, 1273
5. Temple, D. L. and Sam, J. (1970). *J. Heterocycl. Chem.*, **7**, 847
6. Zhaparov, T., Agashkin, O. V. and Sulaimankov, K. (1970). *Isv. Akad. Nauk Kirg. S.S.S.R.*, 54
7. Rosazza, J. P., Bobbitt, J. M. and Schwarting, A. E. (1970). *J. Org. Chem.*, **35**, 2564
8. Arata, Y. and Kobayashi, T. (1970). *Chem. Pharm. Bull.*, **18**, 2361
9. Van Binst, G. and Nouls, J. C. (1970). *J. Chem. Soc. C*, 150

10. Terzyan, A. G., Khazhakyan, L. V., Arutyunyan, N. A. and Tatevosyan, G. T. (1971). *Arm. Khim. Zh.*, **24**, 56
11. Foldeak, S. (1971). *Tetrahedron*, **27**, 3465
12. Askam, V. and Deeks, R. H. L. (1970), *Chem. Commun.*, 597
13. Fry, E. M. and Beisler, J. A. (1970). *J. Org. Chem.*, **35**, 2809
14. Allen, M. S., Gaskell, A. J. and Joule, J. A. (1971). *J. Chem. Soc. C*, 736
15. Hesse, K. D. (1970). *Justus Liebigs Ann. Chem.*, **741**, 117
16. Bishop, D. C. and Tucker, M. J. (1970). *J. Chem. Soc. C*, 2184
17. Dyke, S. F., Brown, D. W., Sainsbury, M. and Hardy, G. (1971). *Tetrahedron*, **27**, 3495
18. Shamma, M. and Jones, C. D. (1970). *J. Amer. Chem. Soc.*, **92**, 4943
19. Wiegrebe, W., Sasse, D., Reinhart, H. and Faber, L. (1970). *Z. Naturforsch. B*, **25**, 1408
20. Weise, K. and Zymalkowski, F. (1971). *Tetrahedron Lett.*, 1231
21. Evans, D. A. (1970). *J. Amer. Chem. Soc.*, **92**, 7593
22. Akiba, M. and Ohki, S. (1970). *Chem. Pharm. Bull.*, **18**, 2195
23. Wille, H. J., Pandit, U. K. and Huisman, H. O. (1970). *Tetrahedron Lett.*, 4429
24. Brown, R. E., Meyers, A. I., Trefonas, L. M., Towns, R. L. R. and Brown, J. N. (1971). *J. Heterocycl. Chem.*, **8**, 279
25. Bhacca, N. S. and Williams, D. H. (1965). *Tetrahedron*, **21**, 2021
26. Booth, H. and Lemieux, R. U. (1971). *Can. J. Chem.*, **49**, 778
27. Wolfe, S., Rauk, A., Tel, L. M. and Csizmadia, I. G. (1971). *J. Chem. Soc. B*, 136
28. Crabb, T. A. and Jones, E. R. (1970). *Tetrahedron*, **26**, 1217
29. Nielsen, A. T. (1970). *J. Heterocycl. Chem.*, **7**, 231
30. Schöpf, C., Erwin, G., Hinkel, H. and Höhn, M. (1970). *Justus Liebigs. Ann. Chem.*, **737**, 24
31. Crabb, T. A. and Newton, R. F. (1970). *Tetrahedron*, **26**, 3941
32. Crabb, T. A. and Newton, R. F. (1970). *Tetrahedron*, **26**, 701
33. Chivers, P. J. and Crabb, T. A. (1970). *Tetrahedron*, **26**, 3369
34. Chivers, P. J. and Crabb, T. A. (1970). *Tetrahedron*, **26**, 3389
35. Houlihan, W. J. (1970). *U.S. Pat.*, 3, 526 626
36. Gribble, G. W. (1970). *J. Org. Chem.*, **35**, 1944
37. Crabb, T. A. and Newton, R. F. (1971). *Tetrahedron Lett.*, 3361
38. Crabb, T. A. and Mitchell, J. (1971). *J. Heterocycl. Chem.*, **8**, 721
39. Cahill, R., Crabb, T. A. and Newton, R. F. (1971). *Org. Mag. Res.*, **3**, 263
40. Crabb, T. A. and Newton, R. F. (1970). *Tetrahedron Lett.*, 1551
41. Hibbett, E. P. and Sam, J. (1970). *J. Heterocycl. Chem.*, **7**, 857
42. Edwards, O. E., Paton, J. M., Benn, M. H., Mitchell, R. E., Watanatada, C. and Vohra, K. N. (1971). *Can. J. Chem.*, **49**, 1648
43. Govindachari, T. R. and Viswanathan, N. (1970). *Tetrahedron*, **26**, 715
44. Winterfeld, K. and Hoffstadt, W. (1970). *Arch. Pharm. (Weinheim)*, **303**, 812
45. Speckamp, W. N., Borends, R. J. P., de Gee, A. J. and Huisman, H. O. (1970). *Tetrahedron Lett.*, 383
46. Kutney, J. P., Abdurahman, N., Gletsos, C., Le Quesne, P., Piers, E. and Vlattas, I. (1970). *J. Amer. Chem. Soc.*, **92**, 1727
47. Bocchi, V., Casnati, G. and Gardini, G. P. (1971). *Tetrahedron Lett.*, 683
48. Winterfeldt, E. and Jutta, M. N. (1970). *Chem. Ber.*, **103**, 1174
49. Askam, V. and Deeks, R. H. L. (1970). *J. Chem. Soc. C*, 2245
50. Walker, G. N. and Kempton, R. J. (1971). *J. Org. Chem.*, **36**, 1413
51. Kessar, S. V., Singh, M., Ahuja, V. K. and Lumb, A. K. (1971). *J. Chem. Soc. C*, 262
52. Freed, M. E. and Day, A. R. (1960). *J. Org. Chem.*, **25**, 2108
53. Beim, H. J. and Day, A. R. (1970). *J. Heterocycl. Chem.*, **7**, 355
54. Crabb, T. A. and Newton, R. F. (1968). *Tetrahedron*, **24**, 6327
55. Evnin, A. B., Lam, A. and Blyskal, J. (1970). *J. Org. Chem.*, **35**, 3097
56. Ben-Ishai, D. and Goldstein, E. (1971). *Tetrahedron*, **27**, 3119
57. Smissman, E. E., Chien, P. L. and Robinson, R. A. (1970). *J. Org. Chem.*, **35**, 3655
58. Capuano, L., Welter, M. and Zander, R. (1970). *Chem. Ber.*, **103**, 2394
59. Poludnenko, V. G. and Simonov, A. M. (1970). *Khim. Geterotsikl. Soedin*, 1410
60. Baxter, C. A. R. (1970). *Ger. Offen.* 2, 007 345
61. Benkovic, S. J., Benkovic, P. A. and Chrzanowski, R. (1970). *J. Amer. Chem. Soc.*, **92**, 523
62. Crabb, T. A. and Newton, R. F. (1969). *J. Heterocycl. Chem.*, **6**, 301
63. Houlihan, W. J. (1970). *U.S. Pat.*, 3, 551 441

64. Archer, S. and Schulenberg, J. W. (1971). *U.S. Pat.*, 3, 557 120
65. Nielsen, A. T. (1970). *J. Org. Chem.*, **35**, 2498
66. Chivers, P. J., Crabb, T. A. and Williams, R. O. (1968). *Tetrahedron*, **24**, 6625
67. Hino, T., Katsuko, U. and Akaboshi, S. (1970). *Chem. Pharm. Bull.*, **18**, 384
68. Kessar, S. V., Jit, P., Mundra, K. P. and Lumb, A. K. (1971). *J. Chem. Soc. C*, 266
69. Ohno, M., Spande, T. F. and Witkop, B. (1970). *J. Amer. Chem. Soc.*, **92**, 343
70. Garcia, E. E., Arfaei, A. and Fryer, R. I. (1970). *J. Heterocycl. Chem.*, **7**, 1161
71. Houlihan, W. J. (1970). *U.S. Pat.*, 3, 509 147
72. Harnden, M. R. (1971). *U.S. Pat.*, 3, 563 992
73. Houlihan, W. J. (1970). *Fr. M.*, 7076
74. Bellasio, E. and Testa, E. (1970). *Farmaco Ed. Sci.*, **25**, 305
75. Nakamura, A. and Kamiya, S. (1970). *Chem. Pharm. Bull.*, **18**, 1526; Nakamura, A. (1970). ibid., **18**, 1426
76. Meinwald, J. and Ottenheym, H. C. J. (1971). *Tetrahedron*, **27**, 3307
77. Yonemitsu, O., Okuna, Y., Kansoka, Y. and Witkop, B. (1970). *J. Amer. Chem. Soc.*, **92**, 5686
78. Takada, T., Kunugi, S. and Ohki, S. (1971). *Chem. Pharm. Bull.*, **19**, 982
79. Corey, E. J., McCaulby, R. J. and Sachdev, H. S. (1970). *J. Amer. Chem. Soc.*, **92**, 2476
80. Bellasio, E. and Gallo, G. G. (1970). *Farm. Ed. Sci.*, **25**, 295
81. Zimmer, G. and Bose, D. (1970). *Arch. Pharm. (Weinheim)*, **303**, 218
82. Speckamp, W. N., Dijkink, J. and Huisman, H. O. (1970). *Chem. Commun.*, 197
83. Speckamp, W. N., Dijkink, J. and Huisman, H. O. (1970). *Chem. Commun.*, 196
84. Dekkers, A. W. J. D., Speckamp, W. N. and Huisman, H. O. (1971). *Tetrahedron Lett.*, 489
85. Sasaki, T., Eguchi, S. and Kiriyama, T. (1971). *Tetrahedron*, **27**, 893
86. Hammer, C. F. and Craig, J. H. (1971). *J. Heterocycl. Chem.*, **8**, 411
87. Bobbit, J. M. and Shibuya, S. (1970). *J. Org. Chem.*, **35**, 1181
88. Mitsuhashi, K. and Shiotani, S. (1970). *Japan Pat.*, **7**, 007 750
89. Stetter, H. and Schoeps, J. (1970). *Chem. Ber.*, **103**, 205
90. Eliel, E. L. (1970). *Acc. Chem. Res.*, **3**, 1; *Bull. Soc. Chim. Fr.*, 517
91. Olah, G. A. and Szilagyi, P. J. (1971). *J. Org. Chem.*, **36**, 1121
92. Stapp, P. R. and Drake, C. A. (1971). *J. Org. Chem.*, **36**, 522
93. Descotes, G., Sinou, D. and Martin, J. C. (1970). *Bull. Soc. Chim. Fr.*, 3730
94. Gelin, M., Bahurel, Y. and Descotes, G. (1970). *Bull. Soc. Chim. Fr.*, 3723
95. Bahurel, Y., Lissac-Cahu, M., Descotes, G., Gelin, M., Delmau, J. and Duplan, J. C. (1970). *Bull. Soc. Chim. Fr.*, 4006
96. Zefirov, N. S. and Shekhtman, N. M. (1970). *Zh. Org. Chim.*, **6**, 863
97. De Hoog, A. J. and Havinga, E. (1970). *Rec. Trav. Chim.*, **89**, 972
98. Srivasta, R. M. and Brown, R. K. (1970). *Can. J. Chem.*, **48**, 2334
99. Zamojski, A., Chmielewski, M. and Konowal, A. (1970). *Tetrahedron*, **26**, 183
100. Giudici, T. A. and Bruice, T. C. (1970). *J. Org. Chem.*, **35**, 2386
101. Bucci, P., Lippi, G. and Macchia, B. (1970). *J. Org. Chem.*, **35**, 913
102. Fisch, M., Smallcombe, S., Gramain, J. C., McKervey, M. A. and Anderson, J. E. (1970). *J. Org. Chem.*, **35**, 1886
103. Stapp, P. R. and Randall, J. C. (1970). *J. Org. Chem.*, **35**, 2948
104. Black, R. M. and Gill, G. B. (1971). *Chem. Commun.*, 172
105. Ganter, C. and Wicker, K. (1970). *Helv. Chim. Acta.*, **53**, 1693
106. Zefirov, N. S., Blagoveshchenskii, V. S., Kazimirchik, I. V. and Yakovleva, O. P. (1971). *Zh. Org. Khim.*, **7**, 594
107. Lambert, J. B., Craig, E. M. and Bailey, D. S. (1971). *Chem. Commun.*, 316
108. Brown, M. D., Cook, M. J., Hutchinson, B. J. and Katritzky, A. R. (1971). *Tetrahedron*, **27**, 593
109. Gautier, J. A., Miocque, M., Plat, M., Moskowitz, H. and Blanc-Guenee, J. (1970). *Tetrahedron Lett.*, 895
110. Johnson, P. Y. and Berchtold, G. A. (1970). *J. Org. Chem.*, **35**, 584
111. Dronov, V. I. and Krivonogov, V. P. (1970). *Khim. Geterotsikl. Soedin*, **9**, 1185
112. Kashman, Y. and Kaufman, E. D. (1971). *Tetrahedron*, **27**, 3437
113. Jones, R. A. Y.. Katritzky, A. R., Richards, A. C., Wyatt, R. J., Bishop, R. J. and Sutton, L. E..(1970). *J. Chem. Soc. B*, 127

114. Jones, R. A. Y., Katritzky, A. R., Richards, A. C. and Wyatt, R. J. (1970). *J. Chem. Soc. B*, 122
115. Lambert, J. B., Bailey, D. S. and Michel, B. F. (1970). *Tetrahedron Lett.*, 691
116. Price, C. C. (1971). *Tetrahedron Lett.*, 4527
117. Yonezawa, T., Morishima, I. and Ohmori, Y. (1970). *J. Amer. Chem. Soc.*, **92**, 1267
118. Morishima, I., Okada, K., Ohashi, K. and Yonezawa, T. (1971). *Chem. Commun.*, 33
119. Brown, M. D., Cook, M. J., Desimoni, G. and Katritzky, A. R. (1970). *Tetrahedron*, **26**, 5281
120. Lehn, J. M. and Wagner, J. (1970). *Chem. Commun.*, 415
121. Fedeli, W., Mazza, F. and Vaciago, A. (1970). *J. Chem. Soc. B*, 1219
122. Jones, R. A. Y., Katritzky, A. R. and Mente, P. G. (1970). *J. Chem. Soc. B*, 1210
123. Delpuech, J.-J. and Deschamps, M. N. (1970). *Tetrahedron*, **26**, 2723
124. Leviston, P. G., McKenna, J., McKenna, J. M., Melia, R. A. and Pratt, J. C. (1970). *Chem. Commun.*, 587
125. Stolow, R. D., Lewis, D. I. and D'Angelo, P. A. (1970). *Tetrahedron*, **26**, 5831
126. Stein, M.-L., Chiurdoglu, G., Ottinger, R., Reisse, J. and Christol, H. (1971). *Tetrahedron*, **27**, 411
127. Mistryukov, E. A. and Smirnova, G. N. (1971). *Tetrahedron*, **27**, 375
128. Forrest, P. and Ray, S. (1970). *Chem. Commun.*, 1537
129. Hassan, M. M. A. and Casy, A. F. (1970). *Tetrahedron*, **26**, 4517
130. Mistryukov, E. A., Katvalyan, G. T. and Smirnova, G. N. (1970). *Izv. Akad. Nauk. SSSR*, 1131
131. Lyle, R. E. and Spicer, C. K. (1970). *Tetrahedron Lett.*, 1133
132. Lyle, R. E., Carle, K. R., Ellefson, C. R. and Spicer, C. K. (1970). *J. Org. Chem.*, **35**, 802
133. Boehme, H. and Dehmel, H. (1971). *Arch. Pharm. (Weinheim)*, **304**, 397
134. Boehme, H. and Delmel, H. (1971). *Arch. Pharm. (Weinheim)*, **304**, 407
135. Möhrle, H. and Weber, H. (1971). *Tetrahedron*, **27**, 3241
136. Iorio, M. A. and Miraglia, M. (1971). *Tetrahedron*, **37**, 4983
137. Barbieri, W., Bernardi, L. and Maggione, P. (1970). *Chimica e l'Industria*, **52**, 240
138. Kimoto, S., Okanoto, M., Uneo, M., Ohta, S., Nakamura, M. and Niiya, Y. (1970). C. (1971). *Chem. Commun.*, 644
139. Crabb, T. A. and Newton, R. F. (1970). *Chem. Commun.*, 1123
140. Bărbulescu, N. and Potmischil, F. (1970). *Rev. Chim. (Bucharest)*, **21**, 385
141. Bărbulescu, N. and Potmischil, F. (1970). *Rev. Roum. Chim.*, **15**, 1601
142. Bărbulescu, N. and Potmischil, F. (1970). *Justus Liebigs Ann. Chem.*, **735**, 132
143. Moskovkina, T. V. and Tilichenko, M. N. (1970). *Khim. Farm. Zh.*, **4**, 28
144. Wall, R. T. (1970). *Tetrahedron*, **26**, 2107
145. Speckamp, W. N., Dijkink, J., Dekkers, A. W. J. D. and Huisman, H. O. (1971). *Tetrahedron*, **27**, 3143
146. Esposito, G., Furstoss, R. and Waegell, B. (1971). *Tetrahedron Lett.*, 895.
147. Ninomiya, I., Naito, T., Higuchi, S. and Mori, T. (1971). *Chem. Commun.*, 457
148. Heusler, K. (1970). *Tetrahedron Lett.*, 97; Perelman, D., Sicsic, S. and Welvart, Z. (1970). *ibid.*, 103; Dubé, S. and Deslongchamps, P. (1970). *ibid.*, 101
149. Chen, F. P. and Jesaitis, R. G. (1970). *Chem. Commun.*, 1533
150. Jones, R. A. Y., Katritzky, A. R. and Snarey, M. (1970). *J. Chem. Soc. B*, 131
151. Riddell, F. G. and Williams, D. A. R. (1971). *Tetrahedron Lett.*, 2073
152. Eliel, E. L., Kopp, L. D., Dennis, J. E. and Slayton, A. E. (1971). *Tetrahedron Lett.*, 3409
153. Armarego, W. L. F. and Kobayashi, T. (1970). *J. Chem. Soc. C*, 1597
154. Armarego, W. L. F. and Kobayashi, T. (1971). *J. Chem. Soc. C*, 2502
155. Jones, R. A. Y., Katritzky, A. R. and Trepanier, D. L. (1971). *J. Chem. Soc. B*, 1300
156. Halls, P. J., Jones, R. A. Y., Katritzky, A. R., Snarey, M. and Trepanier, D. L. (1971). *J. Chem. Soc. B*, 1320
157. Crabb, T. A. and Judd, S. I. (1970). *Org. Mag. Res.*, **2**, 317
158. Riddell, F. G. (1971). *J. Chem. Soc. B*, 1028
159. Pihlaja, K. and Pasanen, P. (1970). *Acta Chem. Scand.*, **24**, 2257
160. Allingham, Y., Crabb, T. A. and Newton, R. F. (1971). *Org. Mag. Res.*, **3**, 37
161. Pasanen, P. and Pihlaja, K. (1971). *Tetrahedron Lett.*, 4515
162. Vedjes, E. and Fuchs, P. L. (1971). *J. Org. Chem.*, **36**, 366
163. Jones, R. A. Y., Katritzky, A. R., Ostercamp, D. L., Record, K. A. F. and Richards, A. C. (1971). *Chem. Commun.*, 644

164. Jones, R. A. Y., Katritzky, A. R. and Scattergood, R. (1971). *Chem. Commun.*, 644
165. Feuer, H. and Brown, F. (1970). *J. Org. Chem.*, **35**, 1468
166. Zefirov, N. S., Blagoveshchensky, V. S., Kazimirchik, I. V. and Surova, N. S. (1971). *Tetrahedron*, **27**, 3111
167. Jankowski, K. and Coulombe, R. (1971). *Tetrahedron Lett.*, 991
168. Gelas, J. (1971). *Tetrahedron Lett.*, 509
169. Harpp, D. N. and Gleason, J. G. (1971). *J. Org. Chem.*, **36**, 1314
170. Durst, T. (1971). *Tetrahedron Lett.*, 4171
171. Seyferth, D., Mai, V. A. and Gordon, M. E. (1970). *J. Org. Chem.*, **35**, 1993
172. Dana, G. and Zysman, A. (1970). *Bull. Soc. Chim. Fr.*, 1957
173. Bailey, D. M. and Johnson, R. E. (1970). *J. Org. Chem.*, **35**, 3574
174. Paquette, L. A. and Storm, P. C. (1970). *J. Amer. Chem. Soc.*, **92**, 4295
175. Cope, A. C., McKervey, M. A., Weinshenker, N. M. and Kinnel, R. B. (1970). *J. Org. Chem.*, **35**, 2918
176. Weinges, K. and Wiesenhuetter, A. (1971). *Justus Liebigs Ann. Chem.*, **746**, 70
177. Nesmeyanov, N. A. and Kalyavin, V. A. (1971). *Zh. Org. Khim.*, **7**, 608
178. Mock, W. L. (1970). *J. Amer. Chem. Soc.*, **92**, 6918
179. Wilder, P. W. and Gratz, R. F. (1970). *J. Org. Chem.*, **35**, 3295
180. Kellogg, R. M. (1971). *J. Amer. Chem. Soc.*, **93**, 2344
181. Rigau, J. J., Bacon, C. C. and Johnson, C. R. (1970). *J. Org. Chem.*, **35**, 3655
182. Jones, A. R. (1971). *Chem. Commun.*, 1042
183. Surzur, J.-M., Nouguier, R., Crozet, M-P. and Dupuy, C. (1971). *Tetrahedron Lett.*, 2035
184. Lehn, J. M. and Wagner, J. (1970). *Tetrahedron*, **26**, 4227
185. Lambert, J. B., Oliver, W. L. and Packard, B. S. (1971). *J. Amer. Chem. Soc.*, **93**, 933
186. Rauk, A., Leland, C. A. and Mislow, K. (1970). *Angew. Chem. Internat. Ed.*, **9**, 400; Lehn, J. (1970). *Fortschr. Chem. Forschung.*, **15**, 311; Lambert, J. B. (1971). *Topics in Stereochemistry*, **6**, 19 (Wiley-Interscience: New York)
187. Kessler, H. (1970). *Angew. Chem. Internat. Edn.*, **9**, 219
188. Wong, C. M., Buccini, J., Schwenk, R. and Te Raa, J. (1971). *Can. J. Chem.*, **49**, 639
189. Maia, H. L., Orrell, K. G. and Rydon, H. N. (1971). *Chem. Commun.*, 1209
190. Pachaly, P. (1971). *Chem. Ber.*, **104**, 412
191. Pachaly, P. (1971). *Chem. Ber.*, **104**, 429
192. Iqbal, A. F. M. (1971). *Tetrahedron Lett.*, 3381
193. Doumaux, A. R. and Trecker, D. J. (1970). *J. Org. Chem.*, **35**, 2121
194. Fraser, R. R. and Swingle, R. B. (1970). *Can. J. Chem.*, **48**, 2065
195. Chow, Y. L., Perry, R. A., Menon, B. C. and Chen, S. C. (1971). *Tetrahedron Lett.*, 1545
196. Chow, Y. L., Perry, R. A. and Menon, B. C. (1971). *Tetrahedron Lett.*, 1549
197. Furstoss, R., Teissier, P. and Waegell, B. (1970). *Chem. Commun.*, 384
198. Portoghese, P. S. and Turcotte, J. G. (1971). *Tetrahedron*, **27**, 961
199. Perie, J., Laval, J. P., Roussel, J. and Lattes, A. (1971). *Tetrahedron Lett.*, 4399
200. Paquette, L. A. and Haluska, R. J. (1970). *J. Org. Chem.*, **35**, 132
201. Paquette, L. A. and Kelly, J. F. (1971). *J. Org. Chem.*, **36**, 442
202. Sasaki, T., Eguchi, S. and Kiriyama, T. (1971). *Tetrahedron Lett.*, 2651
203. Smith, A. and Utley, J. H. P. *J. Chem. Soc. C*, 1
204. Shaw, D., Smith, A. and Utley, J. H. P. (1970). *J. Chem. Soc. B*, 1161
205. Smith, A. and Utley, J. H. P. (1971). *J. Chem. Soc. B*, 1201
206. Wittekind, R. R. and Lazarus, S. (1970). *J. Heterocycl. Chem.*, **7**, 1241
207. Ban, Y., Kinoshita, H., Murakami, S. and Oishi, T. (1971). *Tetrahedron Lett.*, 3687
208. Wilson, G. E., Huang, M. G. and Bovey, F. A. (1970). *J. Amer. Chem. Soc.*, **92**, 5907
209. Pihlaja, K. (1970). *Suomen Kemistilehti*, **43**, 143
210. Glue, S., Kay, I. T. and Kipps, M. R. (1970). *Chem. Commun.*, 1158
211. Willy, W. E., Binsch, G. and Eliel, E. L. (1970). *J. Amer. Chem. Soc.*, **92**, 5394
212. Rommelaere, Y. and Anteunis, M. (1970). *Bull. Soc. Chim. Belges*, **79**, 11
213. Inch, T. D. and Williams, N. (1970). *J. Chem. Soc. C*, 263
214. Riddell, F. G. and Robinson, M. J. T. (1971). *Tetrahedron*, **27**, 4163
215. Cort, L. A. and Stewart, R. A. (1971). *J. Chem. Soc. C*, 1386
216. Hartgerink, J. W., van der Laan, L. C. J., Engberts, J. B. F. N. and de Boer, Th. J. (1971). *Tetrahedron*, **27**, 4323
217. Allrand, J. P., Cogne, A., Gagnaire, D. and Robert, J. B. (1971). *Tetrahedron*, **27**, 2453

218. Grée, R. and Carrié, R. (1971). *Tetrahedron Lett.*, 4117
219. Raban, M., Freeman, B. J., Carlson, E. H., Banucci, E. and LeBel, N. A. (1970). *J. Org. Chem.*, **35**, 1496
220. Padwa, A. and Gruber, R. (1970). *J. Org. Chem.*, **35**, 1781
221. Dobson, R. M., Hammen, P. D. and Davis, R. A. (1971). *J. Org. Chem.*, **36**, 2693
222. Elguero, J., Jacquier, R. and Tizane, D. (1970). *Bull. Soc. Chim. Fr.*, 1121
223. Elguero, J., Jacquier, R. and Tizane, D. (1971). *Tetrahedron*, **27**, 123
224. Dittli, C., Elguero, J. and Jacquier, R. (1971). *Bull. Soc. Chim. Fr.*, 1038

11
Macromolecules Derived from Pyrroles

K. M. SMITH
University of Liverpool

Abbreviations used in structural formulae

$$V = -CH{=}CH_2$$
$$PR = -CH_2CH_2CO_2R$$
$$PH = -CH_2CH_2CO_2H$$
$$PMe = -CH_2CH_2CO_2Me$$

11.1 INTRODUCTION

This chapter will review advances in the chemistry of macromolecules derived either synthetically or biosynthetically from pyrrolic intermediates. Review articles concerning individual aspects of this area, such as metallopor-phyrins[1,2], haemoglobin[3], vitamin B_{12}[4], and bile pigments[5], have been published, as well as a general survey[6] of the field as a whole*. In addition, details of lectures describing approaches towards the total synthesis of vitamin B_{12} have appeared[7,8].

11.2 PORPHYRINS

11.2.1 Ring synthesis

Protoporphyrin-IX (1a), the biosynthetic precursor of the haemoproteins and plant pigments has been a target for synthetic chemists for some time; full details of the synthesis through the now familiar[6] *a*- and *b*-oxobilane[9] and *a,c*-biladiene[10] routes have been published. A preliminary report[11] of a highly efficient approach to protoporphyrin-IX (1a) and certain specifically deuterated derivatives (required for contact shift n.m.r. studies of haemo-proteins) has described the condensation of the pyrromethanes (2) and (3), catalysed by toluene *p*-sulphonic acid, in a MacDonald-type[6] cyclisation, followed by aerial oxidation to the porphyrin state. The vinyl groups were generated by base-promoted elimination of hydrogen chloride from the bis(2-chloroethyl)porphyrin zinc chelate. Of particular interest is the observa-tion that deuterium can be incorporated with high efficiency into the α- and γ-*meso* positions if the cyclisation of (2) and (3) is carried out in the presence of deuteriotoluene *p*-sulphonic acid; this has obvious merits for the

---

*Nomenclature used in this chapter is as defined in reference 6.

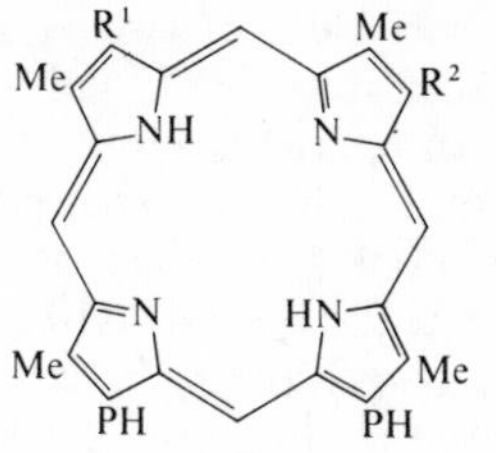

Structure (1):

(a) $R^1 = R^2 = V$
(b) $R^1 = V; R^2 = PH$
(c) $R^1 = PH; R^2 = V$
(d) $R^1 = R^2 = H$
(e) $R^1 = R^2 = Et$

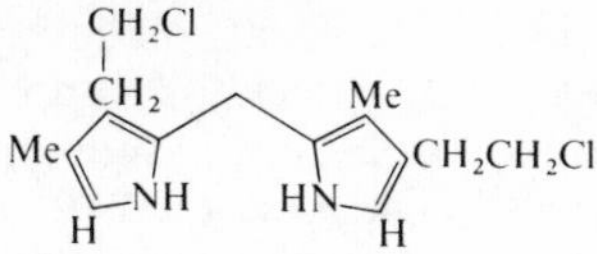

Structures (2), 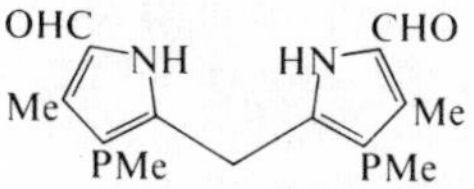 (3):

$V = -CH=CH_2$
$PR = CH_2CH_2CO_2R$

synthesis of tritiated porphyrins (of the type-III or -IX substituent orientation) for biosynthetic investigations.

A tricarboxylic porphyrin, found in the Harderian gland of animals such as the rat, has been named *harderoporphyrin* and characterised as (1b) through the synthesis[12] (by the *b*-oxobilane route) of the two possible isomers (1b) and (1c), and comparison of these with the natural product. The biological significance of harderoporphyrin is discussed in Section 11.11.1. Additional

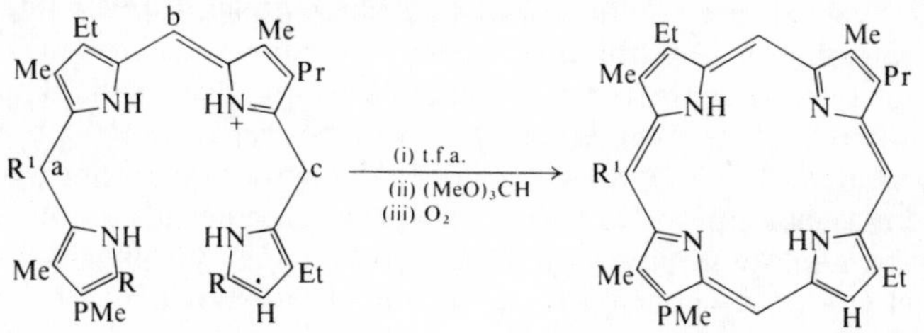

Structure (4) → (i) t.f.a. (ii) $(MeO)_3CH$ (iii) O_2 → Structure (5)

(4) $R = CO_2Bu^t$

(5)
(a) $R^1 = Me$
(b) $R^1 = Et$

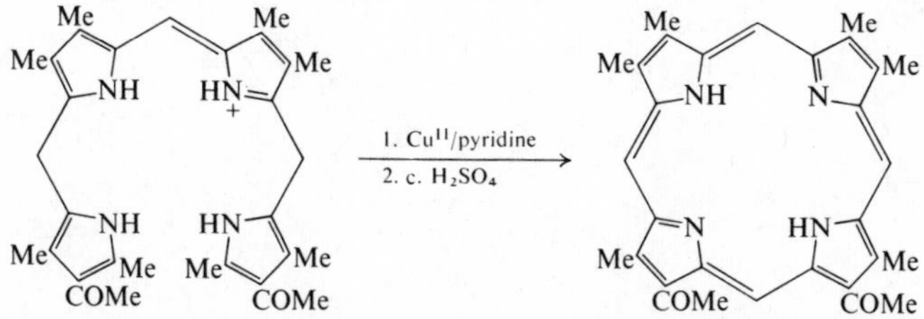

Structure (6) → 1. Cu^{II}/pyridine 2. c. H_2SO_4 → Structure (7)

applications of the *a*- and *b*-oxobilane routes have been published[13], along with a detailed study of the protection of pyrrole rings during porphyrin synthesis.

b-Bilenes (e.g. (4)) have been exploited increasingly as intermediates in the construction of porphyrins. Electron-withdrawing groups sited on the internal rings of t-butyl *b*-bilene-1',8'-dicarboxylates have been shown[14] to cause the production of porphyrin mixtures due to randomisation promoted by the increased electrophilicity of the *b*-position. However, this general route has been successfully applied[15] to the synthesis of two isomerically pure phylloporphyrins (5) thought[16] to be related to the *Chlorobium* chlorophylls (660)*. The *meso*-methylporphyrin (5a) (prepared from the bilene (4a) by treatment with trifluoroacetic acid and then trimethyl orthoformate and trichloroacetic acid) was identical with the phylloporphyrin obtained by degradation of chlorophyll (660) fraction 4, but the *meso*-ethyl homologue (5b) (from bilene (4b)) could not be identified with fraction 3, or any of the other (660) series fractions. Both of the structures allocated to *Chlorobium* chlorophyll (660) fractions which feature *meso*-ethyl groups[16] must now be regarded with suspicion; these structures were in any case difficult to justify on biogenetic grounds.

1′,8′-Dimethyl-*b*-bilenes (e.g. (6)) are known[17] to give porphyrins (via the copper chelates) when oxidatively cyclised with copper salts, but earlier work[18] has indicated that this route gives prohibitively low yields when the *β*-positions of the terminal rings are substituted with electron-withdrawing groups. However, Clezy and Liepa[19] have shown that 1′,8′-dimethyl-*b*-bilenes (6) can be oxidatively cyclised to the appropriate copper porphyrins in 20% yield through the agency of copper acetate in pyridine; the production of a labile complex with copper(II) and pyridine is suggested. Treatment of the copper chelate with concentrated sulphuric acid gives the free porphyrin (7); these relatively severe conditions are a disadvantage of this route. Similar yields of porphyrin were obtained[19] using the copper acetate–pyridine procedure when it was applied to 1′,8′-dimethyl-*a,c*-biladiene salts. The Russian Schools have used this procedure regularly in the preparation of porphyrins[20].

The synthesis of *α,β,γ,δ-meso*-tetraphenylporphyrin (8) from pyrrole and benzaldehyde has interested research workers for decades, both from the mechanistic standpoint and as a simple route to the porphyrin macrocycle for model studies. A porphyrinogen (9) has been shown[21] to be the initial

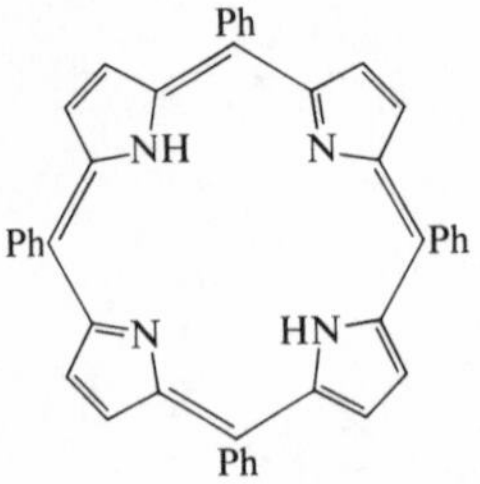

(8) 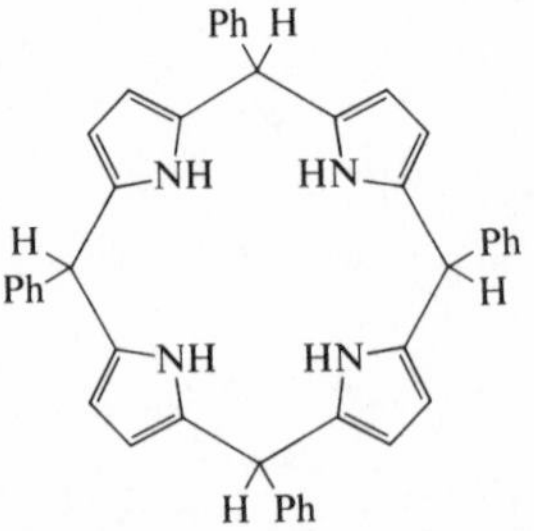   (9) 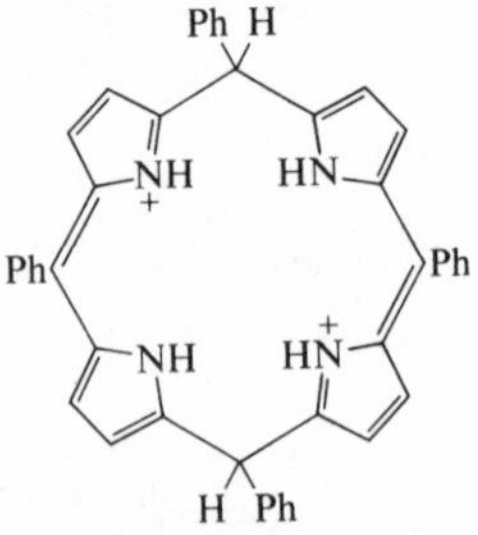   (10)

stage of the polymerisation; this undergoes acid catalysed autoxidation to (8) via a porphodimethene (10). Improvements in the yield of aetioporphyrin-I (11a), a more useful porphyrin for model studies, have been reported[22]. Treatment of the pyrromethene perbromide (12) (readily available in one step from kryptopyrrole (13)) in boiling formic acid gives the porphyrin (11a)

---

*The figures in parentheses refer to the absorption maximum (nm) of the pigments in their visible spectra measured in ether.

in yields between 50 and 60%. An analogue (11b) of aetioporphyrin-I has been prepared[23] from the corresponding pyrromethenes in a succinic acid melt; the yield was, however, minimal.

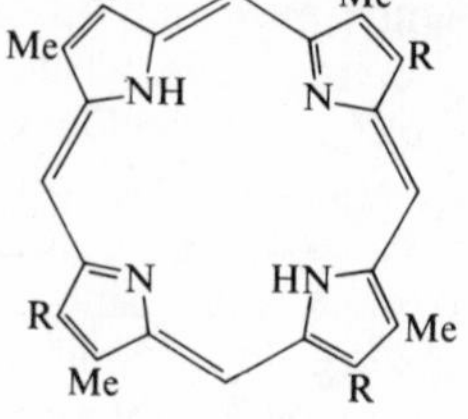

(11)

(a) R = Et
(b) R = $C_{18}H_{37}$

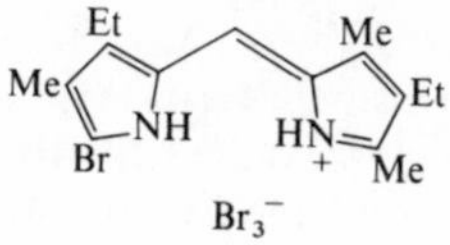

(12)

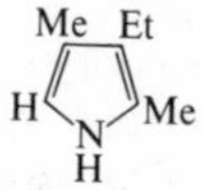

(13)

11.2.2 Reactions and properties

Early workers reported a by-product in the alkylation of aetioporphyrin-I (11a) with methyl iodide and formulated it as a methiodide of undefined structure. Two groups have recently re-examined this work, but in the octa-

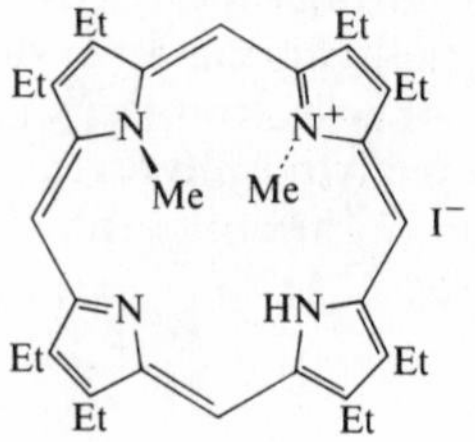

(14)

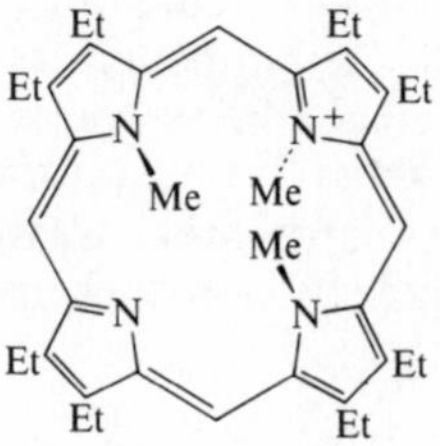

(15)

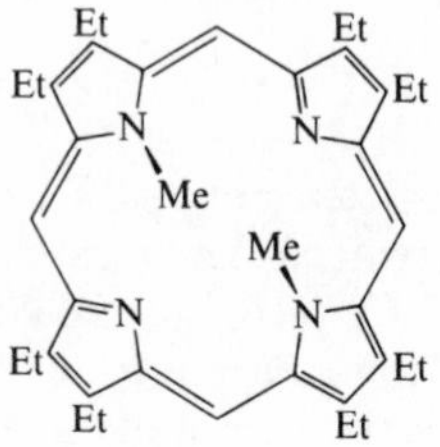

(16)

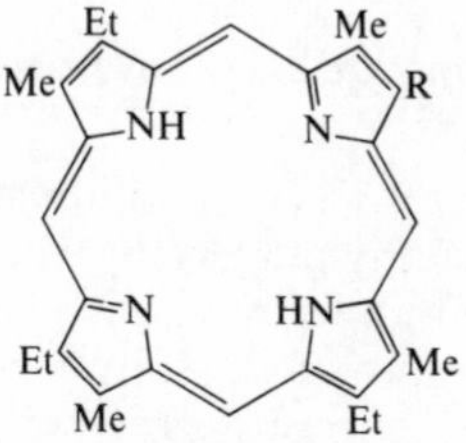

(17)

(a) R = COCl
(b) R = $COCHN_2$
(c) R = $CO(CH_2)_7Me$
(d) R = $CH(CH_2)_7Me$
$\quad\quad\;\; |$
$\quad\quad\; OH$

ethylporphyrin series, and the by-product has been identified[24, 25] as the N,N'-dimethylporphyrin salt (14). The *trans-* disposed methyl groups were sited on adjacent rather than opposite nitrogen atoms on the basis of n.m.r.

assignments of the *meso*-protons[24, 25] and on partial resolution of the D-camphor-10-sulphonate[25]. Such methylated porphyrins must be experiencing considerable strain and it is therefore all the more surprising that treatment of octaethylporphyrin with methyl fluorosulphonate at 100 °C gives[26] a *N,N',N''*-trimethylporphyrin (15) which must exhibit considerable deviations from planarity of the macrocyclic ring. The *N*-methyl groups were tentatively assigned to the *trans,trans*-configuration. On keeping, (15) decomposes to the *N,N'*-dimethyloctaethylporphyrin (16), which could be prepared directly (presumably via (15)) by carrying out the methylation with methyl fluorosulphonate in boiling chloroform. A novel zinc-ion-catalysed demethylation of *N*-methylaetioporphyrin has been reported[27].

Treatment of the porphyrin acid chloride (17a) with diazomethane leads[28] to the first reported porphyrinyl diazoketone (17b); this compound was surprisingly stable and survived chromatography. With an excess of triheptyl borane, the diazoketone (17b) was transformed into the porphyrin ketone (17c) which was reduced with sodium borohydride to the alcohol (17d), completing a series of reactions designed to be applicable to the construction of the 2-side chain of haem-*a*, the prosthetic group of cytochrome-*c* oxidase.

The reaction of benzoyl peroxide with octaethylporphyrin under nitrogen gives[29] side-chain and *meso*-benzoyloxy derivatives. Repetition of the reaction under oxygen almost eliminated the side-chain substituted products.

Electrochemical reduction of deuteroporphyrin-IX (1d) and *meso*-porphyrin-IX (1e) dimethyl esters in dimethylformamide shows[30] three discrete waves; the first two are reversible one-electron steps and the third an irreversible two-electron process. Tetraphenylporphyrin (8) shows[31] four well defined polarographic waves. The mass spectra of porphyrins and their derivatives have been further investigated[32].

11.3   METALLOPORPHYRINS

11.3.1   Preparation from porphyrins

A very wide range of metalloporphyrins has been prepared[33] from the treatment of octaethylporphyrin with metal acetylacetonates, usually in phenol as solvent. Complexes with Mg, Al, Sc, Ti, V, Cr, Mn, Fe, Co, Ni, Cu, Zn, In, and Zr ions were successfully prepared. Existing methods for complex formation have been improved experimentally by the use of divalent metal salts in refluxing dimethyl formamide[34]. The thallium(III) complexes of several porphyrins are obtained[35] by treatment of the free porphyrin with thallium trifluoroacetate; the metal atom has a nuclear spin of $\frac{1}{2}$ and the proton magnetic resonance characteristics of these metalloporphyrins have been reported[35, 36]. A number of studies of the mechanism of divalent metal ion incorporation into the porphyrin ligand have been carried out[2, 37] and also kinetic studies of the acid catalysed solvolyses of zinc porphyrins and *N*-methyl porphyrins[38]. The zinc-ion-catalysed demethylation of an *N*-methyl porphyrin has been mentioned earlier[27].

11.3.2   Reactions and properties

Many significant advances in the chemistry of metalloporphyrins have been reported in the past 2 years. Certainly the most remarkable of these has been reported by Perutz, who has published his solution to the problems of the fundamental working of haemoglobin, the protein responsible for the transport of oxygen in the red blood cells. When one of the four haem groups in the haemoglobin molecule takes up oxygen, the 'co-operative' effect results in the uptake of oxygen by the three remaining haem units becoming exponential; this is essential in order to avoid suffocation. The mechanism which 'triggers' the co-operative effect has mystified researchers for a long time and this has now been solved by Perutz[39] as a result of a lifelong interest in this molecule. The binding of oxygen to the high-spin iron(II) atom causes it to switch over to the low-spin electronic configuration, thereby causing a shrinkage of the iron atomic radius by some 13%. This allows the metal ion to slip into the plane of the porphyrin ring, pulling with it the coordinated imidazole ring of a globin histidine residue, which accordingly moves *c*. 1 Å nearer to the porphyrin ring plane:

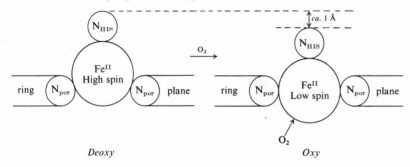

The relatively minor change from high- to low-spin iron(II) is therefore amplified by the haem unit to produce a large-scale rearrangement of the haemoglobin molecule involving also the breaking of salt bridges between the polypetide chains of neighbouring sub-units. Constraints within the molecule are relieved by this latter process, causing the high-spin iron atoms in the remaining haem units to begin to revert to low spin of their own accord, in the absence of oxygen but thereby increasing their affinity for oxygen. In simple terms, the salt bridges of the deoxyhaemoglobin molecule keep the iron atom spring-loaded in the high-spin form and this is released by the first molecule of oxygen to bind with a sub-unit and then further released with the additional molecules of oxygen. With this new understanding it is now possible to study abnormal forms of haemoglobin and even predict the clinical implications of particular genetic mutations and aberrations[40].

Several research groups are currently interested in the construction of synthetic biochemical models[41] in the hope that these will aid in the interpretation of many of the physical properties of haemoproteins. Thus the synthesis[42] and properties[43] of haem models from mesohaem sulphuric anhydrides, histidine and methionine have been reported, as well as similar compounds derived from haemin and histidine containing peptides; spectroscopic evidence showed[44] that the degree of interaction between imidazole and the iron atom was dependent upon the length of the peptide chain. A cyclophane porphyrin has also been synthesised recently, albeit in low yield[45].

A method for calculation of the conformations of metalloporphyrins in solution has been reported[46], and is based on the high-field shift of the coordinated extraplanar ligand (pyridine) in the n.m.r. spectrum. The linear relationship between the paramagnetic chemical shifts of the protons of the peripheral methyl groups in low-spin *bis*-pyridine iron(III) protoporphyrin-IX complexes and the basicity of the coordinated pyridine has been demonstrated[47]. Other n.m.r. spectra of iron[48] and related metal complexes[49] have been reported and analysed. Several infrared[50], Raman[51] and visible absorption spectral[52] studies of metalloporphyrins have appeared.

Tin(II) phthalocyanine has been shown[53] by x-ray crystallographic techniques to have the tin atom lying 1.1 Å out of the plane. In contrast, the tin atom of dichlorotin(IV)phthalocyanine is in the plane of the macrocyclic ring, although the ring is crumpled[54]. The chemistry of phthalocyanines will not be discussed in this review, but these results are relevant because they have been extended to an x-ray study[55] of dichlorotin(IV)octaethylporphyrin (18). Phthalocyanines are known to have a smaller 'hole' than porphyrins, and (18) is shown[55] to be planar with the tin(IV) ion in the porphyrin plane. This finding agrees with the predictions (concerning bond strain in the porphyrin ring) of Collins and Hoard[56] who suggest that a metal ion will lie in the plane of the porphyrin ring if the metal–ring nitrogen distance is less than 2.11 Å. Highly unstable tin(II) porphyrins have been isolated for

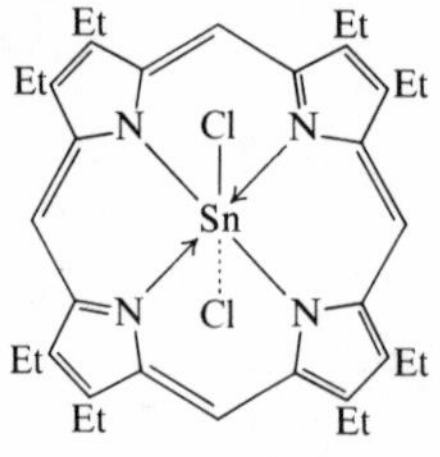

(18)

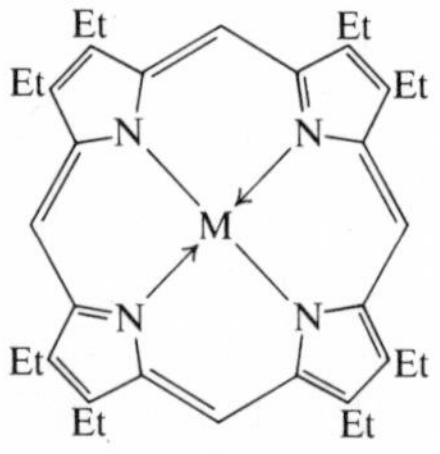

(19)

(a) M = Sn
(b) M = Pb

(20) M = Sc.OCOMe
(21) M = Tl(H$_2$O)OH

the first time[57]; n.m.r. and visible absorption spectroscopy of tin(II) octaethylporphyrin (19a) indicate that the tin ion may be far out of the plane. This could in any case be predicted from the x-ray data for tin(IV) octaethylporphyrin (18)[55] and the much greater ionic radius of tin(II) compared with tin(IV). One interesting property of (19a) and the analogous lead chelate (19b) can be found in the n.m.r. spectra[57] which indicate nonequivalence of the methylene protons, presumably due to the metal being out of the plane of the porphyrin ring. A similar phenomenon has been reported in scandium(III) octaethylporphyrin acetate (20)[33] and in aquothallium(III)octaethylporphyrin hydroxide (21)[35, 36]. In the latter case a full n.m.r. analysis was performed, indicating an ABXR$_3$ $(CH_2 \cdot Tl \cdot CH_3)$ system[36]. Dihydroxytin(IV)octaethylporphyrin (22) has been converted to the chlorin

(23) via a chloroform adduct and treatment of this with sodium borohydride and hydrochloric acid[58]. Tin(IV) porphyrins can also be photo-reduced with

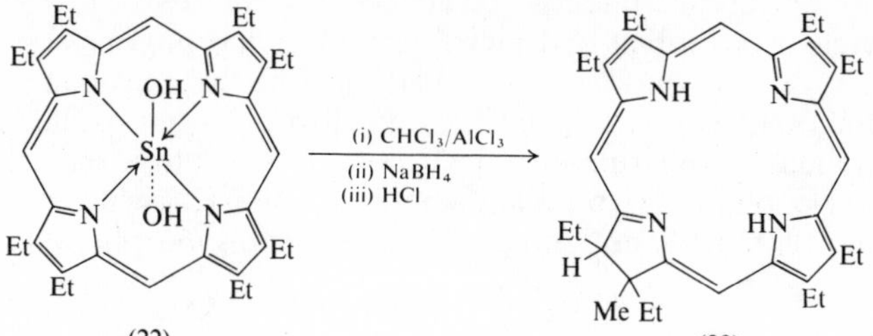

(22)   (i) CHCl₃/AlCl₃ (ii) NaBH₄ (iii) HCl   (23)

$SnCl_2 \cdot H_2O$ in pyridine to give[57] firstly the tin(IV) chlorin and subsequently the *vic*-tetrahydroporphyrin or bacteriochlorin.

A monomer–dimer (oxo-bridged) equilibrium has been characterised[59] in the water soluble $\alpha,\beta,\gamma,\delta$-*meso*-tetra(*p*-sulphophenyl)porphyrin ferric complex (FeTPPS):

$$2 \overset{\displaystyle OH_2}{\underset{\displaystyle |}{FeTPPS}} \underset{pH<9}{\overset{pH>9}{\rightleftharpoons}} O(FeTPPS)_2 + 2H^+ + H_2O$$

This equilibrium has also been assigned to the non-sulphonated species. Ligand binding[60] and lability has featured in many interesting publications in the metalloporphyrin area; earlier work[61] had revealed that the porphyrin ligand labilises cobalt(III) in its substitution reactions and this work has been extended to the examination of a chromium(III) system. It is well known that Cr^{III} complexes are very inert towards substitution, and a labilising effect on the part of the porphyrin ligand (*c.* $10^3–10^4$ times faster than normal chromium complexes) was noted[62]. A case of metal 'shuttling' between imidazole nitrogens in a ruthenium porphyrin imidazole has been reported[63]:

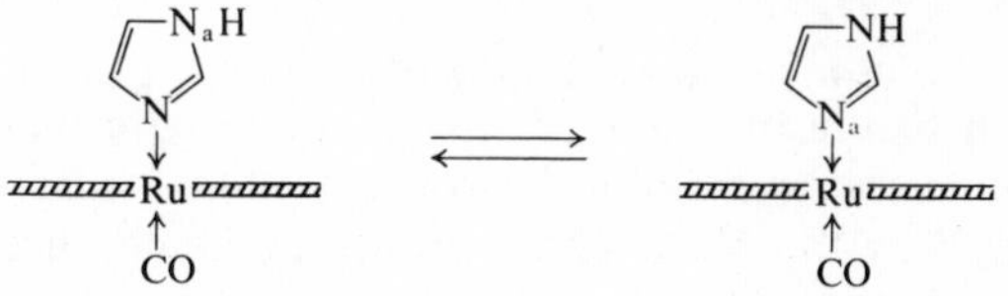

An intra- rather than inter-molecular process was favoured (accounting for the term 'shuttling') on the basis of n.m.r. assignments. While accepting the lability of the imidazole ligand, the conclusions regarding 'shuttling' have been criticised, both on the basis of the n.m.r. assignments[64] and analogy with a similar ruthenium–porphyrin system[65]. Two new unusual porphyrin chelates, illustrating additional categories of porphyrin geometry[1], are an example of a porphyrin bridging two rhenium atoms and a rhenium porphyrin having three axial ligands[66].

Electrophilic substitution of porphyrins and metalloporphyrins has been further examined[67–69] and some deuteration anomalies in the literature have been pointed out[69]. An interesting conclusion[69] is that the rate of electrophilic attack on a metalloporphyrin is a function of the axial ligands as well as the metal and its oxidation state.

The full experimental papers concerning the π-cation radicals[70-72] and di-cations[71,72] of metalloporphyrins have now appeared; magnesium, zinc and copper porphyrins undergo reversible one- and two-electron abstractions whereas iron, cobalt and nickel tetraphenylporphyrins showed three successive reversible one-electron oxidations, the first involving the metal ion and the other two the ligand[72]. Controlled potential oxidation of zinc *meso*-tetraphenylporphyrin generates[73] initially the cation radical which is stable in nucleophilic solvents and then the di-cation (24) which is a powerful electrophile; with methanol, this gave[73] the metalloisoporphyrin (25). Use of

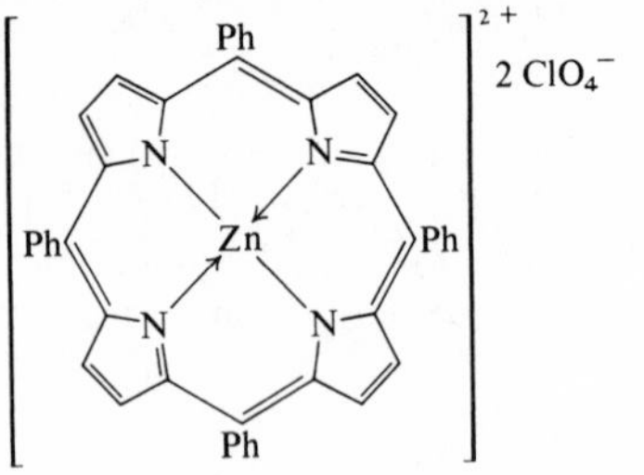

(24) 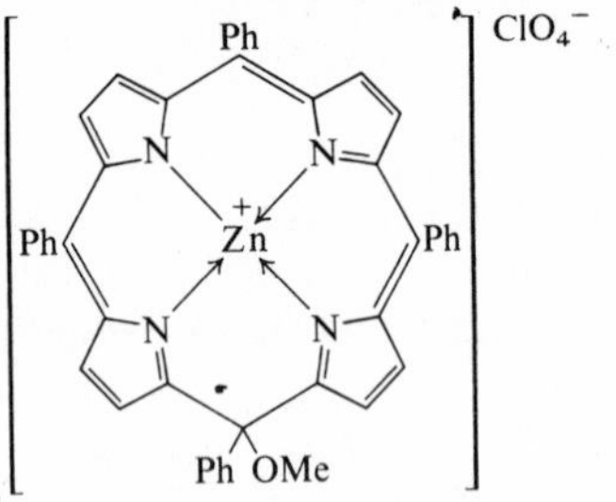   (25)

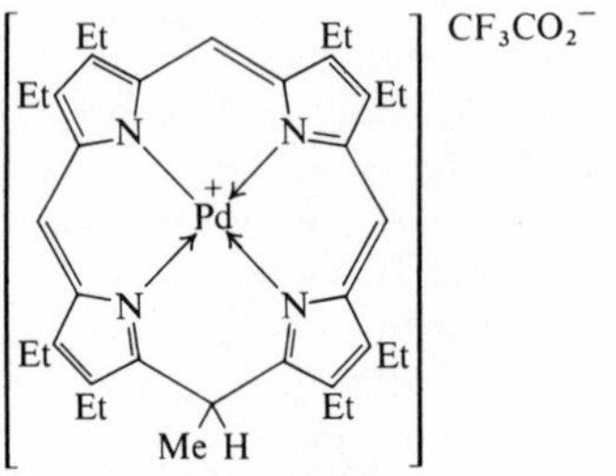

(26) 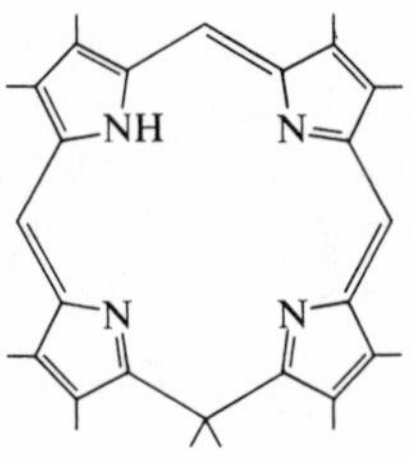   (27)

water in place of methanol gave the corresponding hydroxyisoporphyrin, and this reaction has been extended to the copper and magnesium tetraphenylporphyrin cases. A metalloisoporphyrin (26) has also been observed spectroscopically[67] by the Nottingham group. The theoretically interesting

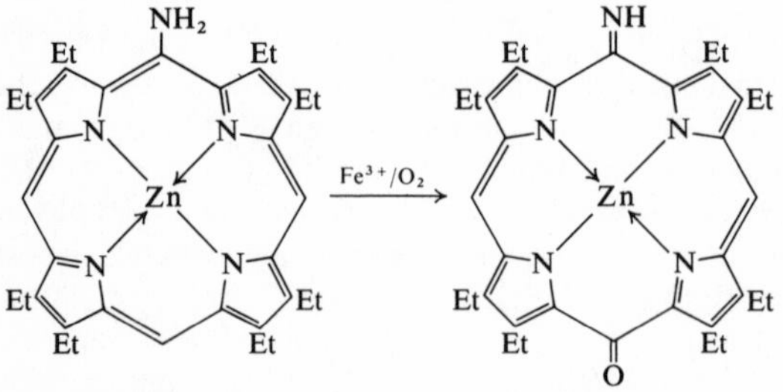

(28)   (29) 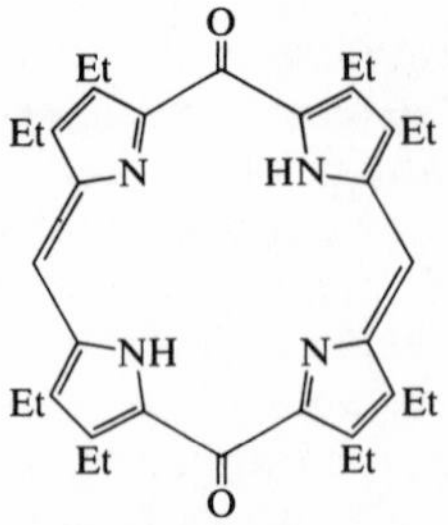   (30)

isoporphyrin free base (27) has not yet been observed. The oxidation of zinc octaethylporphyrin to its cation radical has been extended to the α-amino compound (28) which gives[74] (29) when oxidised in the presence of air. This material could be hydrolysed readily to the $\alpha\gamma$-dioxo-species which was then demetallated to (30). An alternative route to (30) has been reported[35]; treatment of octaethylporphyrin with thallium trifluoroacetate in the presence of trifluoroacetic acid gives (30) via its thallium(III)chelate. Electron spin resonance studies have demonstrated[75] hyperfine interaction of halide ions with metalloporphyrin π-cation radicals. The vapour absorption and high-temperature solution spectra of a number of phthalocyanines, octa-alkylporphyrins and tetraphenylporphyrins have been reported[76].

There has been little argument about the gross structure of haem-*a*, the prosthetic group of cytochrome oxidase. However, there has been much more discussion concerning the precise nature of the 2-side-chain, which on the basis of n.m.r. spectra and other properties has now been given[77] as a *trans,trans*-farnesylethyl group with a labile substituent at C-1' which is lost during isolation. The full structure[77] is therefore that shown in (31).

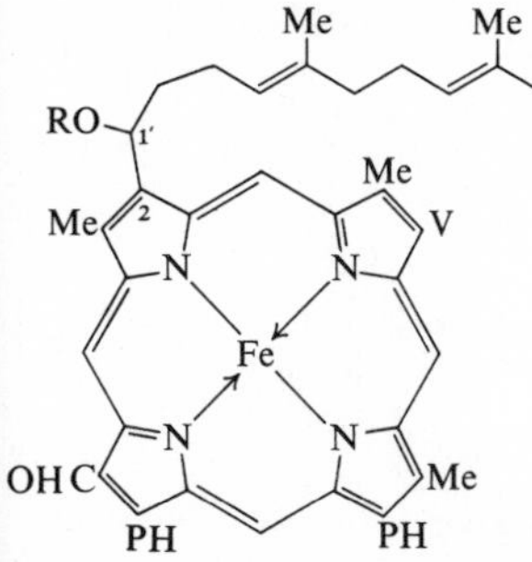

(31) 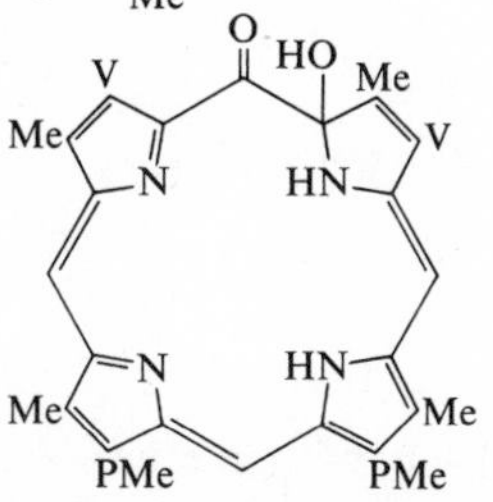   (32) 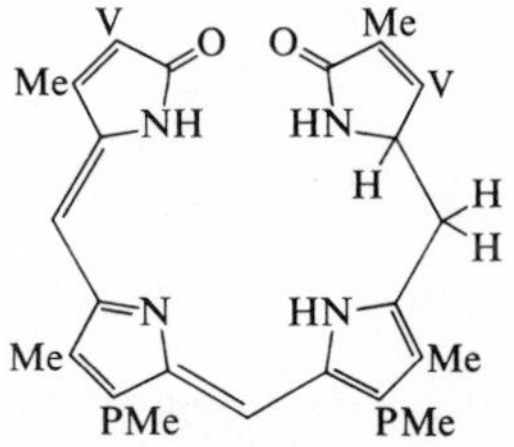   (33)

Photo-oxidation of magnesium protoporphyrin-IX was reported[78] to give an α-hydroxyoxochlorin (32) which decomposed readily to a biliviolin compound (33) with loss of one carbon atom; this was one of very few reported routes to bile pigments from the porphyrin macrocyclic molecule[6]. Different structures for the primary compound (34) and its decomposition product (35) (in the octaethylporphyrin series) are now proposed[79] in the light of more extensive spectral data.

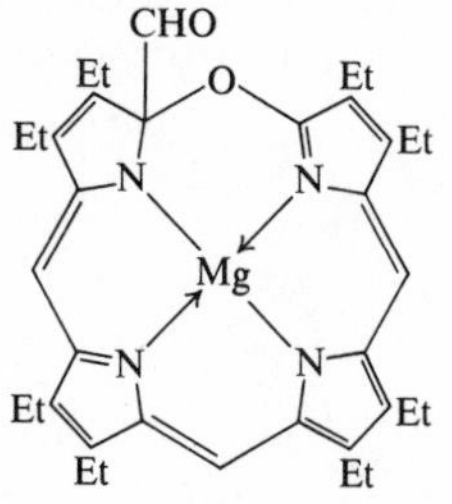

(34) 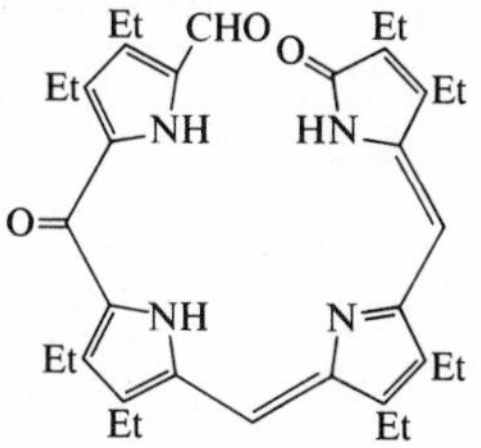   (35)

Formylation of iron(III) porphyrins and chlorins of natural origin via the Vilsmeier procedure has shown[80] the relative reactivities of substitution position to be vinyl > *meso*-position > peripheral position.

Reductive methylation of zinc (octaethylporphyrin gives[81] $\alpha\gamma$-dimethyl-$\alpha\gamma$ dihydro zinc octaethylporphyrin (36) which can be demetallated readily to (37), or else dehydrogenated to the $\alpha\gamma$-dimethylporphyrin (38). With triethylaluminium, octaethylporphyrin gives[82] the aluminium porphyrin hydroxide (39); reduction of octaethylporphyrin with di(isobutyl)aluminium

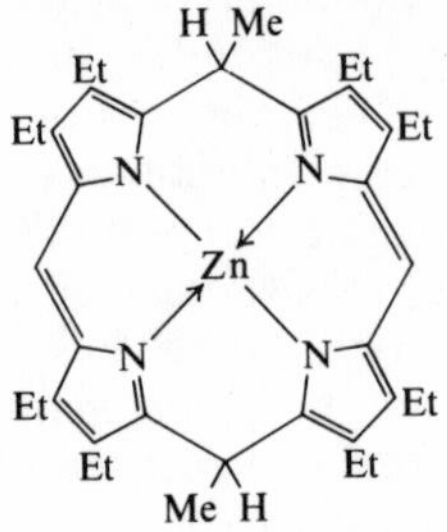

(36) 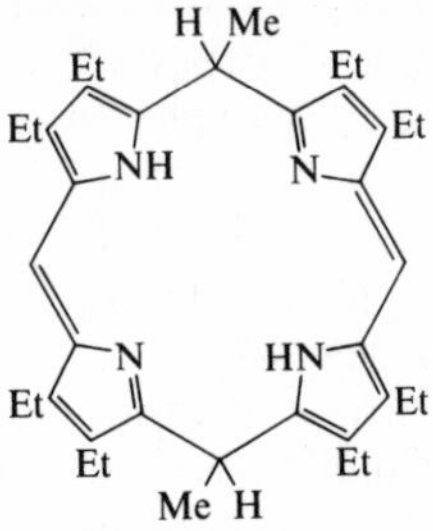   (37)

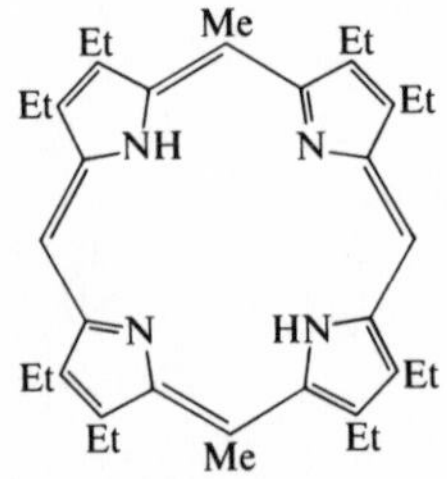

(38) 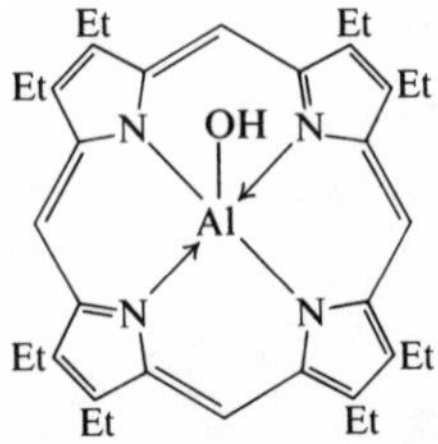   (39)

hydride, followed by work-up with methanol also gives (39) through the $\alpha\gamma$-dihydro metalloporphyrin analogous to (36).

11.4   CHLORINS

11.4.1   Chlorophylls

Chlorophyll-c is found in certain marine algae, brown seaweeds and fresh water diatoms, but not in terrestrial life forms. It has been characterised[83] as the magnesium *porphyrin* chelate (40), bearing an acrylic side-chain which is one of the several notable differences between this pigment and the normal plant chlorophylls. The two components of the mixture can be separated[84]

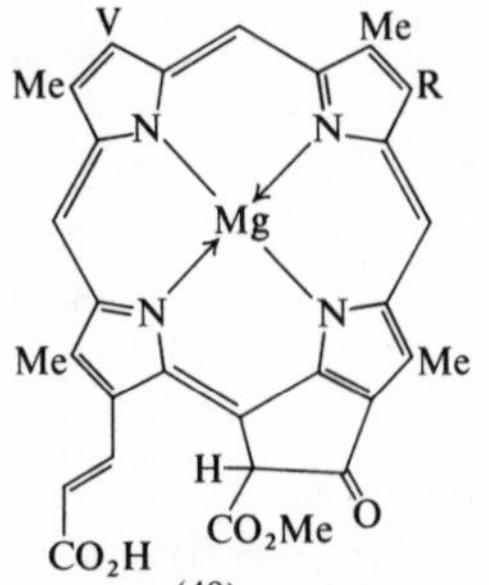

(40)
(a) R = Et
(b) R = V

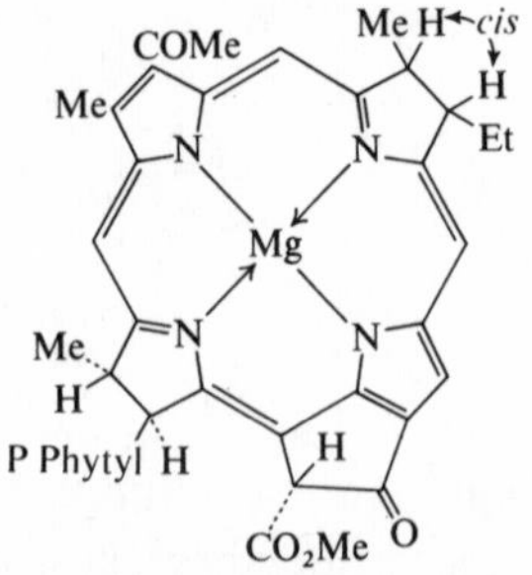

(41)

into chlorophylls-c_1 (40a) and -c_2 (40b) by chromatography on polyethylene; the ratio $c_1 : c_2$ was c. 0.6. Other workers have found[85] the chlorophylls-c from related organisms to be identical with those described above. As an extension of the various absolute stereochemistry assignments of chlorophyll derivatives which have been reported earlier[6], the constitution and absolute stereochemistry of bacteriochlorophyll-b has been given[86] as (41). Inhoffen

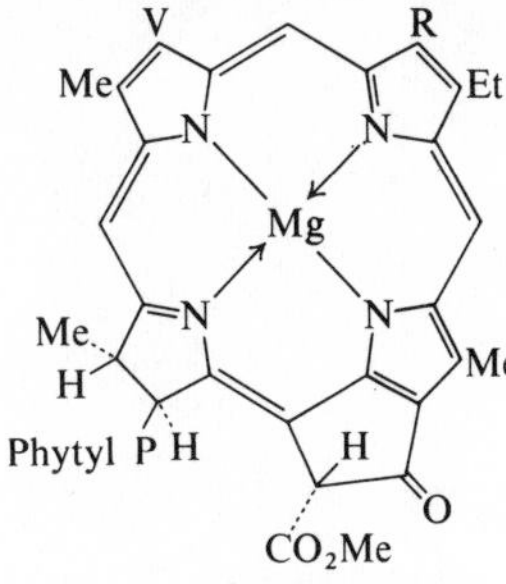

(42)

(a) R = Me
(b) R = CHO

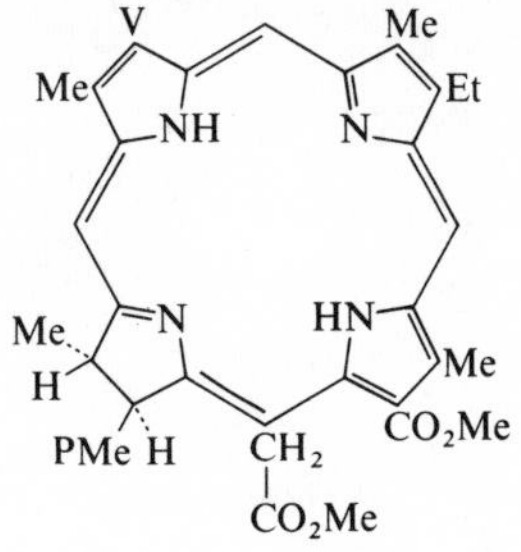

(43)

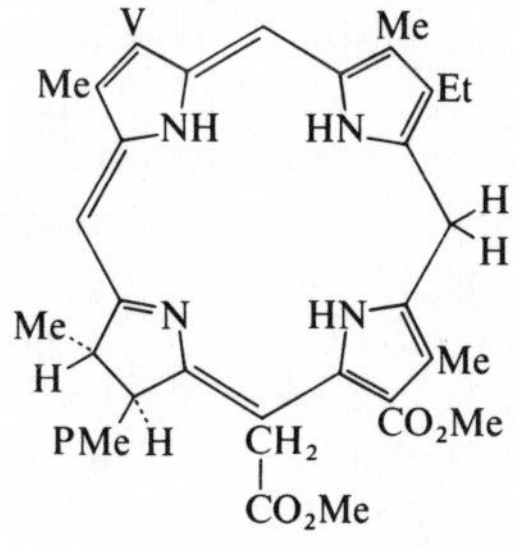

(44)

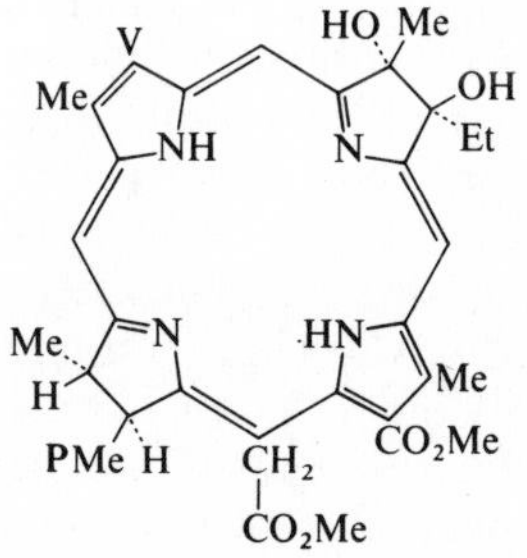

(45)

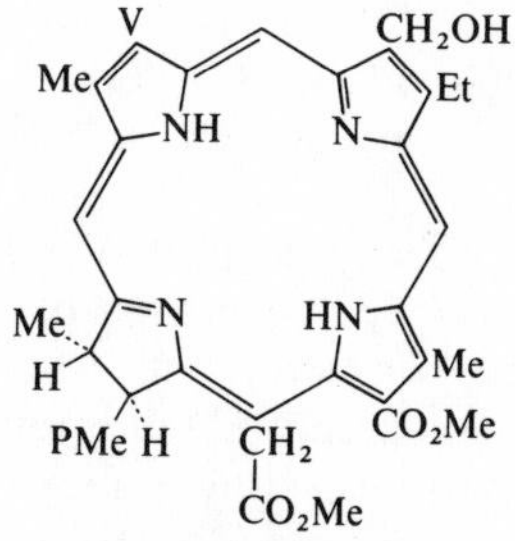

(46)

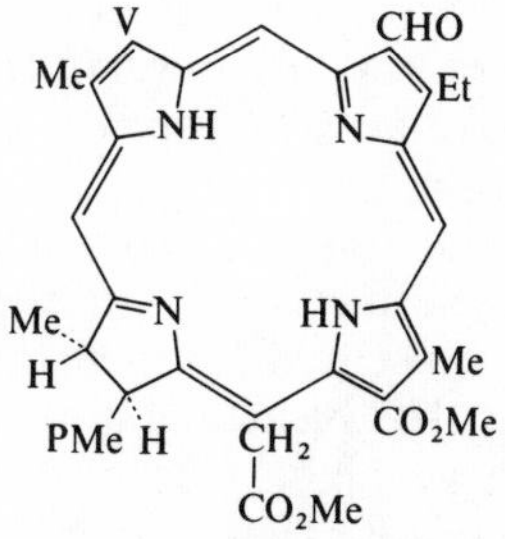

(47)

and co-workers[87] have described the formal total synthesis of chlorophyll-b (42b) based on the famous synthesis of chlorin e_6 trimethyl ester (43) (and formally of chlorophyll-a) by Woodward. Thus electrolysis of chlorin e_6 (43) gave the chlorin-β-phlorin (44) which was photo-oxidised in dioxane–water to the *trans*-dihydroxybacteriochlorin (45). With HCl–water–dioxane, (45) gave (46) which furnished rhodin g_7 trimethyl ester (47) when treated with acetic anhydride in dimethyl sulphoxide. Fischer and Willstätter had earlier transformed rhodin g_7 (47) into chlorophyll-*b* (42b). The photochemical reduction of the zinc phaeoporphyrin (48) into the *cis*-phaeophorbides (49) has been reported[88]; separation of the diastereoisomers of (49) was achieved by

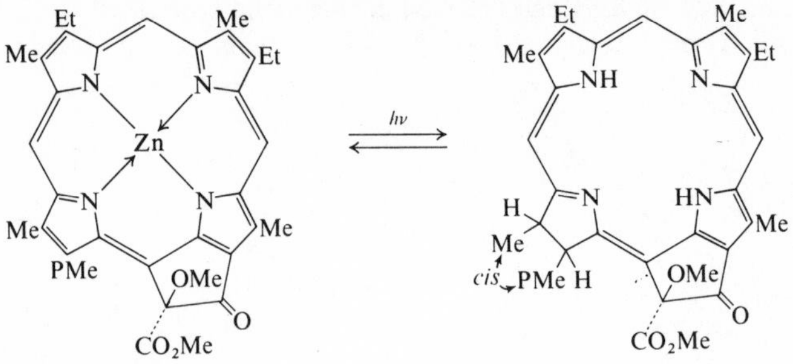

thin-layer chromatography and the relative configurations were determined by n.m.r. spectroscopy.

Oxidised P700, the first photochemical product of Photosystem 1 in green plants, has been shown to contain a cation radical by correlation[89] of optical and e.s.r. studies of the electrochemical oxidation of chlorophyll-*a* to its cation radical with published studies on photosynthesis. This radical is the likely source of the rapidly decaying narrow e.s.r. signal of photosynthesis. Comparisons have been made[90] between Signal 1 and free-radical signals generated in various chlorophyll species *in vitro*; the line-width of Signal 1 has been found to be consistent with unpaired spin delocalisation over an entity $(Chl–H_2O–Chl)^{\overset{+}{\cdot}}$ containing two chlorophyll molecules. The optical spectra of Compounds 1 of horseradish peroxidase and catalase have been compared[91] with π-cation radicals of cobalt octaethylporphyrin, resulting in the proposal that these Compounds 1 contain a π-cation radical of the haem group; explanations have been advanced to account for the missing electron spin resonance signal of Compound 1 of horseradish peroxidase, which is formulated as a quadrivalent iron atom (cf. reference 151) complexed with a porphyrin radical. The reversible one-electron oxidations of certain metallo-chlorins have been examined[92] and the chlorin and porphyrin ligands have been contrasted.

Photoreduction[93] and in particular the photodecomposition of chlorophyll *in vitro* has been studied; chlorophyll solutions are irreversibly bleached by light in the presence of oxygen, the rates for chlorophylls-*a* and -*b* being of the same order. The bleaching process can be interrupted and red intermediates are isolable which have been shown by n.m.r. spectroscopy to lack vinyl and methine protons. These red materials can be oxidised with chromic acid to give methyl ethyl maleimide which is also present in solutions of the fully bleached end product[94].

11.4.2   Other chlorins

Photoreduction of porphyrin tin(IV) complexes with $SnCl_2 \cdot H_2O$ in pyridine leads to tin(IV) chlorins initially and then by further reduction gives tetrahydroporphyrins or bacteriochlorins[57]. Treatment of *trans*-octaethylchlorin (50) with thallium trifluoroacetate leads[95] to the dihydrobiliverdin (51) in 25% yield. The structure with the reduced terminal, rather than internal

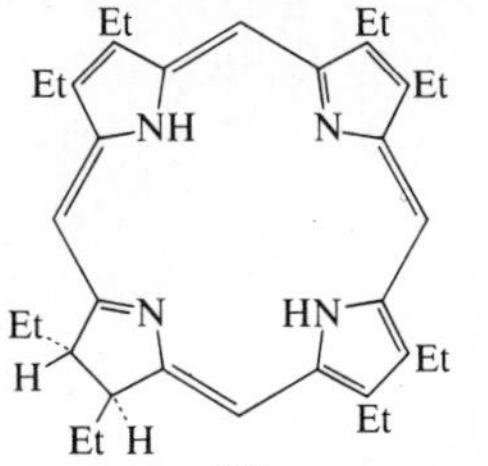

(50) 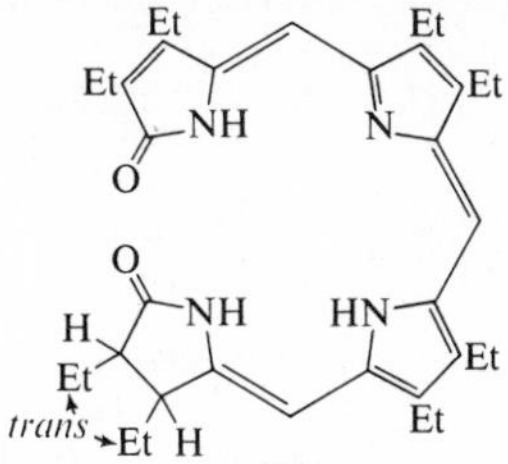   (51)

ring, was confirmed by repetition of the reaction with γ,δ-dideuterio-*trans*-octaethylchlorin and examination of the n.m.r. spectrum of the product, which lacked only one methine proton. The boiling of *meso*-tetraphenyl-chlorin in dimethyl sulphoxide has been shown to be an efficient method for the conversion of this substance into the corresponding porphyrin[96].

11.5 CORROLES

11.5.1 Synthesis

The most recently reported[97] route to the corrole macrocycle is the theoretic-ally interesting extrusion of sulphur from a *meso*-thiaphlorin (54) to give the corrole (55) in yields as high as 60%. This high yield reflects the fact that the

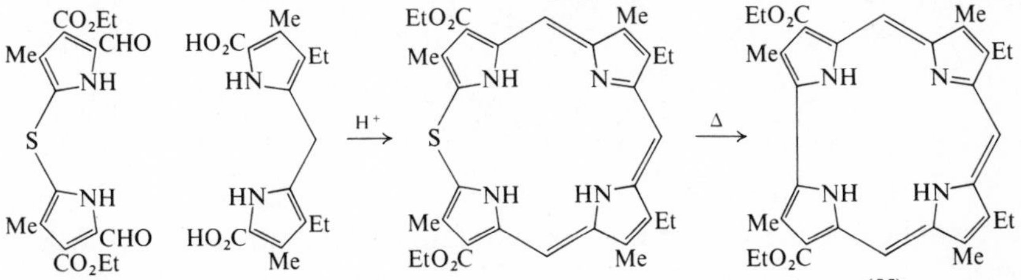

(52)   (53)   (54)   (55)

thermal extrusion is orbital symmetry allowed. The thiaphlorin (54) was obtained by acid-catalysed condensation of the sulphide (52) with the pyrro-methane-5,5'-dicarboxylic acid (53).

11.5.2 Reactions and properties

Alkylation of nickel corrole ambident anions with alkyl halides has been shown[98] to occur at the ring A nitrogen atom; methylation had earlier been thought to take place at the metal atom. Moderate heating of the N-alkylated products (e.g. 56a) causes alkyl group migration from nitrogen to C-3 (giving (57a)) presumably via C-4 through a double [1,5] sigmatropic shift. Bulky alkyl halides react with nickel corrole anions directly at C-3. Alkylation of copper corrole ambident anions leads[99] to an *N*-alkylated product (e.g. 56b) which cannot be rearranged thermally, but rather suffers loss of the *N*-methyl function. In contrast, palladium corrole anions give a mixture of *N*- (56c) and C-3- (57c) methylated compounds with methyl iodide; the ratio of the products can be changed by variation of the solvent. Compound (56c) can be

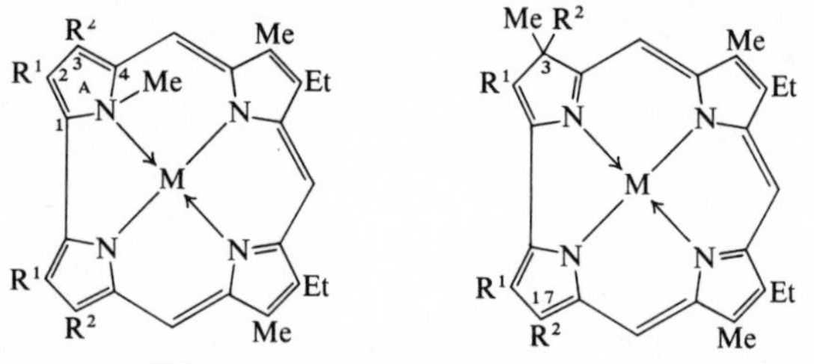

(56)   (57)

(a) M = Ni; R¹ = R² = Me
(b) M = Cu; R¹ = R² = Me
(c) M = Pd; R¹ = Et; R² = Me

thermally isomerised to (57c). Protonation and alkylation of 3,3-disubstituted metallocorroles occurs[99] at C-17. Alkylation of metal-free corroles with methyl iodide and potassium carbonate gives[25] a mixture of two isomeric *N*-methyl corroles (58) and (59) which showed no tendency towards thermal rearrangement[99]. Nickel complexes of these *N*-methyl corroles were in-

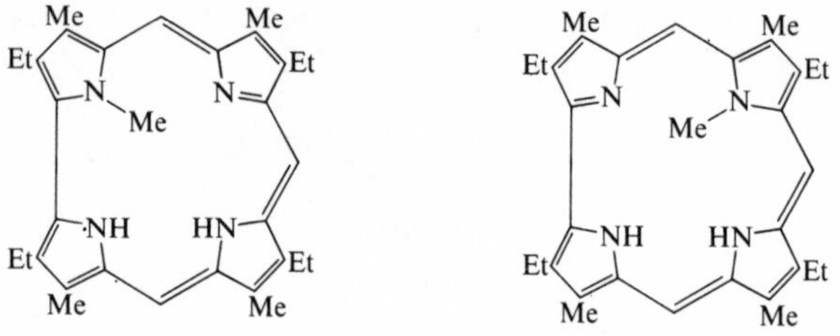

(58)   (59)

accessible, but the copper and palladium chelates were obtained in high yield[25]. A study of the bond lengths and the location of the inner ring protons of the corrole molecule has been reported[100].

11.6 TETRADEHYDROCORRINS

11.6.1 Synthesis, reactions and properties

Full details of the syntheses of tetradehydrocorrin complexes from *a,c*-biladiene salts have appeared[101, 102]. Cyclisation of 1′-methyl-*a,c*-biladienes (60a) in basic solution in the presence of nickel ions gives 1-methyltetrade-hydrocorrin nickel complexes (61); this reaction has been discussed[103] as an electrocyclic process for which conrotation is symmetry allowed. Under the same conditions, 1′,8′-dimethyl-*a,c*-biladienes (60b) or *b*-bilenes (62) can be cyclised[101] to the nickel tetradehydrocorrins (63a, b) which can be hydro-

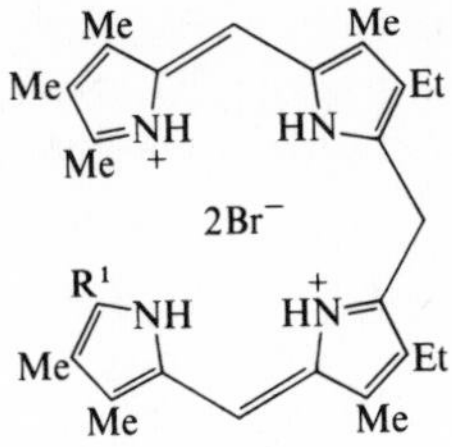

(60)

(a) $R^1 = H$
(b) $R^1 = Me$

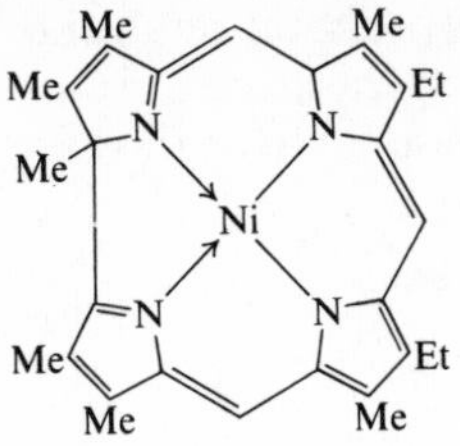

(61)

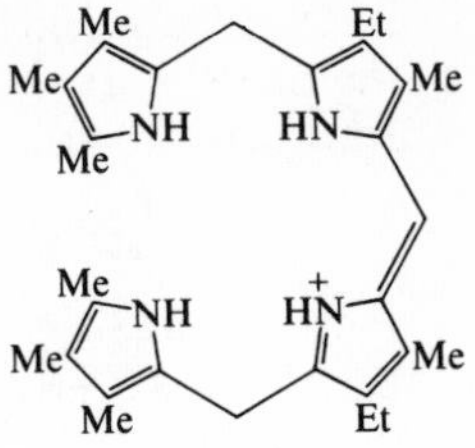

(62)

genated (Raney nickel, 100–160 °C, 100 atm) to the bisdehydrocorrin salts (64). In the absence of 2- and 18-substituents, 1,19-dimethyltetradehydrocorrin salts yield nickel 1,19-dimethylcorrins (65). Dehydrogenation of the corrin

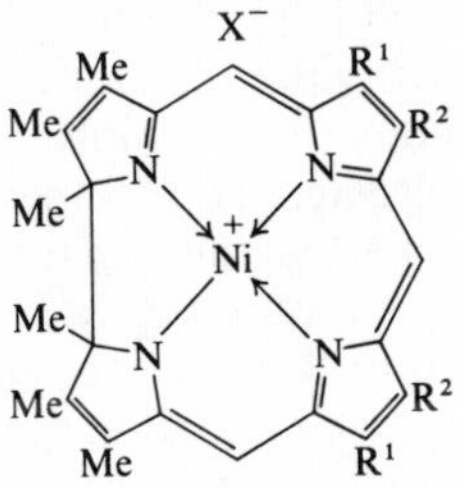

(63)

(a) $R^1 = Me; R^2 = Et; X = ClO_4$
(b) $R^1 = Et; R^2 = Me; X = Cl$
(c) $R^1 = Me; R^2 = Et; X = NO_3$
(d) $R^1 = R^2 = Me; X = ClO_4$
(e) $R^1 = R^2 = Me; X = Br$
(f) $R^1 = R^2 = Me; X = I$

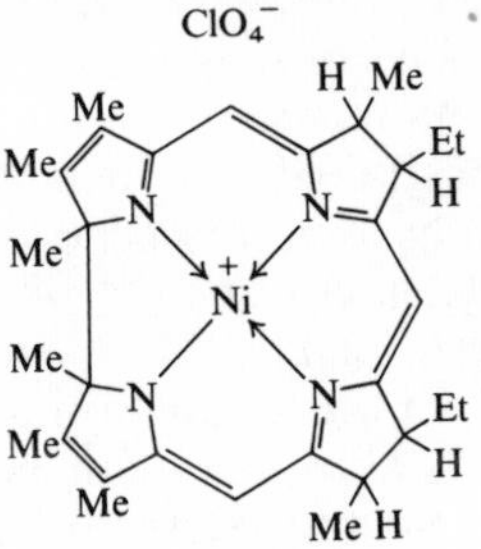

(64)

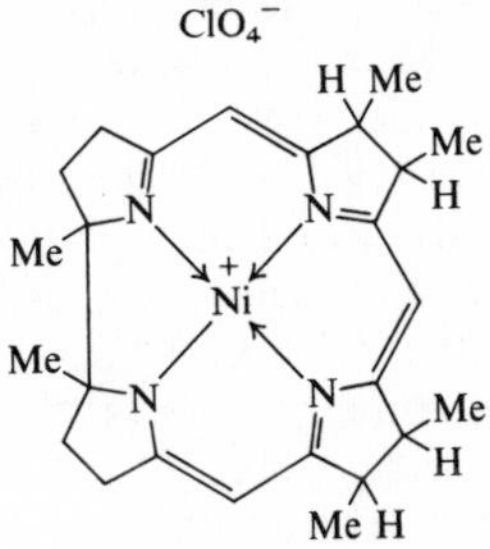

(65)

salt (66) furnishes the bisdehydrocorrin salt (67) and a neutral oxo-macrocycle (68), several properties of which apparently resemble those of the oxophlorins (see Section 11.8). The nickel tetradehydrocorrin (63a) has been partially

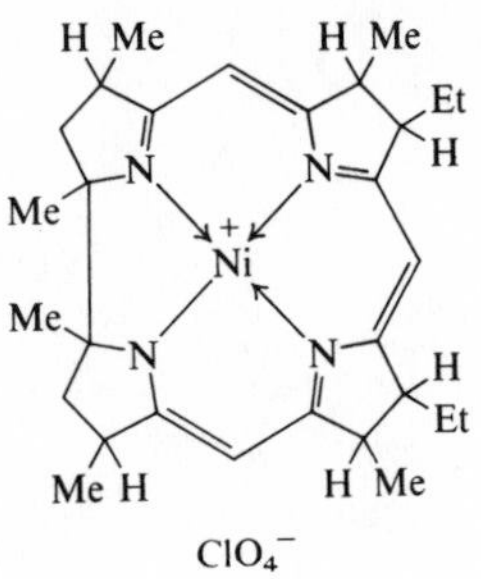

(66)

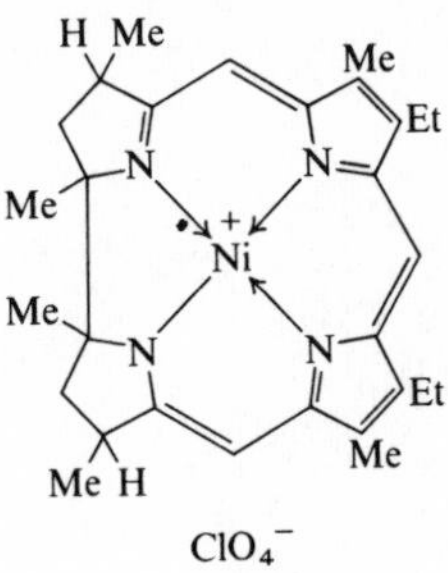

(67)

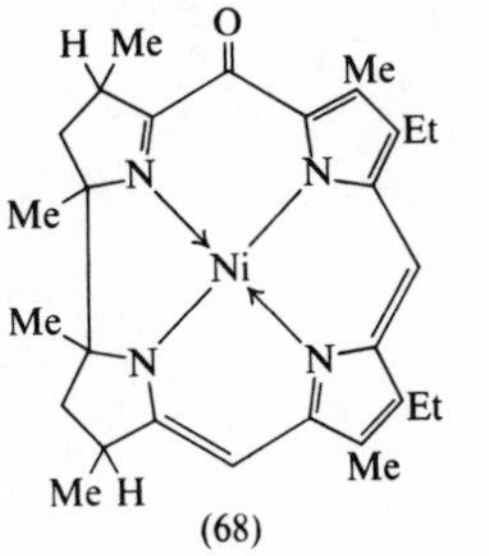

(68)

resolved[103] as its (+)-D-camphorsulphonate, providing evidence for the *trans*- arrangement of the 1- and 19-methyl groups. Such partially resolved salts (63a) are shown[103] to undergo thermal racemisation with a rate dependent upon the nature of the anion. 1-Methyltetradehydrocorrin nickel salts (e.g. 61) rearrange on heating, probably by an allowed sigmatropic mechanism, to the nickel corroles (e.g. 69) featuring a 3,3-*gem*-dialkyl grouping[104]; these substances were originally thought to be the 2,2-*gem*-disubstituted analogues.

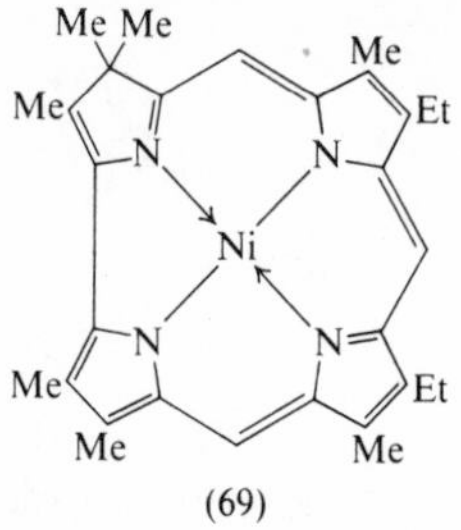

(69) 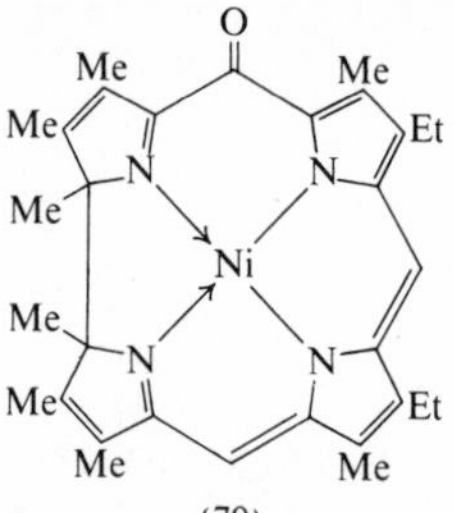   (70)

Substitution reactions of (63c) show[105] the salt to have a strong tendency for transformation into derivatives of the 5-oxo-compound (70). Earlier claims that electrophilic substitution reactions of 1-methyltetradehydrocorrin salts gave only the 5-substituted products (bromination, chlorination, nitration and methylation) have been amended and extended[105], especially in view of the work of Melenteva *et al.*[106] who reported the isolation of 5-nitro, 5,10- and 5,15-dinitro and 5,10,15-trinitro compounds as well as a 3-nitromethyl derivative from the fuming nitric acid nitration of the nickel tetradehydrocorrin salt (63d). Thus, (63c) with methanolic potassium cyanide gave[105] three characterised neutral compounds, bromination gave a 5,15-dibromo derivative and the 5,15-dimethyl derivative was obtained with lithium copper methyl.

Tetradehydrocorrin salts have also been used by the Braunschweig group in their approaches to corrins; treatment of nickel tetradehydrocorrin salts (63d, e) with osmium tetroxide gives a mixture of diols and tetra-ols which can be further hydroxylated to hexa-ols and octa-ols (e.g. (71)); the latter has the chromophore of the corrins[107]. Some of the tetra-ols were rearranged to

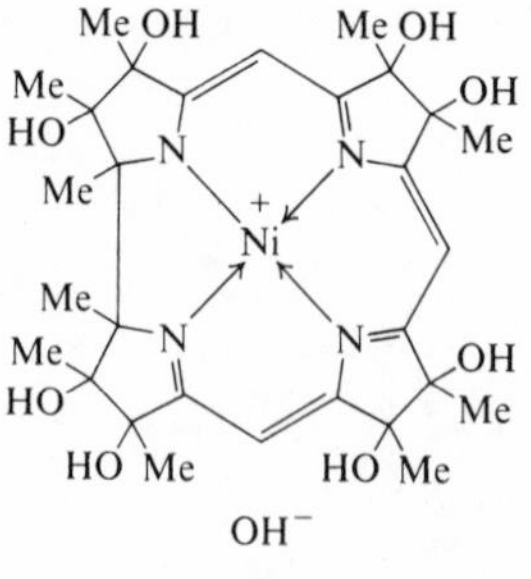

(71) 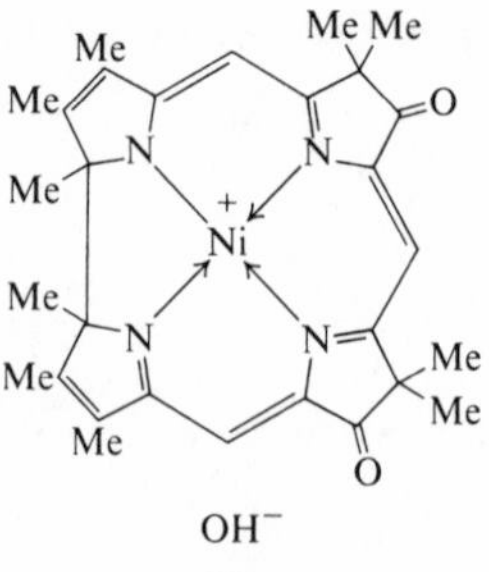   (72)

the corresponding 'geminiketones' (e.g. (72)). The hydroxylations are influenced by substituents at the 2- and 18-positions, which was also a limiting factor in Johnson's reduction of tetradehydrocorrins to corrins[101]. In a different approach[108] compounds such as (73) have been self-condensed to give the corresponding 2,7-diketo-*a,c*-biladienes (74):

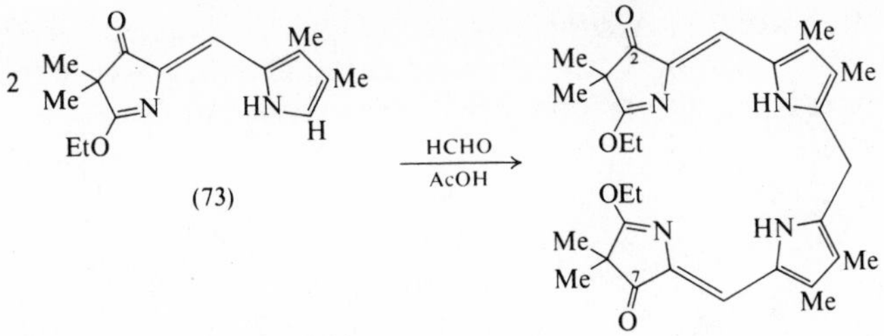

(73)

HCHO
AcOH

(74)

Cyclisation of these substances (74) in the presence of metal ions should give macrocycles without the steric limitations of the earlier route. Reductive methylation of the nickel tetradehydrocorrin iodide (63f) gives[109] (75) through opening of the macrocyclic ring; this material can be oxidised to (76) which in turn gives (77) on reductive methylation.

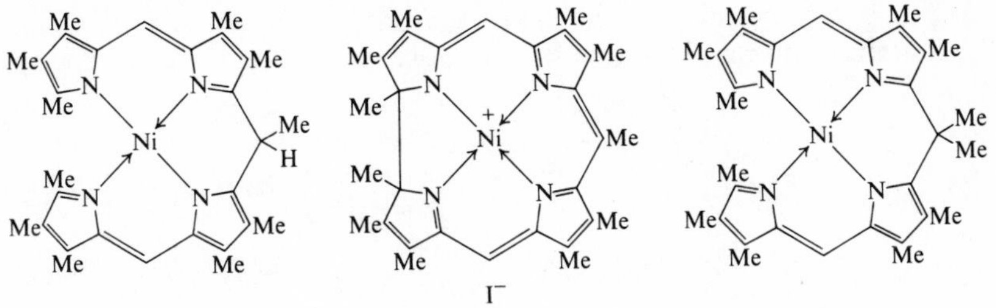

(75)   (76)   (77)

11.7 CORRINS

11.7.1 Synthesis

Routes to corrins from tetradehydrocorrin salts have been reported in Section 11.6.1. The synthesis of nickel(II) and cobalt(II) complexes (79) of 1,19-dimethylcorrin by room temperature hydrogenation of the corresponding tetradehydrocorrins (78) over Raney nickel at 25 atm pressure has been reported[110]. These conditions are markedly milder than those used in the

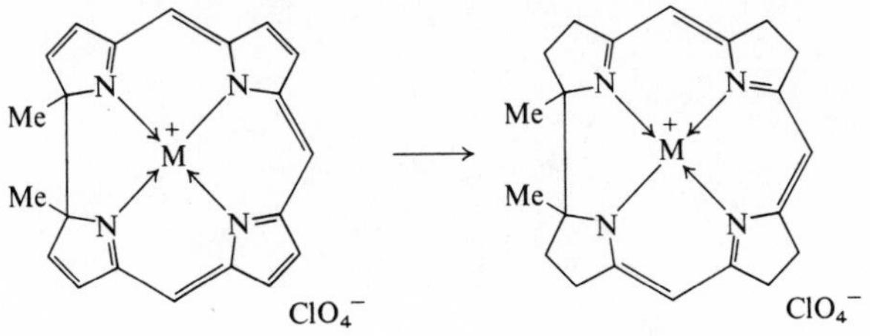

(78)   (79)

M = Co, Ni

synthesis of corrin metal chelates bearing peripheral alkyl substituents[101]. All other synthetic avenues to the corrin macrocycle have used nonpyrrolic intermediates. A simple synthesis of γ-substituted butyrolactams and their conversion into semicorrinoid substances has been published[111]; semicorrins (81) can also be prepared[112] from isoxazoles (80):

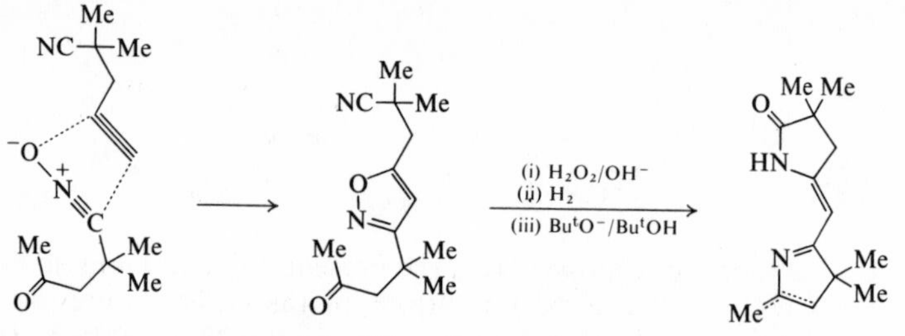

(i) H_2O_2/OH^-
(ii) H_2
(iii) Bu^tO^-/Bu^tOH

(80)   (81)

The key isoxazole intermediate (80) is obtained from a nitrile oxide–acetylene cyclo-addition.

The progress of the Zurich ETH–Harvard groups in their combined efforts in the synthesis of vitamin B_{12} (or more specifically, the cobyrinic acid

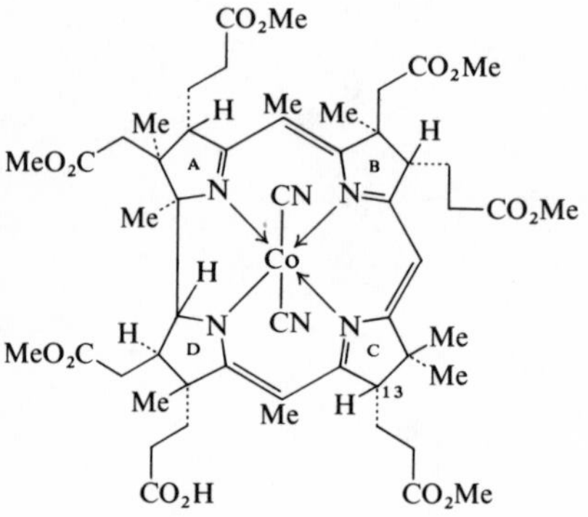

(82) 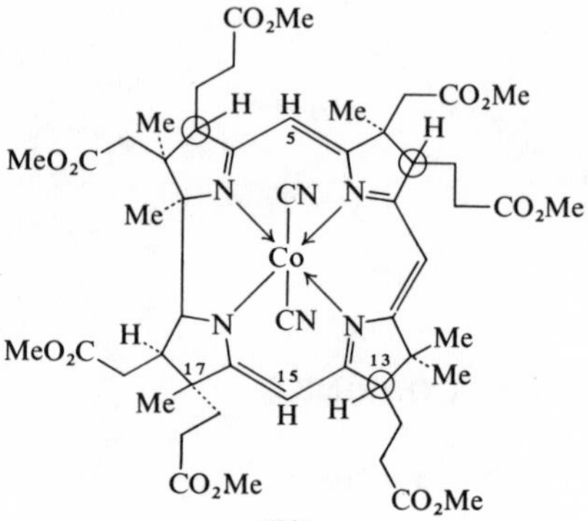   (83)

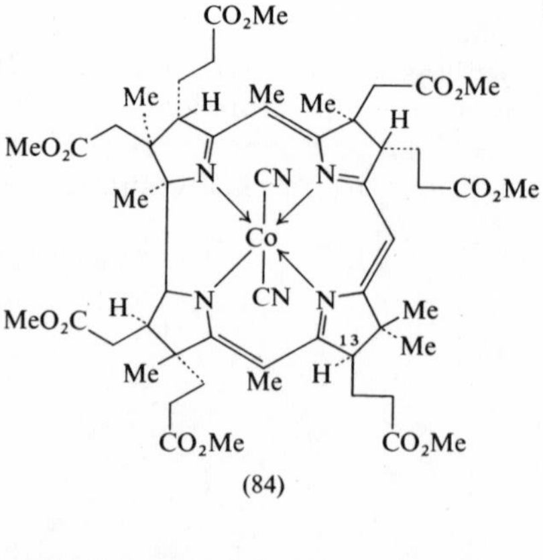

(84)

(82), thereby achieving the formal total synthesis of the vitamin) has recently been reviewed[7,8] by the heads of both groups. Brevity prevents any attempt to discuss the beautiful reactions which are described[7,8]; the furthermost point attained by the combined forces is the corrin (83), leaving only the following problems* to be solved: (a) Separation of the isomers due to the epimeric centres in rings A, B, and C. (b) Differentiation of the 17-propionate function from the esters. (c) Insertion of the 5- and 15-methyl groups. The solution of the first of these problems has been helped by the identification of neovitamin B_{12} compounds as the 13-epimers (e.g. (84)) of the natural series, and in particular by the observation that the neo-series can be equilibrated with the natural compounds under highly acidic conditions[113].

11.7.2   Reactions and properties

Space does not allow the discussion of the massive literature on the chemistry of the cobalt atom of vitamin B_{12} and its derivatives in this review; readers are referred to other recent reviews (e.g. reference 4) for leading references. A series of crystallographic studies on corrinoid metal chelates has appeared[114].

11.8   OXOPHLORINS

11.8.1   Synthesis

The synthesis of a series of electronegatively substituted oxophlorins (e.g. (87a)) has been reported[115] using the highly efficient method[116] based on MacDonald's porphyrin synthesis:

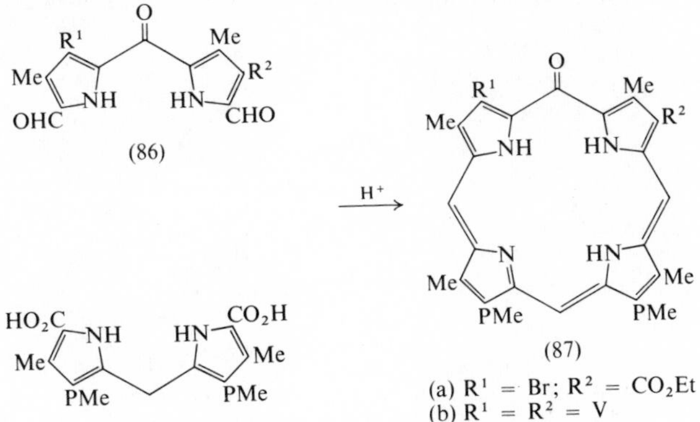

(86)

$$\xrightarrow{\text{H}^+}$$

(87)

(a) $R^1 = Br$; $R^2 = CO_2Et$
(b) $R^1 = R^2 = V$
(c) $R^1 = R^2 = Et$

(85)

Though having certain symmetry limitations, this route has become particularly attractive as a result of improvements in the synthesis of pyrromethanes (e.g. 85)[117] and diformylpyrroketones (e.g. 86)[118]. The acetate and benzoate of α-oxyprotoporphyrin-IX dimethyl ester (87b) have been prepared[119] and their hydrolysis to (87b) studied. Oxidation of the oxophlorin haemin (88a) gave a bile pigment which showed spectroscopic and chromato-

---

*Recent, as yet unpublished, lectures (I.U.P.A.C. Symposium, Boston, July, 1971) have reported the realisation of almost all of these goals.

graphic identity with biliverdin-IXα dimethyl ester (89a). α-Oxymesoporphy-rin-IX dimethyl ester (87c) has been synthesised[120] by the b-oxobilane method;

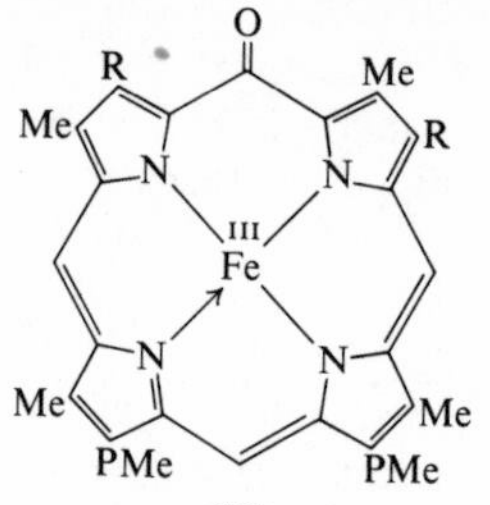

(88)

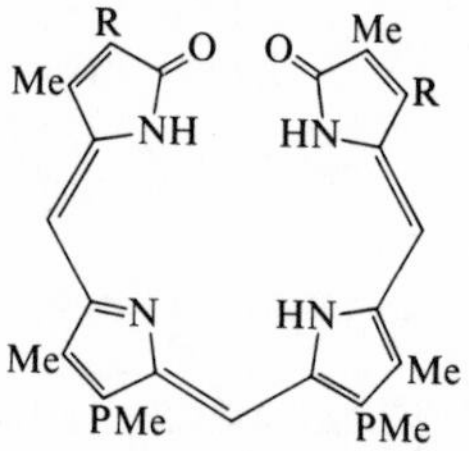

(a) R = V
(b) R = Et

(89)

with oxygen in pyridine the corresponding haemin (88b) gives mesobiliver-din-IXα dimethyl ester (89b).

11.8.2 Reactions and properties

The copper complexes of porphyrins (e.g. (90a)) react[121] with thiocyanogen to give the corresponding meso-thiocyanato-derivatives (90b) which can be demetallated to the free ligand (91a) and then hydrolysed to the sulphur analogue of an oxophlorin with uncertain structure; it is, however, formu-

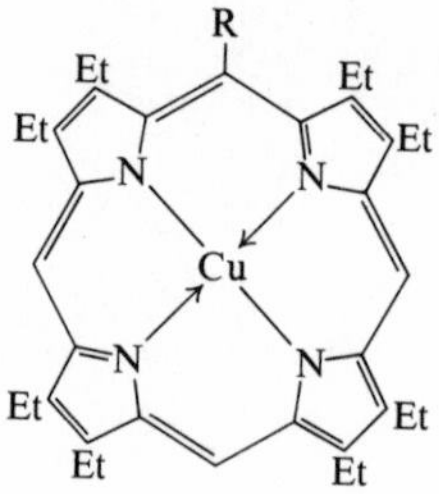

(90)

(a) R = H
(b) R = SCN

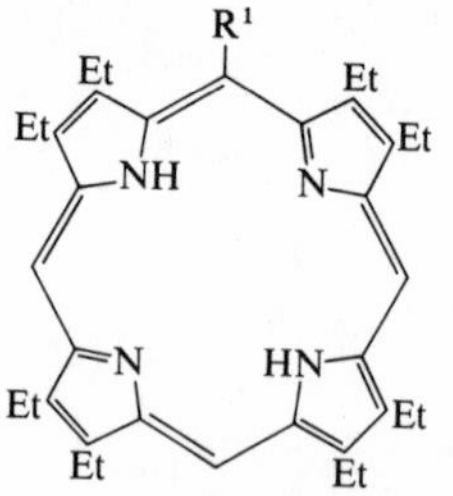

(91)

(a) R^1 = SCN
(b) R^1 = SH

lated as the di-cation of (91b) in acidic solution. A further study of the free radical nature of oxophlorins has appeared[122].

11.9 OPEN-CHAIN PIGMENTS

11.9.1 Prodigiosin

The isolation and characterisation of a cyclic nonylprodigiosin derivative (92) has been reported[123], making an interesting addition to the gradually

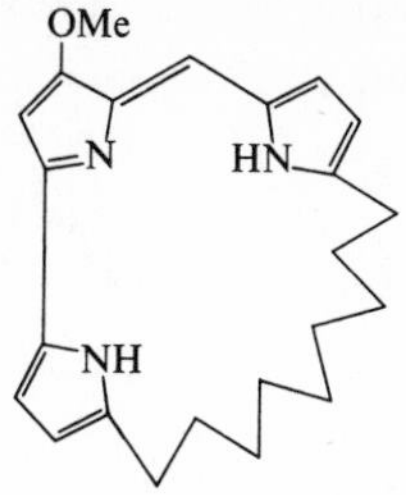

(92)

increasing number of novel prodigiosin derivatives which have been reported in recent times.[6]

11.9.2   Bile pigments

11.9.2.1   Synthesis

The synthesis of optically-active stercobilin-IXα (95) and other optically active analogues has been reported[124], using the classical, and highly effective, condensation of the two optically active oxopyrromethanes (93) and (94).

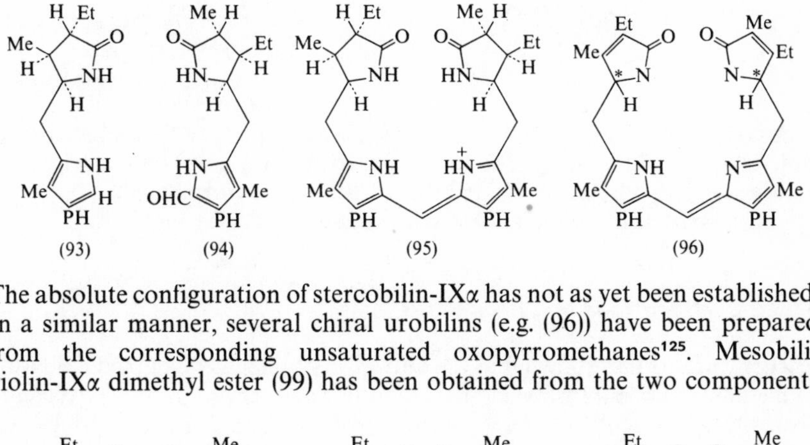

(93)   (94)   (95)   (96)

The absolute configuration of stercobilin-IXα has not as yet been established. In a similar manner, several chiral urobilins (e.g. (96)) have been prepared from the corresponding unsaturated oxopyrromethanes[125]. Mesobiliviolin-IXα dimethyl ester (99) has been obtained from the two components

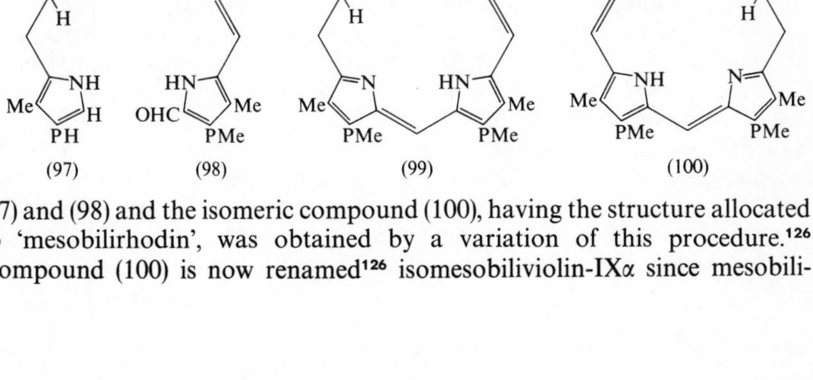

(97)   (98)   (99)   (100)

(97) and (98) and the isomeric compound (100), having the structure allocated to 'mesobilirhodin', was obtained by a variation of this procedure.[126] Compound (100) is now renamed[126] isomesobiliviolin-IXα since mesobili-

rhodin, a red contaminant often observed in traces in crude preparations of mesobiliviolin and i-urobilin has been characterised as (101) or (102)[127].

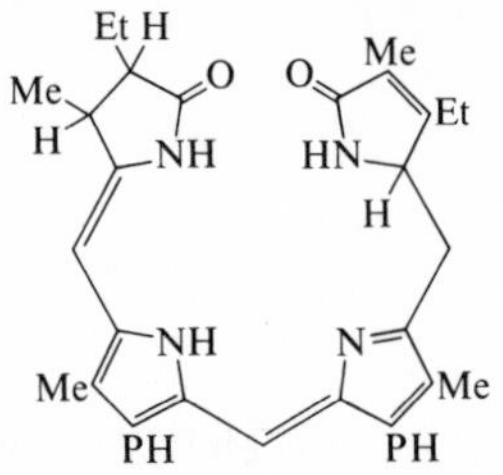

(101) 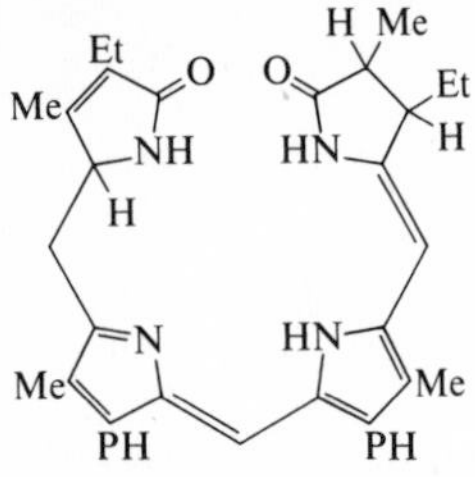　　　　　　　　　　　　　(102)

trans-1,2-Dihydro-octaethylbiliverdin (51) has been prepared by the treatment of *trans*-octaethylchlorin with thallium(III) trifluoroacetate[95] and prolonged treatment with the reagent has been shown to give the bilipurpurin-

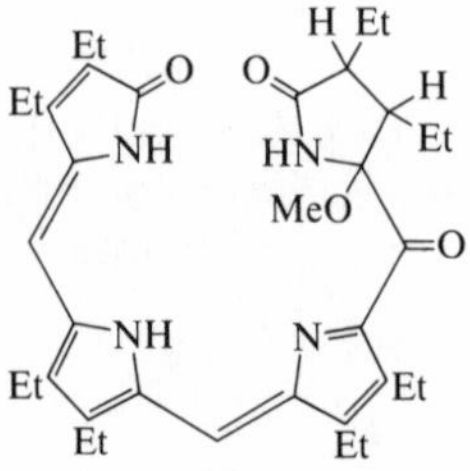

(103) 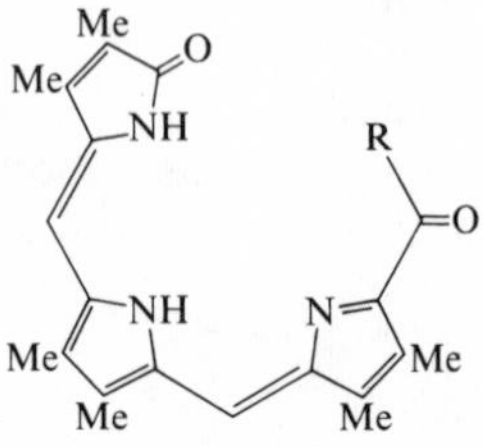　　　　　　　　　　　　　(104)

type compound (103) which has the same chromophore and a visible absorption spectrum similar to those of the tripyrrene compounds (e.g. 104) synthesised by Plieninger[128]. The latter substances form metal complexes having remarkable solubility characteristics.

11.9.2.2　Reactions and properties

Earlier workers had shown that the biliverdin obtained by coupled oxidation of haemin was a mixture of isomers, and these observations have been conclusively confirmed[129] by the separation, isolation and characterisation of all four biliverdin-IX isomers as their methyl esters (105a–108a). The identification of the compounds was carried out by visible, n.m.r., and mass spectroscopy, as well as a mixed melting point with authentic material in the -IXα case. O'Carra and Colleran[130] have also carried out an isomer analysis of phycobilins and bilirubin, as well as the separation and identification of the biliverdin isomers. Dehydrogenation of bilirubin with benzoquinone in acetic acid, followed by methylation gives[131] three isomeric biliverdin dimethyl esters, the two non-rational compounds (109) and (110) arising from a reversible acid-catalysed cleavage of the bilirubin molecule about the central methylene bridge, followed by random recombination. Apparently pure compounds from the ferric chloride oxidation of crude mesobilirubinogen have been shown[132] to be complex mixtures by ascending t.l.c.

The specificity of biliverdin reductase has been examined[133], and the coupled oxidation of myoglobin has been shown[134] to give only biliverdin-IXα

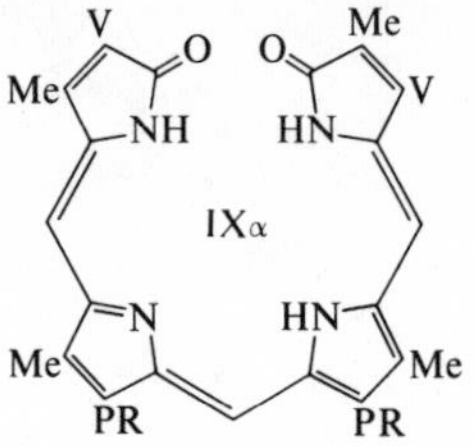

(105)

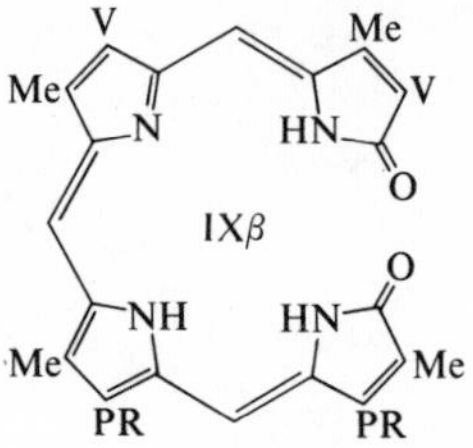

(106)

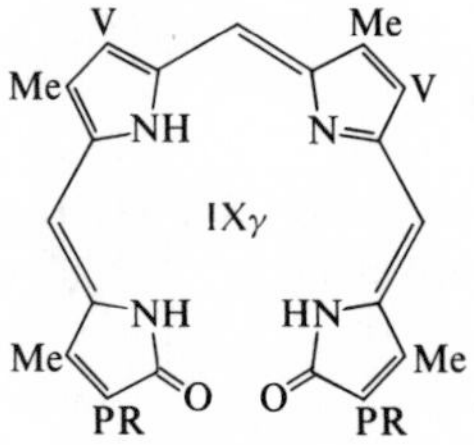

(107)

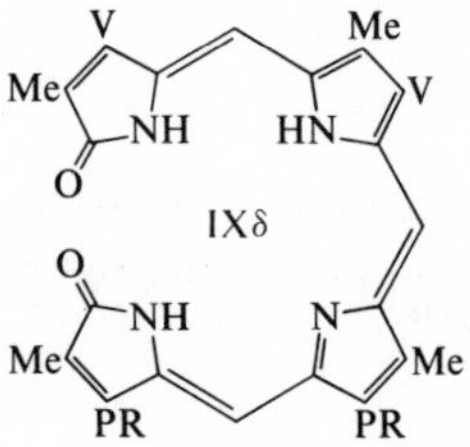

(108)

(a) R = Me
(b) R = H

(105b), while similar *in vitro* treatment of haemoglobin gives largely -IXα (105b) with some of the -IXβ isomer (106b).

Further work on the isolation and spectral characterisation of phycobiliproteins has been reported[135], and also a further study[136] of the structure of

(109)

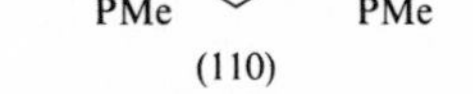

(110)

phycocyanobilin, the prosthetic group of phycocyanin; the generally accepted structure for the bilin, neglecting intramolecular hydrogen bonds, is that shown in (111).

11.10 OTHER MACROCYCLES RELATED TO PORPHYRINS

The preparation of porphyrin analogues (112) containing furan and thiophene as well as pyrrole rings have been described[137]: the spectroscopic similarities of these compounds with the porphyrins supports their formulation as aromatic macrocycles. Only the mono-furan (112a) and mono-thiophene

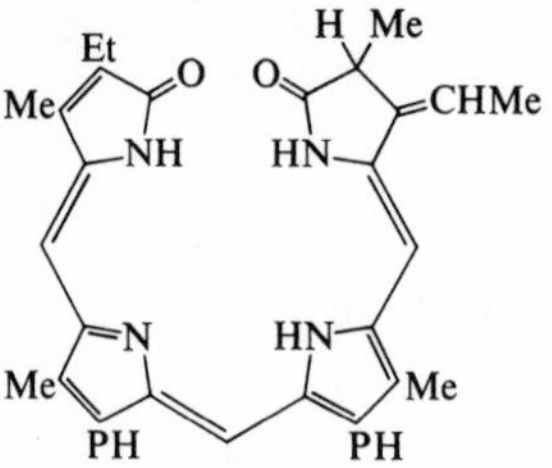

(111) 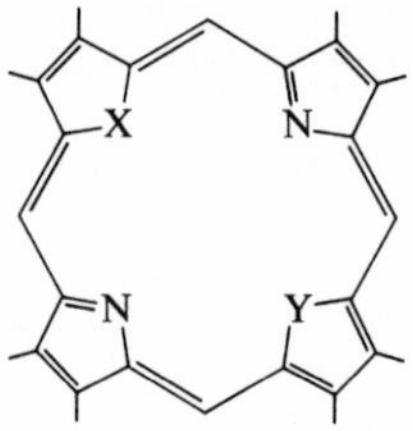   (112)   (113)

(a) X = O; Y = NH
(b) X = S; Y = NH
(c) X = Y = O
(d) X = Y = S
(e) X = O; Y = S

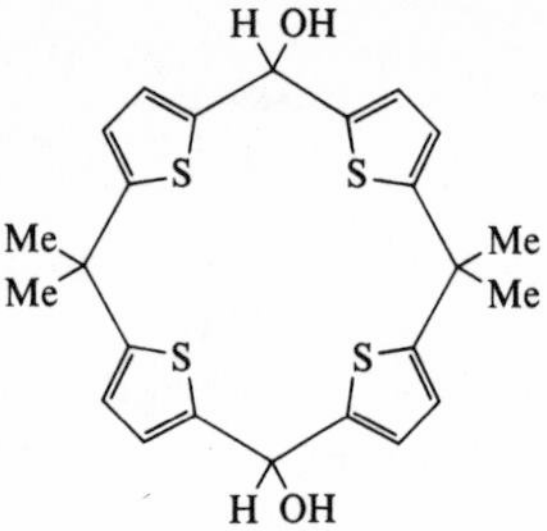

(112b) compounds formed metal complexes under the usual conditions. Routes to thiophene analogues of porphyrinogens (e.g. 113) have also been elaborated[138].

11.11 BIOSYNTHESIS

11.11.1 Porphyrins

An intermediate in the conversion of porphobilinogen (PBG; 114) to uroporphyrinogen-I (115a) by uroporphyrinogen-I synthetase has been identified[139] as the pyrromethane (116a). The enzyme does not catalyse the condensation of two molecules of (116a) to give uroporphyrinogen-I, but in the presence of PBG (114), (116a) is incorporated into uroporphyrinogen-I. Rapoport

(114)

$AR = CH_2CO_2R$

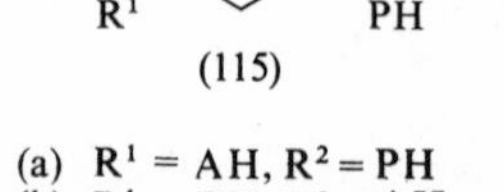

(115)

(a) $R^1 = AH, R^2 = PH$
(b) $R^1 = PH, R^2 = AH$

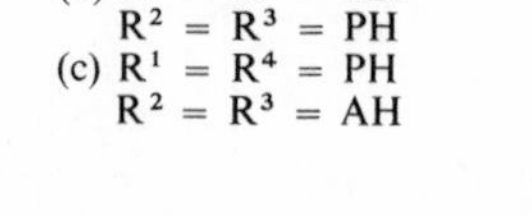

(116)

(a) $R^1 = R^3 = PH$
    $R^2 = R^4 = AH$
(b) $R^1 = R^4 = AH$
    $R^2 = R^3 = PH$
(c) $R^1 = R^4 = PH$
    $R^2 = R^3 = AH$

and co-workers[140] have reported the synthesis of two pyrromethanes (116a, b), neither of which were substrates for either uroporphyrinogen-I synthetase or uroporphyrinogen-III cosynthetase (or the combined enzyme system). In the presence of PBG, the pyrromethane (116a) inhibited PBG consumption, whereas (116b) had only a slight inhibitory effect. In a type-III forming system, (116a) diverted the formation of uroporphyrinogen-III to give uroporphyrinogen-I. It is therefore proposed that the pyrromethane

(116c) is the intermediate of uroporphyrinogen-III (115b) synthesis. This postulate, once confirmed by the synthesis and feeding of (116c) would also have implications in the biosynthesis of vitamin B_{12}, which most workers have assumed shares a pathway with the porphyrins far past the PBG stage[152]. The pyrromethane (116c) features the portion of the future corrin skeleton which contains the 1-, 19-direct link, and positive results would suggest that a rearrangement involving the internuclear carbon atom was involved, possibly of a macrocycle such as a corphin. An alternative bio-synthetic route to the type-III skeleton has been expounded[141], though not yet published. Elegant use of deuterium labelling has shown[141, 142] the conversion of the 2- and 4-propionic acid side-chains of coproporphyrinogen-III to the vinyls of protoporphyrin-IX (1a) to proceed with the retention of three of the methylene protons of each propionic acid residue; this confirms the postulate of Sano that the transformation involves initial hydroxylation, followed by decarboxylative dehydration:

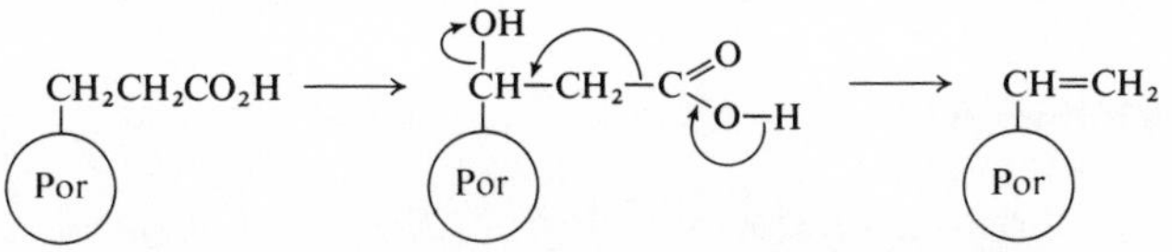

These results have therefore eliminated the possibility that the monoacrylic tripropionic porphyrin (postulated structure (117)) isolated[143] from meconium might be an intermediate in the enzymic conversion of coproporphyrinogen-III to protoporphyrin-IX. However, the isolation and characterisation of

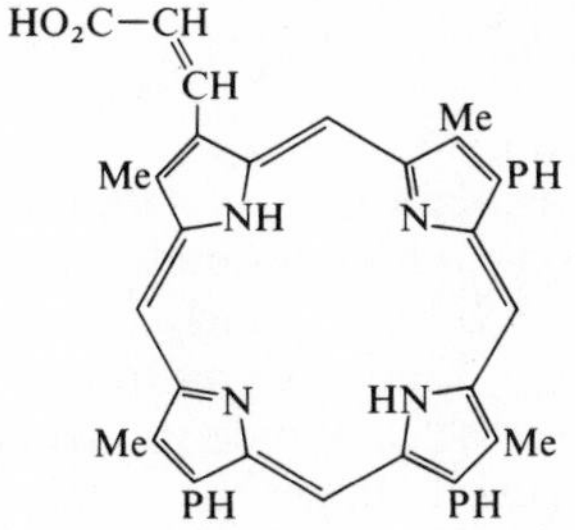

(117)

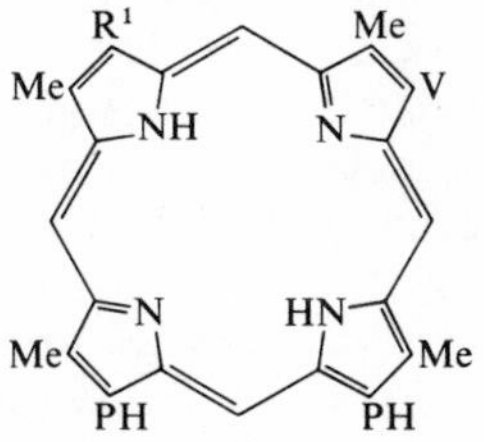

(118)

(a) R^1 = H
(b) R^1 = CHO

harderoporphyrin[12] (1b) (see Section 11.2.1) seems to indicate clearly that the 2-propionic acid side-chain of coproporphyrinogen-III is transformed to vinyl before the 4-propionic residue. This tendency of the 2-side-chains to be degraded before the 4-side-chains is also reflected in the structures of pemptoporphyrin (118a) and *Spirographis* porphyrin (118b).

11.11.2 Chlorins

Incubation of ^{14}C-labelled chlorophyll-*a* with homogenates of soybean leaves has been shown to furnish ^{14}C-labelled chlorophyll-*b*[144], indicating that chlorophyll-*b* is obtained *in vivo* from chlorophyll-*a*, and not through a pathway branched at an earlier stage.

11.11.3 Prodigiosin

Earlier work has shown that proline induces the biosynthesis of prodigiosin; when radioactively-labelled methionine is included with the proline a four-fold increase in the amount of incorporated ^{14}C label has been noted[145]. Labelled methionine in the absence of proline was not incorporated. The origin of certain of the carbon atoms in prodigiosin has been established from a ^{13}C Fourier n.m.r. study of the pigments produced from *Serratia marcescens* incubated with ^{13}C enriched acetate[146]; this represents the first report of the study of a detailed labelling pattern in a metabolite biosynthetically enriched with ^{13}C.

11.11.4 Vitamin B$_{12}$

A cobalt-free corrin has been isolated from *Streptomyces olivaceus* grown under cobalt deficient conditions;[147] in the presence of cobalt ions normal vitamin B$_{12}$ was obtained. The material from *Streptomyces* is considered to be located at an earlier stage in the biosynthetic pathway than the material isolated from *Chromatium* by Toohey[148].

11.11.5 Bile pigments

When tritiated α-oxymesoporphyrin-IX haemin (diacid of 88b) was injected into rats with biliary fistulae it was converted into mesobilirubin-IXα, whereas the corresponding β-oxy-haemin was only poorly converted into bile pigment[149]. It is concluded by analogy, that α-oxyprotoporphyrin-IX haemin (diacid of (88a)) is an intermediate in the catabolism of haemoglobin, undergoing further oxidation to bile pigment under the catalysis of an enzyme of definite specificity. The biosynthesis of phycocyanin from *Cyanidium caldarium* has been examined[150]; δ-aminolaevulinic acid was incorporated exclusively into the bilin which is, therefore, presumably derived from a porphyrin *in vivo*.

References

1. Fleischer, E. B. (1970). *Accounts Chem. Res.*, **3**, 105
2. Hambright, P. (1971). *Coord. Chem. Rev.*, **6**, 247
3. Buse, G. (1971). *Angew. Chem. Int. Ed.*, **10**, 663
4. Stadtman, T. C. (1971). *Science*, **171**, 859
5. Rüdiger, W. (1970). *Angew. Chem. Int. Ed.*, **9**, 473
6. Smith, K. M. (1971). *Quart. Rev. Chem. Soc.*, **25**, 31
7. Eschenmoser, A. (1970). *Quart. Rev. Chem. Soc.*, **24**, 366
8. Woodward, R. B. (1971). *Pure Appl. Chem.*, **25**, 283
9. Carr, R. P., Jackson, A. H., Kenner, G. W. and Sach, G. S. (1971). *J. Chem. Soc. C*, 487

10. Grigg, R., Johnson, A. W. and Roche, M. (1970). *J. Chem. Soc. C*, 1928
11. Gonsalves, A. M. d'A. R., Kenner, G. W. and Smith, K. M. (1971). *Chem. Commun.*, 1304
12. Kennedy, G. Y., Jackson, A. H., Kenner, G. W. and Suckling, C. J. (1970). *FEBS Letters*, **6**, 9; **7**, 205; Kennedy, G. Y. (1970). *Comp. Biochem. Physiol.*, **36**, 21
13. Crook, P. J., Jackson, A. H. and Kenner, G. W. (1971). *J. Chem. Soc. C*, 474
14. Jackson, A. H., Kenner, G. W. and Smith, K. M. (1971). *J. Chem. Soc. C*, 502
15. Cox, M. T., Jackson, A. H. and Kenner, G. W. (1971). *J. Chem. Soc. C*, 1974
16. Holt, A. S., Purdie, J. W. and Wasley, J. W. F. (1966). *Can. J. Chem.*, **44**, 88
17. Johnson, A. W. and Kay, I. T. (1961). *J. Chem. Soc.*, 2418; Grigg, R., Johnson, A. W., Kenyon, R., Math, V. B. and Richardson, K. (1969). *J. Chem. Soc. C*, 176
18. Badger, G. M., Harris, R. L. N. and Jones, R. A. (1964). *Aust. J. Chem.*, **17**, 1013
19. Clezy, P. S. and Liepa, A. J. (1971). *Aust. J. Chem.*, **24**, 1027
20. Mironov, A. F., Rumyantseva, V. D., Modnikova, G. A. and Evstigneeva, R. P. (1970). *Zh. Obshch. Khim.*, **40**, 385; Ponomarev, G. V., Nasr-ala, S. M., Mironov, A. F., Rumyantseva, V. D. and Evstigneeva, R. P. (1970). *Zh. Obshch. Khim.*, **40**, 247
21. Dolphin, D. (1970). *J. Heterocycl. Chem.*, **7**, 275
22. Smith, K. M. (1971). *Tetrahedron Lett.*, 2325
23. Treibs, A. and Schulze, L. (1971). *Ann. Chem.*, **751**, 127
24. Dearden, G. R. and Jackson, A. H. (1970). *Chem. Commun.*, 205
25. Broadhurst, M. J., Grigg, R., Shelton, G. and Johnson, A. W. (1970). *Chem. Commun.*, 231
26. Grigg, R., Sweeney, A., Dearden, G. R., Jackson, A. H. and Johnson, A. W. (1970). *Chem. Commun.*, 1273
27. Shears, B. and Hambright, P. (1970). *Inorg. Nucl. Chem. Lett.*, **6**, 679
28. Jones, R. V. H., Kenner, G. W., Lewis, T. and Smith, K. M. (1971). *Chem. Ind.* (London), 129
29. Bonnett, R. and McDonagh, A. F. (1970). *Chem. Commun.*, 337
30. Peychal-Heiling, G. and Wilson, G. S. (1971). *Anal. Chem.*, **43**, 545
31. Peychal-Heiling, G. and Wilson, G. S. (1971). *Anal. Chem.*, **43**, 550
32. Boylan, D. B. (1970). *Org. Mass Spectrom.*, **3**, 339; Adler, A. D., Green, J. H. and Mautner, M. (1970). *Org. Mass Spectrom.*, **3**, 955; Baker, E. W. (1971). *Chem. Geol.*, **7**, 45; Roxynov, B. V., Mironov, A. F. and Evstigneeva, R. P. (1971). *Khim. Prir. Soedin.*, **7**, 197; Evstigneeva, R. P., Mamaev, V. M. and Ponomarev, G. V. (1971). *Khim. Prir. Soedin.*, **7**, 49
33. Buchler, J. W., Eikelmann, G., Puppe, L., Rohbock, K., Schneehage, H. H. and Weck, D. (1971). *Ann. Chem.*, **745**, 135
34. Adler, A. D., Longo, F. P., Kampas, F. and Kim, J. (1970). *J. Inorg. Nucl. Chem.*, **32**, 2443
35. Smith, K. M. (1971). *Chem. Commun.*, 540
36. Abraham, R. J. and Smith, K. M. (1971). *Tetrahedron Lett.*, 3335
37. Hambright, P. and Fleischer, E. B. (1970). *Inorg. Chem.*, **9**, 1757; Hambright, P. (1970). *J. Inorg. Nucl. Chem.*, **32**, 2449; Burnham, B. F. and Zuckerman, J. J. (1970). *J. Amer. Chem. Soc.*, **92**, 1547; Cabiness, D. K. and Magerum, D. W. (1970). *J. Amer. Chem. Soc.*, **92**, 2151; Kingham, D. J. and Brisbin, S. A. (1970). *Inorg. Chem.*, **9**, 2034; Koifman, O. I. and Berezin, B. D. (1970). *Zh. Obshch. Khim.*, **40**, 2701; Shah, B., Shears, B. and Hambright, P. (1971). *Inorg. Chem.*, **10**, 1828; Koifman, O. I., Golubchikov, O. A., Andrianov, V. G. and Berezin, B. D. (1971). *Izv. Vyssh. Ucheb. Zaved. Khim. Khim. Tekhnol.*, **14**, 519; Berezin, B. D. and Koifman, O. I. (1971). *Zh. Fiz. Khim.*, **45**, 1451
38. Shah, B. and Hambright, P. (1970). *J. Inorg. Nucl. Chem.*, **32**, 3420; Shears, B., Shah, B. and Hambright, P. (1971). *J. Amer. Chem. Soc.*, **93**, 776
39. Perutz, M. F. (1970). *Nature (London)*, **228**, 726, 734; Perutz, M. F. (1971). *New Scientist*, 676; Bolton, W. and Perutz, M. F. (1970). *Nature (London)*, **228**, 551
40. Perutz, M. F. (1971). *New Scientist*, 762; Morimoto, H., Lehmann, H. and Perutz, M. F. (1971). *Nature (London)*, **232**, 408
41. Wang, J. H. (1970). *Accounts Chem. Res.*, **3**, 90
42. Warme, P. K. and Hager, L. P. (1970). *Biochemistry*, **9**, 1599
43. Warme, P. K. and Hager, L. P. (1970). *Biochemistry*, **9**, 1606, 4237, 4244
44. v. d. Heijden, A., Peer, H. G. and v. d. Oord, A. H. A. (1971). *Chem. Commun.*, 369
45. Diekmann, H., Chang, C. K. and Traylor, T. G. (1971). *J. Amer. Chem. Soc.*, **93**, 4068

46. Storm, C. B. (1970). *J. Amer. Chem. Soc.*, **92**, 1423
47. Hill, H. A. O. and Morallee, K. G. (1970). *Chem. Commun.*, 266
48. Boyd, P. D. W. and Smith, T. D. (1971). *Inorg. Chem.*, **10**, 2041; Degani, H. A. and Fiat, D. (1971). *J. Amer. Chem. Soc.*, **93**, 4281; Yamane, T., Wüthrich, K., Shulman, R. G. and Ogawa, S. (1970). *J. Mol. Biol.*, **49**, 197; Sheard, B., Yamane, T. and Shulman, R. G. (1970). *J. Mol. Biol.*, **53**, 35; Shulman, R. G., Wüthrich, K., Yamane, T., Patel, D. J. and Blumberg, W. E. (1970). *J. Mol. Biol.*, **53**, 143; Ogawa, S. and Shulman, R. G. (1971). *Biochem. Biophys. Res. Commun.*, **42**, 9; Ogawa, S., Shulman, R. G., Kynoch, P. A. M. and Lehmann, H. (1970). *Nature (London)*, **225**, 1042. Davis, D. G., Mock, N. H., Lindstrom, T. R., Charachi, S. and Ho, C. (1970). *Biochem. Biophys. Res. Commun.*, **40**, 343
49. Kane, A. R., Sullivan, J. F., Kenny, D. H. and Kenney, M. E. (1970). *Inorg. Chem.*, **9**, 1445. Maskasky, J. E. and Kenney, M. E. (1971). *J. Amer. Chem. Soc.*, **93**, 2062
50. Brackett, G. C., Richards, P. L. and Caughey, W. S. (1971). *J. Chem. Phys.*, **54**, 4383; Bürger, H., Burczyk, K. and Fuhrhop, J.-H. (1971). *Tetrahedron*, **27**, 3257; Dickson, F. E. and Petrakis, L. (1970). *J. Phys. Chem.*, **74**, 2850; Caughey, W. S., Bayne, R. A. and McCoy, S. (1970). *Chem. Commun.*, 950. McCoy, S. and Caughey, W. S. (1970). *Biochemistry*, **9**, 2387; Uenoyama, H. (1971). *Biochim. Biophys. Acta*, **230**, 479; Ogoshi, H., Masai, N. Yoshida, Z., Takemoto, J. and Nakamoto, K. (1971). *Bull. Chem. Soc. Jap.*, **44**, 49; Ogoshi, H. and Yoshida, Z. (1971). *Bull. Chem. Soc. Jap.*, **44**, 1722
51. Bürger, H., Burczyk, K., Buchler, J. W., Fuhrhop, J.-H., Höfler, F. and Schrader, B. (1970). *Inorg. Nucl. Chem. Lett.*, **6**, 171
52. Boucher, L. J. (1970). *J. Amer. Chem. Soc.*, **92**, 2725
53. Friedel, M. K., Hoskins, B. F., Martin, R. L. and Mason, S. A. (1970). *Chem. Commun.*, 400
54. Rogers, D. and Osborn, R. S. (1971). *Chem. Commun.*, 840
55. Cullen, D. L. and Meyer, E. F. (1971). *Chem. Commun.*, 616
56. Collins, D. M. and Hoard, J. L. (1970). *J. Amer. Chem. Soc.*, **92**, 3761
57. Whitten, D. G., Yau, J. C. and Carrol, F. A. (1971). *J. Amer. Chem. Soc.*, **93**, 2291
58. Fuhrhop, J.-H., Lumbantobing, T. and Ullrich, J. (1970). *Tetrahedron Lett.*, 3771
59. Fleischer, E. B., Palmer, J. M., Srivastava, T. S. and Chatterjee, A. (1971). *J. Amer. Chem. Soc.*, **93**, 3162
60. Cole, S. J., Curthoys, G. C. and Magnusson, E. A. (1970). *J. Amer. Chem. Soc.*, **92**, 2991; (1971). *J. Amer. Chem. Soc.*, **93**, 2153; *Aust. J. Chem.*, **24**, 1967; Kirksey, C. H. and Hambright, P. (1970). *Inorg. Chem.*, **9**, 958
61. Fleischer, E. B., Jacobs, S. and Mestichelli, L. (1968). *J. Amer. Chem. Soc.*, **90**, 2527
62. Fleischer, E. B. and Krishnamurthy, M. (1971). *J. Amer. Chem. Soc.*, **93**, 3784
63. Tsutsui, M., Ostfeld, D. and Hoffman, L. M. (1971). *J. Amer. Chem. Soc.*, **93**, 1820
64. Faller, J. W. and Sibert, J. W. (1971). *J. Organometallic Chem.*, **31**, C5
65. Eaton, S. S., Eaton, G. R. and Holm, R. H. (1971). *J. Organometallic Chem.*, **32**, C52
66. Ostfeld, D., Tsutsui, M., Hrung, C. P. and Conway, D. C. (1971). *J. Amer. Chem. Soc.*, **93**, 2548
67. Grigg, R., Sweeney, A. and Johnson, A. W. (1970). *Chem. Commun.*, 1237
68. Bonnett, R. and Brewer, R. (1970). *Tetrahedron Lett.*, 2579
69. Paine, J. B. and Dolphin, D. (1971). *J. Amer. Chem. Soc.*, **93**, 4080
70. Fuhrhop, J.-H. and Mauzerall, D. (1969). *J. Amer. Chem. Soc.*, **91**, 4174
71. Fajer, J., Borg, D. C., Forman, A., Dolphin, D. and Felton, R. H. (1970). *J. Amer. Chem. Soc.*, **92**, 3451
72. Wolberg, A. and Manassen, J. (1970). *J. Amer. Chem. Soc.*, **92**, 2982; (1970). *Inorg. Chem.*, **9**, 2365
73. Dolphin, D., Felton, R. H., Borg, D. C. and Fajer, J. (1970). *J. Amer. Chem. Soc.*, **92**, 743
74. Fuhrhop, J.-H. (1970). *Chem. Commun.*, 781
75. Forman, A., Borg, D. C., Felton, R. H. and Fajer, J. (1971). *J. Amer. Chem. Soc.*, **93**, 2790
76. Edwards, L. and Gouterman, M. (1970). *J. Mol. Spectrosc.*, **33**, 292; Edwards, L., Dolphin, D. and Gouterman, M. (1970). *J. Mol. Spectrosc.*, **35**, 90; Edwards, L., Dolphin, D., Gouterman, M. and Adler, A. D. (1971). *J. Mol. Spectrosc.*, **38**, 16
77. Smythe, G. A. and Caughey, W. S. (1970). *Chem. Commun.*, 809
78. Barrett, J. (1967). *Nature (London)*, **215**, 733
79. Fuhrhop, J.-H. and Mauzerall, D. (1971). *Photochem. Photobiol.*, **13**, 453

80. Nichol, A. W. (1970). *J. Chem. Soc. C*, 903
81. Buchler, J. W. and Puppe, L. (1970). *Ann. Chem.*, **740**, 142
82. Buchler, J. W., Puppe, L. and Schneehage, H. H. (1971). *Ann. Chem.*, **749**, 134
83. Dougherty, R. C., Strain, H. H., Svec, W. A., Uphaus, R. A. and Katz, J. J. (1970). *J. Amer. Chem. Soc.*, **92**, 2826
84. Strain, H. H., Cope, B. T., McDonald, G. N., Svec, W. A. and Katz, J. J. (1971). *Phytochemistry*, **10**, 1109
85. Wasley, J. W. F., Scott, W. T. and Holt, A. S. (1970). *Can. J. Biochem.*, **48**, 376; Budzikiewicz, H. and Taraz, K. (1971). *Tetrahedron*, **27**, 1447; Croft, J. A. and Howden, M. E. H. (1970). *Phytochemistry*, **9**, 901; De-Greef, J. A. and Caubergs, R. (1970). *Naturwissenschaften*, **57**, 673
86. Brockmann, H. and Kleber, I. (1970). *Tetrahedron Lett.*, 2195
87. Inhoffen, H. H., Jäger, P. and Mahlhop, R. (1971). *Ann. Chem.*, **749**, 109
88. Wolf, H. and Scheer, H. (1971). *Angew. Chem. Int. Ed.*, **10**, 866
89. Borg, D. C., Fajer, J., Felton, R. H. and Dolphin, D. (1970). *Proc. Nat. Acad. Sci. U. S.*, **67**, 813
90. Norris, J. R., Uphaus, R. A., Crespi, H. L. and Katz, J. J. (1971). *Proc. Nat. Acad. Sci. U. S.*, **68**, 625
91. Dolphin, D., Forman, A., Borg, D. C., Fajer, J. and Felton, R. H. (1971). *Proc. Nat. Acad. Sci. U. S.*, **68**, 614
92. Fuhrhop, J.-H. (1970). *Z. Naturforsch., B*, **25**, 255
93. Gurinovitch, G. P. and Byteva, I. M. (1970). *Biofizika*, **15**, 602; Shulga, A. M. and Suboch, V. P. (1971). *Biofizika*, **16**, 214
94. Jen, J. J. and Mackinney, G. (1970). *Photochem. Photobiol.*, **11**, 297, 303
95. Cavaleiro, J. A. S. and Smith, K. M. (1971). *Chem. Commun.*, 1384
96. Datta-Gupta, N. and Williams, G. E. (1971). *J. Org. Chem.*, **36**, 2019
97. Broadhurst, M. J., Grigg, R. and Johnson, A. W. (1970). *Chem. Commun.*, 807
98. Grigg, R., Johnson, A. W. and Shelton, G. (1971). *Ann. Chem.*, **746**, 32; Grigg, R., King, T. J. and Shelton, G. (1970). *Chem. Commun.*, 56; Dobinson, G. C. and Mason, R. Unpublished results
99. Grigg, R., Johnson, A. W. and Shelton, G. (1971). *J. Chem. Soc. C*, 2287
100. Dyke, J. M., Hush, N. S., Williams, M. L. and Woolsey, I. S. (1971). *Mol. Phys.*, **20**, 1149
101. Dicker, I. D., Grigg, R., Johnson, A. W., Richardson, K. and v. d. Broek, P. (1971). *J. Chem. Soc. C*, 536
102. Melenteva, T. A., Pekel, N. D. and Berezovskii, V. M. (1970). *Zh. Obshch. Khim.*, **40**, 165
103. Grigg, R., Johnson, A. P., Johnson, A. W. and Smith, M. J. (1971). *J. Chem. Soc. C*, 2457
104. Grigg, R., Johnson, A. W., Richardson, K. and Smith, M. J. (1970). *J. Chem. Soc. C*, 1289
105. Hamilton, A. and Johnson, A. W. (1971). *Chem. Commun.*, 523
106. Melenteva, T. A., Pekel, N. D., Genokhova, N. S. and Berezovskii, V. M. (1970). *Dokl. Akad. Nauk SSSR*, **134**, 591
107. Inhoffen, H. H., Ullrich, J., Hoffmann, H. A., Klinzmann, G. and Scheu, R. (1970). *Ann. Chem.*, **738**, 1
108. Gossauer, A. and Inhoffen, H. H. (1970). *Ann. Chem.*, **738**, 18; Gossauer, A., Miehe, D. and Inhoffen, H. H. (1970). *Ann. Chem.*, **738**, 31
109. Inhoffen, H. H., Buchler, J. W., Puppe, L. and Rohbock, K. (1971). *Ann. Chem.*, **747**, 133
110. Johnson, A. W. and Overend, W. R. (1971). *Chem. Commun.*, 710
111. Stevens, R. V. and Kaplan, M. (1970). *Chem. Commun.*, 822; Stevens, R. V., Christensen, C. G., Edmonson, W. L., Kaplan, M., Reid, E. B. and Wentland, M. P. (1971). *J. Amer. Chem. Soc.*, **93**, 6629
112. Stevens, R. V., DuPree, L. E. and Wentland, M. P. (1970). *Chem. Commun.*, 821; Stevens, R. V., DuPree, L. E., Edmonson, W. L., Magid, L. L. and Wentland, M. P. (1971). *J. Amer. Chem. Soc.*, **93**, 6637
113. Bonnett, R., Godfrey, J. M. and Math, V. B. (1971). *J. Chem. Soc., C*, 3736; Bonnett, R., Godfrey, J. M., Math, V. B., Edmond, E., Evans, H. and Hodder, O. J. R. (1971). *Nature (London)*, **229**, 473
114. Dunitz, J. D. and Meyer, E. F. (1971). *Helv. Chim. Acta*, **54**, 77. Dobler, M. and Dunitz, J. D. (1971). *Helv. Chim. Acta*, **54**, 90; Currie, M. and Dunitz, J. D. (1971). *Helv. Chim. Acta*, **54**, 98
115. Clezy, P. S. and Liepa, A. J. (1970). *Aust. J. Chem.*, **23**, 2461

116. Clezy, P. S., Liepa, A. J. and Smythe, G. A. (1970). *Aust. J. Chem.*, **23**, 603
117. Clezy, P. S. and Liepa, A. J. (1970). *Aust. J. Chem.*, **23**, 2443
118. Clezy, P. S., Liepa, A. J., Nichol, A. W. and Smythe, G. A. (1970). *Aust. J. Chem.*, **23**, 589
119. Clezy, P. S. and Liepa, A. J. (1970). *Aust. J. Chem.*, **23**, 2477
120. Crook, P. J., Jackson, A. H. and Kenner, G. W. (1971). *Ann. Chem.*, **748**, 26
121. Clezy, P. S. and Fookes, C. J. R. (1971). *Chem. Commun.*, 1268
122. Bonnett, R., Dimsdale, M. J. and Sales, K. D. (1970). *Chem. Commun.*, 962
123. Gerber, N. N. (1970). *Tetrahedron Lett.*, 809
124. Plieninger, H. and Ruppert, J. (1970). *Ann. Chem.*, **736**, 43
125. Plieninger, H., Ehl, K. and Tapia, A. (1970). *Ann. Chem.*, **736**, 62
126. Plieninger, H., Ehl, K. and Klinga, K. (1971). *Ann. Chem.*, **743**, 112
127. Rüdiger, W., Köst, H. P., Budzikiewicz, H. and Kramer, V. (1970). *Ann. Chem.*, **738**, 197. O'Carra, P. and Killilea, S. D. (1970). *Tetrahedron Lett.*, 4211
128. Plieninger, H. and Stumpf, K. (1970). *Chem. Ber.*, **103**, 2562
129. Bonnett, R. and McDonagh, A. F. (1970). *Chem. Commun.*, 237
130. O'Carra, P. and Colleran, E. (1970). *J. Chromatogr.*, **50**, 458
131. Bonnett, R. and McDonagh, A. F. (1970). *Chem. Commun.*, 238
132. Stoll, M. S. and Gray, C. H. (1970). *Biochem. J.*, **117**, 271
133. Colleran, E. and O'Carra, P. (1970). *Biochem. J.*, **119**, 16P
134. Colleran, E. and O'Carra, P. (1970). *Biochem. J.*, **119**, 42P
135. Teale, F. W. J. and Dale, R. E. (1970). *Biochem. J.*, **116**, 161
136. Schram, B. L. (1970). *Biochem. J.*, **119**, 15P. Schram, B. L. and Kroes, H. H. (1971). *Eur. J. Biochem.*, **19**, 581
137. Broadhurst, M. J., Grigg, R. and Johnson, A. W. (1971). *J. Chem. Soc. C*, 3681
138. Ahmed, M. and Meth-Cohn, O. (1971). *J. Chem. Soc. C*, 2104
139. Pluscec, J. and Bogorad, L. (1970). *Biochemistry*, **9**, 4736
140. Frydman, B., Reil, S., Valasinas, A., Frydman, R. B. and Rapoport, H. (1971). *J. Amer. Chem. Soc.*, **93**, 2738. See also, Frydman, R. B., Reil, S. and Frydman, B. (1971). *Biochemistry*, **10**, 1154
141. Battersby, A. R. (I.U.P.A.C. Symposium Lecture, Boston, July, 1971)
142. Battersby, A. R. (1971). *Chem. Brit.*, **7**, 297
143. French, J., Nicholson, D. C. and Rimington, C. (1970). *Biochem. J.*, **120**, 393. Rimington, C. (1971). *South African J. Lab. Clin. Med.*, **17**, 187
144. Ellsworth, R. K., Perkins, H. J., Detwiller, J. P. and Lui, K. (1971). *Biochim. Biophys. Acta*, **223**, 275
145. Quadri, S. M. H. and Williams, R. P. (1971). *Biochim. Biophys. Acta*, **230**, 181
146. Cushley, R. J., Anderson, D. R., Lipsky, S. R., Sykes, R. J. and Wasserman, H. H. (1971). *J. Amer. Chem. Soc.*, **93**, 6284
147. Sato, K., Shimzu, S. and Fukui, S. (1970). *Biochem. Biophys. Res. Commun.*, **39**, 170
148. Toohey, J. I. (1965). *Proc. Nat. Acad. Sci. U.S.*, **54**, 934; Toohey, J. I. (1966). *Fed. Proc.*, **25**, 1628
149. Kondo, T., Nicholson, D. C., Jackson, A. H. and Kenner, G. W. (1971). *Biochem. J.* **121**, 601
150. Troxler, R. F., Brown, A., Vester, R. and White, P. (1970). *Science*, **167**, 192; Troxler, R. F. and Brown, A. (1970). *Biochim. Biophys. Acta*, **215**, 503
151. Felton, R. H., Owen, G. S., Dolphin, D. and Fajer, J. (1971). *J. Amer. Chem. Soc.*, **93**, 6332
152. Müller, G. and Dieterle, W. (1971). *Hoppe-Seyler's Z. Physiol. Chem.*, **352**, 143